**MECHANICAL
ENGINEERING
DESIGN**

McGRAW-HILL SERIES IN MECHANICAL ENGINEERING

Jack P. Holman, Southern Methodist University, Consulting Editor

BARRON: Cryogenic Systems
ECKERT: Introduction to Heat and Mass Transfer
ECKERT and DRAKE: Analysis of Heat and Mass Transfer
ECKERT and DRAKE: Heat and Mass Transfer
HAM, CRANE, and ROGERS: Mechanics of Machinery
HARTENBERG and DENAVIT: Kinematic Synthesis of Linkages
HINZE: Turbulence
JACOBSEN and AYRE: Engineering Vibrations
JUVINALL: Engineering Considerations of Stress, Strain, and Strength
KAYS: Convective Heat and Mass Transfer
LICHTY: Combustion Engine Processes
MARTIN: Kinematics and Dynamics of Machines
PHELAN: Dynamics of Machinery
PHELAN: Fundamentals of Mechanical Design
RAVEN: Automatic Control Engineering
SCHENCK: Theories of Engineering Experimentation
SCHLICHTING: Boundary-Layer Theory
SHIGLEY: Dynamic Analysis of Machines
SHIGLEY: Kinematic Analysis of Mechanisms
SHIGLEY: Mechanical Engineering Design
SHIGLEY: Simulation of Mechanical Systems
STOECKER: Refrigeration and Air Conditioning
SUTTON and SHERMAN: Engineering Magnetohydrodynamics

**McGRAW-HILL
BOOK COMPANY**

New York
St. Louis
San Francisco
Auckland
Bogotá
Düsseldorf
Johannesburg
London
Madrid
Mexico
Montreal
New Delhi
Panama
Paris
São Paulo
Singapore
Sydney
Tokyo
Toronto

JOSEPH EDWARD SHIGLEY
Professor of Mechanical Engineering
The University of Michigan

Mechanical Engineering Design

THIRD EDITION

This book was set in Times New Roman.
The editors were B. J. Clark and Douglas J. Marshall;
the production supervisor was Leroy A. Young.
New drawings were done by J & R Services, Inc.
Fairfield Graphics was printer and binder.

Library of Congress Cataloging in Publication Data

Shigley, Joseph Edward.
    Mechanical engineering design.

    (McGraw-Hill series in mechanical engineering)
    Includes index.
    1. Machinery—Design. I. Title.
TJ230.S5 1977   621.8'15   76-18775
ISBN 0-07-056881-2

**MECHANICAL
ENGINEERING
DESIGN**

    7 8 9 0 FGRFGR 7 8 3 2 1 0

CONTENTS

Preface   xi

Acknowledgments   xv

Part 1   Fundamentals of Mechanical Design

1   Introduction   3

1-1   The Phases of Design.   *1-2*   Recognition and Identification.
1-3   The Mathematical Model.   *1-4*   Evaluation and Presentation.
1-5   Design Factors.   *1-6*   The Strength Design Factor.
1-7   Economics.   *1-8*   Systems of Units.   *1-9*   English Systems.
1-10   The International System of Units.   *1-11*   Rules for Use of SI
Units.   *1-12*   Precision and Rounding of Quantities.
1-13   Conversion.

2   Stress Analysis   26

2-1   Stress.   *2-2*   Mohr's Circle.   *2-3*   Mohr's Circle for Three-
dimensional Stress.   *2-4*   Uniform Stress.   *2-5*   Elastic Strain.

2-6 Stress-Strain Relations. 2-7 Shear and Moment in Beams.
2-8 Singularity Functions. 2-9 Normal Stresses in Bending.
2-10 Beams with Unsymmetrical Sections. 2-11 Shear Stresses in
Bending. 2-12 Shear Flow. 2-13 Torsion. 2-14 Thin-walled
Cylinders. 2-15 Stresses in Thick-walled Cylinders. 2-16 Press
and Shrink Fits. 2-17 Thermal Stresses and Strains. 2-18 Curved
Beams. 2-19 Hertz Contact Stresses.

3 Deflection Analysis   94

3-1 Spring Rates. 3-2 Simple Tension, Compression, and Torsion.
3-3 Deflection of Beams. 3-4 Deflections by Use of Singularity
Functions. 3-5 The Method of Superposition. 3-6 The Graphical-
integration Method. 3-7 Strain Energy. 3-8 The Theorem of
Castigliano. 3-9 Deflection of Curved Members. 3-10 Theory of
Columns. 3-11 Column Design. 3-12 The Secant Formula.

4 Statistical Considerations in Design   130

4-1 Permutations and Combinations. 4-2 Probability.
4-3 Probability Theorems. 4-4 Random Variables. 4-5 Sample
and Population. 4-6 The Normal Distribution. 4-7 Sampling
Distributions. 4-8 Population Combinations. 4-9 Dimensioning—
Definitions and Standards. 4-10 Statistical Analysis of Tolerancing.

5 The Strength of Mechanical Elements   162

5-1 Some Remarks on Strength. 5-2 Ductility and Hardness.
5-3 Mechanical Properties. 5-4 The Maximum-Normal Stress
Theory. 5-5 The Maximum-Shear-Stress Theory. 5-6 The
Distortion-Energy Theory. 5-7 Failure of Ductile Materials with
Static Loads. 5-8 Failure of Brittle Materials with Static Loads.
5-9 Fatigue. 5-10 Fatigue Strength and Endurance Limit.
5-11 Finite-Life Strength. 5-12 Cumulative Fatigue Damage.
5-13 Endurance-Limit Modifying Factors. 5-14 Surface Finish.
5-15 Size Effects. 5-16 Reliability. 5-17 Temperature Effects.
5-18 Stress Concentration. 5-19 Miscellaneous Effects.
5-20 Fluctuating Stresses. 5-21 Fatigue Strength Under Fluctuating
Stresses. 5-22 Torsional Fatigue Strength. 5-23 Fatigue Failure
Due to Combined Stresses. 5-24 Surface Strength.

Part 2   Design of Mechanical Elements

6 The Design of Screws, Fasteners, and Connections   227

6-1 Thread Standards and Definitions. 6-2 The Mechanics of
Power Screws. 6-3 Thread Stresses. 6-4 Threaded Fasteners.

6-5 Preloading of Bolts. *6-6* Torque Requirements. *6-7* Bolt Strength and Preload. *6-8* Selection of the Nut. *6-9* Fatigue Loading. *6-10* Bolted and Riveted Joints Loaded in Shear. *6-11* Centroids of Bolt Groups. *6-12* Shear of Bolts and Rivets Due to Eccentric Loading. *6-13* Keys, Pins, and Retainers.

7   Welded, Brazed, and Bonded Joints   274

7-1 Welding. *7-2* Butt and Fillet Welds. *7-3* Torsion in Welded Joints. *7-4* Bending in Welded Joints. *7-5* The Strength of Welded Joints. *7-6* Resistance Welding. *7-7* Bonded Joints.

8   Mechanical Springs   295

8-1 Stresses in Helical Springs. *8-2* Deflection of Helical Springs. *8-3* Extension Springs. *8-4* Compression Springs. *8-5* Spring Materials. *8-6* Fatigue Loading. *8-7* Helical Torsion Springs. *8-8* Belleville Springs. *8-9* Miscellaneous Springs. *8-10* Critical Frequency of Helical Springs. *8-11* Energy-Storage Capacity.

9   Antifriction Bearings   319

9-1 Bearing Types. *9-2* Bearing Life. *9-3* Bearing Load. *9-4* Selection of Ball and Straight Roller Bearings. *9-5* Selection of Tapered Roller Bearings. *9-6* Lubrication. *9-7* Enclosure. *9-8* Shaft and Housing Details.

10   Lubrication and Journal Bearings   347

10-1 Types of Lubrication. *10-2* Viscosity. *10-3* Petroff's Law. *10-4* Stable Lubrication. *10-5* Thick-Film Lubrication. *10-6* Hydrodynamic Theory. *10-7* Design Factors. *10-8* The Relation of the Variables. *10-9* Temperature and Viscosity Considerations. *10-10* Optimization Techniques. *10-11* Pressure-Fed Bearings. *10-12* Heat Balance. *10-13* Bearing Design. *10-14* Bearing Types. *10-15* Thrust Bearings. *10-16* Boundary Lubrication. *10-17* Bearing Materials. *10-18* Design of Boundary-lubricated Bearings.

11   Spur Gears   398

11-1 Nomenclature. *11-2* Conjugate Action. *11–3* Involute Properties. *11-4* Fundamentals. *11-5* Contact Ratio. *11-6* Interference. *11-7* The Forming of Gear Teeth. *11-8* Tooth Systems. *11-9* Gear Trains. *11-10* Force Analysis. *11-11* Tooth Stresses. *11-12* Estimating Gear Size. *11-13* Tooth Fatigue Stress. *11-14* Bending Strength. *11-15* Factor of Safety. *11-16* Surface Durability. *11-17* Surface Fatigue Strength. *11-18* Heat

Dissipation. *11-19* Gear Materials. *11-20* Gear-Blank Design. *11-21* Involute Splines.

12 Helical, Worm, and Bevel Gears   456

12-1 Parallel Helical Gears—Kinematics. *12-2* Helical Gears—Tooth Proportions. *12-3* Helical Gears—Force Analysis. *12-4* Helical Gears—Strength Analysis. *12-5* Crossed-helical Gears. *12-6* Worm Gearing—Kinematics. *12-7* Worm Gearing—Force Analysis. *12-8* Power Rating of Worm Gearing. *12-9* Straight Bevel Gears—Kinematics. *12-10* Bevel Gears—Force Analysis. *12-11* Bevel Gearing—Bending Stress and Strength. *12-12* Bevel Gearing—Surface Durability. *12-13* Spiral Bevel Gears.

13 Shafts   504

13-1 Introduction. *13-2* Design for Static Loads. *13-3* Reversed Bending and Steady Torsion. *13-4* The Soderberg Approach. *13-5* The General Bi-axial Stress Problem. *13-6* The Sines Approach. *13-7* The Kececioglu Approach. *13-8* Formulas for Stress-Concentration Factors.

14 Clutches, Brakes, and Couplings   524

14-1 Statics. *14-2* Internal-expanding Rim Clutches and Brakes. *14-3* External-contracting Rim Clutches and Brakes. *14-4* Band-Type Clutches and Brakes. *14-5* Frictional-contact Axial Clutches. *14-6* Cone Clutches and Brakes. *14-7* Miscellaneous Clutches and Couplings. *14-8* Friction Materials. *14-9* Energy Considerations. *14-10* Heat Dissipation.

15 Flexible Mechanical Elements   555

15-1 Belts. *15-2* Flat-Belt Drives. *15-3* V Belts. *15-4* Roller Chain. *15-5* Rope Drives. *15-6* Wire Rope. *15-7* Flexible Shafts.

16 A Systems Approach   578

16-1 The Mathematical Model. *16-2* Lumped Systems. *16-3* The Dynamic Response of a Distributed System. *16-4* The Dynamic Response of a Lumped System. *16-5* Modeling the Elasticities. *16-6* Modeling the Masses and Inertias. *16-7* Modeling Friction and Damping. *16-8* Mathematical Models for Shock Analysis. *16-9* Stress and Deflection Due to Impact. *16-10* Cam Systems. *16-11* Designing with the Programmable Calculator.

Answers to Selected Problems   627

Appendix 634

A-1 Standard SI Prefixes. *A-2* Conversion of English Units to SI Units. *A-3* Conversion of SI Units to English Units.
A-4 Preferred SI Units for Bending Stress and Torsion Stress.
A-5 Preferred SI Units for Axial Stress and Direct Shear Stress.
A-6 Preferred SI Units for the Deflection of Beams. *A-7* Physical Constants of Materials. *A-8* Properties of Structural Shapes—Equal Angles. *A-9* Properties of Structural Shapes—Unequal Angles.
A-10 Properties of Round Tubing. *A-11* Properties of Structural Shapes—Channels. *A-12* Shear, Moment, and Deflection of Beams.
A-13 Ordinates of the Standard Normal Distribution. *A-14* Areas Under the Standard Normal Distribution Curve. *A-15* Limits and Fits for Cylindrical Parts. *A-16* American Standard Pipe.
A-17 Mechanical Properties of Steels. *A-18* Mechanical Properties of Wrought Aluminum Alloys. *A-19* Mechanical Properties of Aluminum Alloy Castings. *A-20* Typical Properties of Gray Cast Iron. *A-21* Typical Properties of Some Copper-base Alloys.
A-22 Typical Mechanical Properties of Wrought Stainless Steels.
A-23 Typical Properties of Magnesium Alloys. *A-24* Decimal Equivalents of Wire and Sheet-Metal Gauges. *A-25* Charts of Theoretical Stress-Concentration Factors K_t. Greek Alphabet.
A-26 Dimensions of Round-Head Machine Screws.
A-27 Dimensions of Hexagon-Head Cap Screws. *A-28* Dimensions of Finished Hexagon Bolts. *A-29* Dimensions of Finished Hexagon and Hexagon Jam Nuts. *A-30* Properties of Sections. *A-31* Mass and Mass Moments of Inertia of Geometric Shapes.

Index 681

PREFACE

This book has been written for engineering students who are beginning a course of study in mechanical engineering design. Such students will have acquired a set of engineering *tools* consisting, essentially, of mathematics, computer languages, and the ability to use the English language to express themselves in the spoken and written forms. Mechanical design involves a great deal of geometry, too, therefore another useful tool is the ability to sketch and draw the various configurations which arise. Students will also have studied a number of basic engineering *sciences*, including physics, engineering mechanics, materials and processes, and the thermal-fluid sciences. These, the tools and sciences, constitute the foundation for the practice of engineering, and so, at this stage of undergraduate education, it is appropriate to introduce the professional aspects of engineering. These professional studies should integrate and use the tools and the sciences in the accomplishment of an engineering objective. The pressures upon the undergraduate curricula today require that we do this in the most efficient manner. Most engineering educators are agreed that mechanical design integrates and utilizes a greater number of the tools and the sciences than any other professional study. Mechanical design is also the very core of other professional and design types of studies in mechanical engineering. Thus studies in mechanical design seem to be the most effective and economical method of starting the student in the practice of mechanical engineering.

Books, like cars, always seem to grow larger and larger with each new edition. And books have been subjected to the same inflationary pressures as automobiles have in recent years, producing ever higher prices for them. In this edition I have attempted to buck this trend by trimming the fat and the frills wherever possible. The result is a leaner, more concise book that will cost the student less than one having more pages.

One of the principal reasons for writing a new edition now is the urgent need to introduce the International System of Units (SI) and the attendant rules into mechanical design studies. Accordingly about 50 percent of the illustrative examples and problems for student solution are expressed in SI units. In the case of gears, there are no metric standards in this country and so the examples and problems are mostly in English units.

It is especially important to note that the International System of Units (SI) is presented and used in this book *exactly* in accordance with the rules and recommendations as given in the National Bureau of Standards Special Publication 330, 1974 edition. To do otherwise could lead to great confusion because of the proliferation of units.

The pocket electronic calculator came along just in time to aid in the introduction of SI units into engineering. The scientific notation capability is just what is needed. But the calculator can be expected to affect the teaching and practice of design in many other ways. We can expect to see a decline in the use of charts, graphs, graphical computations, and tabular material in the very near future. Examples of this are to be found in Section 5-11, where it is found that the S-N diagram is no longer required, and in Sections 8-5 and 9-2.

There are a number of other features of this edition which should be noted. Singularity functions are introduced in Chapter 2 and are used there for shear and moment diagrams and in Chapter 3 for deflection analysis. Improved material on column analysis and design is presented in Chapter 3. In Chapter 6 additional material on both fatigue loading and shear loading of bolted joints is presented. Chapter 7 is a new chapter devoted to welded, brazed, and bonded joints. A feature of this chapter is the analytical approach used in the stress analysis of joints loaded in shear, torsion, and bending. Chapter 9 on antifriction bearings has been completely rewritten, and it contains new material on bearing life, reliability, bearing load, and tapered roller bearings. Chapter 13 on shafts has also been largely rewritten. It now contains the Sines and the Kececioglu methods of shaft design and analysis.

One cannot help but be gratified by the reception accorded the previous editions of this book. Nevertheless they did contain many unclear or glossed-over passages, examples that served no useful purposes, illustrations that did not illustrate, and problems for student solution that were poorly phrased and not well organized. So another of my objectives in writing this edition has been to remedy all these deficiencies. The illustrations and text have been checked very carefully and redone or rewritten where desirable. Most of the illustrative examples have been replaced or revised. And most of the problems for student homework are new ones. But a few problems were retained from previous editions because they did such an excellent job of permitting self-instruction of the student. Most of these have been renumbered in this edition.

It is of course very pleasing to learn that previous editions of this book are used so much by design engineers in the practice of their profession. In many places, for this reason, the text addresses the practicing engineer separately from the student, recognizing that the engineer's needs are somewhat different and that more facilities may be available to him or her for use in the entire design process. The decisions made by an engineer in solving a design problem will depend upon those facilities, and hence may vary to some extent from one industry or one engineering department to another. The student, however, wants to get the right answer, which is the answer obtained by the professor. It has not been difficult to achieve this dual objective. The entire problem is explained to the practicing engineer so that, considering problem-solving constraints, an optimum approach can be chosen. And an appropriate course of action is specified or suggested for students. This method is of more use to students because it contains fewer ambiguities and optional courses of action for their purposes, and of greater value to practicing engineers because it provides them with the options.

I especially want to encourage the users of this edition to send me their comments and suggestions. Every chapter of this edition has been influenced by the users who have taken the time to write me concerning their use of the previous editions, and I am indeed appreciative of their interest.

JOSEPH EDWARD SHIGLEY

ACKNOWLEDGMENTS

The author expresses his gratitude for suggestions to:

Robert W. Adamson, *California State Polytechnic College, San Luis Obispo, California*

Charles W. Allen, *California State University, Chico, California*

Rolin F. Barrett, *North Carolina State University, Raleigh, North Carolina*

W. K. Bodger, *Fresno State College, Fresno, California*

O. M. Browne, Jr., *University of Washington, Seattle, Washington*

Milton A. Chace, *The University of Michigan, Ann Arbor, Michigan*

Frederick A. Costello, *University of Delaware, Newark, Delaware*

Joseph Datsko, *The University of Michigan, Ann Arbor, Michigan*

Winston M. Dudley, *Sacramento State College, Sacramento, California*

G. A. Fazekas, *University of Houston, Houston, Texas*

Ferdinand Freudenstein, *Columbia University, New York, New York*

Franklin D. Hart, *North Carolina State University, Rayleigh, North Carolina*

Robert C. Juvinall, *The University of Michigan, Ann Arbor, Michigan*

William C. Kieling, *University of Washington, Seattle, Washington*

W. A. Kleinhenz, *University of Minnesota, Minneapolis, Minnesota*

Charles Lipson, *The University of Michigan, Ann Arbor, Michigan*

Robert A. Lucas, *Lehigh University, Bethlehem, Pennsylvania*

Charles R. Mischke, *Iowa State University, Ames, Iowa*

Larry D. Mitchell, *Virginia Polytechnic Institute and State University, Blacksburg, Virginia*

Charles Nuckolls, *Florida Technical University, Orlando, Florida*

Charles B. O'Toole, *Pennsylvania State University, McKeesport Campus, McKeesport, Pennsylvania*

Dan R. Rankin, *California State College, Los Angeles, California*

George N. Sandor, *Rensselaer Polytechnic Institute, Troy, New York*

Arthur W. Sear, *California State College, Los Angeles, California*

Walter L. Starkey, *Ohio State University, Columbus, Ohio*

Ralph I. Stevens, *University of Iowa, Iowa City, Iowa*

Ward O. Winer, *Georgia Institute of Technology, Atlanta, Georgia*

JOSEPH EDWARD SHIGLEY

**MECHANICAL
ENGINEERING
DESIGN**

PART ONE

Fundamentals of Mechanical Design

INTRODUCTION

This book is a study of the decision-making processes which mechanical engineers use in the formulation of plans for the physical realization of machines, devices, and systems. These decision-making processes are applicable to the entire field of engineering design—not just to mechanical engineering design. To understand them, to apply them to practical situations, and to make them pay off, however, require a set of circumstances, a particular situation, or a vehicle, so to speak. In this book we have therefore chosen the field of mechanical engineering as the vehicle for the application of these decision-making processes.

Mechanical design means the design of things and systems of a mechanical nature—machines, products, structures, devices, and instruments. For the most part mechanical design utilizes mathematics, the materials sciences, and the engineering-mechanics sciences.

Mechanical engineering design includes all mechanical design, but it is a broader study because it includes all the disciplines of mechanical engineering, such as the thermal-fluids sciences, too. Aside from the fundamental sciences which are required, the first studies in mechanical engineering design are in mechanical design, and hence this is the approach taken in this book.

The book is divided into two parts. Part 1 is concerned with the fundamentals of decision making, the mathematical and analytical tools we will require, and the actual subject matter to be employed in using these tools. Occasionally you may encounter a familiar subject. This is included so that you can review it, if necessary, but more importantly, to establish the nomenclature for use in the more advanced portions of the book, for continuity, and as a reference source.

In Part 2 the fundamentals are applied to many typical design situations which arise in the design or selection of the elements of mechanical systems. An attempt has been made to arrange Part 2 so that the basic or more common elements are studied first. In this manner, as you become familiar with the design of single elements, you can begin to put them together to form complete machines or systems. Thus, with respect to a whole system, Part 2 becomes progressively more comprehensive. The intent, therefore, is that Part 2 should be studied chapter by chapter in the order in which it is presented.

1-1 THE PHASES OF DESIGN

The total design process is of interest to us in this chapter. How does it begin? Does the engineer simply sit down at his desk with a blank sheet of paper? And, as he jots down some ideas, what happens next? What factors influence or control the decisions which have to be made? Finally, then, how does this design process end?

This complete process, from start to finish, is often outlined as in Fig. 1-1. The process begins with a *recognition of a need* and a decision to do something about it. After many iterations, the process ends with the *presentation* of the plans for satisfying the need. In the next several sections we shall examine these steps in the design process in detail.

1-2 RECOGNITION AND IDENTIFICATION

Sometimes, but not always, design begins when an engineer recognizes a need and decides to do something about it. Recognition of the need and phrasing it in so many words often constitute a highly creative act because the need may be only a vague discontent, a feeling of uneasiness, or a sensing that something is not right. Recognition is usually triggered by a particular adverse circumstance or a set of random circumstances which arise almost simultaneously. It is evident, too, that a sensitive person, one who is easily disturbed by things, is more likely to recognize a need—and also more likely to do something about it. And for this reason sensitive people are more creative.

The need is usually not evident at all, as we have indicated. For example, the need to do something about a food-packaging machine may be indicated by the noise level, by the variation in package weight, and by slight but perceptible variations in the quality of the packaging or wrap.

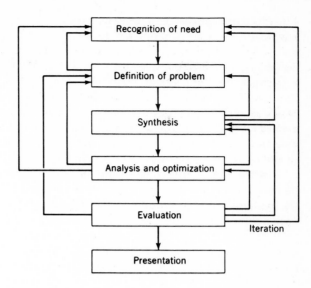

FIGURE 1-1
The phases of design.

A need is easily recognized after someone else has stated it. Thus the need in this country for cleaner air and water, for more parking facilities in the cities, for better public transportation systems, and for faster traffic flow has become quite evident.

There is a distinct difference between the statement of the need and the *identification of the problem* which follows this statement (Fig. 1-1). The problem is more specific. If the need is for cleaner air, the problem might be that of reducing the dust discharge from power-plant stacks, or reducing the quantity of irritants from automotive exhausts, or means for quickly extinguishing forest fires.

Definition of the problem must include all the specifications for the thing that is to be designed. The specifications are the input and output quantities, the characteristics and dimensions of the space the thing must occupy and all the limitations on these quantities. We can regard the thing to be designed as something in a black box. In this case we must specify the inputs and outputs of the box together with their characteristics and limitations. The specifications define the cost, the number to be manufactured, the expected life, the range, the operating temperature, and the reliability. Obvious items in the specifications are the speeds, feeds, temperature limitations, maximum range, expected variations in the variables, and dimensional and weight limitations.

There are many implied specifications which result either from the designer's particular environment or from the nature of the problem itself. The manufacturing processes which are available, together with the facilities of a certain plant, constitute restrictions on a designer's freedom, and hence are a part of the implied specifications. A small plant, for instance, may not own cold-working machinery. Knowing this, the designer selects other metal-processing methods which can be performed in the plant. The labor skills available and the competitive situation also constitute implied specifications.

Anything which limits the designer's freedom of choice is a specification. Many materials and sizes are listed 'in supplier's catalogs, for instance, but these are not all easily available and shortages frequently occur. Furthermore, inventory economics require that a manufacturer stock a minimum number of materials and sizes.

After the problem has been defined and a set of written and implied specifications has been obtained, the next step in design, as shown in Fig. 1-1, is the synthesis of an optimum solution. Now synthesis cannot take place without both analysis and optimization because the system under design must be analyzed to determine whether the performance complies with the specifications. The analysis may reveal that the system is not an optimum one. If the design fails either or both of these tests, the synthesis procedure must begin again.

1-3 THE MATHEMATICAL MODEL

We have noted, and we shall do so again and again, that design is an iterative process in which we proceed through several steps, evaluate the results, and then return to an earlier phase of the procedure. Thus we may synthesize several components of a system, analyze and optimize them, and return to synthesis to see what effect this has on the remaining parts of the system. Both analysis and optimization require that we construct or devise abstract models of the system which will admit some form of mathematical analysis. We call these models *mathematical models*. In creating them it is our hope that we can find one which will simulate the real physical system very well.

All real physical systems are complex. Creating a mathematical model of the system means that we are simplifying the system to the point where it can be analyzed. The term *rigid body* is one such idealization. There are no rigid bodies to be found in nature. The term *concentrated force* is another idealization. When we use it we are assuming that the area through which the force acts is relatively small. As we shall see, there are many other kinds of idealizations which have to be used.

The nature of the problem, its economics, the computational facilities available, and the ability and working time of the engineer all play a key role in the formulation of the model. Some problems are so important that they require a very sophisticated model, and the economics of the problem may justify such a model. At other times the nature of the problem and its economics indicate that high-quality engineering is neither necessary nor desired. In such a case the engineer may elect to create a simple model in order to obtain results in a short period of time.

1-4 EVALUATION AND PRESENTATION

As indicated in Fig. 1-1, *evaluation* is a significant phase of the total design process. Evaluation is the final proof of a successful design, which usually involves the testing of a prototype in the laboratory. Here we wish to discover if the design really satisfies the need or needs. Is it reliable? Will it compete successfully with similar products? Is it economical to manufacture and to use? Is it easily maintained and adjusted? Can a profit be made from its sale or use?

Communicating the design to others is the final, vital step in the design process. Undoubtedly many great designs, inventions, and creative works have been lost to mankind simply because the originators were unable or unwilling to explain their accomplishments to others. Presentation is a selling job. The engineer, when presenting a new solution to administrative, management, or supervisory persons, is attempting to sell or to prove to them that this solution is a better one. Unless this can be done successfully, the time and effort spent on obtaining the solution have been largely wasted.

One who sells a new idea also sells oneself. If one is repeatedly successful in selling ideas, designs, new solutions, and the like, to management, then one begins to receive salary increases and promotions; in fact, this is how one climbs the ladder to success. Basically, there are only three means of communication available to us. These are the *written*, the *oral*, and the *graphical* forms. Therefore the successful engineer will be technically competent and versatile in *all three forms* of communication. A technically competent person who lacks ability in any one of these forms is severely handicapped. If ability in all three forms is lacking, no one will ever know how competent that person is!

The three forms of communication, writing, speaking, and drawing, are *skills*, that is, abilities which can be developed or acquired by any reasonably intelligent person. Skills are acquired only by practice—hour after monotonous hour of it. Musicians, athletes, surgeons, typists, writers, dancers, aerialists, and artists, for example, are skillful because of the number of hours, days, weeks, months, and years they have practiced. Nothing worthwhile in life can be achieved without work, often tedious, dull, and monotonous, and lots of it; and engineering is no exception.

Ability in writing can be acquired by writing letters, reports, memos, papers, and articles. It does not matter whether or not the articles are published—the practice is the important thing. Ability in speaking can be obtained by participating in fraternal, civic, church, and professional activities. This participation provides abundant opportunities for practice in speaking. To acquire drawing ability, pencil sketching should be employed to illustrate every idea possible. The written or spoken word often requires study for comprehension, but pictures are readily understood and should be used freely.

The competent engineer should not be afraid of the possibility of not succeeding in a presentation. In fact, occasional failure should be expected because failure or criticism seems to accompany every really creative idea. There is a great deal to be learned from a failure, and the greatest gains are obtained by those willing to risk defeat. In the final analysis, the real failure would lie in deciding not to make the presentation at all.

The purpose of this section is to note the importance of presentation as the final step in the design process. Thus, no matter whether you are planning a presentation to your teacher or to your employer, you should communicate thoroughly and clearly, for this is the payoff. Helpful information on report writing, public speaking, and sketching or drafting is available from countless sources, and you should take advantage of these aids.

1-5 DESIGN FACTORS

Sometimes the strength of an element is an important consideration in the determination of the geometry and the dimensions of the element. In such a situation we say that *strength* is an important design factor.

When we use the expression *design factor* we are referring to some characteristic or consideration which influences the design of the element or, perhaps, the entire system. Usually a number of these design factors have to be considered in any given design situation.* Sometimes one of these will turn out to be critical, and when it is satisfied, the other factors no longer need to be considered. As an example, the following list of design factors must often be considered:

1	Strength	*12*	Noise
2	Reliability	*13*	Styling
3	Thermal considerations	*14*	Shape
4	Corrosion	*15*	Size
5	Wear	*16*	Flexibility
6	Friction	*17*	Control
7	Processing	*18*	Stiffness
8	Utility	*19*	Surface finish
9	Cost	*20*	Lubrication
10	Safety	*21*	Maintenance
11	Weight	*22*	Volume

Some of these factors have to do directly with the dimensions, the material, the processing, and the joining of the elements of the system. Other factors affect the configuration of the total system. We shall be giving our attention to these factors, as well as many others, throughout the book.

In this book you will be faced with a great many design situations in which engineering fundamentals must be applied, usually in a mathematical approach, to resolve the problem or problems. This is completely correct and appropriate in

* In design literature the expression "design factor" is used by some writers to designate the ratio of the strength of an element to the internal stresses created by the external forces which act upon the element.

an academic atmosphere where the need is actually to utilize these fundamentals in the resolution of professional problems. To keep the correct perspective, however, it should be observed that in many design situations the important design factors are such that no calculations or experiments are necessary in order to define an element or a system. Students, especially, are often confounded when they run into situations in which it is virtually impossible to make a single calculation and yet an important design decision must be made. These are not extraordinary occurrences at all; they happen every day. This point is made here so that you will not be misled into believing that there is a rational mathematical approach to every design decision.

1-6 THE STRENGTH DESIGN FACTOR

Strength is a *property* of a material or of a mechanical element. The strength of an element depends upon the choice, the treatment, and the processing of the material. Consider, for example, a shipment of 1000 springs. We can associate a *strength* S_i with the *i*th spring. The *stress* in this spring is zero, however, until it is assembled into a machine. After assembly, external forces are applied to the spring which result in stresses whose magnitudes depend upon the geometry of the spring and are independent of the material and its processing. If the spring is removed from the machine unharmed, the stress will again be zero, but the strength S_i remains as one of the properties of the spring. So remember, *stress* is something that happens to a part because of the application of a force. But *strength* is an inherent property of that part built in by the use of a particular material and process.

In this book the capital letter S is used to denote strength with appropriate subscripts to designate the kind of strength.

Many of the static strengths to be used depend upon information gathered from the standard tensile test. This test uses a specimen machined to stated dimensions. The original area and length of the gauge used to measure the strains (usually 2 in in English units) are recorded before the test is begun. The specimen is slowly loaded in tension while the load and strain are observed. At the conclusion of the test the results are plotted as a *stress-strain diagram* (Figs. 1-2 and 1-3).

Point A in Fig. 1-2a is called the *proportional limit*. This is the point at which the stress-strain diagram first begins to deviate from a straight line. Point B is called the *elastic limit*. No permanent set will be observable in the specimen if the load is removed at this point. Between A and B the diagram is not a perfectly straight line even though the specimen is elastic. Thus *Hooke's law*, which states that stress is proportional to strain, applies only up to the proportional limit.

During the tension test, many materials reach a point at which the strain begins to increase very rapidly without a corresponding increase in stress. This is called the *yield point*: point C in Fig. 1-2a. Not all materials have a yield point that is so easy to find. For this reason, *yield strength* is often defined by an *offset method*

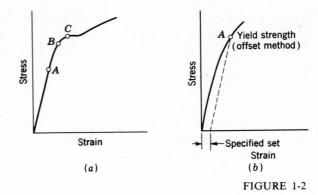

(a)

(b)

FIGURE 1-2

as shown in Fig. 1-2*b*. Such a yield strength corresponds to a definite amount of permanent set, usually 0.2 or 0.5 percent of the original gauge length.

The *ultimate*, or *tensile*, *strength* is the maximum stress reached on the stress-strain diagram (Fig. 1-3*a*, point *B*). Some materials exhibit a downward trend after the maximum stress is reached. Others, such as cast iron and some of the high-strength steels, fracture when the stress-strain diagram is still rising.

Compression tests are more difficult to make, and the geometry of the test specimens differs from the geometry of those used in tension tests. The reason for this is that the specimen may buckle during testing or it may be difficult to get the stresses distributed evenly. Other difficulties occur because ductile materials will bulge after yielding. However, the results can be plotted on a stress-strain diagram too, and the same strength definitions can be applied. For many materials the compressive strengths are about the same as the tensile strengths. When substantial differences occur, however, as is the case with the cast irons, the tensile and compressive strengths should be stated separately.

Torsional strengths are found by twisting round bars and recording the

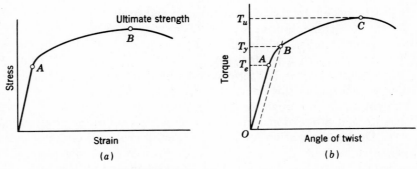

(a)

(b)

FIGURE 1-3
(a) Tension test of a ductile material; (b) torque-twist diagram.

torque and the twist angle. The results are then plotted as a *torque-twist diagram* as shown in Fig. 1-3*b*. By using the torsional stress equations to be presented in Chap. 2 the elastic limit corresponding to torque T_e, and the torsional yield strength, corresponding to T_y, may be found. Point *C* on the diagram corresponds to the ultimate torsional strength S_{su}. However, this is not a true maximum stress because the core of a bar so twisted will still be in an elastic state. For this reason, the strength S_{su}, corresponding to T_u, is properly called the *modulus of rupture*.

In accordance with accepted practice, we shall employ the Greek letters σ and τ in this book to designate the normal and shear stresses, respectively.

The strength design factor, called *factor of safety*, is a factor used to evaluate the safeness of a member. Let a member be subjected to some applied load *F*. We assume that *F* is a very general term and that it can be either a force, torque, or a moment, for example. If *F* is increased, eventually it will become so large that any additional small increase would permanently impair the ability of the member to perform its proper function. If we designate this limiting value of *F* as F_u, then the factor of safety is defined as

$$n = \frac{F_u}{F}$$

In many cases stress is linearly proportional to load. Under these conditions factor of safety is

$$n = \frac{S}{\sigma} \quad \text{or} \quad n = \frac{S_s}{\tau} \qquad (1\text{-}1)$$

When the stress becomes equal to the strength, $n = 1$ and there is no safety at all. Consequently the term *margin of safety* is frequently used. Margin of safety is defined by the equation

$$m = n - 1 \qquad (1\text{-}2)$$

The terms factor of safety and margin of safety are widely used in industrial practice, and the meaning and intent of these terms is clearly understood. However, the strength of an element is a statistically varying quantity, and the stress, too, is variable. For this reason a factor of safety $n > 1$ *does not preclude failure*. Because of this correlation between the degree of hazard and *n*, some authorities prefer to use the term "design factor" instead. As long as you understand the intent, either term is correct. As indicated earlier, "design factor," in this book, is used to describe the various decision-making considerations. *Factor of safety*, here, is the factor *n* of Eq. (1-1).

All stress equations may be represented by the general equation

$$\sigma = Cf(x_1, x_2, \ldots, x_i)F(F_1, F_2, \ldots, F_j) \qquad (1\text{-}3)$$

where *C* = a constant
  x_i = dimensions of part to be designed
  F_j = external forces or loads applied to part to be designed

Equations for stress in this book will always be written in this form. In analysis, of course, the equations can be solved directly for the stresses. In design, the right-hand side contains the forces which are usually known and the dimensions which are usually to be determined. In writing a stress equation, always set it up in this form, substitute in it the known quantities, and leave the appropriate symbols, designating the dimensions to be found, on the right of the equals sign. Now, in place of σ, substitute S/n on the left-hand side of the equation, giving

$$\frac{S}{n} = Cf(x_1, x_2, \ldots, x_i)F(F_1, F_2, \ldots, F_j) \qquad (1\text{-}4)$$

Both the strength S and the factor of safety n will have been previously determined, and the equation can now be solved for the dimensions.

Safety can also be obtained by the use of a so-called allowable stress. An *allowable stress* is the maximum safe stress that is permitted in an element. Such stresses may be defined by a construction code or by a factor of safety. Thus an alternative approach is to write

$$\sigma_{\text{all}} = \frac{S}{n} \qquad \text{or} \qquad \tau_{\text{all}} = \frac{S_s}{n} \qquad (1\text{-}5)$$

where σ_{all} and τ_{all} are the allowable normal and shear stresses respectively.

Sometimes the allowable stresses are specified by a construction code. For example, the AISC manual* specifies the allowable tensile stress for structural steel, rivets, bolts, and weld metal as 20 kpsi. The code does not state what factor of safety was used to obtain this stress nor what strength was used. However, since these codes do specify the grade of steel, it is possible to work backward from the known strength and determine what factor of safety was used.

1-7   ECONOMICS

The cost design factor plays such an important role in the design decision process that one could spend as much time in studying it as in the study of design itself. Here we shall introduce only a few general approaches and simple rules.

First, observe that nothing can be said in an absolute sense concerning costs. Materials and labor usually show an increasing cost from year to year. But the costs of processing the materials can be expected to exhibit a decreasing trend because of the use of automated machine tools. The cost of manufacturing a single product will vary from one city to another and from one plant to another because of the differences in overhead, labor, freight differentials, and slight manufacturing variations.

---

* "Steel Construction," p. 5-16, American Institute of Steel Construction (AISC), New York. This manual is republished frequently, and the latest edition can be obtained from any bookseller.

Use Standard Sizes

This is a first principle of cost reduction. An engineer who specifies a G10350* bar of hot-rolled steel $2\frac{1}{8}$-in square, called a hot-rolled square, has added cost to the product provided a 2- or $2\frac{1}{4}$-in square, both of which are standard, would do equally well. The $2\frac{1}{8}$-in size can be obtained by special order, or by rolling or machining a $2\frac{1}{4}$-in bar, but these approaches add cost to the product. Of course, millimetre sizes are used in SI.

To assure that standard sizes are specified, the designer must have access to stock lists of the materials he employs. These are available in libraries or can be obtained directly from the suppliers.

A further word of caution regarding the selection of standard sizes of materials is necessary. Although a great many sizes are usually listed in catalogs, they are not all readily available. Some sizes are used so infrequently that they are not stocked. A rush order for such sizes may mean more expense and delay. Thus one should also have access to a list of preferred sizes.

There are many purchased parts, such as motors, pumps, bearings, and fasteners, which are specified by designers. In the case of these, too, the designer should make a special effort to specify parts which are readily available. If they are made and sold in large quantities, they usually cost considerably less than the odd sizes. The cost of ball bearings, for example, depends more upon the quantity of production by the bearing manufacturer than upon the size of the bearing.

Use Large Tolerances

Among the effects of design specifications on costs, those of tolerances are perhaps most significant. Tolerances in design influence the producibility of the end product in many ways, from necessitating additional steps in processing to rendering a part completely impractical to produce economically. Tolerances cover dimensional variation and surface-roughness range and also the variation in mechanical properties resulting from heat treatment and other processing operations.

Parts having large tolerances can often be produced by machines with higher production rates. And the labor cost will be smaller if skilled workmen are not required because of larger tolerances. Also, fewer of such parts will be rejected in the inspection process, and they are usually easier to assemble.

Breakeven Points

Sometimes it happens that, when two or more design approaches are compared for cost, the choice between the two depends upon another set of conditions, such as the quantity of production, the speed of the assembly lines, or some other condition. There then occurs a point corresponding to equal cost which is called the *breakeven point*.

* See Table A-17.

As an example, consider a situation in which a certain part can be manufactured at the rate of 25 parts per hour on an automatic screw machine or 10 parts per hour on a hand screw machine. Let us suppose, too, that the setup time for the automatic is 3 h and that the labor cost for either machine is $20 per h, including overhead. Figure 1-4 is a graph of the cost versus production by the two methods. The breakeven point corresponds to 50 parts. If the desired production is greater than 50 parts, the automatic machine should be used.

Cost Estimates

There are many ways of obtaining relative cost figures so that two or more designs can be roughly compared. A certain amount of judgment may be required in some instances. For example, we can compare the relative value of two automobiles by comparing the dollar cost per pound of weight. Another way to compare the cost of one design with another is simply to count the number of parts. The design having the smallest number of parts is likely to cost less.

Many other cost estimators can be used, depending upon the application, such as area, volume, horsepower, torque, capacity, speed, and various performance ratios.

Value Engineering

A method of evaluating several proposed designs using a systematic approach called *value analysis*, or *value engineering*, is often very useful and may even point

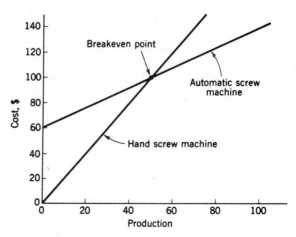

FIGURE 1-4

the way to new design approaches. In this method* *value* is defined as a numerical ratio, the ratio of the *function,* or *performance,* to the *cost.*

As an elementary introduction to the methods of value analysis let us analyze the ball bearing of Fig. 1-5. The analysis is carried out in tabular form and is illustrated by Table 1-1. Since only bearing manufacturers have access to actual costs, the analysis which follows is a hypothetical situation. The steps proceed from left to right across the table. These are:

1 Name of part
2 Brief statement or summary of function
3 Estimate of percent of total performance contributed by this function
4 Cost of part
5 Percent of total cost
6 Value, the ratio of percent performance to percent cost

The reasoning behind the analysis shown in Table 1-1 is as follows:

As we shall learn in Chap. 9, the outer and inner rings provide raceways for the balls. Also, the bore of the inner ring and the outside diameter of the outer ring are very accurately ground so as to provide an excellent fit when the completed bearing is assembled into a machine. Further, the faces of these two rings often bear against other parts to permit the bearing to resist axial loads and to maintain alignment of rotating parts.

The function of the balls is to support the load and, by their rolling action, to reduce friction.

If the balls were not kept apart by the separator, they would roll against one another and generate a larger frictional torque. The separator also assures the same number of balls in the load-carrying zone at all times.

As noted in the table, the rivets secure the two separator halves together.

Note that the value of the separator is equal to that of the balls even though the function of the separator is a secondary one. It would therefore be poor economy to attempt to eliminate the separator. A considerable saving might be made, however, if the value of the outer ring could be increased.

* See D. Henry Edel, Jr., "Introduction to Creative Design," pp. 111–119, Prentice-Hall, Inc., Englewood Cliffs, N.J., 1967.

Table 1-1 VALUE ANALYSIS OF A BALL BEARING

Name of part	Function	% performance	Cost, $	% cost	Value
Outer ring	Provides ball race	20	2.35	47	0.425
Inner ring	Provides ball race	20	2.00	40	0.500
Balls	Supports load	50	0.50	10	5.000
Separator	Separates balls	10	0.10	2	5.000
Rivets	Fastens separator halves	0	0.05	1	0
Total		100	5.00	100	

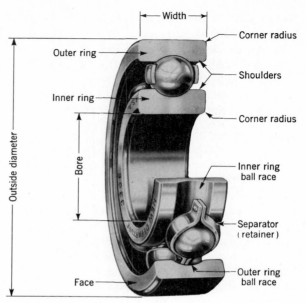

FIGURE 1-5
Nomenclature of a ball bearing. (*Courtesy of New-Departure-Hyatt Division, General Motors Corporation.*)

1-8 SYSTEMS OF UNITS

Newton's first two laws may be summarized by the equation

$$\mathbf{F} = m\mathbf{a} \qquad (1\text{-}6)$$

which is called the equation of particle motion. In this equation $\mathbf{a}$ is the acceleration experienced by a particle of mass m when it is acted upon by the force $\mathbf{F}$. Both $\mathbf{F}$ and $\mathbf{a}$ are vector quantities.

An important use of Eq. (1-6) occurs in the standardization of systems of units. Let us employ the following symbols to designate units as follows:

Force, F
Mass, M
Length, L
Time, T

These symbols are to stand for any unit we might choose to employ. Thus, possible choices for the symbol L could be inches, or kilometres, or miles. The symbols F, M, L, and T are not numbers, but they may be substituted into Eq. (1-6) as if they were. The equality sign then implies that the symbols on one side are equivalent to those on the other. Making the indicated substitution then gives

$$F = MLT^{-2} \qquad (1\text{-}7)$$

because the acceleration **a** has units of length divided by time squared. Equation (1-7) expresses an equivalency between the four units, force, mass, length, and time. We are free to choose units for three of these and then the units used for the fourth depend upon the first three. It is for this reason that the first three units chosen are called the *basic units* while the fourth is called the *derived unit*.

When force, length, and time are chosen as the basic units, the mass is the derived unit and the system which results is called a *gravitational system of units*.

When mass, length, and time are chosen as the basic units, force is the derived unit and the system which results is called an *absolute system of units*.

1-9 ENGLISH SYSTEMS

In the English-speaking countries the *British foot-pound-second system* (FPS) and the *inch-pound-second system* (IPS) are the two standard gravitational systems most used by engineers. In the FPS system the unit of mass is

$$M = \frac{FT^2}{L} = \frac{(\text{pound-force})(\text{second})^2}{\text{foot}} = \text{lbf·s}^2/\text{ft} = \text{slug} \qquad (1\text{-}8)$$

Thus, length, time, and force are the three basic units in the FPS gravitational system. The unit of length is the foot which is abbreviated as ft; but to save space in illustrations, length in feet will be designated by a single superior mark according to conventional practice. Thus a distance of 16 ft in the text will appear as 16′ on illustrations.

The unit of time in the FPS system is the second, abbreviated as s.

The unit of force in the FPS system is the pound, more properly called the *pound-force*. We shall seldom abbreviate this unit as lbf; the abbreviation lb is permissible, since we shall be dealing only with English gravitational systems. It is interesting to know that the abbreviation lb for pound comes from Libra the balance, the seventh sign of the zodiac, which is represented as a pair of scales.

In some branches of engineering it is useful to represent 1000 lb as a kilopound and to abbreviate it as kip. Many authors add the letter s to obtain the plural but to be consistent with the practice of using only singular units we shall not do so here. Thus both 1 kip and 3 kip are appropriate.

Finally we note in Eq. (1-8) that the derived unit of mass in the FPS gravitational system is the lb·s²/ft and that this unit is called a *slug*; there is no abbreviation. Note that a centered period, instead of a hyphen, is used between lb and s².

The unit of mass in the IPS gravitational system is

$$M = \frac{FT^2}{L} = \frac{(\text{pound-force})(\text{second})^2}{\text{inch}} = \text{lb·s}^2/\text{in} \qquad (1\text{-}9)$$

In this system the unit of length is the inch and is abbreviated as in without being followed by a period. Conventional practice on drawings is to use double superior marks to mean inches. Thus eight inches will be written as 8″ on drawings and illustrations and as 8 in in the text.

The weight of an object is the force exerted upon it by gravity. Designating the weight as W and the acceleration due to gravity as g, Eq. (1-6) becomes

$$W = mg \qquad (1\text{-}10)$$

In the FPS system, standard gravity is

$$g = 32.1740 \text{ ft/s}^2$$

For most cases this is rounded off to 32.2. Thus the weight of a mass of one slug in the FPS system is

$$W = mg = (1 \text{ slug})(32.2 \text{ ft/s}^2) = 32.2 \text{ lb}$$

In the IPS system, standard gravity is 386.088 or about 386 in/s². Thus, in this system, a unit mass weighs

$$W = (1 \text{ lbf·s}^2/\text{in})(386 \text{ in/s}^2) = 386 \text{ lbf}$$

1-10   THE INTERNATIONAL SYSTEM OF UNITS

The official name of the International System of Units is Le Système International d'Unités, abbreviated SI. Don't call it the "SI system" because the S in SI means "system."

SI is an absolute system. The basic units are the metre, the kilogram-mass, and the second.* The unit of force is derived and is called a newton to distinguish it from the kilogram which, as indicated, is the unit of mass. The units of the newton are

$$F = \frac{ML}{T^2} = \frac{(\text{kilogram})(\text{metre})}{(\text{second})^2} = \text{kg·m/s}^2$$

The abbreviation for a newton is N.

The seven SI base units, with their symbols, are shown in Table 1-2. These are dimensionally independent. Lower-case letters are used for the symbols unless they are derived from a proper name, then a capital is used for the first letter of the symbol. Note that the unit of mass uses the prefix kilo; this is the only base unit having a prefix.

---

* In the past Americans and the English have differed on the correct spelling of the words metre and kilogram. The English used metre and kilogramme. The usual American spelling was meter and kilogram. According to the latest edition of the National Bureau of Standards Special Publication 330, the spellings "metre" and "kilogram" are used and recommended in the hope of securing worldwide uniformity in the English spelling of the names of the units of the International System.

Table 1-2 shows that the SI unit of temperature is the kelvin. The Celsius temperature scale (once called Centigrade) is not a part of SI but a difference of one degree on the Celsius scale equals one kelvin.

A second class of SI units comprises the derived units, many of which have special names. Table 1-3 is a list of those we shall find most useful in our work in this book.

The radian (symbol rad) is a supplemental unit in SI for a plane angle.

A series of names and symbols to form multiples and submultiples of SI units have been established to provide an alternative to the writing of powers of 10. Table A-1 includes these prefixes and symbols.

Table 1-2 SI BASE UNITS

Quantity	Name	Symbol
Length	metre	m
Mass	kilogram	kg
Time	second	s
Electric current	ampere	A
Thermodynamic temperature	kelvin	K
Amount of matter	mole	mol
Luminous intensity	candela	cd

Table 1-3 EXAMPLES OF SI DERIVED UNITS*

Quantity	Unit	SI symbol	Formula
Acceleration	metre per second squared		$m \cdot s^{-2}$
Angular acceleration	radian per second squared		$rad \cdot s^{-2}$
Angular velocity	radian per second		$rad \cdot s^{-1}$
Area	square metre		m^2
Circular frequency	radian per second	ω	$rad \cdot s^{-1}$
Density	kilogram per cubic metre		$kg \cdot m^{-3}$
Energy	joule	J	$N \cdot m$
Force	newton	N	$kg \cdot m \cdot s^{-2}$
Force couple	newton metre		$N \cdot m$
Frequency	hertz	Hz	s^{-1}
Power	watt	W	$J \cdot s^{-1}$
Pressure	pascal	Pa	$N \cdot m^{-2}$
Quantity of heat	joule	J	$N \cdot m$
Speed	revolution per second		s^{-1}
Stress	pascal	Pa	$N \cdot m^{-2}$
Torque	newton metre		$N \cdot m$
Velocity	metre per second		$m \cdot s^{-1}$
Volume	cubic metre		m^3
Work	joule	J	$N \cdot m$

* In this book, negative exponents are seldom used; thus circular frequency, for example, would be expressed in rad/s.

1-11 RULES FOR USE OF SI UNITS

The International Bureau of Weights and Measures (BIPM), the international standardizing agency for SI, has established certain rules and recommendations for the use of SI. These are intended to eliminate differences which occur between various countries of the world in scientific and technical practices.

Number Groups

Numbers having four or more digits are placed in groups of three and separated by a space instead of a comma. However, the space may be omitted for the special case of numbers having four digits. A period is used as a decimal point. These recommendations avoid the confusion caused by certain European countries in which a comma is used as a decimal point, and by the English use of a centered period. Examples of correct and incorrect usage are as follows:

   1924 or 1 924 but not 1,924
   0.1924 or 0.192 4 but not 0.192,4
   192 423.618 50 but not 192,423.61850

The decimal point should always be preceded by a zero for numbers less than unity.

Using Prefixes

The multiple and submultiple prefixes in steps of 1000 only are recommended (Table A-1). This means that length can be expressed in mm, m, or km, but not in cm,* unless a valid reason exists.

When SI units are raised to a power the prefixes are also raised to the same power. This means that km^2 is

$$km^2 = (1000\ m)^2 = (1000)^2\ m^2 = 10^6\ m^2$$

Similarly mm^2 is

$$(0.001\ m)^2 = (0.001)^2\ m^2 = 10^{-6}\ m^2$$

When raising prefixed units to a power it is permissible, though not often convenient, to use the nonpreferred prefixes such as cm^2 or dm^3.

Except for the kilogram, which is a base unit, prefixes should not be used in the denominators of derived units. Thus the meganewton per square metre MN/m^2 is satisfactory but the newton per square millimetre N/mm^2 is not to be used. Note that this recommendation avoids a proliferation of derived units.

Double prefixes should not be used. Thus instead of millimillimetres mmm, use micrometres μm.

---

* The American Society of Mechanical Engineers (ASME) in the third edition of the Orientation and Guide for Use of Metric Units states on p. 11 that centimetres should not be used on drawings. Though its use is not recommended, the centimetre is such a conveniently sized unit that many people expect its use to continue.

Do not use an equals sign unless the equalities are true. Here is a true equality

$$14\,240 \text{ m} = 14.240\ E + 03 \text{ m}$$

But 14 240 m is *not* equal to 14.240 km though they are equivalent. Therefore do not confuse equivalency with equality.

Mass, Force, and Weight

The so-called metric countries may find it more difficult to transfer from their form of metric engineering units than the English-speaking countries. The reason for this is that the greatest difference is in the use of the newton as the unit of force instead of the kilogram-force.

A basic change in thinking about force and mass is required. Mass, measured in kilograms, is the amount of matter contained in physical objects. But weight, measured in newtons, is the gravitational force acting upon an object at a specified location. Thus, it is necessary to discontinue the current practice of indicating the quantity of material in an object by specifying its weight. When weight is given, then the location and the corresponding acceleration due to gravity should also be specified.

Standard gravity is 9.806 m/s^2 or about 9.80 m/s^2. Thus, the weight of a one kilogram mass at Washington, D.C., near sea level, is

$$W = mg = (1 \text{ kg})(9.80 \text{ m/s}^2) = 9.80 \text{ N}$$

It is convenient to remember that an apple weighs about 1 N.

Volume

The litre, symbol l, is used as a special name for the cubic decimetre dm^3, but its use for high-accuracy measurements is not recommended. Since the volume of a litre is approximately the same as that of a U.S. liquid quart, a conveniently sized unit, the litre will probably have widespread usage when SI is used universally.

Especial care must be taken with the symbol l for a litre to avoid confusing it with the number 1.

1-12 PRECISION AND ROUNDING OF QUANTITIES

The use of the electronic calculator having eight or ten digits or more in the display can easily lead to the appearance of great precision. Consider, for example, a computation which leads to the result $F = 142.047$ lb. This number contains six significant figures and so the implication exists that the result is precise to all six figures. However, if the data entered into the computation consisted of numbers containing only three significant figures, then the answer $F = 142.047$ lb indicates

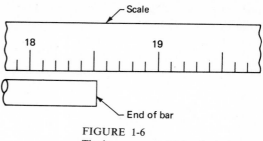

FIGURE 1-6
The bar measures 18.5 units in length.

a precision which does not exist. To express results with a nonexistent precision is misleading to say the least, but it is also dishonest, and may even be dangerous. So, in this section, we want to examine the means of learning the precision of a number, and then specify methods of rounding off a result to the correct precision.

When objects are counted, the result is always exact. When you purchase a dozen rolls at the bakery, you expect to receive exactly 12 rolls—no more, no less. But, as opposed to counting, measurements are always approximations. For example, in weighing yourself recently, you may have found your weight to be, say, 123 lb. This does not mean 123.000 lb because the three zeros after the decimal point imply that you were able to read the scales to a thousandth of a pound. Thus, a weight of 123 lb simply means that it is closer to 123 lb than it is to either 122 or 124.

The degree of precision obtained from a measurement depends upon a lot of factors, one of which is just how much precision is really needed. Usually great accuracy is expensive because such measurements must be taken with more expensive instruments and by more skillful workers. Unneeded precision is wasteful of both time and money.

The implied precision of any value is plus or minus one-half of the last significant digit used to state the value. Figure 1-6 shows a bar which measures 18.5 units in length. The length is closer to 18.5 units than it is to either 18.4 or to 18.6. Three significant figures were used to express this result. The precision of this measurement is ±0.05 because the result would have been expressed as 18.5 units for all lengths between 18.45 and 18.55.

A value written as 18.50 contains four significant figures because the zero must be counted too. The value 18.50 implies that the measurement is closer to 18.50 than it is to either 18.49 or to 18.51. The precision is ±0.005.

The measurement 0.0018 contains two significant figures because the zeros are only used to locate the decimal point. If the measurement is stated in the form 1.8 E-03 the two significant figures are evident. Then note that 1.80 E-03, which is the same as 0.001 80 has three significant figures.

The precision of a value such as 18 000 cannot be determined. The measurement could have been made to the nearest 1000, in which case there are only two significant figures. In such cases knowledge of how the measurement was made will usually reveal the precision. In this book ambiguous cases, such as this one, are arbitrarily assumed to have a precision equivalent to three significant figures. Note that if the value is stated as $1.80\ E + 04$, it is clear that there are three significant figures.

Now suppose we have some value, say 142.507, which we know to be accurate only to three significant figures. We therefore wish to eliminate the insignificant figures. This process is called *rounding off* a value. Here are the rules:

1 Retain the last digit unchanged if the first digit discarded is less than 5. For example 234.315 rounds to 234.3 if four significant figures are desired, or to 234 if three significant figures are desired.

2 Increase the last digit retained by one unit if the first digit discarded is greater than 5, or if it is a 5 followed by at least one digit other than zero. For example, 14.6 rounds off to 15; 14.501 rounds off to 15.

3 Retain the last digit unchanged if it is an even number and the first digit discarded is exactly 5 followed only by zeros. Thus 1.45 and 1.450 both round off to 1.4 if two significant figures are desired.

4 Increase the last digit by one unit if it is an odd number and the first digit discarded is exactly 5 followed only by zeros. Thus 1.55 rounds off to 1.6 and 15.50 rounds off to 16.

Computations involving multiplication and division should be rounded after the computation has been performed. The answer should then be rounded off such that it contains no more significant digits than are contained in the least precise value. Thus, the product

$$(1.68)(104.2) = 175.056$$

should be rounded off to 175 because the value 1.68 has three significant digits. Similarly

$$\frac{1.68}{104.2} = 0.016\ 122\ 8$$

should be rounded to 0.0161 for the same reason.

The rules are slightly different for computations involving addition and subtraction. Before performing the operation, round off all values to one significant digit farther to the right than that of the least accurate number. Then perform the addition or subtraction. Finally, round off the answer so that it has the same number of significant figures as are in the least accurate number. For example, suppose we wish to sum

$$A = 104.2 + 1.687 + 13.46$$

Of these, the number 104.2 is the least accurate. Therefore, using the round-off rules, we first compute

$$A = 104.2 + 1.69 + 13.46 = 119.35$$

Then we round off the result to $A = 119.4$ using rule 4. Note that the value 1.687 was rounded off to 1.69 before the terms were summed.

1-13 CONVERSION

Tables A-2 and A-3 can be used for converting from English units to SI or vice versa. Actually, it shouldn't be necessary to convert from one system to another very often. If the data for a problem are given in SI units, then the problem should be solved using SI, making conversions unnecessary.

Infrequently, data necessary to solve a problem will not be available in SI. Sizes, ratings, weights, and other appropriate data relating to manufactured items in the English-speaking countries have customarily been given only in English units. Examples of this practice are the unit weights and dimensions of structural-steel shapes, such as I beams and angles, and the power output and torque ratings of motors and engines. Thus when data are available only in English units, a conversion to SI is necessary to solve the problem in SI.

Referring to Table A-2 we see that to convert from feet to metres requires that we multiply feet by the factor 0.305 to get metres. This means that the units of the conversion constant are in m/ft. When converting from one system of units to another, always insert the units into the equation beside each value. If the equation is correctly structured, the units will cancel to give the desired result.

For example, let us convert the dimension $h = 8.50$ ft to SI units. The equation is

$$h = (8.50 \text{ ft})\left(0.305 \frac{\text{m}}{\text{ft}}\right) = 2.59 \text{ m}$$

Notice that the symbol ft appears in both the numerator and denominator and cancels, leaving only the desired unit.

Some conversions require more steps, but the procedure is the same. Concrete, made of gravel, weighs about 140 lb/ft^3. What is its weight in SI units?

$$w = \left(140 \frac{\text{lb}}{\text{ft}^3}\right)\left(4.45 \frac{N}{\text{lb}}\right)\left(\frac{1}{0.305 \text{ m/ft}}\right)^3\left(\frac{1}{1000 \ N/kN}\right) = 22.0 \text{ kN/m}^3$$

The second term in this equation converts pounds to newtons. The third term converts cubic feet to cubic metres; note that the conversion must be made in the denominator. The fourth term converts the result to kilonewtons per cubic metre.

One of the problems that arises in the use of SI is that the number magnitudes differ so from those in the English system. For example, the modulus of elasticity of nickel steel is 30 Mpsi in the English system and 207 GPa in SI (see Table A-7). And an area of 1 in^2 in the English system becomes

$$A = (1 \text{ in}^2)\left(0.0254 \frac{\text{m}}{\text{in}}\right)^2\left(1000 \frac{\text{mm}}{\text{m}}\right)^2 = 645 \text{ mm}^2$$

in square millimetres, or

$$A = (1 \text{ in}^2)\left(0.0254 \frac{\text{m}}{\text{in}}\right)^2 = 64.5(10)^{-3} \text{ m}^2$$

in square metres. This is why electronic calculators having scientific notation capability are particularly valuable when solving problems in SI.

Tables A-4 to A-6 provide an additional aid when solving for stresses and deflections of beams in SI. These tables give the prefix of the result for various combinations of prefixed terms which are substituted into the stress and deflection formulas.

PROBLEMS

1-1 A heat-treated high-carbon steel was tested in tension. The original diameter of the specimen was 12.5 mm and the final diameter was 11.6 mm. Using a 50-mm gauge length, the following data were obtained:

Load kN	Elongation mm	Load kN	Elongation mm
25.9	0.051	88.6	0.508
36.0	0.071	93.4	0.711
46.6	0.092	98.8	1.016
54.5	0.107	107.2	1.524
70.8	0.152	113.5	2.032
74.8	0.203	117.5	2.541
80.5	0.305	121.0	3.048
85.0	0.406	123.3	3.560

Find the 0.2 percent offset yield strength, the ultimate strength, and the percentage elongation.

1-2 Convert the following to appropriate SI units:
(*a*) The area of a ½-acre lot.
(*b*) 30 psi tire pressure.
(*c*) 2400 rpm.
(*d*) The mass of an automobile weighing 3200 lbf.
(*e*) The displacement of a 425-in³ engine.
(*f*) 2 U.S. qt and 1 pt.
(*g*) 55 mph.
(*h*) 15 gal of gasoline.
(*i*) 14.7 psi atmospheric pressure.

2

STRESS ANALYSIS

The public today is clamoring for faster, cleaner, safer, and quieter machines. And design engineers, in their efforts to satisfy these demands with products at a lower cost, are being sued in the courts throughout the land for designing things that are unsafe, that are dangerous to human life. One of the real problems we shall face in this book is that of relating the strength of an element to the external loads which are imposed upon it. These external loads cause internal stresses in the element. To avoid product liability the design engineer must have positive assurance that these stresses will never exceed the strength. In this chapter we shall concern ourselves with the determination of these stresses. Then, in other chapters to follow, we shall introduce the concept of strength and begin the process of relating strength and stress to achieve safety.

It is worth observing that the stresses which are calculated are only as reliable as are the external loads and geometries which are used in the determination. Once a product has been manufactured and placed in the hands of the consumer or user, the manufacturer has relinquished control over that product, and to some extent over the external loads to which it may be subjected. It is for this reason, among others, that testing and experimental proving are so important in the design and development of a new machine or device.

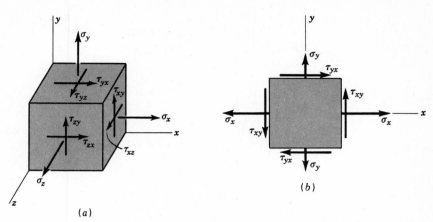

FIGURE 2-1

2-1 STRESS

Figure 2-1a is a general three-dimensional stress element, showing three normal stresses σ_x, σ_y, and σ_z, all positive; and six shear stresses τ_{xy}, τ_{yx}, τ_{yz}, τ_{zy}, τ_{zx}, and τ_{xz}, also all positive. The element is in static equilibrium and hence

$$\tau_{xy} = \tau_{yx} \qquad \tau_{yz} = \tau_{zy} \qquad \tau_{zx} = \tau_{xz} \qquad (2\text{-}1)$$

Outwardly directed normal stresses are considered as tension and positive. Shear stresses are positive if they act in the positive direction of a reference axis. The first subscript of a shear stress component is the coordinate normal to the element face. The shear stress component is parallel to the axis of the second subscript. The negative faces of the element will have shear stresses acting in the opposite direction; these are also considered as positive.

Figure 2-1b illustrates a state of *biaxial* or *plane stress*, which is the more usual case. For this state, only the normal stresses will be treated as positive or negative. The senses of the shear stress components will be specified by the clockwise (cw) or counterclockwise (ccw) convention. Thus, in Fig. 2-1b, τ_{xy} is ccw and τ_{yx} is cw.

2-2 MOHR'S CIRCLE

One of the most challenging problems in design is that of relating the strength of a mechanical element to the internal stresses which are produced by the external loads. Usually we have only a single value for the strength, such as the yield strength, but several stress components. One of the problems to be faced in a future chapter is how to take a stress element like Fig. 2-1b and relate it to a single strength magnitude to achieve safety. This section is the first step in the solution.

Suppose the element of Fig. 2-1b is cut by an oblique plane at angle ϕ to the x axis as shown in Fig. 2-2. This section is concerned with the stresses σ and τ which act upon this oblique plane. By summing the forces caused by all the stress components to zero, the stresses σ and τ are found to be*

$$\sigma = \frac{\sigma_x + \sigma_y}{2} + \frac{\sigma_x - \sigma_y}{2} \cos 2\phi + \tau_{xy} \sin 2\phi \qquad (2\text{-}2)$$

$$\tau = -\frac{\sigma_x - \sigma_y}{2} \sin 2\phi + \tau_{xy} \cos 2\phi \qquad (2\text{-}3)$$

Differentiating the first equation with respect to ϕ and setting the result equal to zero gives

$$\tan 2\phi = \frac{2\tau_{xy}}{\sigma_x - \sigma_y} \qquad (2\text{-}4)$$

Equation (2-4) defines two particular values for the angle 2ϕ, one of which defines the maximum normal stress σ_1 and the other, the minimum normal stress σ_2. These two stresses are called the *principal stresses*, and their corresponding directions, the *principal directions*. The angle ϕ, between the principal directions is 90°.

In a similar manner, we differentiate Eq. (2-3), set the result equal to zero, and obtain

$$\tan 2\phi = -\frac{\sigma_x - \sigma_y}{2\tau_{xy}} \qquad (2\text{-}5)$$

Equation (2-5) defines the two values of 2ϕ at which the shear stress τ is maximum.

It is interesting to note that Eq. (2-4) can be written in the form

$$2\tau_{xy} \cos 2\phi = (\sigma_x - \sigma_y) \sin 2\phi$$

or

$$\sin 2\phi = \frac{2\tau_{xy} \cos 2\phi}{\sigma_x - \sigma_y} \qquad (a)$$

Now substitute Eq. (a) for $\sin 2\phi$ in Eq. (2-3). We obtain

$$\tau = -\frac{\sigma_x - \sigma_y}{2} \frac{2\tau_{xy} \cos 2\phi}{\sigma_x - \sigma_y} + \tau_{xy} \cos 2\phi = 0 \qquad (2\text{-}6)$$

Equation (2-6) states that the shear stress associated with both principal directions is zero.

---

* For the complete force analysis see Joseph E. Shigley, "Applied Mechanics of Materials," pp. 259–261, McGraw-Hill Book Company, New York, 1976.

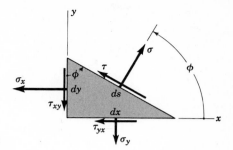

FIGURE 2-2

Solving Eq. (2-5) for sin 2ϕ, in a similar manner, and substituting the result in Eq. (2-2) yields

$$\sigma = \frac{\sigma_x + \sigma_y}{2} \qquad (2\text{-}7)$$

Equation (2-7) tells us that the two normal stresses associated with the directions of the two maximum shear stresses are equal.

Formulas for the two principal stresses can be obtained by substituting the angle 2ϕ from Eq. (2-4) into Eq. (2-2). The result is

$$\sigma_1, \sigma_2 = \frac{\sigma_x + \sigma_y}{2} \pm \sqrt{\left(\frac{\sigma_x - \sigma_y}{2}\right)^2 + \tau_{xy}^2} \qquad (2\text{-}8)$$

In a similar manner the two maximum shear stresses are found to be

$$\tau_{\max} = \pm \sqrt{\left(\frac{\sigma_x - \sigma_y}{2}\right)^2 + \tau_{xy}^2} \qquad (2\text{-}9)$$

But this may not be the maximum shear stress when the z direction is considered too. See Sec. 2-3.

A graphical method for expressing the relations developed in this section, called *Mohr's circle diagram*, is a very effective means of visualizing the stress state at a point and keeping track of the directions of the various components associated with plane stress. In Fig. 2-3 we create a coordinate system with normal stresses plotted along the abscissa, and shear stresses plotted as the ordinates. On the abscissa, tensile (positive) normal stresses are plotted to the right of the origin O and compressive (negative) normal stresses to the left.

Using the stress state of Fig. 2-1b, we plot Mohr's circle diagram (Fig. 2-3) by laying off σ_x as OA, τ_{xy} as AB, σ_y as OC, and τ_{yx} as CD. The line DEB is the diameter of Mohr's circle with center at E on the σ axis. Point B represents the stress coordinates $\sigma_x \tau_{xy}$ on the x faces, and point D the stress coordinates $\sigma_y \tau_{yx}$ on the y faces. Thus EB corresponds to the x axis and ED to the y axis. The angle 2ϕ, measured counterclockwise from EB to ED, is $180°$, which corresponds to $\phi = 90°$, measured counterclockwise from x to y, on the stress element of Fig. 2-1b.

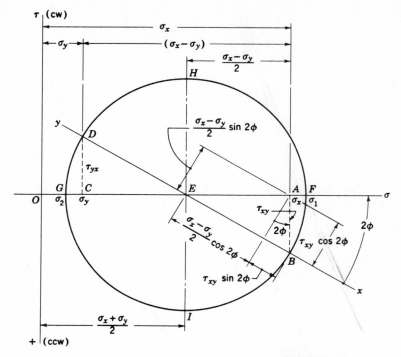

FIGURE 2-3
Mohr's circle diagram.

The maximum principal normal stress σ_1 occurs at F, and the minimum principal normal stress σ_2 at G. The two maximum shear stresses, one clockwise and one counterclockwise, occur at H and I, respectively.

You should demonstrate for yourself that the geometry of Fig. 2-3 satisfies all the relations developed in this section.

Though Eqs. (2-8) and (2-4) can be solved directly for the principal stresses and directions, a semigraphical approach is easier and quicker and offers fewer opportunities for error. This method is illustrated by the following example:

EXAMPLE 2-1   The stress element shown in Fig. 2-4 has $\sigma_x = 80$ MPa and $\tau_{xy} = 50$ MPa cw.* Find the principal stresses and directions and show these on a stress element correctly aligned with respect to the xy system. Draw another stress element to show τ_{max}, find the corresponding normal stresses, and label the drawing completely.

---

* Any stress components such as σ_y, τ_{zx}, etc., that are not given in a problem are always taken as zero.

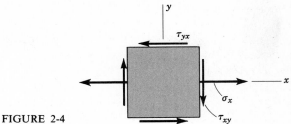

FIGURE 2-4

SOLUTION   We shall construct Mohr's circle corresponding to the given data, and then read the results directly from the diagram. You can construct this diagram with compass and scales and find the required information with the aid of scales and protractor, if you choose to do so. Such a complete graphical approach is perfectly satisfactory and sufficiently accurate for most purposes.

In the semigraphical approach used here, we first make an approximate freehand sketch of Mohr's circle and then use the geometry of the figure to obtain the desired information.

Draw the σ and τ axes first, in Fig. 2-5a, and locate $\sigma_x = 80$ MPa along the σ axis. Then, from σ_x, locate $\tau_{xy} = 50$ MPa in the cw direction of the τ axis to establish point A. Corresponding to $\sigma_y = 0$, locate $\tau_{yx} = 50$ MPa in the ccw direction along the τ axis to obtain point D. The line AD forms the diameter of the required circle which can now be drawn. The intersection of the circle with the σ axis defines σ_1 and σ_2 as shown. The x axis is the line CA; the y axis is the line CD. Now, noting the triangle ABC, indicate on the sketch the length of the legs AB and BC, as 50 and 40 MPa, respectively. The length of the hypotenuse AC is

$$\tau_{max} = \sqrt{(50)^2 + (40)^2} = 64.0 \text{ MPa}$$

and this should be labeled on the sketch too. Since intersection C is 40 MPa from the origin, the principal stresses are now found to be

$$\sigma_1 = 40 + 64 = 104 \text{ MPa} \qquad \sigma_2 = 40 - 64 = -24 \text{ MPa}$$

The angle 2ϕ from the x axis cw to σ_1 is

$$2\phi = \tan^{-1} \frac{50}{40} = 51.3°$$

If your calculator has rectangular-to-polar conversion, you should always use it to obtain τ_{max} and 2ϕ.

To draw the principal stress element (Fig. 2-5b), sketch the x and y axes parallel to the original axes (Fig. 2-4). The angle ϕ on the stress element must be measured in the *same* direction as is the angle 2ϕ on Mohr's circle. Thus, from x measure 25.7° (half of 51.3°) clockwise to locate the σ_1 axis. The stress element can now be completed and labeled as shown.

The two maximum shear stresses occur at points E and F in Fig. 2-5a. The two normal stresses corresponding to these shear stresses are each 40 MPa, as indicated. Point E is 38.7° ccw from point A on Mohr's circle. Therefore, in

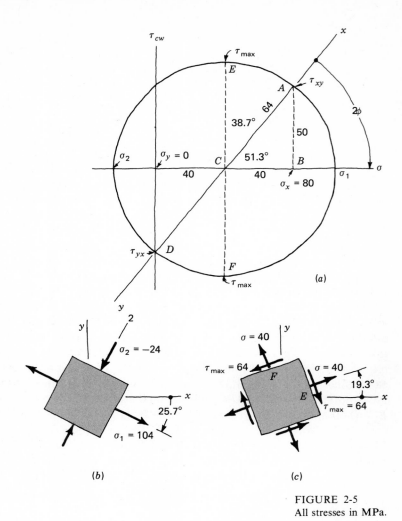

FIGURE 2-5
All stresses in MPa.

Fig. 2-5c, draw a stress element oriented 19.3° (half of 38.7°) ccw from x. The element should then be labeled with magnitudes and directions as shown.

In constructing these stress elements it is important to indicate the x and y directions of the original reference system. This completes the link between the original machine element and the orientation of its principal stresses.   ////

2-3 MOHR'S CIRCLE FOR THREE-DIMENSIONAL STRESS

The general case of three-dimensional stress, or *triaxial stress*, is illustrated by Fig. 2-1a, as we have already learned. As in the case of plane or biaxial stress, a

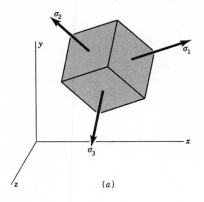
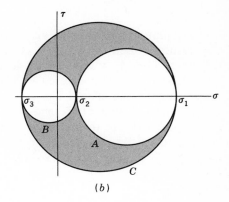

FIGURE 2-6
Mohr's circles for triaxial stress.

particular orientation of the stress element occurs in space for which all shear-stress components are zero. When an element has this particular orientation, the normals to the faces correspond to the principal directions and the normal stresses associated with these faces are the principal stresses. Since there are six faces, there are three principal directions and three principal stresses σ_1, σ_2, and σ_3.

In our studies of plane stress we were able to specify any stress state σ_x, σ_y, and τ_{xy} and find the principal stresses and principal directions. But six components of stress are required to specify a general state of stress in three dimensions, and the problem of determining the principal stresses and directions is much more difficult. It turns out that it is rarely necessary in design, and so we shall not investigate the problem in this book. The process involves finding the three roots to the cubic

$$\sigma^3 - (\sigma_x + \sigma_y + \sigma_z)\sigma^2 + (\sigma_x\sigma_y + \sigma_x\sigma_z + \sigma_y\sigma_z - \tau_{xy}^2 - \tau_{yz}^2 - \tau_{zx}^2)\sigma$$
$$- (\sigma_x\sigma_y\sigma_z + 2\tau_{xy}\tau_{yz}\tau_{zx} - \sigma_x\tau_{yz}^2 - \sigma_y\tau_{zx}^2 - \sigma_z\tau_{xy}^2) = 0 \qquad (2\text{-}10)$$

Having solved Eq. (2-10) for a given state of stress, one might obtain a principal stress element like that of Fig. 2-6a.

In plotting Mohr's circles for triaxial stress, the principal stresses are arranged so that $\sigma_1 > \sigma_2 > \sigma_3$. Then the result appears as in Fig. 2-6b. The stress coordinates $\sigma_N \tau_N$ for any arbitrarily located plane will always lie within the shaded area.

It frequently happens, in the case of plane stress, that the two principal stresses σ_1 and σ_2 have the same sign. In this situation construction of a single Mohr's circle through σ_1 and σ_2 will *not* give τ_{max}. This can easily be seen by constructing three such circles, using $\sigma_3 = 0$ for the third principal stress.

2-4 UNIFORM STRESS

The assumption of a uniform distribution of stress is frequently made in design. The result is then often called *pure tension, pure compression,* or *pure shear,* depending upon how the external load is applied to the body under study. The word *simple* is sometimes used instead of "pure" to indicate that there are no other complicating effects. The tension rod is typical. Here a tension load F is applied through pins at the ends of the bar. The assumption of uniform stress means that if we cut the bar at a section remote from the ends, and remove the lower half, we can replace its effect by applying a uniformly distributed force of magnitude σA to the cut end. So the stress σ is said to be uniformly distributed. It is calculated from the equation

$$\sigma = \frac{F}{A} \qquad (2\text{-}11)$$

This assumption of uniform stress distribution requires that:

1 The bar be straight and of a homogeneous material.
2 The line of action of the force coincide with the centroid of the section.
3 The section be taken remote from the ends and from any discontinuity or abrupt change in cross section.

The same equation and assumptions hold for simple compression. A slender bar in compression may, however, fail by buckling, and this possibility must be eliminated from consideration before Eq. (2-11) is used.*
Use of the equation

$$\tau = \frac{F}{A} \qquad (2\text{-}12)$$

for a body, say a bolt, in shear assumes a uniform stress distribution too. It is very difficult in practice to obtain a uniform distribution of shear stress; the equation is included because occasions do arise in which this assumption is made.

2-5 ELASTIC STRAIN

When a straight bar is subjected to a tensile load, the bar becomes longer. The amount of stretch, or elongation, is called *strain*. The elongation per unit length of the bar is called *unit strain*. In spite of these definitions, however, it is customary to use the word "strain" to mean "unit strain" and the expression "total strain" to mean total elongation, or deformation, of a member. Using this custom here, the expression for strain is

$$\epsilon = \frac{\delta}{l} \qquad (2\text{-}13)$$

where δ is the total elongation (total strain) of the bar within the length l.

* See Sec. 3-10.

Shear strain γ is the change in a right angle of a stress element subjected to pure shear.

Elasticity is that property of a material which enables it to regain its original shape and dimensions when the load is removed. Hooke's law states that, within certain limits, the stress in a material is proportional to the strain which produced it. An elastic material does not necessarily obey Hooke's law, since it is possible for some materials to regain their original shape without the limiting condition that stress be proportional to strain. On the other hand, a material which obeys Hooke's law is elastic. For the condition that stress is proportional to strain, we can write

$$\sigma = E\epsilon \qquad (2\text{-}14)$$

$$\tau = G\gamma \qquad (2\text{-}15)$$

where E and G are the constants of proportionality. Since the strains are dimensionless numbers, the units of E and G are the same as the units of stress. The constant E is called the *modulus of elasticity*. The constant G is called the *shear modulus of elasticity*, or sometimes, the *modulus of rigidity*. Both E and G, however, are numbers which are indicative of the stiffness or rigidity of the materials. These two constants represent fundamental properties.

By substituting $\sigma = F/A$ and $\epsilon = \delta/l$ into Eq. (2-14) and rearranging, we obtain the equation for the total deformation of a bar loaded in axial tension or compression.

$$\delta = \frac{Fl}{AE} \qquad (a)$$

Experiments demonstrate that when a material is placed in tension, there exists not only an axial strain, but also a lateral strain. Poisson demonstrated that these two strains were proportional to each other within the range of Hooke's law. This constant is expressed as

$$\mu = -\frac{\text{lateral strain}}{\text{axial strain}} \qquad (2\text{-}16)$$

and is known as *Poisson's ratio*. These same relations apply for compression, except that a lateral expansion takes place instead.

The three elastic constants are related to each other as follows:

$$E = 2G(1 + \mu) \qquad (2\text{-}17)$$

2-6 STRESS-STRAIN RELATIONS

Stress, being an artificial concept, cannot be experimentally measured. There are many experimental techniques, though, which can be used to measure strain. Thus, if the relationship between stress and strain is known, the stress state at a point can be calculated after the state of strain has been measured.

We define the *principal strains* as the strains in the direction of the principal stresses. It is true that the shear strains are zero, just as the shear stresses are zero, on the faces of an element aligned in the principal directions. From Eq. (2-16) the three principal strains corresponding to a state of uniaxial stress are

$$\epsilon_1 = \frac{\sigma_1}{E} \qquad \epsilon_2 = -\mu\epsilon_1 \qquad \epsilon_3 = -\mu\epsilon_1 \qquad (2\text{-}18)$$

The minus sign is used to indicate compressive strains. Note that, while the stress state is uniaxial, the strain state is triaxial.

Biaxial Stress

For the case of biaxial stress we give σ_1 and σ_2 prescribed values and let σ_3 be zero. The principal strains can be found from Eq. (2-16) if we imagine each principal stress to be acting separately and then combine the results by superposition. This gives

$$\epsilon_1 = \frac{\sigma_1}{E} - \frac{\mu\sigma_2}{E}$$

$$\epsilon_2 = \frac{\sigma_2}{E} - \frac{\mu\sigma_1}{E} \qquad (2\text{-}19)$$

$$\epsilon_3 = -\frac{\mu\sigma_1}{E} - \frac{\mu\sigma_2}{E}$$

Equations (2-19) give the principal strains in terms of the principal stresses. But the usual situation is the reverse. To solve for σ_1, multiply ϵ_2 by μ, and add the first two equations together. This produces

$$\epsilon_1 + \mu\epsilon_2 = \frac{\sigma_1}{E} - \frac{\mu\sigma_2}{E} + \frac{\mu\sigma_2}{E} - \frac{\mu^2\sigma_1}{E}$$

Solving for σ_1 gives

$$\sigma_1 = \frac{E(\epsilon_1 + \mu\epsilon_2)}{1 - \mu^2} \qquad (2\text{-}20)$$

Similarly,
$$\sigma_2 = \frac{E(\epsilon_2 + \mu\epsilon_1)}{1 - \mu^2} \qquad (2\text{-}21)$$

In solving these equations, remember that a tensile stress or strain is treated as positive and that compression is negative.

Triaxial Stress

The state of stress is said to be *triaxial* when none of the three principal stresses is zero. For this, the principal strains are

$$\epsilon_1 = \frac{\sigma_1}{E} - \frac{\mu\sigma_2}{E} - \frac{\mu\sigma_3}{E}$$

$$\epsilon_2 = \frac{\sigma_2}{E} - \frac{\mu\sigma_1}{E} - \frac{\mu\sigma_3}{E} \qquad (2\text{-}22)$$

$$\epsilon_3 = \frac{\sigma_3}{E} - \frac{\mu\sigma_1}{E} - \frac{\mu\sigma_2}{E}$$

In terms of the strains, the principal stresses are

$$\sigma_1 = \frac{E\epsilon_1(1 - \mu) + \mu E(\epsilon_2 + \epsilon_3)}{1 - \mu - 2\mu^2}$$

$$\sigma_2 = \frac{E\epsilon_2(1 - \mu) + \mu E(\epsilon_1 + \epsilon_3)}{1 - \mu - 2\mu^2} \qquad (2\text{-}23)$$

$$\sigma_3 = \frac{E\epsilon_3(1 - \mu) + \mu E(\epsilon_1 + \epsilon_2)}{1 - \mu - 2\mu^2}$$

Values of Poisson's ratio for various materials are given in Table A-7.

2-7 SHEAR AND MOMENT IN BEAMS

Figure 2-7a *shows a beam supported by reactions* R_1 *and* R_2 *and loaded by the* concentrated forces F_1, F_2, and F_3. The direction chosen for the y axis is the clue to the sign convention for the forces. F_1, F_2, and F_3 are negative because they act in the negative y direction; R_1 and R_2 are positive.

   If the beam is cut at some section location at $x = x_1$, and the left-hand portion removed as a free body, an *internal shear force* V and *bending moment* M must act on the cut surface to assure equilibrium. The shear force is obtained by summing the forces to the left of the cut section. The bending moment is the sum of the moments of the forces to the left of the section taken about an axis through the section. Shear force and bending moment are related by the equation

$$V = \frac{dM}{dx} \qquad (2\text{-}24)$$

   Sometimes the bending is caused by a distributed load. Then, the relation between shear force and bending moment may be written

$$\frac{dV}{dx} = \frac{d^2M}{dx^2} = -w \qquad (2\text{-}25)$$

where w is a downward acting load of w units of force per unit length.

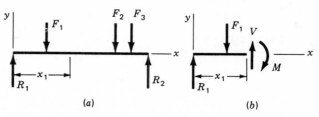

(a)

(b)

FIGURE 2-7

The sign conventions used for bending moment and shear force in this book are shown in Fig. 2-8.

The loading w of Eq. (2-25) is uniformly distributed. A more general distribution can be defined by the equation

$$q = \lim_{\Delta x \to 0} \frac{\Delta F}{\Delta x}$$

where q is called the *load intensity*; thus $q = -w$.

Equations (2-24) and (2-25) reveal additional relations if they are integrated. Thus, if we integrate between, say, x_A and x_B, we obtain

$$\int_{V_A}^{V_B} dV = \int_{x_A}^{x_B} q \, dx = V_B - V_A \qquad (2\text{-}26)$$

which states that *the change in shear force from A to B is equal to the area of the loading diagram between x_A and x_B.*

In a similar manner

$$\int_{M_A}^{M_B} dM = \int_{x_A}^{x_B} V \, dx = M_B - M_A \qquad (2\text{-}27)$$

which states that *the change in moment from A to B is equal to the area of the shear-force diagram between x_A and x_B.*

EXAMPLE 2-2 Figure 2-9a shows a beam OC 18 in long loaded by concentrated forces of 50 lb at A and 70 lb at B. Find the reactions R_1 and R_2 and sketch the shear-force and moment diagrams.

SOLUTION Summing moments about O gives $R_2 = 42.2$ lb. Then, summing forces in the y direction yields $R_1 = 77.8$ lb.

To plot the shear-force diagram (Fig. 2-9b) begin at the origin and work to the right. The shear force is $+77.8$ lb from O to A. At A it is reduced by 50 lb to $+27.8$ lb, and it continues with this value to B. At B it is reduced again by 70 lb to -42.2 lb. The positive reaction of 42.2 lb at C brings the shear force back to zero at the end.

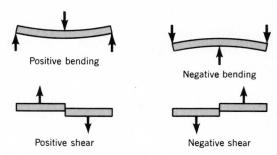

FIGURE 2-8
Sign conventions for bending and shear.

To obtain the moment diagram (Fig. 2-9c), remember that the change in moment between any two points is equal to the area of the shear-force diagram between those points. As indicated by the loading diagram, which is a free-body diagram of the entire beam, the moment at O is zero. Therefore the moment at A is the area of the shear diagram between O and A. Thus

$$M_A = (77.8)(4) = 311.2 \text{ lb} \cdot \text{in}$$

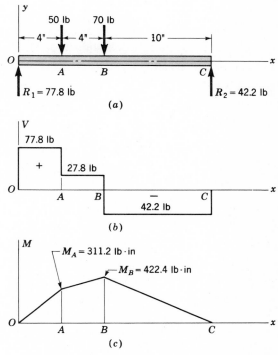

FIGURE 2-9

Note that this could also be obtained by multiplying the reaction R_1 by the distance from O to A.

The change in moment from A to B is the area of the shear diagram between A and B. Thus

$$M_B - M_A = (27.8)(4) = 111.2 \text{ lb·in}$$

and

$$M_B = 111.2 + 311.2 = 422.4 \text{ lb·in}$$

This result can also be obtained by taking moments about B as follows:

$$M_B = -(77.8)(8) + (50)(4) = -422.4 \text{ lb·in}$$

Here we have obeyed the clockwise-counterclockwise rule for moments. Though the bending moment is negative using this rule, it is positive according to the convention of Fig. 2-8.

2-8 SINGULARITY FUNCTIONS

The five singularity functions defined in Table 2-1 constitute a useful and easy means of integrating across discontinuities. By their use, general expressions for shear force and bending moment in beams can be written when the beam is loaded by concentrated forces or moments. As shown in the table, the concentrated moment and force functions are zero for all values of x except the particular value $x = a$. The unit step, ramp, and parabolic functions are zero only for values of x that are less than a. The integration properties shown in the table constitute a part of the mathematical definition too. The examples which follow show how these functions are used.*

EXAMPLE 2-3   Derive expressions for the loading, shear-force, and bending-moment diagrams for the beam of Fig. 2-10.

SOLUTION   Using Table 2-1 and $q(x)$ for the loading function, we find

$$q = R_1 \langle x \rangle^{-1} - F_1 \langle x - a_1 \rangle^{-1} - F_2 \langle x - a_2 \rangle^{-1} + R_2 \langle x - l \rangle^{-1} \tag{1}$$

Next, we use Eq. (2-26) to get the shear force. Note that $V = 0$ at $x = -\infty$.

$$V = \int_{-\infty}^{x} q \, dx = R_1 \langle x \rangle^0 - F_1 \langle x - a_1 \rangle^0 - F_2 \langle x - a_2 \rangle^0$$
$$+ R_2 \langle x - l \rangle^0 \tag{2}$$

A second integration, in accordance with Eq. (2-27), yields

$$M = \int_{-\infty}^{x} V \, dx = R_1 \langle x \rangle^1 - F_1 \langle x - a_1 \rangle^1 - F_2 \langle x - a_2 \rangle^1$$
$$+ R_2 \langle x - l \rangle^1 \tag{3}$$

---

* For additional examples see S. H. Crandall, Norman C. Dahl, and Thomas J. Lardner (eds.), "An Introduction to the Mechanics of Solids," 2d ed., pp. 164–172, McGraw-Hill Book Company, New York, 1972.

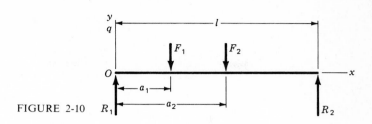

FIGURE 2-10

Table 2-1 SINGULARITY FUNCTIONS

Function	Graph of $f_n(x)$	Meaning
Concentrated moment	$\langle x - a \rangle^{-2}$	$\langle x - a \rangle^{-2} = 0 \qquad x \neq a$ $\displaystyle\int_{-\infty}^{x} \langle x - a \rangle^{-2} \, dx = \langle x - a \rangle^{-1}$
Concentrated force	$\langle x - a \rangle^{-1}$	$\langle x - a \rangle^{-1} = 0 \qquad x \neq a$ $\displaystyle\int_{-\infty}^{x} \langle x - a \rangle^{-1} \, dx = \langle x - a \rangle^{0}$
Unit step	$\langle x - a \rangle^{0}$	$\langle x - a \rangle^{0} = \begin{cases} 0 & x < a \\ 1 & x \geq a \end{cases}$ $\displaystyle\int_{-\infty}^{x} \langle x - a \rangle^{0} \, dx = \langle x - a \rangle^{1}$
Ramp	$\langle x - a \rangle^{1}$	$\langle x - a \rangle^{1} = \begin{cases} 0 & x < a \\ x - a & x \geq a \end{cases}$ $\displaystyle\int_{-\infty}^{x} \langle x - a \rangle^{1} \, dx = \dfrac{\langle x - a \rangle^{2}}{2}$
Parabolic	$\langle x - a \rangle^{2}$	$\langle x - a \rangle^{2} = \begin{cases} 0 & x < a \\ (x - a)^{2} & x \geq a \end{cases}$ $\displaystyle\int_{-\infty}^{x} \langle x - a \rangle^{2} \, dx = \dfrac{\langle x - a \rangle^{3}}{3}$

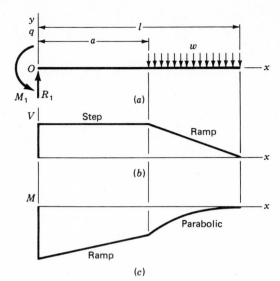

FIGURE 2-11
(a) Loading diagram for a beam cantilevered at O; (b) shear-force diagram;
(c) bending-moment diagram.

The reactions R_1 and R_2 can be found by taking a summation of moments and forces as usual or they can be found by noting that the shear force and bending moment must be zero everywhere except in the region $0 \le x \le l$. This means that Eq. (2) should give $V = 0$ at x slightly larger than l. Thus

$$R_1 - F_1 - F_2 + R_2 = 0 \qquad (4)$$

Since the bending moment should also be zero in the same region, we have from Eq. (3),

$$R_1 l - F_1(l - a_1) - F_2(l - a_2) = 0 \qquad (5)$$

Equations (4) and (5) can now be solved for the reactions R_1 and R_2.   ////

EXAMPLE 2-4   Figure 2-11a shows the loading diagram for a beam cantilevered at O and having a uniform load w acting on the portion $a \le x \le l$. Derive the shear force and moment relations. M_1 and R_1 are the support reactions.

SOLUTION   Following the procedure of Example 2-3 we find the loading function to be

$$q = -M_1\langle x \rangle^{-2} + R_1\langle x \rangle^{-1} - w\langle x - a \rangle^0 \qquad (1)$$

Then integrating successively gives

$$V = \int_{-\infty}^{x} q\, dx = -M_1\langle x \rangle^{-1} + R_1\langle x \rangle^0 - w\langle x - a \rangle^1 \qquad (2)$$

$$M = \int_{-\infty}^{x} V\, dx = -M_1\langle x \rangle^0 + R_1\langle x \rangle^1 - \frac{w}{2}\langle x - a \rangle^2 \qquad (3)$$

The reactions are found by making x slightly larger than l because both V and M are zero in this region. Equation (2) will then give

$$-M_1(0) + R_1 - w(l - a) = 0 \qquad (4)$$

which can be solved for R_1. From Eq. (3) we get

$$-M_1 + R_1 l - \frac{w}{2}(l - a)^2 = 0 \qquad (5)$$

which can be solved for M_1. Figures 2-11b and 2-11c show the shear-force and bending-moment diagrams.   ////

2-9 NORMAL STRESSES IN BENDING

In deriving the relations for the normal bending stresses in beams, we make the following idealizations:

1 The beam is subjected to pure bending; this means that the shear force is zero, and that no torsion or axial loads are present.
2 The material is isotropic and homogeneous.
3 The material obeys Hooke's law.
4 The beam is initially straight with a cross section that is constant throughout the beam length.
5 The beam has an axis of symmetry in the plane of bending.
6 The proportions of the beam are such that it would fail by bending rather than by crushing, wrinkling, or sidewise buckling.
7 Cross sections of the beam remain plane during bending.

In Fig. 2-12a we visualize a portion of a beam acted upon by the positive bending moment M. The y axis is the axis of symmetry. The x axis is coincident with the *neutral axis* of the section, and the xz plane, which contains the neutral axes of all cross sections, is called the *neutral plane*. Elements of the beam coincident with this plane have zero strain. The location of the neutral axis with respect to the cross section has not yet been defined.

Application of the positive moment will cause the upper surface of the beam to bend downward, and the neutral axis will then be curved, as shown in Fig. 2-12b. Because of the curvature, a section AB originally parallel to CD, since the beam was straight, will rotate through the angle $d\phi$ to $A'B'$. Since AB and $A'B'$ are both straight lines, we have utilized the assumption that plane sections remain plane during bending. If we now specify the radius of curvature of the neutral axis as ρ, the length of a differential element of the neutral axis as ds, and the angle subtended by the two adjacent sides CD and $A'B'$ as $d\phi$, then, from the definition of curvature, we have

$$\frac{1}{\rho} = \frac{d\phi}{ds} \qquad (a)$$

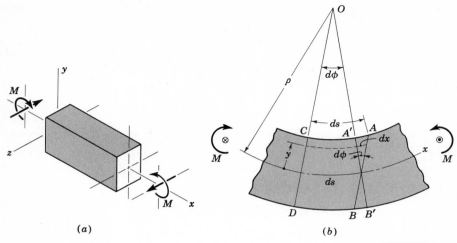

(a) (b)

FIGURE 2-12

As shown in Fig. 2-12b, the deformation of a "fiber" at distance y from the neutral axis is

$$dx = y \, d\phi \qquad (b)$$

The strain is the deformation divided by the original length, or

$$\epsilon = -\frac{dx}{ds} \qquad (c)$$

where the negative sign indicates compression. Solving Eqs. (a), (b), and (c) simultaneously gives

$$\epsilon = -\frac{y}{\rho} \qquad (d)$$

Thus the strain is proportional to the distance y from the neutral axis. Now, since $\sigma = E\epsilon$, we have for the stress

$$\sigma = -\frac{Ey}{\rho} \qquad (e)$$

We are now dealing with pure bending, which means that there are no axial forces acting on the beam. We can state this in mathematical form by summing all the horizontal forces acting on the cross section and equating this sum to zero. The force acting on an element of area dA is $\sigma \, dA$; therefore

$$\int \sigma \, dA = -\frac{E}{\rho} \int y \, dA = 0 \qquad (f)$$

Equation (f) defines the location of the neutral axis. The moment of the area about the neutral axis is zero, and hence the neutral axis passes through the centroid of the cross-sectional area.

Next we observe that equilibrium requires that the internal bending moment created by the stress σ must be the same as the external moment M. In other words,

$$M = \int y\sigma \, dA = \frac{E}{\rho} \int y^2 \, dA \qquad (g)$$

The second integral in Eq. (g) is the moment of inertia of the area about the z axis. That is,

$$I = \int y^2 \, dA \qquad (2\text{-}28)$$

This is called the *area moment of inertia;* since area cannot truly have inertia, it should not be confused with *mass moment of inertia.*

If we next solve Eqs. (g) and $(2\text{-}28)$ and rearrange them, we have

$$\frac{1}{\rho} = \frac{M}{EI} \qquad (2\text{-}29)$$

This is an important equation in the determination of the deflection of beams, and we shall employ it in Chap. 3. Finally, we eliminate ρ from Eqs. (e) and $(2\text{-}29)$ and obtain

$$\sigma = -\frac{My}{I} \qquad (2\text{-}30)$$

Equation $(2\text{-}30)$ states that the bending stress σ is directly proportional to the distance y from the neutral axis and the bending moment M, as shown in Fig. 2-13. It is customary to designate $c = y_{\max}$, to omit the negative sign, and to write

$$\sigma = \frac{Mc}{I} \qquad (2\text{-}31)$$

where it is understood that Eq. $(2\text{-}31)$ gives the maximum stress. Then tensile or compressive maximum stresses are determined by inspection when the sense of the moment is known.

Equation $(2\text{-}31)$ is often written in the two alternative forms

$$\sigma = \frac{M}{I/c} \qquad \sigma = \frac{M}{Z} \qquad (2\text{-}32)$$

where $Z = I/c$ is called the *section modulus.*

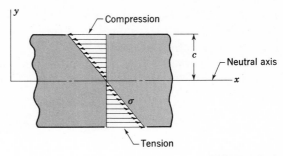

FIGURE 2-13

EXAMPLE 2-5 A beam having a T section with the dimensions shown in Fig. 2-14 is subjected to a bending moment of 14 kip·in. Locate the neutral axis and find the maximum tensile and compressive bending stresses. The sign of the moment produces tension at the top surface.

SOLUTION The area of the composite section is $A = 3.25$ in^2. Now divide the T section into two rectangles numbered 1 and 2 and sum the moments of these areas about the top edge. We then have

$$3.25c_1 = (\tfrac{1}{2})(3)(\tfrac{1}{4}) + (\tfrac{1}{2})(3\tfrac{1}{2})(2\tfrac{1}{4})$$

and hence $c_1 = 1.327$ in. Therefore $c_2 = 4 - 1.327 = 2.673$ in. Note that we have chosen not to round off these values.

Next we calculate the moment of inertia of each rectangle about its own centroidal axis. Using Table A-30, we find for the top rectangle

$$I_1 = \frac{bh^3}{12} = \frac{(3)(\tfrac{1}{2})^3}{12} = 0.0321 \text{ in}^4$$

Similarly, for the bottom rectangle,

$$I_2 = \frac{bh^3}{12} = \frac{(\tfrac{1}{2})(3\tfrac{1}{2})^3}{12} = 1.787 \text{ in}^4$$

We now employ the parallel-axis theorem to obtain the moment of inertia of the composite figure about its own centroidal axis. This theorem states

$$I_x = I_{cg} + Ad^2$$

where I_{cg} is the moment of inertia of the area about its own centroidal axis and I_x is the moment of inertia about any parallel axis a distance d removed. For the top rectangle, the distance is

$$d_1 = c_1 - \tfrac{1}{4} = 1.327 - 0.25 = 1.077 \text{ in}$$

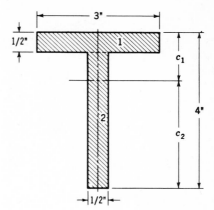

FIGURE 2-14

And for the bottom rectangle,

$$d_2 = c_2 - 1\tfrac{3}{4} = 2.673 - 1.75 = 0.923 \text{ in}$$

Using the parallel-axis theorem twice, we now find

$$I = (I_1 + A_1 d_1^2) + (I_2 + A_2 d_2^2)$$
$$= [0.0312 + (1.5)(1.077)^2] + [1.787 + (1.75)(0.923)^2] = 5.048 \text{ in}^4$$

Finally then, the maximum tensile stress, which occurs at the top surface, is found to be

$$\sigma = \frac{Mc_1}{I} = \frac{14(10)^3(1.327)}{5.048} = 3680 \text{ psi}$$

Similarly, the maximum compressive stress at the lower surface is found to be

$$\sigma = -\frac{Mc_2}{I} = -\frac{14(10)^3(2.673)}{5.048} = -7400 \text{ psi} \qquad ////$$

2-10 BEAMS WITH UNSYMMETRICAL SECTIONS

The relations developed in Sec. 2-9 can also be applied to beams having unsymmetrical sections provided the plane of bending coincides with one of the two principal axes of the section. We have found that the stress at a distance y from the neutral axis is

$$\sigma = -\frac{Ey}{\rho} \qquad (a)$$

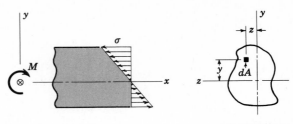

FIGURE 2-15

Therefore the force on the element of area dA in Fig. 2-15 is

$$\sigma \, dA = -\frac{Ey}{\rho} \, dA$$

Taking moments of this force about the y axis and integrating across the section gives

$$M_y = \int \sigma z \, dA = -\frac{E}{\rho} \int yz \, dA \qquad (b)$$

We recognize the second integral in Eq. (b) as the *product of inertia I_{yz}*. If the bending moment is in the plane of one of the principal axes, then

$$I_{yz} = \int yz \, dA = 0 \qquad (c)$$

and with this restriction the relations hold for any cross-sectional shape. In design, of course, this does not come about unless the designer designs the parts which connect to, and transfer the load into, the beam such that the plane of these bending loads is the same as one of the principal planes of the beam.

2-11   SHEAR STRESSES IN BENDING

When the shear force in a beam is not zero, a shear stress is developed of magnitude*

$$\tau = \frac{V}{Ib} \int_{y_1}^{c} y \, dA \qquad (2\text{-}33)$$

This stress occurs at distance y_1 from the neutral surface as shown in Fig. 2-16. In this equation b is the width of the surface at y_1, V is the shear force over the entire section, and I is the moment of inertia of the section about the neutral axis.

The integral, in Eq. (2-33), is the first moment of the area of the vertical face between y_1 and c about the neutral axis. This moment is usually designated as Q.

---

* For the complete development, see Joseph E. Shigley, "Applied Mechanics of Materials," pp. 149–158, McGraw-Hill Book Company, New York, 1976.

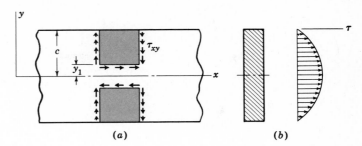

FIGURE 2-16

Thus

$$Q = \int_{y_1}^{c} y \, dA \qquad (2\text{-}34)$$

With this simplification, Eq. (2-33) may be written as

$$\tau = \frac{VQ}{Ib} \qquad (2\text{-}35)$$

Equation (2-33) shows that the shear stress is maximum when $y_1 = 0$ and zero when $y_1 = c$ for a rectangular section. Thus the shear stress is zero at the top and bottom of the beam, where the bending stress is maximum, and the shear stress is maximum at the neutral axis, where the bending stress (normal stress) is zero. Since horizontal shear stress is always accompanied by vertical shear stress, the distribution can be diagrammed as shown in Fig. 2-16a.

Figure 2-16b shows the shear-stress distribution for a rectangular section. Substitution of I for such a section in Eq. (2-33), with $y_1 = 0$, gives the maximum

$$\tau_{max} = \frac{3V}{2A} \qquad (2\text{-}36)$$

For a solid circular beam, Eq. (2-33) gives

$$\tau_{max} = \frac{4V}{3A} \qquad (2\text{-}37)$$

and for a hollow circular section,

$$\tau_{max} = \frac{2V}{A} \qquad (2\text{-}38)$$

A good approximation for structural W and S shapes is given by*

$$\tau_{max} = \frac{V}{A_w} \qquad (2\text{-}39)$$

where A_w is the area of the web only.

---

* The symbols W, S, and C are used in the structural industry to designate hot-rolled shapes. The symbol W designates a wide-flange beam, S an eye-beam, and C a channel section.

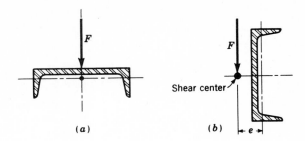

FIGURE 2-17
(a) Channel section has only one axis of symmetry; (b) to avoid twisting, the plane of the loads must pass through the shear center.

In the case of certain nonsymmetrical cross sections, the plane of the bending moment must pass through the *shear center* if twisting of the beam is to be avoided (Fig. 2-17). This problem usually arises when thin open sections are used for beams, and under these conditions local failure or buckling should also be investigated. Methods of locating the shear center are beyond the scope of this book.*

EXAMPLE 2-6   Determine a diameter for the solid round shaft, 18 in long, shown in Fig. 2-18. The shaft is supported by self-aligning bearings at the ends. Mounted upon the shaft are a V-belt sheave, which contributes a radial load of 400 lb to the shaft, and a gear, which contributes a radial load of 150 lb. Both loads are in the same plane and have the same directions. The bending stress is not to exceed 10 kpsi.

SOLUTION   As indicated in Chap. 1, certain assumptions are necessary. In this problem we decide on the following conditions:

1   The weight of the shaft is neglected.
2   Since the bearings are self-aligning, the shaft is assumed to be simply supported and the loads and bearing reactions to be concentrated.
3   The normal bending stress is assumed to govern the design.

The assumed loading diagram is shown in Fig. 2-19a. Taking moments about C,

$$\sum M_C = -18R_1 + (12)(400) + (4)(150) = 0$$
$$R_1 = 300 \text{ lb}$$

Next

$$\sum M_0 = -(6)(400) - (14)(150) + 18R_2 = 0$$
$$R_2 = 250 \text{ lb}$$

* See Crandall, Dahl, and Lardner, op. cit., p. 470.

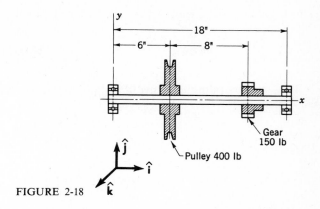

FIGURE 2-18

This statics problem can also be solved by vectors. We define the unit vector triad **ijk** associated with the xyz axes, respectively. Then position vectors from the origin O to loading points A, B, and C are

$$\mathbf{r}_A = 6\mathbf{i} \qquad \mathbf{r}_B = 14\mathbf{i} \qquad \mathbf{r}_C = 18\mathbf{i}$$

Also, the force vectors are

$$\mathbf{R}_1 = R_1\,\mathbf{j} \qquad \mathbf{F}_A = -400\mathbf{j} \qquad \mathbf{F}_B = -150\mathbf{j} \qquad \mathbf{R}_2 = R_2\,\mathbf{j}$$

Then taking moments about O yields the vector equation

$$\mathbf{r}_A \times \mathbf{F}_A + \mathbf{r}_B \times \mathbf{F}_B + \mathbf{r}_C \times \mathbf{R}_2 = 0 \qquad (1)$$

The terms in Eq. (1) are called *vector cross products*. The cross product of the two vectors

$$\mathbf{A} = x_A\mathbf{i} + y_A\mathbf{j} + z_A\mathbf{k} \qquad \mathbf{B} = x_B\mathbf{i} + y_B\mathbf{j} + z_B\mathbf{k}$$

is

$$\mathbf{A} \times \mathbf{B} = (y_A z_B - z_A y_B)\mathbf{i} + (z_A x_B - x_A z_B)\mathbf{j} + (x_A y_B - y_A x_B)\mathbf{k}$$

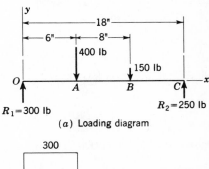

(a) Loading diagram

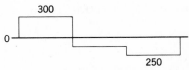

FIGURE 2-19 (b) Shearing force diagram

But this relation is more conveniently viewed as the determinant

$$\mathbf{A} \times \mathbf{B} = \begin{vmatrix} \mathbf{i} & \mathbf{j} & \mathbf{k} \\ x_A & y_A & z_A \\ x_B & y_B & z_B \end{vmatrix}$$

Thus we compute the terms in Eq. (1) as follows:

$$\mathbf{r}_A \times \mathbf{F}_A = \begin{vmatrix} \mathbf{i} & \mathbf{j} & \mathbf{k} \\ 6 & 0 & 0 \\ 0 & -400 & 0 \end{vmatrix} = -2400\mathbf{k}$$

$$\mathbf{r}_B \times \mathbf{F}_B = \begin{vmatrix} \mathbf{i} & \mathbf{j} & \mathbf{k} \\ 14 & 0 & 0 \\ 0 & -150 & 0 \end{vmatrix} = -2100\mathbf{k}$$

$$\mathbf{r}_C \times \mathbf{R}_2 = \begin{vmatrix} \mathbf{i} & \mathbf{j} & \mathbf{k} \\ 18 & 0 & 0 \\ 0 & R_2 & 0 \end{vmatrix} = 18R_2\,\mathbf{k}$$

Substituting these three terms into Eq. (1) and solving the resulting algebraic equation gives $R_2 = 250$ lb. Hence $\mathbf{R}_2 = 250\mathbf{j}$. Next we write

$$\mathbf{R}_1 = -\mathbf{F}_A - \mathbf{F}_B - \mathbf{R}_2 = -(-400\mathbf{j}) - (-150\mathbf{j}) - 250\mathbf{j} = 300\mathbf{j}\text{ lb}$$

In this instance the vector approach is more laborious. It is probably an advantage, however, for three-dimensional problems and also for programming the digital computer.

The next step is to draw the shear-force diagram (Fig. 2-19b). The maximum bending moment occurs at the point of zero shear force. Its value is

$$M = (300)(6) = 1800\text{ lb·in}$$

In view of the assumptions, Eq. (2-32) applies. We first compute the section modulus. From Table A-30,

$$\frac{I}{c} = \frac{\pi d^3}{32} = 0.0982d^3 \qquad (2)$$

Then, using Eq. (2-32),

$$\sigma = \frac{M}{I/c} = \frac{1800}{0.0982d^3}$$

Substituting $\sigma = 10\,000$ psi and solving for d yields

$$d = \sqrt[3]{\frac{1800}{(0.0982)(10\,000)}} = 1.22\text{ in}$$

Therefore we select $d = 1\tfrac{1}{4}$ in for the shaft diameter.

////

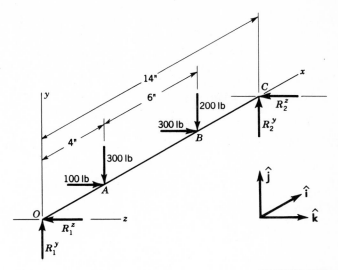

FIGURE 2-20

EXAMPLE 2-7 For the loading shown in Fig. 2-20 find the reactions R_1 and R_2 and the value and location of the maximum bending moment.

SOLUTION Examination of the forces at A and B in Fig. 2-20 indicates that the reactions R_1^y and R_2^y will be in the positive y direction, and the reactions R_1^z and R_2^z will be in the negative z direction, and they are so shown. In writing equations, however, it is convenient to assume positive directions for all unknown quantities.

First, using the algebraic approach, and taking moments about the z axis, we have

$$\sum M^z = -(4)(300) - (10)(200) + 14R_2^y = 0$$
$$R_2^y = 228 \text{ lb}$$
$$\sum M^y = -(4)(100) - (10)(300) - 14R_2^z = 0$$
$$R_2^z = -243 \text{ lb}$$

To obtain R_1^y and R_1^z, sum the forces in the y and z directions. This gives

$$\sum F^y = R_1^y - 300 - 200 + 228 = 0$$
$$R_1^y = 272 \text{ lb}$$
$$\sum F^z = R_1^z + 100 + 300 - 243 = 0$$
$$R_1^z = -157 \text{ lb}$$

Using the vector approach to accomplish the same purpose requires that we define the position vectors

$$\mathbf{r}_A = 4\mathbf{i} \qquad \mathbf{r}_B = 10\mathbf{i} \qquad \mathbf{r}_C = 14\mathbf{i}$$

Also, expressing the forces and reactions in vector form gives

$$\mathbf{R}_1 = R_1^y\,\mathbf{j} + R_1^z\mathbf{k} \qquad \mathbf{F}_A = -300\mathbf{j} + 100\mathbf{k} \qquad \mathbf{F}_B = -200\mathbf{j} + 300\mathbf{k}$$

$$\mathbf{R}_2 = R_2^y\,\mathbf{j} + R_2^z\,\mathbf{k}$$

Then, summing moments about the origin gives the vector equation

$$\sum \mathbf{M}_0 = \mathbf{r}_A \times \mathbf{F}_A + \mathbf{r}_B \times \mathbf{F}_B + \mathbf{r}_C \times \mathbf{R}_2 = 0 \qquad (1)$$

The terms for Eq. (1) are easily computed by expressing the cross products as determinants as in Example 2-6. The results are

$$\mathbf{r}_A \times \mathbf{F}_A = -400\mathbf{j} - 1200\mathbf{k}$$

$$\mathbf{r}_B \times \mathbf{F}_B = -3000\mathbf{j} - 2000\mathbf{k}$$

$$\mathbf{r}_C \times \mathbf{R}_2 = -14R_2^z\,\mathbf{j} + 14R_2^y\,\mathbf{k}$$

Substituting these in Eq. (1) and solving for $\mathbf{R}_2$ gives

$$\mathbf{R}_2 = 228\mathbf{j} - 243\mathbf{k}$$

Then

$$|\mathbf{R}_2| = 333 \text{ lb}$$

Next, for equilibrium,

$$\mathbf{R}_1 + \mathbf{F}_A + \mathbf{F}_B + \mathbf{R}_2 = 0 \qquad (2)$$

The solution to this equation is

$$\mathbf{R}_1 = 272\mathbf{j} - 157\mathbf{k} \qquad |\mathbf{R}_1| = 314 \text{ lb}$$

The maximum bending moment will occur either at point A or at point B. For both points we must take moments about axes that are parallel to the y and z axes. For point A, we have

$$M^y = -(272)(4) = -1090 \text{ lb·in} \qquad M^z = -(157)(4) = -628 \text{ lb·in}$$

and so the resultant moment at A is

$$M_A = \sqrt{(1090)^2 + (628)^2} = 1260 \text{ lb·in}$$

For point B,

$$M^z = -(272)(10) + (300)(6) = -920 \text{ lb·in}$$

$$M^y = -(157)(10) + (100)(6) = -970 \text{ lb·in}$$

and so

$$M_B = \sqrt{(920)^2 + (970)^2} = 1340 \text{ lb·in}$$

These computations can be verified by drawing two shear-force diagrams,

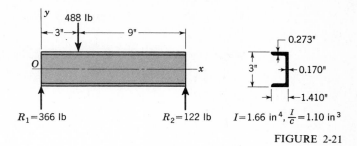

FIGURE 2-21

one for the xz plane and one for the yz plane, and computing the two corresponding moment diagrams. ////

EXAMPLE 2-8   A beam 12 in in length is to support a load of 488 lb acting 3 in from the left end, as in Fig. 2-21. Basing the design only on the bending stress, a designer has selected a 3-in A97075-O aluminum channel, the dimensions of which are shown in the figure too. The design may not be safe, however, since the designer did not consider the direct shear due to bending. If the shear stress is also included, the maximum combined stress is likely to occur near the point of load application and where the web joins the flange. Assume the load passes through the shear center, and compute the principal stresses.

SOLUTION   The loading, shear-force, and bending-moment diagrams are shown in Fig. 2-22a. To simplify the calculations we assume a cross section with square corners (Fig. 2-22b). As indicated on the shear-force diagram, the maximum shear stress occurs just to the left of the 488-lb load. We wish to determine

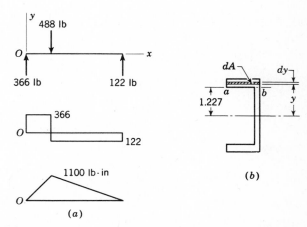

FIGURE 2-22
(a) Loading, shear, and moment diagrams; (b) assumed cross section.

the normal and shear stresses along the plane ab of Fig. 2-22b. The normal stress is

$$\sigma_x = -\frac{My}{I} = -\frac{(1100)(1.227)}{1.66} = -814 \text{ psi}$$

For the shear stress we are going to use Eq. (2-33). The terms needed are $dA = 1.410 \, dy$, $y_1 = 1.227$ in, and $c = 1.500$ in. We then have

$$Q = \int_{y_1}^{c} y \, dA = \int_{1.227}^{1.500} y(1.410) \, dy = 0.521 \text{ in}^3$$

Then
$$\tau_{xy} = 0.521 \frac{V}{Ib} = \frac{(0.521)(366)}{(1.66)(0.170)} = 675 \text{ psi}$$

When σ_x and τ_{xy} are plotted on Mohr's circle, it is found that $\sigma_1 = 375$ psi and $\sigma_2 = -1200$ psi. This compares with the maximum normal stress at the outer surface of the channel of

$$\sigma = \pm \frac{Mc}{I} = \pm \frac{1100(1.5)}{1.66} = \pm 994 \text{ psi}$$

Thus, in this example, it is important to consider the contribution made by the direct shear stress to the design.   ////

2-12 SHEAR FLOW

In determinining the shear stress in a beam, the dimension b is not always measured parallel to the neutral axis. The beam sections shown in Fig. 2-23 show how to measure b in order to compute the static moment Q. Always remember that it is the tendency of the shaded area to slide relative to the unshaded area, which causes the shear stress. Thus, the shear stress at any section is always obtained by selecting the minimum b for that section.

Another useful idea in the analysis of beams is the concept of *shear flow*. By merely removing the dimension b from Eq. (2-35) we get the shear flow q as

$$q = VQ/I \qquad (2\text{-}40)$$

where q is in force units per unit of length of the beam at the section under consideration. Basically, the shear flow is simply the shear force per unit length at the section defined by $y = y_1$. When the shear flow is known, the shear stress is determined by the equation

$$\tau = q/b \qquad (2\text{-}41)$$

EXAMPLE 2-9   Figure 2-24 illustrates an aluminum box beam. It is built up by riveting two aluminum plates $\frac{1}{4}$ in thick to a pair of cold-formed 6 in by 1 in

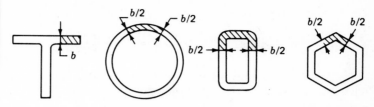

FIGURE 2-23

channels. The beam is subjected to a shear force of 1700 lb, which is constant along the length of the beam. If the $\frac{1}{4}$-in aluminum rivets used have an allowable shear load of 500 lb each, what should be the spacing a of the rivets? Neglect the bearing stresses.

SOLUTION   It is necessary first to find the moment of inertia of the built-up section. The easiest way to do this is to find the moment of inertia of the enclosing rectangle, and, from this, to subtract the moments of inertia of the three empty rectangular spaces. The enclosing rectangle is 4 in wide and $6\frac{1}{2}$ in high. The central space is 2 in wide by 6 in high. The two side spaces are each $\frac{3}{4}$ in wide by $5\frac{1}{2}$ in high. Thus, the moment of inertia of the built-up section is

$$I = \frac{4(6\frac{1}{2})^3}{12} - \frac{2(6)^3}{12} - 2\left[\frac{(\frac{3}{4})(5\frac{1}{2})^3}{12}\right] = 34.7 \text{ in}^4$$

We are going to solve this problem by using enough rivets to successfully resist the shear flow between the plate and the two channels. Thus, Q is the area of a plate times its centroidal distance. Therefore

$$Q = A\bar{y} = (4)(\tfrac{1}{4})(3\tfrac{1}{4} - \tfrac{1}{8}) = 3.125 \text{ in}^3$$

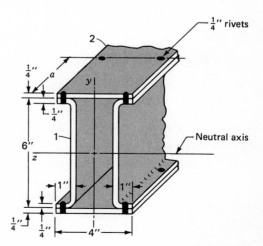

FIGURE 2-24

where $\bar{y}$ is the centroidal distance. Then, using Eq. (2-40), we get

$$q = \frac{VQ}{I} = \frac{1700(3.125)}{34.7} = 153 \text{ lb/in}$$

This means that 153 lb of shear force is to be transferred from the plate to the channels for every inch of beam length. Each pair of rivets will transfer 1000 lb into the channels. Therefore, the rivet spacing must be

$$a = \frac{1000}{153} = 6.54 \text{ in} \qquad Ans.$$

In other words, there must be a pair of rivets every $6\frac{1}{2}$ in along the length of the beam on both the top and bottom.

A similar approach can be used for built-up sections made by spot welding or cementing. ////

2-13 TORSION

Twisting of a member by an external twisting moment or torque, is called torsion. The angle of twist for a round bar is

$$\theta = \frac{Tl}{GJ} \qquad (2\text{-}42)$$

where T = torque
  l = length
  G = modulus of rigidity
  J = polar area moment of inertia

For a solid round bar, the shear stress is zero at the center and maximum at the surface. The distribution is proportional to the radius ρ and is

$$\tau = \frac{T\rho}{J} \qquad (2\text{-}43)$$

Designating r as the radius to the outer surface, we have

$$\tau_{\text{max}} = \frac{Tr}{J} \qquad (2\text{-}44)$$

The assumptions used in the analysis are:

1 The bar is acted upon by a pure torque, and the sections under consideration are remote from the point of application of the load and from a change in diameter.
2 Adjacent cross sections originally plane and parallel remain plane and parallel after twisting, and any radial line remains straight.
3 The material obeys Hooke's law.

Equation (2-44) applies only to circular sections. For a solid round section,

$$J = \frac{\pi d^4}{32} \qquad (2\text{-}45)$$

where d is the diameter of the bar. For a hollow round section,

$$J = \frac{\pi}{32}(d^4 - d_i^4) \qquad (2\text{-}46)$$

where d_i is the inside diameter, often referred to as the ID.

In using Eq. (2-44) it is often necessary to obtain the torque T from a consideration of the horsepower and speed of a rotating shaft. For convenience, three forms of this relation are

$$P = \frac{2\pi T n}{(33\,000)(12)} = \frac{FV}{33\,000} = \frac{Tn}{63\,000} \qquad (2\text{-}47)$$

$$T = \frac{63\,000P}{n} \qquad (2\text{-}48)$$

where P = horsepower
$\quad T$ = torque, lb·in
$\quad n$ = shaft speed, rpm
$\quad F$ = force, lb
$\quad V$ = velocity, fpm

If SI units are used, the applicable equation is

$$P = T\omega \qquad (2\text{-}49)$$

where P = power, W
$\quad T$ = torque, N·m
$\quad \omega$ = angular velocity, radians/s

Determination of the torsional stresses in noncircular members is a difficult problem, generally handled experimentally using a soap-film or membrane analogy, which we shall not consider here. Timoshenko and MacCullough,* however, give the following approximate formula for the maximum torsional stress in a rectangular section:

$$\tau_{max} = \frac{T}{wt^2}\left(3 + 1.8\,\frac{t}{w}\right) \qquad (2\text{-}50)$$

In this equation w and t are the width and thickness of the bar, respectively; they cannot be interchanged since t must be the shortest dimension. For thin plates in torsion, t/w is small and the second term may be neglected. The equation is also

---

* S. Timoshenko and Gleason H. MacCullough, "Elements of Strength of Materials," 3d ed., p. 265, D. Van Nostrand Company, Inc., New York, 1949. See also F. R. Shanley, "Strength of Materials," p. 509, McGraw-Hill Book Company, New York, 1957.

approximately valid for equal-sided angles; these can be considered as two rectangles each of which is capable of carrying half the torque.

2-14 THIN-WALLED CYLINDERS

When the wall thickness of a cylindrical pressure vessel is about one-tenth, or less, of its radius, the stress which results from pressurizing the vessel may be assumed to be uniformly distributed across the wall thickness. When this assumption is made, the vessel is called a *thin-walled* pressure vessel. The stress state in tanks, pipe, and hoops may also be determined using this assumption.

In Fig. 2-25 the internal pressure p is exerted on the sides of a cylinder of thickness t and internal diameter D. The load tending to separate the two halves of a unit length of the cylinder is pD. This load is resisted by the *tangential stress*, also called *hoop stress*, acting uniformly over the stressed area. We then have $pD = 2t\sigma_t$, or

$$\sigma_t = \frac{pD}{2t} \qquad (2\text{-}51)$$

where σ_t is the tangential stress.

In a closed cylinder a longitudinal stress σ_l will exist because of the pressure upon the ends of the vessel. The force acting upon the ends $p(\pi D^2/4)$ must be equated to the longitudinal stress times the area over which the stress acts. Thus

$$p\frac{\pi D^2}{4} = \sigma_l(\pi D t)$$

and so

$$\sigma_l = \frac{pD}{4t} \qquad (2\text{-}52)$$

2-15 STRESSES IN THICK-WALLED CYLINDERS

A thick-walled cylinder subjected to external or internal pressure, or both, has radial and tangential stress with values which are dependent upon the radius of the element under consideration. A thick-walled cylinder may also be stressed longitudinally. In design, gun barrels, high-pressure hydraulic cylinders, and pipes carrying fluids at high pressures, must all be considered as thick-walled cylinders.

In determining the radial stress σ_r and the tangential stress σ_t, we make use of the assumption that the longitudinal elongation is constant around the circumference of the cylinder. In other words, a right section of the cylinder remains plane after stressing.

Referring to Fig. 2-26, we designate the inside radius of the cylinder by a, the outside radius by b, the internal pressure by p_i, and the external pressure by p_o.

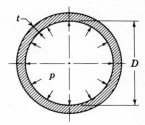

FIGURE 2-25
Tangential, or hoop, stress is caused by
internal pressure.

Let us consider the equilibrium of an infinitely thin semicircular ring cut from the cylinder at radius r and having a unit length. Taking a summation of forces in the vertical direction equal to zero, we have

$$2\sigma_t \, dr + 2\sigma_r r - 2(\sigma_r + d\sigma_r)(r + dr) = 0 \qquad (a)$$

Simplifying and neglecting higher-order quantities gives

$$\sigma_t - \sigma_r - r\frac{d\sigma_r}{dr} = 0 \qquad (b)$$

Equation (b) relates the two unknowns σ_t and σ_r, but we must obtain a second relation in order to evaluate them. The second equation is obtained from the assumption that the longitudinal deformation is constant. Since both σ_t and σ_r are positive for tension, Eq. (2-22) can be written

$$\epsilon_l = -\frac{\mu\sigma_t}{E} - \frac{\mu\sigma_r}{E} \qquad (c)$$

where μ is Poisson's ratio and ϵ_l is the longitudinal unit deformation. Both of these are constants, and so Eq. (c) can be rearranged in the form

$$-\frac{E\epsilon_l}{\mu} = \sigma_t + \sigma_r = 2C_1 \qquad (d)$$

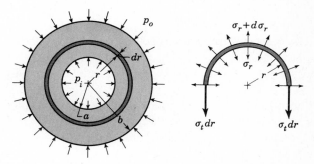

FIGURE 2-26
A thick-wall cylinder.

Next, solving Eqs. (b) and (d) to eliminate σ_t produces

$$r\frac{d\sigma_r}{dr} + 2\sigma_r = 2C_1 \tag{e}$$

Multiplying Eq. (e) by r gives

$$r^2\frac{d\sigma_r}{dr} + 2r\sigma_r = 2rC_1 \tag{f}$$

We note that

$$\frac{d}{dr}(r^2\sigma_r) = r^2\frac{d\sigma_r}{dr} + 2r\sigma_r \tag{g}$$

Therefore

$$\frac{d}{dr}(r^2\sigma_r) = 2rC_1 \tag{h}$$

which, when integrated, gives

$$r^2\sigma_r = r^2C_1 + C_2 \tag{i}$$

where C_2 is a constant of integration. Solving for σ_r, we obtain

$$\sigma_r = C_1 + \frac{C_2}{r^2} \tag{j}$$

Substituting this value in Eq. (d), we find

$$\sigma_t = C_1 - \frac{C_2}{r^2} \tag{k}$$

To evaluate the constants of integration, note that, at the boundaries of the cylinder,

$$\sigma_r = \begin{cases} -p_i & \text{at } r = a \\ -p_o & \text{at } r = b \end{cases}$$

Substituting these values in Eq. (j) yields

$$-p_i = C_1 + \frac{C_2}{a^2} \qquad -p_o = C_1 + \frac{C_2}{b^2} \tag{l}$$

The constants are found by solving these two equations simultaneously. This gives

$$C_1 = \frac{p_i a^2 - p_o b^2}{b^2 - a^2} \qquad C_2 = \frac{a^2 b^2(p_o - p_i)}{b^2 - a^2} \tag{m}$$

Substituting these values into Eqs. (j) and (k) yields

$$\sigma_t = \frac{p_i a^2 - p_o b^2 - a^2 b^2(p_o - p_i)/r^2}{b^2 - a^2} \tag{2-53}$$

$$\sigma_r = \frac{p_i a^2 - p_o b^2 + a^2 b^2(p_o - p_i)/r^2}{b^2 - a^2} \tag{2-54}$$

In the above equations positive stresses indicate tension and negative stresses compression.

Let us now determine the stresses when the external pressure is zero. Substitution of $p_o = 0$ in Eqs. (2-53) and (2-54) gives

$$\sigma_t = \frac{a^2 p_i}{b^2 - a^2}\left(1 + \frac{b^2}{r^2}\right) \qquad (2\text{-}55)$$

$$\sigma_r = \frac{a^2 p_i}{b^2 - a^2}\left(1 - \frac{b^2}{r^2}\right) \qquad (2\text{-}56)$$

These equations are plotted in Fig. 2-27 to show the distribution of stresses over the wall thickness. The maximum stresses occur at the inner surface, where $r = a$. Their magnitudes are

$$\sigma_t = p_i \frac{b^2 + a^2}{b^2 - a^2} \qquad (2\text{-}57)$$

$$\sigma_r = -p_i \qquad (2\text{-}58)$$

The stresses in the outer surface of a cylinder subjected only to external pressure are found similarly. They are

$$\sigma_t = -p_o \frac{b^2 + a^2}{b^2 - a^2} \qquad (2\text{-}59)$$

$$\sigma_r = -p_o \qquad (2\text{-}60)$$

2-16   PRESS AND SHRINK FITS

When two cylindrical parts are assembled by shrinking or press-fitting one part upon another, a contact pressure is created between the two parts. The stresses resulting from this pressure may easily be determined with the equations of the preceding section.

Figure 2-28 shows two cylindrical members which have been assembled with a shrink fit. A contact pressure p exists between the members at radius b, causing radial stresses $\sigma_r = -p$ in each member at the contacting surfaces. From Eq. (2-59), the tangential stress at the outer surface of the inner member is

$$\sigma_{it} = -p\frac{b^2 + a^2}{b^2 - a^2} \qquad (2\text{-}61)$$

In the same manner, from Eq. (2-57), the tangential stress at the inner surface of the outer member is

$$\sigma_{ot} = p\frac{c^2 + b^2}{c^2 - b^2} \qquad (2\text{-}62)$$

These equations cannot be solved until the contact pressure is known. In obtaining a shrink fit, the diameter of the male member is made larger than the diameter

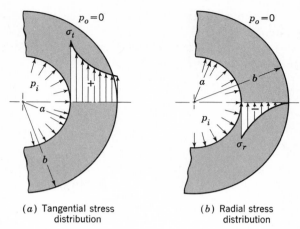

(a) Tangential stress
distribution

(b) Radial stress
distribution

FIGURE 2-27
Distribution of stresses in a thick-walled cylinder subjected to internal pressure.

of the female member. The difference in these dimensions is called the interference and is the deformation which the two members must experience. Since these dimensions are usually known, the deformation should be introduced in order to evaluate the stresses. Let

$$\delta_o = \text{increase in radius of hole}$$

and
$$\delta_i = \text{decrease in radius of inner cylinder}$$

The tangential strain in the outer cylinder at the inner radius is

$$\epsilon_{ot} = \frac{\text{change in circumference}}{\text{original circumference}}$$

or
$$\epsilon_{ot} = \frac{2\pi(b + \delta_o) - 2\pi b}{2\pi b} = \frac{\delta_o}{b}$$

and so
$$\delta_o = b\epsilon_{ot} \qquad (a)$$

but since
$$\epsilon_{ot} = \frac{\sigma_{ot}}{E_o} - \frac{\mu\sigma_{or}}{E_o}$$

then, from Eqs. (2-61) and (2-62), we have

$$\delta_o = \frac{bp}{E_o}\left(\frac{c^2 + b^2}{c^2 - b^2} + \mu\right) \qquad (b)$$

This is the increase in radius of the outer cylinder. In a similar manner the decrease in radius of the inner cylinder is found to be

$$\delta_i = -\frac{bp}{E_i}\left(\frac{b^2 + a^2}{b^2 - a^2} - \mu\right) \qquad (c)$$

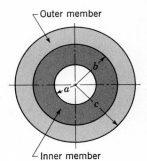

Outer member

b

a

c

FIGURE 2-28 Inner member

Then the total deformation δ is

$$\delta = \delta_o - \delta_i = \frac{bp}{E_o}\left(\frac{c^2 + b^2}{c^2 - b^2} + \mu\right) + \frac{bp}{E_i}\left(\frac{b^2 + a^2}{b^2 - a^2} - \mu\right) \qquad (2\text{-}63)$$

This equation can be solved for the pressure p when the interference δ is given. If the two members are of the same material, $E_o = E_i = E$ and the relation simplifies to

$$p = \frac{E\delta}{b}\left[\frac{(c^2 - b^2)(b^2 - a^2)}{2b^2(c^2 - a^2)}\right] \qquad (2\text{-}64)$$

Substitution of this value of p in Eqs. (2-61) and (2-62) will then give the tangential stresses at the inner surface of the outer cylinder and at the outer surface of the inner cylinder. In addition, Eq. (2-63) or (2-64) can be employed to obtain the value of p for use in the general equations [Eqs. (2-53) and (2-54)] in order to obtain the stress at any point in either cylinder.

Assumptions

In addition to the assumptions both stated and implied by the development, it is necessary to assume that both members have the same length. In the case of a hub which has been press-fitted to a shaft, this assumption would not be true, and there would be an increased pressure at each end of the hub. It is customary to allow for this condition by the employment of a stress-concentration factor. The value of this factor depends upon the contact pressure and the design of the female member, but its theoretical value is seldom greater than 2.

EXAMPLE 2-10 A tube, such as a gun barrel, has nominal dimensions of 1 in ID × 2 in OD, over which a second tube having nominal dimensions of 2 in ID × 3 in OD is to be shrink-fitted. The material is steel. It is desired to fit these two members together to cause a stress of $\sigma_t = 10$ kpsi at the inner surface of the outer member.

(a) Find the required original dimensions of the members.
(b) Determine the resulting stress distributions.

SOLUTION   (a) Using Eq. (2-62), we solve for the contact pressure. Thus

$$p = \sigma_{ot}\frac{c^2 - b^2}{c^2 + b^2} = (10\,000)\frac{(1.5)^2 - (1)^2}{(1.5)^2 + (1)^2} = 3850 \text{ psi}$$

Solving Eq. (2-64) for the deformation, we obtain

$$\delta = \frac{bp}{E}\frac{2b^2(c^2 - a^2)}{(c^2 - b^2)(b^2 - a^2)}$$

$$= \frac{(1)(3850)}{(30)(10)^6}\frac{(2)(1)^2[(1.5)^2 - (0.5^2)]}{[(1.5)^2 - (1)^2][(1)^2 - (0.5)^2]}$$

$$= 0.000\,548 \text{ in}$$

The dimensions selected are as follows:

Outside diameter of inner member $= 2.000\,00$ in

Inside diameter of outer member $= 2.000\,00 - (2)(0.000\,548)$

$$= 1.998\,904 \text{ in}$$

(b)   We shall determine the stress distribution in the outer member first. This is a cylinder subjected to an internal pressure of $p_i = 3850$ psi and an external pressure of $p_o = 0$ psi. Equations (2-55) and (2-56) apply. In these equations $a = 1$ in and $b = 1.5$ in, and they are to be solved for various values of r between 1 and 1.5 in. By using $r = 1.1$ in, a sample calculation is as follows:

$$\sigma_t = \frac{a^2 p_i}{b^2 - a^2}\left(1 + \frac{b^2}{r^2}\right) = \frac{(1)^2(3850)}{(1.5)^2 - (1)^2}\left[1 + \frac{(1.5)^2}{(1.1)^2}\right]$$

$$= 8810 \text{ psi}$$

$$\sigma_r = \frac{a^2 p_i}{b^2 - a^2}\left(1 - \frac{b^2}{r^2}\right) = \frac{(1)^2(3850)}{(1.5)^2 - (1)^2}\left[1 - \frac{(1.5)^2}{(1.1)^2}\right]$$

$$= -2650 \text{ psi}$$

These, and other results obtained in the same manner, are shown in Table 2-2.

Table 2-2 TANGENTIAL AND RADIAL STRESSES IN THE OUTER MEMBER

Radius r, in	Tangential stress σ_t, psi	Radial stress σ_r, psi
1.0	10 000	−3 850
1.1	8 810	−2 650
1.2	7 900	−1 740
1.3	7 180	−950
1.4	6 630	−460
1.5	6 160	0

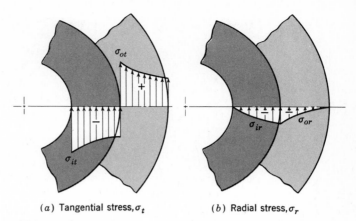

(a) Tangential stress, σ_t (b) Radial stress, σ_r

FIGURE 2-29
Distribution of radial and tangential stresses in shrink-fitted members.

Equations (2-53) and (2-54) are used for the inner member. When $p_i = 0$ and $p_o = p$, these equations reduce to

$$\sigma_t = -\frac{pb^2}{b^2 - a^2}\left(1 + \frac{a^2}{r^2}\right) \qquad (2\text{-}65)$$

$$\sigma_r = -\frac{pb^2}{b^2 - a^2}\left(1 - \frac{a^2}{r^2}\right) \qquad (2\text{-}66)$$

Upon applying this set of equations to this example, $a = 0.5$ in, $b = 1$ in, and r varies from 0.5 to 1 in. The contact pressure is 3850 psi as before. The results are shown in Table 2-3, and the stress distribution for both members is plotted in Fig. 2-29.   ////

2-17 THERMAL STRESSES AND STRAINS

When the temperature of an unrestrained body is uniformly increased, the body expands, and the normal strain is

$$\epsilon_x = \epsilon_y = \epsilon_z = \alpha(\Delta T) \qquad (2\text{-}67)$$

Table 2-3 TANGENTIAL AND RADIAL STRESSES IN THE INNER MEMBER

Radius r, in	Tangential stress σ_t, psi	Radial stress σ_r, psi
0.50	− 10 300	0
0.60	− 8 720	− 1 570
0.70	− 7 750	− 2 520
0.80	− 7 160	− 3 130
0.90	− 6 730	− 3 560
1.00	− 6 420	− 3 850

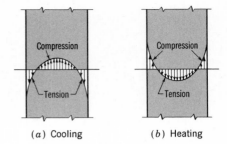

FIGURE 2-30
Thermal stresses in an infinite slab during
heating and cooling.

(a) Cooling   (b) Heating

where α is the coefficient of thermal expansion and ΔT is the temperature change, in degrees. In this action the body experiences a simple volume increase with the components of shear strain all zero.

If a straight bar is restrained at the ends so as to prevent lengthwise expansion and then is subjected to a uniform increase in temperature, a compressive stress will develop because of the axial constraint. The stress is

$$\sigma = \epsilon E = \alpha(\Delta T)E \qquad (2\text{-}68)$$

In a similar manner, if a uniform flat plate is restrained at the edges and also subjected to a uniform temperature rise, the compressive stress developed is given by the equation

$$\sigma = \frac{\alpha(\Delta T)E}{1 - \mu} \qquad (2\text{-}69)$$

The stresses represented by Eqs. (2-68) and (2-69), though due to temperature, are not thermal stresses inasmuch as they result from the fact that the edges were restrained. A *thermal stress* is one which arises because of the existence of a *temperature gradient* in a body.

Figure 2-30 shows the internal stresses within a slab of infinite dimensions during heating and cooling. During cooling, the maximum stress is the surface

Table 2-4 COEFFICIENTS OF THERMAL EXPANSION (LINEAR MEAN COEFFICIENTS FOR THE TEMPERATURE RANGE 0–100°C)

Material	Celsius scale	Fahrenheit scale
Aluminum	$23.9(10)^{-6}$	$13.3(10)^{-6}$
Brass, cast	$18.7(10)^{-6}$	$10.4(10)^{-6}$
Carbon steel	$10.8(10)^{-6}$	$6.0(10)^{-6}$
Cast iron	$10.6(10)^{-6}$	$5.9(10)^{-6}$
Magnesium	$25.2(10)^{-6}$	$14.0(10)^{-6}$
Nickel steel	$13.1(10)^{-6}$	$7.3(10)^{-6}$
Stainless steel	$17.3(10)^{-6}$	$9.6(10)^{-6}$
Tungsten	$4.3(10)^{-6}$	$2.4(10)^{-6}$

tension. At the same time, force equilibrium requires a compressive stress at the center of the slab. During heating, the external surfaces are hot and tend to expand but are restrained by the cooler center. This causes compression in the surface and tension in the center as shown.

Table 2-4 lists approximate values of the coefficient α for various engineering materials.

2-18 CURVED BEAMS

The distribution of stress in a curved beam is determined by using the following assumptions:

1 The cross section has an axis of symmetry in a plane along the length of the beam.
2 Plane cross sections remain plane after bending.
3 The modulus of elasticity is the same in tension as in compression.

Unlike a straight beam, we shall find that the neutral axis and the centroidal axis of a curved beam are not coincident, and also that the stress does not vary linearly from the neutral axis. The notation shown in Fig. 2-31 is defined as follows:

$$r_o = \text{radius of outer fiber}$$

$$r_i = \text{radius of inner fiber}$$

$$h = \text{depth of section}$$

$$c_o = \text{distance from neutral axis to outer fiber}$$

$$c_i = \text{distance from neutral axis to inner fiber}$$

$$r = \text{radius of neutral axis}$$

$$\bar{r} = \text{radius of centroidal axis}$$

$$e = \text{distance from centroidal axis to neutral axis}$$

To begin, we define the element *abcd* by the angle ϕ. A bending moment M causes section *bc* to rotate through $d\phi$ to *b'c'*. The strain on any fiber at distance ρ from the center O is

$$\epsilon = \frac{\delta}{l} = \frac{(r - \rho)\, d\phi}{\rho\phi} \tag{a}$$

The normal stress corresponding to this strain is

$$\sigma = \epsilon E = \frac{E(r - \rho)\, d\phi}{\rho\phi} \tag{b}$$

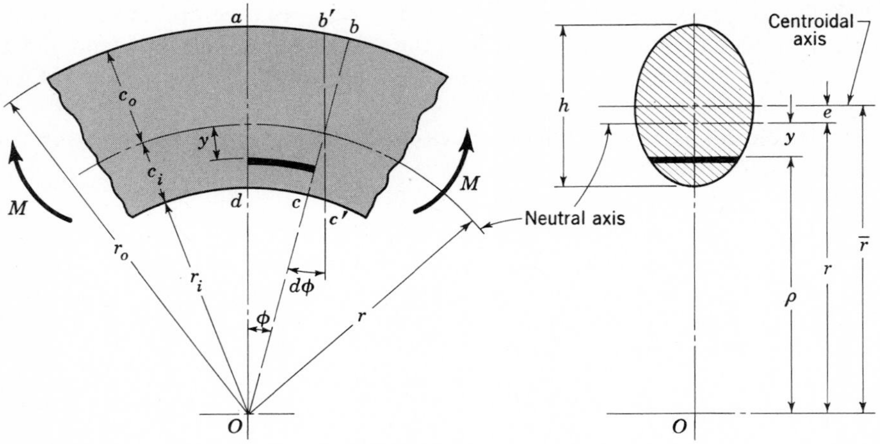

FIGURE 2-31
Note that y is positive in the direction toward point O.

Since there are no axial external forces acting on the beam, the sum of the normal forces acting on the section must be zero. Therefore

$$\int \sigma \, dA = E \frac{d\phi}{\phi} \int \frac{(r - \rho) \, dA}{\rho} = 0 \qquad (c)$$

Now arrange Eq. (c) in the form

$$E \frac{d\phi}{\phi} \left(r \int \frac{dA}{\rho} - \int dA \right) = 0 \qquad (d)$$

and solve the expression in parentheses. This gives

$$r \int \frac{dA}{\rho} - A = 0$$

or

$$r = \frac{A}{\int \frac{dA}{\rho}} \qquad (2\text{-}70)$$

This important equation is used to find the location of the neutral axis with respect to the center of curvature O of the cross section. The equation indicates that the neutral and the centroidal axes are not coincident.

Our next problem is to determine the stress distribution. We do this by balancing the external applied moment against the internal resisting moment. Thus, from Eq. (b),

$$\int (r - \rho)(\sigma \, dA) = E \frac{d\phi}{\phi} \int \frac{(r - \rho)^2 \, dA}{\rho} = M \qquad (e)$$

Since $(r - \rho)^2 = r^2 - 2\rho r + \rho^2$, Eq. ($e$) can be written in the form

$$M = E\frac{d\phi}{\phi}\left(r^2 \int \frac{dA}{\rho} - r \int dA - r \int dA + \int \rho \, dA\right) \qquad (f)$$

Note that r is a constant; then compare the first two terms in parentheses with Eq. (d). These terms vanish, and we have left

$$M = E\frac{d\phi}{\phi}\left(-r \int dA + \int \rho \, dA\right) \qquad (g)$$

The first integral in this expression is the area A, and the second is the product $\bar{r}A$. Therefore

$$M = E\frac{d\phi}{\phi}(\bar{r} - r)A = E\frac{d\phi}{\phi}eA \qquad (h)$$

Now, using Eq. (b) once more, and rearranging, we finally obtain

$$\sigma = \frac{My}{Ae(r - y)} \qquad (2\text{-}71)$$

This equation shows that the stress distribution is hyperbolic. The maximum stresses occur at the inner and outer fibers and are

$$\sigma_i = \frac{Mc_i}{Aer_i} \qquad \sigma_o = \frac{Mc_o}{Aer_o} \qquad (2\text{-}72)$$

These equations are valid for pure bending. In the usual and more general case, such as a crane hook, the U frame of a press, or the frame of a clamp, the bending moment is due to forces acting to one side of the cross section under consideration. In this case the bending moment is computed about the *centroidal axis, not* the neutral axis. Also, an additional axial tensile or compressive stress must be added to the bending stress given by Eqs. (2-71) and (2-72) to obtain the resultant stress acting on the section.

   An optimum section for a curved beam can be designed by relating the inner and outer fiber stresses given by Eq. (2-72) in the same ratio as the tensile and compressive strengths of the beam material. Thus, if the material is steel, the tensile and compressive strengths are nearly equal and we can design the section so that $\sigma_i = \sigma_o$ and arrive at an optimum geometry. Note that this yields the simple relation

$$\frac{c_i}{c_o} = \frac{r_i}{r_o} \qquad (i)$$

which, however, is not so simple to implement. An approach to optimization is to decide on the form of the section and to program Eq. (2-70) and other necessary relations for digital computation. Then, using time sharing with man-machine interaction, simply punch in various trial values of the geometry until Eq. (i) is satisfied. If batch processing is used, one of the standard iteration techniques can be employed with Eq. (*i*) as the STOP or END command.

EXAMPLE 2-11   A rectangular-section beam has the dimensions $b = 1$ in, $h = 3$ in and is subjected to a pure bending moment of 20 000 lb·in so as to produce compression of the inner fiber. Find the stresses for the following geometries:

(*a*) The beam is straight.
(*b*) The centroidal axis has a radius of curvature of 15 in.
(*c*) The centroidal axis has a radius of curvature of 3 in.

SOLUTION

a.

$$\frac{I}{c} = \frac{bh^2}{6} = \frac{(1)(3)^2}{6} = 1.5 \text{ in}^3$$

$$\sigma_{max} = \frac{M}{I/c} = \frac{20\,000}{1.5} = 13\,300 \text{ psi}$$

b. We must integrate Eq. (2-70) first. Designating $dA = b\,d\rho$, we have

$$r = \frac{A}{\int \frac{1}{\rho}\,dA} = \frac{bh}{\int_{r_i}^{r_o} \frac{b}{\rho}\,d\rho} = \frac{h}{\ln \dfrac{r_o}{r_i}} \qquad (2\text{-}73)$$

Now, $r_o = 15 + 1.5 = 16.5$ and $r_i = 15 - 1.5 = 13.5$ in. Solving Eq. (2-73) gives

$$r = \frac{3}{\ln\,(16.5/13.5)} = 14.950 \text{ in}$$

Then $e = \bar{r} - r = 15.00 - 14.950 = 0.050$ in. Also $c_o = 1.550$ in and $c_i = 1.450$ in. Equations (2-72) give

$$\sigma_i = -\frac{Mc_i}{Aer_i} = -\frac{(20\,000)(1.450)}{(3)(0.050)(13.5)} = -14\,330 \text{ psi}$$

$$\sigma_o = \frac{Mc_o}{Aer_o} = \frac{(20\,000)(1.550)}{(3)(0.050)(16.5)} = 12\,530 \text{ psi}$$

c. We have $r_o = 3 + 1.5 = 4.5$ in and $r_i = 3 - 1.5 = 1.5$ in. Using Eq. (2-73) again, we obtain

$$r = \frac{3}{\ln\,(4.5/1.5)} = 2.731 \text{ in}$$

So, $e = \bar{r} - r = 3 - 2.731 = 0.269$ in, $c_o = 1.769$ in, and $c_i = 1.231$ in. We have, from Eqs. (2-72),

$$\sigma_i = -\frac{Mc_i}{Aer_i} = -\frac{(20\,000)(1.231)}{(3)(0.269)(1.5)} = -20\,400 \text{ psi}$$

$$\sigma_o = \frac{Mc_o}{Aer_o} = \frac{(20\,000)(1.769)}{(3)(0.269)(4.5)} = 9750 \text{ psi}$$

It is interesting to note that the difference in stresses for the straight beam and the beam having a 15-in radius of curvature is only about $7\frac{1}{2}$ percent. Such an error may be acceptable in certain design situations. It is here that you must utilize your own judgment in making the decision whether to treat the member as straight or as curved. Only your own time will be saved by assuming the beam is straight. If the radius of curvature is large compared with the section depth, and if your boss is nagging you for a quick result, you might well decide to increase the factor of safety by a modest amount, and compute the stress using the straight-beam assumption. Of course, you should not risk this course of action in the classroom or on the job without being aware of the inaccuracies involved.   ////

Curved-Beam Formulas

Sections most frequently encountered in the stress analysis of curved beams are shown in Fig. 2-32. Formulas for the rectangular section were developed in Example 2-11, but they are repeated here for convenience.

$$\bar{r} = r_i + \frac{h}{2} \tag{2-74}$$

$$r = \frac{h}{\ln{(r_o/r_i)}} \tag{2-75}$$

For the trapezoidal section in Fig. 2-32b, the formulas are

$$\bar{r} = r_i + \frac{h}{3}\frac{b_i + 2b_o}{b_i + b_o} \tag{2-76}$$

$$r = \frac{A}{b_o - b_i + [(b_i r_o - b_o r_i)/h]\ln{(r_o/r_i)}} \tag{2-77}$$

For the T section in Fig. 2-32c we have

$$\bar{r} = r_i + \frac{b_i c_1^2 + 2b_o c_1 c_2 + b_o c_2^2}{2(b_o c_2 + b_i c_1)} \tag{2-78}$$

$$r = \frac{b_i c_1 + b_o c_2}{b_i \ln{[(r_i + c_1)/r_i]} + b_o \ln{[r_o/(r_i + c_1)]}} \tag{2-79}$$

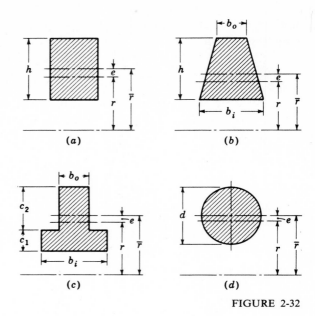

FIGURE 2-32

The equations for the solid round section of Fig. 2-32d are

$$\bar{r} = r_i + \frac{d}{2} \qquad (2\text{-}80)$$

$$r = \frac{d^2}{4(2\bar{r} - \sqrt{4\bar{r}^2 - d^2})} \qquad (2\text{-}81)$$

Note, very particularly, that Eq. (2-81) gives the radius of curvature of the neutral axis. Do not confuse the result with the radius of the section!

The formulas for other sections can be obtained by performing the integration indicated in Eq. (2-70). Cold forming of beams into curved beams introduces distortion of the section such that the resulting section has a different geometry than the unbent section. This distortion, of course, is greatest in the case of rectangular or circular tubing. For all these, as well as for complex sections, a numerical or graphical integration of Eq. (2-70) may be required if accurate results are desired.

2-19 HERTZ CONTACT STRESSES

A state of triaxial stress seldom arises in design. An exception to this occurs when two bodies having curved surfaces are pressed against one another. When this happens, point or line contact changes to area contact, and the stress developed in both bodies is three-dimensional. Contact-stress problems arise in the contact of a wheel and a rail, a cam and its follower, mating gear teeth, and in the action of rolling bearings. To guard against the possibility of surface failure, in such cases, it is necessary to have means of computing the stress states which result from loading one body against another.

When two solid spheres of diameters d_1 and d_2 are pressed together with a force F, a circular area of contact of radius a is obtained. Specifying E_1, μ_1 and E_2, μ_2 as the respective elastic constants of the two spheres, the radius a is given by the equation

$$a = \sqrt[3]{\frac{3F}{8} \frac{[(1 - \mu_1^2)/E_1] + [(1 - \mu_2^2)/E_2]}{(1/d_1) + (1/d_2)}} \qquad (2\text{-}82)$$

The pressure within each sphere has a semi-elliptical distribution, as shown in Fig. 2-33. The maximum pressure occurs at the center of the contact area and is

$$p_{\max} = -\frac{3F}{2\pi a^2} \qquad (2\text{-}83)$$

Equations (2-82) and (2-83) are perfectly general and also apply to the contact of a sphere and a plane surface or to a sphere and an internal spherical surface. For a plane surface use $d = \infty$. For an internal surface the diameter is expressed as a negative quantity.

Thomas and Hoersch* have calculated the stresses for varying depths below the contact surface. Referring to Fig. 2-33, we define a three-dimensional right-handed coordinate system with the origin in the center of the contact area and the z coordinate in the direction of the contact force and defining the depth of any stress element below the contact surface. Then, according to Thomas and Hoersch, the stresses in the x, y, and z directions are principal stresses. We first define a factor A by the equation

$$A = \frac{4a}{\pi} \frac{(1/d_1) + (1/d_2)}{[(1 - \mu_1^2)/E_1] + [(1 - \mu_2^2)/E_2]} \qquad (2\text{-}84)$$

Then, for any stress element on the z axis, the stresses are

$$\sigma_x = \sigma_y = A\left[(1 + \mu)\left(\frac{z}{a}\cot^{-1}\frac{z}{a} - 1\right) + \frac{1}{2}\frac{a^2}{a^2 + z^2}\right] \qquad (2\text{-}85)$$

$$\sigma_z = -A\left(\frac{a^2}{a^2 + z^2}\right) \qquad (2\text{-}86)$$

---

* H. R. Thomas and V. A. Hoersch, Stresses Due to the Pressure of One Elastic Solid upon Another, *Univ. Ill. Eng. Expt. Sta. Bull.* 212, 1930. See also Harold A. Rothbart (ed.), "Mechanical Design and Systems Handbook," pp. 13-3–13-13, McGraw-Hill Book Company, New York, 1964, and M. F. Spotts, "Mechanical Design Analysis," pp. 166–171, Prentice-Hall, Inc., Englewood Cliffs, N.J., 1964.

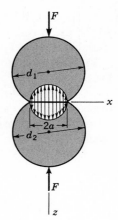

FIGURE 2-33
Pressure distribution between two spheres.

The equations are valid for either sphere, but in Eq. (2-85), the value of Poisson's ratio used must correspond with the sphere under consideration. Note that the equations would be even more complicated if the stress state at points off the z axis were to be determined because the x and y coordinates would also have to be included.

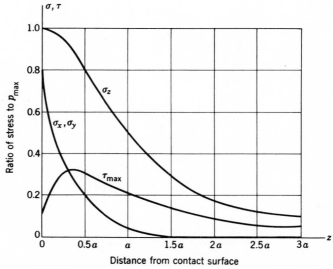

FIGURE 2-34
Magnitude of the stress components below the surface as a function of the maximum pressure for contacting spheres. Note that the maximum shear stress is slightly below the surface and is approximately $0.3p_{max}$.

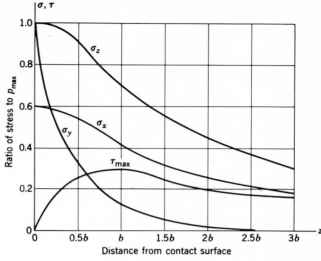

FIGURE 2-35

Magnitude of the stress components below the surface as a function of the maximum pressure for contacting cylinders. Of the three shear stresses the largest is parallel to the *zy* plane. Note that it is about $0.3p_{max}$ and that it occurs slightly below the contacting surface.

Mohr's circles for the stress state described by Eqs. (2-85) and (2-86) are a point and two coincident circles. Since $\sigma_x = \sigma_y$, $\tau_{xy} = 0$, and

$$\tau_{xz} = \tau_{yz} = \frac{\sigma_x - \sigma_z}{2} = \frac{\sigma_y - \sigma_z}{2} \qquad (2\text{-}87)$$

Figure 2-34 is a plot of Eqs. (2-85) to (2-87) for a distance of $3a$ below the surface. Note that τ reaches a maximum value slightly below the surface. It is the opinion of many authorities that this maximum shear stress is responsible for the surface fatigue failure of contacting elements. The explanation is that a crack originates at the point of maximum shear stress below the surface and progresses to the surface and that the pressure of the lubricant flowing into the crack wedges the chip loose.

When the contacting surfaces are cylindrical, the area of contact is a narrow rectangle of half-width b given by the equation

$$b = \sqrt{\frac{2F}{\pi l} \frac{[(1 - \mu_1^2)/E_1] + [(1 - \mu_2^2)/E_2]}{(1/d_1) + (1/d_2)}} \qquad (2\text{-}88)$$

where l is the length of the contact area, and the remaining quantities have the same meaning as before. The pressure has an elliptical distribution across the width $2b$, and the maximum pressure is

$$p_{max} = -\frac{2F}{\pi b l} \qquad (2\text{-}89)$$

Equations (2-88) and (2-89) apply to a cylinder and a plane surface by making $d = \infty$ for the plane surface. They also apply to the contact of a cylinder and an internal cylindrical surface; in this case d is made negative.

To picture the stress state, select the origin of a reference system at the center of the contact area with x parallel to the cylindrical axes, y perpendicular to the plane formed by the two cylinder axes, and z in the plane of the contact force. Then, for stress elements on the z axis, three principal stresses σ_x, σ_y, and σ_z exist, all different. Figure 2-35 is a plot of these stresses for depths to $3b$ below the contact surface. Three different shear stresses exist; the largest of these is

$$\tau_{yz} = \frac{\sigma_y - \sigma_z}{2} \qquad (2\text{-}90)$$

This component, labeled τ_{max}, is also plotted in Fig. 2-35, and it is seen that it reaches a maximum value slightly below the surface just as in the case of contacting spheres.

PROBLEMS

Section 2-2

2-1 For each of the following stress states construct Mohr's circle diagram, and find the principal stresses and maximum shear stresses. Draw a stress element properly oriented from the x axis to show the principal stresses and their directions. Calculate the angle between the x axis and σ_1.

   (a) $\sigma_x = 10$ kpsi, $\sigma_y = -4$ kpsi
   (b) $\sigma_x = 10$ kpsi, $\tau_{xy} = 4$ kpsi cw
   (c) $\sigma_x = -2$ kpsi, $\sigma_y = -8$ kpsi, $\tau_{xy} = 4$ kpsi ccw
   (d) $\sigma_x = 10$ kpsi, $\sigma_y = 5$ kpsi, $\tau_{xy} = 1$ kpsi cw

2-2 The same as Prob. 2-1 for the following:

   (a) $\sigma_x = -17$ kpsi, $\sigma_y = -4$ kpsi, $\tau_{xy} = 7$ kpsi cw
   (b) $\sigma_x = -10$ kpsi, $\sigma_y = -12$ kpsi, $\tau_{xy} = 6$ kpsi ccw
   (c) $\sigma_y = 4$ kpsi, $\tau_{xy} = 8$ kpsi cw
   (d) $\sigma_x = 4$ kpsi, $\sigma_y = 12$ kpsi, $\tau_{xy} = 7$ kpsi ccw

2-3 For each of the following stress states, sketch Mohr's circle, compute the principal stresses and directions, and show these on a stress element properly oriented with respect to the xy reference:

   (a) $\sigma_x = -30$ MPa, $\sigma_y = -60$ MPa, $\tau_{xy} = 30$ MPa ccw
   (b) $\sigma_x = -90$ MPa, $\sigma_y = 10$ MPa, $\tau_{xy} = 40$ MPa cw
   (c) $\sigma_x = 75$ MPa, $\sigma_y = 15$ MPa, $\tau_{xy} = 32$ MPa ccw
   (d) $\sigma_y = 120$ MPa, $\tau_{xy} = 60$ MPa ccw

2-4 The same as Prob. 2-3 for the following stress components:

   (a) $\sigma_x = 84$ MPa, $\sigma_y = -42$ MPa, $\tau_{xy} = 96$ MPa cw

(b) $\sigma_x = -50$ MPa, $\sigma_y = 60$ MPa, $\tau_{xy} = 80$ MPa ccw

(c) $\sigma_x = -60$ MPa, $\tau_{xy} = 40$ MPa cw

(d) $\sigma_x = 20$ MPa, $\sigma_y = -60$ MPa, $\tau_{xy} = 40$ MPa cw

Sections 2-3 to 2-6

2-5 A steel rod 80 mm long and 15 mm in diameter is acted upon by a compressive load of 175 kN. The material is carbon steel. Find:

(a) The compressive stress

(b) The axial strain

(c) The deformation

(d) The increase in diameter of the rod

2-6 Draw Mohr's three-circle diagram, and find the three principal stresses and the maximum shear stress for the following conditions:

(a) Pure tension

(b) Pure compression

(c) Pure torsion

(d) Equal tension in the x and y directions

2-7 A steel rod $\frac{1}{2}$ in in diameter and 40 in long carries a tensile load of 3000 lb.

(a) Find the stress and elongation of the rod.

(b) If this rod is to be replaced by an aluminum rod which is to have the same elongation, what should be its diameter?

(c) Calculate the stress in the aluminum rod.

2-8 A press is to be so designed that the elongation of the two steel tension members A, as shown, does not exceed $\frac{1}{64}$ in.

(a) The press is to be rated at 10 kip. Determine the diameters of the rods.

(b) If a maximum stress of 20 kpsi is permitted, are the rods safe?

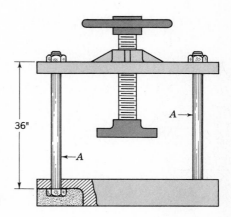

36"

PROBLEM 2-8

2-9 A built-up connecting rod consists of three steel bars, each $\frac{1}{4}$ in thick × $1\frac{1}{4}$ in wide as shown in the figure. During assembly it was found that one of the bars measured only 31.997 in between pin centers, the other two bars measuring exactly 32.000 in. Determine the stress in each bar after assembly but before an external load is applied.

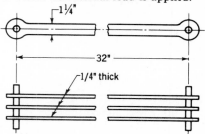

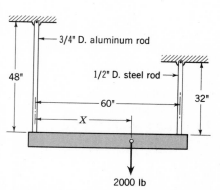

PROBLEM 2-9

2-10 An aluminum rod, $\frac{3}{4}$ in in diameter × 48 in long, and a steel rod, $\frac{1}{2}$ in in diameter × 32 in long, are spaced 60 in apart and fastened to a horizontal beam which carries a 2000-lb load, as shown in the figure. The beam is to remain horizontal after the load is applied. For the purposes of this problem we assume the beam is weightless and absolutely rigid.

(*a*) Find the location X of the load.
(*b*) Find the stress in each rod.

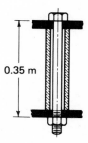

PROBLEM 2-10

2-11 A steel bolt, having a nominal diameter of 20 mm and a pitch of 2.5 mm (pitch is the distance from thread to thread in the axial direction), and an aluminum tube 40 mm OD by 22 mm ID, act as a spacer for two plates, as shown in the figure. The nut is pulled snug and then given a one-third additional turn. Find the resulting stress in the bolt and in the tube, neglecting the deformation of the plates.

PROBLEM 2-11

2-12 Develop expressions for the strain in the x and y directions when the tensile stresses σ_x and σ_y are equal.

2-13 Electrical strain gauges mounted on the surface of a loaded machine part were aligned in the direction of the principal stresses and gave the following values for the strains:

$$\epsilon_1 = 0.0021 \text{ m/m} \qquad \epsilon_2 = -0.000\,76 \text{ m/m}$$

Determine the three principal stresses and the unknown normal strain. What assumptions were necessary? Use $E = 207$ GPa and $\mu = 0.292$.

Sections 2-7, 2-8

2-14, 2-15 Find the reactions at the supports and plot the shear-force and bending-moment diagrams for each of the beams shown in the figure.

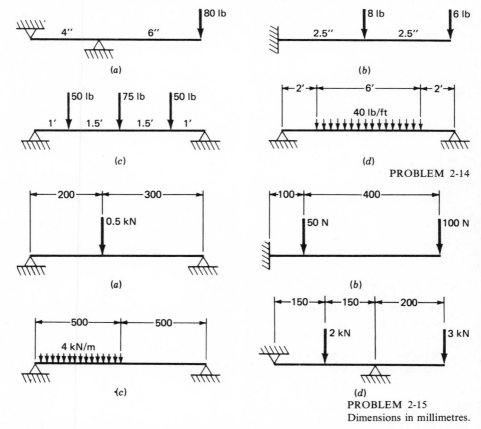

2-16 to 2-19 Using singularity functions, find general expressions for the loading, shear-force, and moment for each beam shown in the figure and compute the reactions at the supports. Where possible check your results with Table A-12.

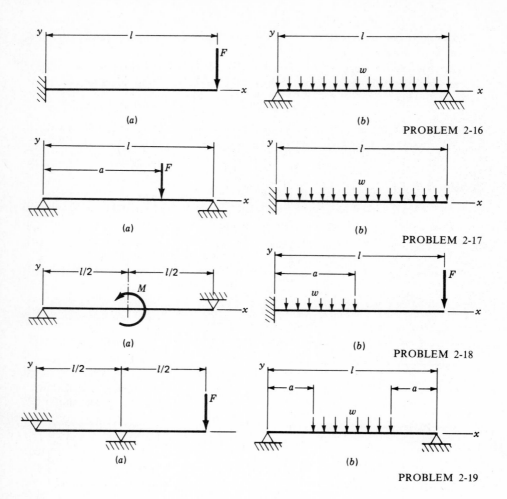

(a)

(b)

PROBLEM 2-16

(a)

(b)

PROBLEM 2-17

(a)

(b)

PROBLEM 2-18

(a)

(b)

PROBLEM 2-19

Sections 2-9, 2-10

2-20 The figure shows a rubber roll used for sizing cloth. Bearings *A* and *B* support the rotating shaft that is loaded uniformly by another similar roll and subjects both rolls to a uniformly distributed load of 125 lb/in of length. Use a hollow shaft having a wall thickness one-tenth of the outside diameter. Find the dimensions of the shaft cross section based upon an allowable normal bending stress of 16 kpsi.

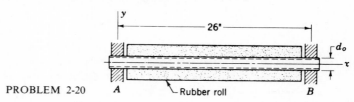

PROBLEM 2-20   A   Rubber roll   B

2-21 The figure shows a shaft mounted in bearings at *A* and *B*. The forces acting on the gear and pulley that are keyed to the shaft cause bending. Assume the loads are concentrated, use an allowable normal bending stress of 24 kpsi, and find the dimensions of a standard-size hollow round shaft, having an ID about three-fourths of the OD, to safely resist the loads.

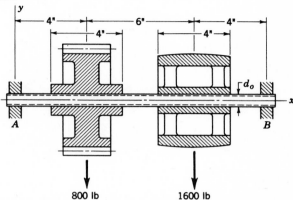

PROBLEM 2-21   800 lb   1600 lb

2-22 Cast iron has a strength in compression of about three to four times the strength in tension, depending upon the grade. Design an optimum T section for a cast-iron beam using a uniform thickness such that the compressive stress will be related to the tensile stress by a factor of 4.

2-23 The shaft shown in the figure is supported in bearings at *O* and *C* and is subjected to bending loads due to the force components acting at *A* and *B*.

(*a*) Sketch two moment diagrams, one for each plane of bending, and compute the moment components at *A* and *B*. Also compute the maximum bending moment.

(*b*) The shaft is to have a uniform diameter and be made of cold-drawn UNS G10350 steel. If a factor of safety of 2.3, based on the yield strength, is to be used, find the allowable normal bending stress in SI units.

(*c*) Find the safe shaft diameter to the nearest millimetre.

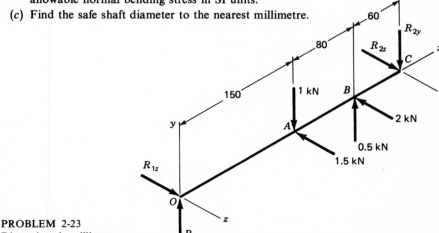

PROBLEM 2-23
Dimensions in millimetres.

2-24 The figure illustrates an overhanging shaft supported in bearings at O and B and loaded by forces at A and C. If the shaft has a diameter of 15 mm, find the maximum normal bending stress and its location.

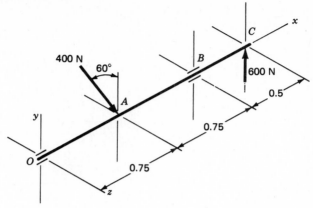

PROBLEM 2-24
Dimensions in metres.

Sections 2-11 to 2-13

2-25 The figure shows a round shaft of diameter d and length l loaded in bending by a force F acting in the center of the span. Three stress elements identified in section A-A are: B, on top and in the xy plane; C, on the front side and in the xz plane; and D, on the bottom and in the xy plane. Draw each of these stress elements properly oriented with respect to xyz, show the stresses which act upon them, and write the stress formulas in terms of F and the geometry of the bar.

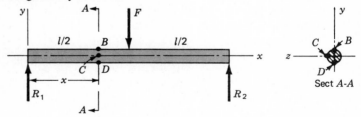

PROBLEM 2-25

2-26 A steel shaft $\frac{7}{8}$ in in diameter and 22 in long between bearings supports two loads as shown.
   (a) Sketch the shear-force and bending-moment diagrams and compute the shear forces and bending moments at points where the loads are applied.
   (b) Draw a stress element corresponding to point A on the shaft and compute and show on this element all stresses that act.
   (c) The same as in b for a stress element at B.

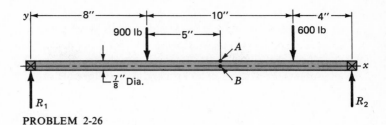

PROBLEM 2-26

2-27 The figure shows a beam made by spot-welding two steel strips on to back-to-back cold-formed channels. Determine the spacing of each pair of welds if the shear force on the beam is 8 kN and if the allowable shear load for each spot-weld is 7 kN.

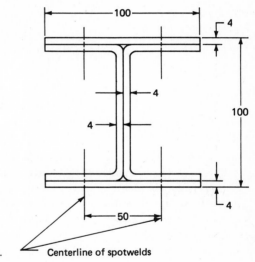

PROBLEM 2-27
Dimensions in millimetres.

Centerline of spotwelds

2-28 The figure shows a box beam made by gluing four boards, each finished to 25 by 150 mm, together. The beam is to support a shear force of 7.4 kN in the xy plane. Find the shear stress in the glued joints.

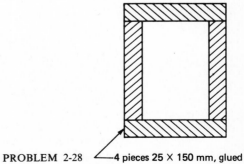

PROBLEM 2-28   4 pieces 25 × 150 mm, glued

2-29 The figure shows a shaft mounted in bearings at *A* and *D* and having pulleys at *B* and *C*. The forces shown acting on the pulley surfaces represent the belt pulls on the tight and loose sides of the pulleys. Compute the torque applied to the shaft by the pulley at *C* and by the pulley at *B*. Find an appropriate diameter for the shaft using an allowable normal stress of 16 kpsi and/or an allowable shear stress of 12 kpsi.

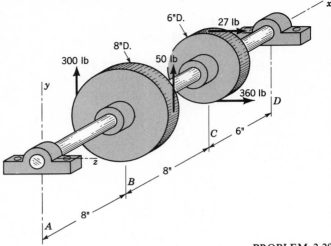

PROBLEM 2-29

2-30 Illustrated in the figure is a shaft having two pulleys and with bearings (not shown) at *A* and *B*. The belt pulls and their directions are shown for each pulley. The coordinate system may be identified by the superscripts on the bearing reaction vectors. Find the safe diameter of the shaft using an allowable normal stress of 24 kpsi and an allowable shear stress of 12 kpsi.

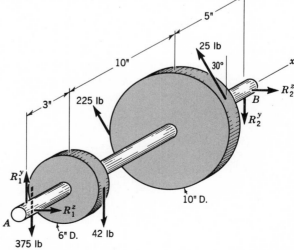

PROBLEM 2-30
The partially concealed 375-lb force vector is the belt pull on the 6-in pulley.

2-31 The figure is a free-body diagram of a steel shaft supported in bearings at O and B resulting in bearing reactions R_1 and R_2. The shaft is loaded by the forces $F_A = 4.5$ kN and $F_C = 1.8$ kN, and by the equal and opposite torques $T_A = -T_C = 600$ N·m.

(*a*) Draw the shear-force and bending-moment diagrams and compute values corresponding to points O, A, B, and C.

(*b*) Locate a stress element on the bottom (minus y side) of the shaft at A. Find all the stress components which act. Do not look through the shaft to see this element; look from underneath!

(*c*) Using the stress element of *b*, draw Mohr's circle diagram, find both principal stresses, and label the locations of σ_1, σ_2, τ_{max}, both reference axes, and the angle from the x axis to the axis containing σ_1.

(*d*) Draw the principal stress element correctly oriented from a horizontal x axis and label it completely, with angles and stresses.

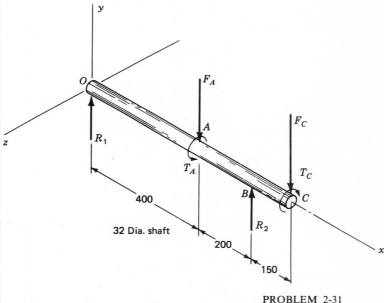

PROBLEM 2-31
Dimensions in millimetres.

2-32 The stresses are to be computed at two points on the cantilever shown in the figure. These are: the stress element which is shown at A on top of the bar and parallel to the xz plane; and the one at B that is on the front side of the bar and parallel to the xy plane. The forces are $F = 0.55$ kN, $P = 8$ kN, and $T = 30$ N·m. Draw both stress elements, label the axes and the stress components using proper magnitudes and directions.

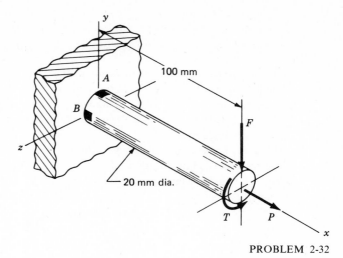

PROBLEM 2-32

2-33 The figure shows the crankshaft and flywheel of a one-cylinder air compressor. In use, a part of the power stored in the flywheel is used to produce a portion of the piston force *P*. In this problem you are to assume that the entire piston force *P* results from the torque of 600 N·m delivered to the crankshaft by the flywheel. A stress element is to be located on the top surface of the crankshaft at *A*, 100 mm from the left bearing. The sides of the element are parallel to the *xz* axes.

(*a*) Compute the stress components that act at *A*.

(*b*) Find the principal stresses and their directions for the element at *A*.

(*c*) Make a sketch of the principal stress element and orient it correctly with reference to the *x* and *z* axes. Label completely.

(*d*) Sketch another stress element correctly oriented to show the maximum shear stress and the corresponding normal stresses. Label this element too.

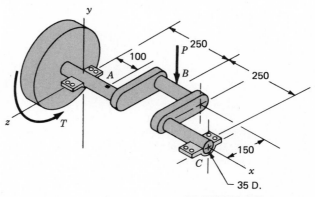

PROBLEM 2-33
Dimensions in millimetres.

2-34 A 100-mm-diameter solid circular shaft can carry a torsion of T_s N·m without exceeding a certain maximum shear stress. What proportion of this torque can a hollow shaft having a wall thickness of 10 mm and the same OD carry? Both shafts are to have the same maximum shear stress. Compute the ratio of the mass of the solid shaft to that of the hollow shaft.

2-35 The figure illustrates a bevel gear keyed to a 1-in-diameter shaft. The shaft is supported by a combined radial and thrust bearing at D and a bearing which takes pure radial load at F. Torque to the shaft is applied through the spur gear at E. In a force analysis of a bevel gear, the equivalent forces acting at the large end of the teeth are often used. In this problem the resultant force at the pitch line and at the large end of the teeth is specified in three components: a thrust force $W_T = 129$ lb, acting in the negative x direction; a tangential force $W = 500$ lb, acting in the negative z direction (the coordinate system is right-handed, with the z direction coming out of the paper); and a radial force $W_R = 129$ lb, acting in the negative y direction. The portion of the shaft to the right of the bearing F may be treated as a cantilever. Then the effect of these forces is to cause bending, compression, and torsion of the shaft. Neglect direct shear, and calculate all the stresses which act on an element at A, an element at B, and an element at C. Draw each stress element, show the stresses acting on it, and calculate the principal stresses and the maximum shear stress in each case.

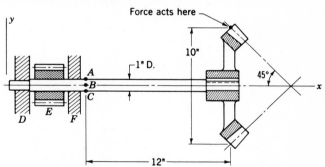

PROBLEM 2-35

2-36 The bit brace shown in the figure is used to apply a torque of 150 lb·in and a thrust of 50 lb to the bit, in boring a hole.

   (*a*) Make a three-dimensional abstract drawing of the brace and calculate and show all the reactions.

   (*b*) Select stress elements at the top, bottom, and two sides at section *A-A* and compute all the stresses acting on these elements.

   (*c*) The bit used has a shank diameter of $\frac{1}{4}$ in where it projects from the chuck at section *B-B*. If the distance from *B-B* to the surface of the wood is 3 in, find the largest of the normal stresses at this section when the plane of the brace is horizontal.

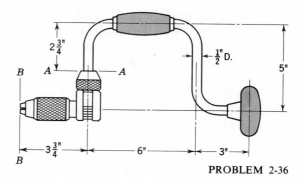

PROBLEM 2-36

2-37 The figure shows a crank loaded by a force $F = 300$ lb which causes twisting and bending of a $\frac{3}{4}$-in-diameter shaft fixed to a support at the origin of the reference system. In actuality, the support may be an inertia which we wish to rotate, but for the purposes of a stress analysis we can consider this as a statics problem.

(a) Draw separate free-body diagrams of the shaft AB and the arm BC, and compute the values of all forces, moments, and torques which act. Label the directions of the coordinate axes on these diagrams.

(b) Compute the maximum torsional stress and the maximum bending stress in the arm BC and indicate where these act.

(c) Locate a stress element on the top surface of the shaft at A, and calculate all the stress components which act upon this element.

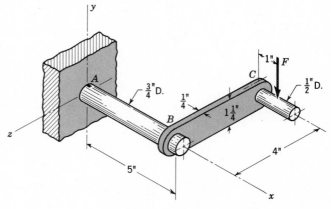

PROBLEM 2-37

Sections 2-14 to 2-16

2-38 A light pressure cylinder is made of an aluminum alloy. This cylinder has a $3\frac{1}{2}$-in OD, a 0.065-in wall thickness, and material properties $E = 10.3$ Mpsi and $\mu = 0.334$. If the internal pressure is 1860 psi, what are the principal normal strains on an element on the circumference?

2-39 A magnesium tube is 5 in in OD and has a wall thickness of $\frac{1}{2}$ in. The tubing is used as a pressure vessel to hold a fluid at an internal pressure of 4 kpsi. Calculate the radial- and tangential-stress components and the three principal normal strains at the outer and inner radii.

2-40 A cylinder is 300 mm OD by 200 mm ID and is subjected to an external pressure of 140 MPa. The longitudinal stress is zero. What is the maximum shear stress and at what radius does it occur?

2-41 Derive the relations for the stress components in a thin-wall spherical pressure vessel.

2-42 A gun barrel is assembled by shrinking an outer barrel over an inner barrel so that the maximum principal stress equals 70 percent of the yield strength of the material. Both members are of steel and have as their properties a yield strength $S_y = 78$ kpsi, $E = 30$ Mpsi, and $\mu = 0.292$. The nominal radii of the barrels are $\frac{3}{16}$, $\frac{3}{8}$, and $\frac{9}{16}$ in.
    (a) Plot the stress distribution for both parts of the barrels.
    (b) What value of interference should be used in assembly?
    (c) When the gun is fired, an internal pressure of 40 kpsi is created. Plot the resulting stress distributions.

2-43 A cylinder is 25 mm ID by 50 mm OD and is subjected to an internal pressure of 150 MPa. Find the tangential stress at the inner and outer surfaces.

2-44 Find the tangential stress at the inner and outer surfaces of the cylinder of Prob. 2-43 when there is an external pressure of 150 MPa and the internal pressure is zero.

2-45 A steel tire $\frac{3}{8}$ in thick is to be shrunk over a cast-iron rim 16 in in diameter and 1 in thick. Calculate the inside diameter to which the tire must be bored in order to induce a hoop stress of 20 kpsi in the tire. The rim is made of grade No. 30 cast iron having $E = 13$ Mpsi and $\mu = 0.211$.

Sections 2-18, 2-19

2-46 Plot the distribution of stresses across section A-A of the crane hook shown in the figure if the load F is 5 kip.

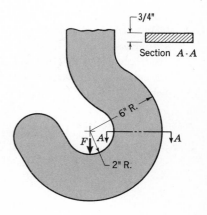

PROBLEM 2-46

2-47 Find the stress at the inner and outer fibers at section *A-A* of the frame shown in the figure if $F = 500$ lb.

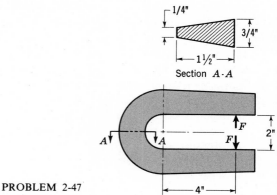

Section *A-A*

PROBLEM 2-47

2-48 The figure is an illustration of the No. 3029T mayfly hook No. 14, 2X long, TDE. It is made of forged Sheffield 0.95 carbon wire. The first number is simply the catalog number. The size is No. 14, and the designation 2X long means that the shank is longer than standard and corresponds to that of a No. 12 hook (two sizes larger). TDE means turned-down eye; if the leader is properly attached, this provides a means of giving a straight pull (no eccentricity) from the leader into the shank. This wire has a tensile strength of 375 kpsi and a yield strength of approximately 340 kpsi. What pull would have to be applied to the hook to cause yielding?

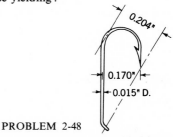

PROBLEM 2-48

2-49 The eye bolt shown in the figure is made of a cold-drawn low-carbon steel having a yield strength of 62 kpsi. In a complete analysis of the strength of this fastener, the stripping strength of the threads would have to be considered, but in this problem you are to consider only the possibility of failure in the eye. Therefore, based on the yield strength, find the maximum axial pull that can be applied. Next find the maximum bending load that can be applied at *A* if the bolt is supported at *B*.

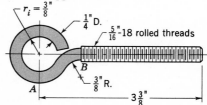

PROBLEM 2-49

2-50 In this problem you are to find the stresses in the U frame of the coping saw shown in the figure. To solve the problem you need only know that the saw blade is assembled with a tension of 65 lb and that the material of the frame is cold-drawn UNS G10180 steel. However, you may be interested to know that the blade has a small pin at each end which engages the slotted jaws. Jaw *B* has a long threaded shank which mates with internal threads in the handle of the saw. Thus, unscrewing the handle permits the two arms of the U to spread apart, relieves the tension on the blade, and allows the blade to be removed and replaced with a new one, or rotated about its axis to a more favorable cutting position relative to the work.

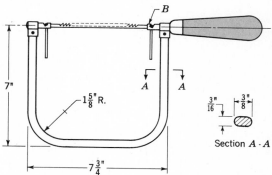

PROBLEM 2-50

2-51 A steel wheel $\frac{1}{4}$ in wide and 3 in in diameter supports a load of 400 lb while rolling without slipping on a flat surface of steel. Using Fig. 2-35, find the distance in inches below the wheel surface at which the maximum shear stress occurs. Draw a stress element for this point, aligned in the direction of the principal stresses, and compute σ_x, σ_y, and σ_z using Fig. 2-35. Draw Mohr's three-circle diagram and find all three maximum shear stresses. Values of the elastic constants are given in Table A-7.

3

DEFLECTION ANALYSIS

A structure or mechanical element is said to be *rigid* when it does not bend, or deflect, or twist too much when an external force, moment, or torque is applied. But if the movement due to the external disturbance is large, the member is said to be *flexible*. The words rigidity and flexibility are qualitative terms which depend upon the situation. Thus the floor of a building which bends only 0.1 in because of the weight of a machine placed upon it would be considered very rigid if the machine were heavy. But a surface plate which bends 0.01 in because of its own weight would be considered too flexible.

Deflection analysis enters into design situations in many ways. A snap ring, or retaining ring, must be flexible enough so that it can be bent, without permanent deformation, and assembled; and then it must be rigid enough to hold the assembled parts together. In a transmission the gears must be supported by a rigid shaft. If the shaft bends too much, that is, if it is too flexible, the teeth will not mesh properly, resulting in excessive impact, noise, wear, and early failure. In rolling sheet or strip steel to prescribed thicknesses, the rolls must be crowned, that is, curved, so that the finished product will be of uniform thickness. Thus, to design the rolls it is necessary to know exactly how much they will bend when a sheet of steel is rolled between them. Sometimes mechanical elements must be designed to

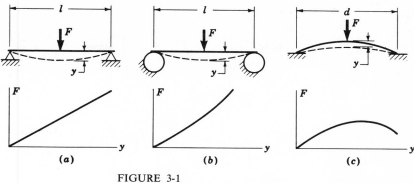

FIGURE 3-1
(a) A linear spring; (b) a stiffening spring; (c) a softening spring.

have a particular force-deflection characteristic. The suspension system of an automobile, for example, must be designed within a very narrow range to achieve an optimum bouncing frequency for all conditions of vehicle loading because the human body is comfortable only within a limited range of frequencies.

3.1 SPRING RATES

Elasticity is that property of a material which enables it to regain its original configuration after having been deformed. A *spring* is a mechanical element which exerts a force when deformed. Figure 3-1a shows a straight beam of length l simply supported at the ends and loaded by the transverse force F. The deflection y is linearly related to the force, as long as the elastic limit of the material is not exceeded, as indicated by the graph. This beam can be described as a linear spring.

In Fig. 3-1b a straight beam is supported on two cylinders such that the length between supports decreases as the beam is deflected by the force F. A larger force is required to deflect a short beam than a long one, and hence the more this beam is deflected, the stiffer it becomes. Also, the force is not linearly related to the deflection, and hence this beam can be described as a *nonlinear stiffening spring*.

Figure 3-1c is a dish-shaped round disk. The force necessary to flatten the disk increases at first and then decreases as the disk approaches a flat configuration, as shown by the graph. Any mechanical element having such a characteristic is called a *nonlinear softening spring*.

If we designate the general relationship between force and deflection by the equation

$$F = F(y) \qquad (a)$$

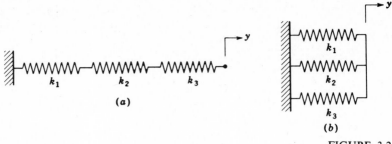

FIGURE 3-2

then *spring rate* is defined as

$$k(y) = \lim_{\Delta y \to 0} \frac{\Delta F}{\Delta y} = \frac{dF}{dy} \qquad (3\text{-}1)$$

where y must be measured in the direction of F and at the point of application of F. Most of the force-deflection problems encountered in this book are linear, as in Fig. 3-1a. For these, k is a constant, also called the *spring constant*; consequently Eq. (3-1) is written

$$k = \frac{F}{y} \qquad (3\text{-}2)$$

We might note that Eqs. (3-1) and (3-2) are quite general and apply equally well for torques and moments provided angular measurements are used for y. For linear displacements the units of k are often lb per in or N per m, and for angular displacements lb·in per radian or N·m per radian.

It is frequently desirable to represent springs abstractly as zigzag lines when they occur in combinations. Figure 3-2a shows three springs in series. We can replace these three springs by a single equivalent spring of stiffness k by noting that the deflection of the equivalent spring must equal the sum of the deflections of the three series springs. Thus

$$y = \frac{F}{k} = \frac{F}{k_1} + \frac{F}{k_2} + \frac{F}{k_3}$$

or

$$k = \frac{1}{(1/k_1) + (1/k_2) + (1/k_3)} \qquad (3\text{-}3)$$

The force required to deflect the three parallel springs of Fig. 3-2b is

$$F = k_1 y + k_2 y + k_3 y$$

These springs can be replaced by a single equivalent spring whose constant is

$$k = \frac{F}{y} = k_1 + k_2 + k_3 \qquad (3\text{-}4)$$

We shall find the ideas presented here useful in the analysis of complex structures made up of many parts and for other purposes. These methods are also useful in the simulation of dynamic systems where the force-deflection relationships are necessary to define the mathematical model of the system.

3-2 SIMPLE TENSION, COMPRESSION, AND TORSION

The relation for the total extension or deformation of a uniform bar has already been developed in Sec. 2-5, Eq. (*a*). It is repeated here for convenience.

$$\delta = \frac{Fl}{AE} \qquad (3\text{-}5)$$

This equation does not apply to a long bar loaded in compression if there is a possibility of buckling (Sec. 3-10). Using Eqs. (3-2) and (3-5), we see that the spring constant of an axially loaded bar is

$$k = \frac{AE}{l} \qquad (3\text{-}6)$$

The angular deflection of a uniform round bar subjected to a twisting moment T was given in Eq. (2-42), and is

$$\theta = \frac{Tl}{GJ} \qquad (3\text{-}7)$$

where θ is in radians. If we multiply Eq. (3-7) by $180/\pi$ and substitute $J = \pi d^4/32$ for a solid round bar, we obtain

$$\theta = \frac{585Tl}{Gd^4} \qquad (3\text{-}8)$$

where θ is in degrees.

Equation (3-7) can be rearranged to give the torsional spring rate as

$$k = \frac{T}{\theta} = \frac{GJ}{l} \qquad (3\text{-}9)$$

3-3 DEFLECTION OF BEAMS

Beams deflect a great deal more than axially loaded members, and the problem of bending probably occurs more often than any other loading problem in design. Shafts, axles, cranks, levers, springs, brackets, and wheels, as well as many other elements, must often be treated as beams in the design and analysis of mechanical structures and systems. The subject of bending, however, is one which you should have studied as preparation for reading this book. It is for this reason that we include here only a brief review to establish the nomenclature and conventions to be used throughout this book.

In Sec. 2-9 we developed the relation for the curvature of a beam subjected to a bending moment M [Eq. (2-29)]. The relation is

$$\frac{1}{\rho} = \frac{M}{EI} \qquad (3\text{-}10)$$

where ρ is the radius of curvature. From studies in mathematics we also learn that the curvature of a plane curve is given by the equation

$$\frac{1}{\rho} = \frac{d^2y/dx^2}{[1 + (dy/dx)^2]^{3/2}} \qquad (3\text{-}11)$$

where the interpretation is that y is the deflection of the beam at any point x along its length. The slope of the beam at any point x is

$$\theta = \frac{dy}{dx} \qquad (a)$$

For many problems in bending the slope is very small, and for these the denominator of Eq. (3-11) can be taken as unity. Equation (3-10) can then be written

$$\frac{M}{EI} = \frac{d^2y}{dx^2} \qquad (b)$$

Noting Eqs. (2-24) and (2-25) and successively differentiating Eq. (b) yields

$$\frac{V}{EI} = \frac{d^3y}{dx^3} \qquad (c)$$

$$\frac{q}{EI} = \frac{d^4y}{dx^4} \qquad (d)$$

It is convenient to display these relations in a group as follows:

$$\frac{q}{EI} = \frac{d^4y}{dx^4} \qquad (3\text{-}12)$$

$$\frac{V}{EI} = \frac{d^3y}{dx^3} \qquad (3\text{-}13)$$

$$\frac{M}{EI} = \frac{d^2y}{dx^2} \qquad (3\text{-}14)$$

$$\theta = \frac{dy}{dx} \qquad (3\text{-}15)$$

$$y = f(x) \qquad (3\text{-}16)$$

The nomenclature and conventions are illustrated by the beam of Fig. 3-3. Here, a beam of length $l = 20$ in is loaded by the uniform load $w = 80$ lb per in of beam

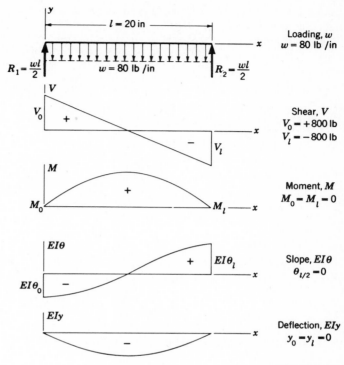

FIGURE 3-3

length. The x axis is positive to the right, and the y axis positive upward. All quantities, loading, shear, moment, slope, and deflection, have the same sense as y; they are positive if upward, negative if downward.

The values of the quantities at the ends of the beam, where $x = 0$ and $x = l$, are called their *boundary values*. For this reason the beam problem is often called a *boundary-value problem*. The reactions $R_1 = R_2 = +800$ lb and the shear forces $V_0 = +800$ lb and $V_l = -800$ lb are easily computed using the methods of Chap. 2. The bending moment is zero at each end because the beam is simply supported. Note that the beam-deflection curve must have a negative slope at the left boundary and a positive slope at the right boundary. This is easily seen by examining the deflection of Fig. 3-3. The magnitude of the slope at the boundaries is, as yet, unknown; because of symmetry, however, the slope is known to be zero at the center of the beam. We note, finally, that the deflection is zero at each end.

3-4   DEFLECTIONS BY USE OF SINGULARITY FUNCTIONS

Many methods have been used to integrate the beam equations; these can be found in nearly any basic book on solid mechanics or mechanics of materials. Because they are precise and quite general we shall present only two approaches in this book, graphical integration and integration using singularity functions (see Sec. 2-8 and Table 2-1). The following examples are illustrations of the use of singularity functions:

EXAMPLE 3-1 As a first example, we choose the beam of Table A-12-6, which is a simply supported beam having a concentrated load F not in the center. Writing Eq. (3-12) for this loading gives

$$EI\frac{d^4y}{dx^4} = q = -F\langle x - a\rangle^{-1} \qquad 0 < x < l \qquad (1)$$

Note that the reactions R_1 and R_2 do not appear in this equation, as they did in Chap. 2, because of the range chosen for x. If we now integrate from 0 to x, not $-\infty$ to x, according to Eq. (3-13) we get

$$EI\frac{d^3y}{dx^3} = V = -F\langle x - a\rangle^0 + C_1 \qquad (2)$$

Using Eq. (3-14) this time we integrate again and obtain

$$EI\frac{d^2y}{dx^2} = M = -F\langle x - a\rangle^1 + C_1 x + C_2 \qquad (3)$$

At $x = 0$, $M = 0$ and Eq. (3) gives $C_2 = 0$. At $x = l$, $M = 0$ and Eq. (3) gives

$$C_1 = \frac{F(l - a)}{l} = \frac{Fb}{l}$$

Substituting C_1 and C_2 into Eq. (3) gives

$$EI\frac{d^2y}{dx^2} = M = \frac{Fbx}{l} - F\langle x - a\rangle^1 \qquad (4)$$

Note that we could have obtained this equation by summing moments about a section a distance x from the origin. Next, integrate Eq. (4) twice in accordance with Eqs. (3-15) and (3-16). This yields

$$EI\frac{dy}{dx} = EI\theta = \frac{Fbx^2}{2l} - \frac{F\langle x - a\rangle^2}{2} + C_3 \qquad (5)$$

$$EIy = \frac{Fbx^3}{6l} - \frac{F\langle x - a\rangle^3}{6} + C_3 x + C_4 \qquad (6)$$

The constants of integration C_3 and C_4 are evaluated using the two boundary conditions $y = 0$ at $x = 0$, and $y = 0$ at $x = l$. The first condition, substituted into Eq. (6), gives $C_4 = 0$. The second condition, substituted into Eq. (6), yields

$$0 = \frac{Fbl^2}{6} - \frac{Fb^3}{6} + C_3 l$$

whence

$$C_3 = -\frac{Fb}{6l}(l^2 - b^2)$$

Upon substituting C_3 and C_4 into Eq. (6) we obtain the deflection relation as

$$y = \frac{F}{6EIl}[bx(x^2 + b^2 - l^2) - l\langle x - a\rangle^3] \tag{7}$$

Compare Eq. (7) with the two deflection equations in Table A-12-6 and note the use of singularity functions enables us to express the entire relation with a single equation.   ////

EXAMPLE 3-2   Find the deflection relation, the maximum deflection, and the reactions for the statically indeterminate beam shown in Fig. 3-4a using singularity functions.

SOLUTION   The loading diagram and approximate deflection curve are shown in Fig. 3-4b. Based upon the range $0 < x < l$, the loading equation is

$$q = R_2\langle x - a\rangle^{-1} - w\langle x - a\rangle^0 \tag{1}$$

We now integrate this equation four times in accordance with Eqs. (3-12) to (3-16). The results are

$$V = R_2\langle x - a\rangle^0 - w\langle x - a\rangle^1 + C_1 \tag{2}$$

$$M = R_2\langle x - a\rangle^1 - \frac{w}{2}\langle x - a\rangle^2 + C_1 x + C_2 \tag{3}$$

$$EI\theta = \frac{R_2}{2}\langle x - a\rangle^2 - \frac{w}{6}\langle x - a\rangle^3 + \frac{C_1}{2}x^2 + C_2 x + C_3 \tag{4}$$

$$EIy = \frac{R_2}{6}\langle x - a\rangle^3 - \frac{w}{24}\langle x - a\rangle^4 + \frac{C_1}{6}x^3 + \frac{C_2}{2}x^2 + C_3 x + C_4 \tag{5}$$

As in the previous example, the constants are evaluated by applying appropriate boundary conditions. First we note from Fig. 3-4b that both $EI\theta = 0$ and $EIy = 0$ at $x = 0$. This gives $C_3 = 0$ and $C_4 = 0$. Next we observe that at $x = 0$ the shear force is the same as the reaction. So Eq. (2) gives $V(0) = R_1 = C_1$. Since the deflection must be zero at the reaction R_2 where $x = a$, we have from Eq. (5) that

$$\frac{C_1}{6}a^3 + \frac{C_2}{2}a^2 = 0 \qquad \text{or} \qquad C_1\frac{a}{3} + C_2 = 0 \tag{6}$$

Also the moment must be zero at the overhanging end where $x = l$; Eq. (3) then gives

$$R_2(l - a) - \frac{w}{2}(l - a)^2 + C_1 l + C_2 = 0$$

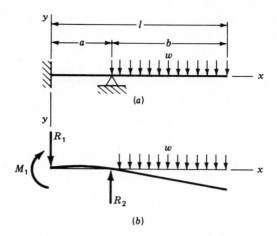

FIGURE 3-4

Simplifying and noting that $l - a = b$ gives

$$C_1 a + C_2 = -\frac{wb^2}{2} \qquad (7)$$

Equations (6) and (7) are now solved simultaneously for C_1 and C_2. The results are

$$C_1 = R_1 = -\frac{3wb^2}{4a} \qquad C_2 = \frac{wb^2}{4}$$

With R_1 known, we can sum the forces in the y direction to zero to get R_2. Thus

$$R_2 = -R_1 + wb = \frac{3wb^2}{4a} + wb = \frac{wb}{4a}(4a + 3b)$$

The moment reaction M_1 is obtained from Eq. (3) with $x = 0$. This gives

$$M(0) = M_1 = C_2 = \frac{wb^2}{4}$$

The complete equation of the deflection curve is obtained by substituting the known terms for R_2 and the constants C_i into Eq. (5). The result is

$$EIy = \frac{wb}{24a}(4a + 3b)\langle x - a \rangle^3 - \frac{w}{24}\langle x - a \rangle^4 - \frac{wb^2 x^3}{8a} + \frac{wb^2 x^2}{8} \qquad (8)$$

The maximum deflection occurs at the free end of the beam at $x = l$. By making this substitution into Eq. (8) and manipulating the resulting expression, one finally obtains

$$y_{\text{max}} = -\frac{wb^3 l}{8EI} \qquad (9)$$

Examination of the deflection curve of Fig. 3-4b reveals that the curve will have a zero slope at some point between R_1 and R_2. By replacing the constants in Eq. (4), setting $\theta = 0$, and solving the result for x we readily find that the slope is zero at $x = 2a/3$. The corresponding deflection at this point can be obtained by using this value of x in Eq. (8). The result is

$$y(x = 2a/3) = \frac{wa^2b^2}{54EI} \qquad (10)$$

////

3-5 THE METHOD OF SUPERPOSITION

While an engineer may enjoy the challenge of solving beam problems and it may sharpen his problem-solving capability, a designer usually has more important and more urgent things to do. If someone else has already solved the problem, the practicing engineer can save his employer money and increase the flow of cash to the stockholders by using worked-out solutions. The solutions to bending problems which occur most frequently in design are included in Table A-12. Solutions to still other beam-deflection problems may be found in various handbooks. If the solution to a particular problem cannot be found, it still may be possible to obtain one using the method of superposition. This method can be employed for all linear force-deflection problems, that is, those in which the force and deflection are linearly related. The *method of superposition* uses the principle that the deflection at any point in a beam is equal to the sum of the deflections caused by each load acting separately. Thus, if a beam is bent by three separate forces, the deflection at a particular point is the sum of three deflections, one for each force.

3-6 THE GRAPHICAL-INTEGRATION METHOD

It often happens that the geometry, or the method of loading a beam, makes the deflection problem so difficult that it is impractical to solve it by classical methods. Under such conditions one can employ numerical integration using desk calculators or electronic computers or by graphical integration. Though of limited accuracy, the graphical approach is fast, and it provides a good physical understanding of what is happening. For many purposes the accuracy is completely satisfactory.

Graphical integration can be explained by reference to Fig. 3-5; the function is plotted in a and the integral in b. Three graphical scales are necessary, one for the dependent variable, one for the independent variable, and one for the integral. Let us denote x as the independent variable and y as the dependent variable. Then the integral of the function $y = f(x)$ between, say, $x = a$ and $x = b$ is simply the area below the curve $y = f(x)$ between ordinates erected to x at a and b. In graphical integration we choose the two ordinates close together and average

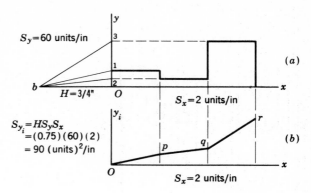

$S_y = 60$ units/in

(a)

b $H = 3/4"$ O

$S_x = 2$ units/in

$S_{y_i} = HS_yS_x$
$\quad = (0.75)(60)(2)$
$\quad = 90 \text{ (units)}^2/\text{in}$

(b)

$S_x = 2$ units/in

FIGURE 3-5

them so that the procedure consists merely in finding a graphical scheme which will enable us to draw lines whose slopes are proportional to the average ordinate in each interval.

To integrate the function of Fig. 3-5a choose the line Ob, called the *pole distance* and designated as H, any convenient length, say $\frac{3}{4}$ in. Project the heights of the various rectangles to the y axis and draw lines connecting these intersections with the pole b. Each line, b-1, b-2, and b-3, has a slope proportional to the height of the corresponding rectangle. The integral in Fig. 3-5b is obtained by drawing lines Op parallel to b-1, pq parallel to b-2, and qr parallel to b-3. The scale of the integral is

$$S_{y_i} = HS_xS_y \qquad (3\text{-}17)$$

where H = pole distance, in
$\quad S_x$ = scale of x in units of x per inch
$\quad S_y$ = scale of y in units of y per inch
$\quad S_{y_i}$ = scale of the integral in units of x times units of y per inch

EXAMPLE 3-3 This method of solution is now applied to the problem of obtaining the deflection of a beam (Fig. 3-6). In this figure is a stepped shaft (1) originally drawn one-fourth size which is further reduced for reproduction reasons. Loading diagram (2) shows a load of 50 lb per in extending over a distance of 10 in of the shaft length. The reactions at the bearings have been calculated and are indicated as R_1 and R_2. The shearing-force diagram (3) has been obtained from the loading diagram, using the conditions of static equilibrium. The moment diagram (4) was obtained by graphically integrating the shearing-force diagram. The construction is shown, and also the calculation for the scale of the moment diagram.

The next step is to obtain the numerical values of the moment at selected points along the shaft. This is done by scaling the diagram. The values of the

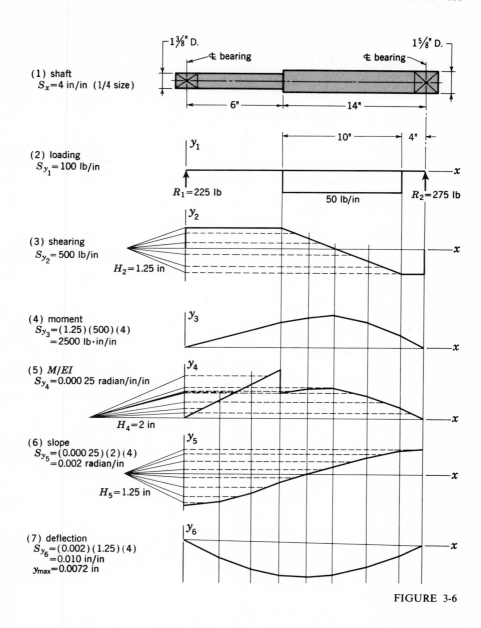

(1) shaft
$S_x = 4$ in/in (1/4 size)

$1\frac{3}{8}$" D.

$1\frac{5}{8}$" D.

$\textup{\textcentoldstyle}$ bearing

$\textup{\textcentoldstyle}$ bearing

6"

14"

10"

4"

(2) loading
$S_{y_1} = 100$ lb/in

y_1

x

$R_1 = 225$ lb

50 lb/in

$R_2 = 275$ lb

(3) shearing
$S_{y_2} = 500$ lb/in

y_2

$H_2 = 1.25$ in

x

(4) moment
$S_{y_3} = (1.25)(500)(4)$
$= 2500$ lb·in/in

y_3

x

(5) M/EI
$S_{y_4} = 0.000\,25$ radian/in/in

y_4

$H_4 = 2$ in

x

(6) slope
$S_{y_5} = (0.000\,25)(2)(4)$
$\quad = 0.002$ radian/in

y_5

$H_5 = 1.25$ in

x

(7) deflection
$S_{y_6} = (0.002)(1.25)(4)$
$\quad = 0.010$ in/in
$y_{max} = 0.0072$ in

y_6

x

FIGURE 3-6

moment of inertia are then calculated for each diameter. The moments are next divided by the products of the modulus of elasticity ($E = 30$ Mpsi) and the moments of inertia, and these values are plotted (5) to obtain the M/EI diagram. (If the moment of inertia is constant, this operation may be performed after the deflection curve is obtained. In case this is done, note that the deflection curve then becomes the *yEI* curve.) The *M/EI* diagram is now integrated twice to obtain the deflection curve (7). In integrating the slope diagram (6) it is necessary to guess at the location of zero slope, that is, the position at which to place the *x* axis. Should this guess be wrong, and it usually is, the deflection curve will not close with a horizontal line. The line should be drawn so as to close the deflection curve, and measurements of deflection made in the vertical direction. (Do not measure perpendicular to the closing line unless it is horizontal.) The correct location of zero slope is found as follows: Draw a line parallel to the closing line and tangent to the deflection curve. The point of tangency is the point of zero slope, and this is also the location of the maximum deflection. ////

3-7 STRAIN ENERGY

The external work done on an elastic member in deforming it is transformed into *strain*, or *potential energy*. The potential energy stored by a member when it is deformed through a distance *y* is the average force times the deflection, or

$$U = \frac{F}{2} y = \frac{F^2}{2k} \qquad (3\text{-}18)$$

Equation (3-18) is general in the sense that the force *F* also means torque, or moment, provided, of course, that consistent units are used for *k*. By substituting appropriate expressions for *k*, formulas for strain energy for various simple loadings may be obtained. For simple tension and compression, for example, we employ Eq. (3-6) and obtain

$$U = \frac{F^2 l}{2AE} \qquad (3\text{-}19)$$

For torsion we use Eq. (3-9); the torsional strain energy is then found to be

$$U = \frac{T^2 l}{2GJ} \qquad (3\text{-}20)$$

To obtain an expression for the strain energy due to direct shear, consider the element with one side fixed in Fig. 3-7*a*. The force *F* places the element in pure shear, and the work done is

$$U = \frac{F\delta}{2} \qquad (a)$$

Now, since the shear strain is

$$\gamma = \frac{\delta}{l} = \frac{\tau}{G} = \frac{F}{AG}$$

then

$$U = \frac{F^2 l}{2AG} \qquad (3\text{-}21)$$

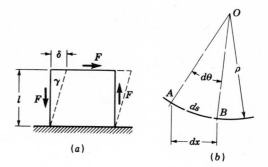

FIGURE 3-7

(a)

(b)

The strain energy stored in a beam or lever by bending may be obtained by referring to Fig. 3-7b. Here AB is a section of the elastic curve of length ds having a radius of curvature ρ. The strain energy stored in this section of the beam is

$$dU = \frac{M}{2}\, d\theta \qquad (b)$$

Now, since $\rho\, d\theta = ds$, we have

$$dU = \frac{M\, ds}{2\rho} \qquad (c)$$

Next, using Eq. (3-10) to eliminate ρ,

$$dU = \frac{M^2\, ds}{2EI} \qquad (d)$$

The strain energy in an entire beam is obtained by adding the energies in all the elemental sections. For small deflections $ds \approx dx$, and so

$$U = \int \frac{M^2\, dx}{2EI} \qquad (3\text{-}22)$$

A useful relation can be obtained by dividing Eqs. (3-19) and (3-20) by the volume lA. We then obtain the two following expressions for the strain energy per unit volume:

$$u = \frac{\sigma^2}{2E} \qquad u = \frac{\tau^2}{2G} \qquad (3\text{-}23)$$

Suppose we wish to design a member to store a large amount of energy (this problem often arises in the design of springs). Then these expressions tell us that the material ought to have a high strength, because σ appears in the numerator, and amazingly, a *low* modulus of elasticity, because E appears in the denominator.

Equation (3-22) gives the strain energy due to pure bending. Most problems encountered in design are not pure bending, but the shear is so small that it is neglected. However, one ought to be able to calculate the shear strain energy for himself and then decide whether to neglect it or not. So let us select a rectangular-section beam of width b and depth h subjected to a vertical shear force F. Using Eq. (2-31) with $I = bh^3/12$ and $dA = b\,dy$, we find the shear stress to be

$$\tau = \frac{V}{Ib}\int_y^{h/2} y\,dA = \frac{3V}{2bh^3}(h^2 - 4y^2) \qquad (e)$$

Taking an element of volume $dv = b\,dy\,dx$ and using Eq. (3-23) for the shear energy gives

$$dU = \frac{b\tau^2}{2G}\,dy\,dx \qquad (f)$$

If we now substitute τ from Eq. (e) into Eq. (f) and integrate with respect to y, we find

$$dU = \frac{9V^2\,dx}{8Gbh^6}\int_{-h/2}^{+h/2}(h^4 - 8h^2y^2 + 16y^4)\,dy = \frac{3V^2\,dx}{5Gbh}$$

and hence

$$U = \frac{3}{5}\int_0^l \frac{V^2\,dx}{Gbh} \qquad (3\text{-}24)$$

Note that this expression holds only for a rectangular cross section and gives the strain energy due to transverse shear. The total energy must include that due to bending too.

A cantilever having a concentrated load F on the free end has a constant shear force $V = -F$, and Eq. (3-24) yields

$$U = \frac{3F^2l}{5Gbh} \qquad (3\text{-}25)$$

Juvinall* states that the constant $\frac{3}{5}$ should be replaced by the approximate value of $\frac{1}{2}$ for other cross sections. Popov† shows that the strain energy due to shear in a cantilever is less than 1 percent of the total when the beam length is ten or more times the beam depth. Thus, except for very short beams, the strain energy given by Eqs. (3-24) and (3-25) is negligible.

As another example of the use of Eq. (3-24), take a simply supported beam having a uniformly distributed load w. The equation for the shear force is

$$V = \frac{wl}{2} - wx \qquad (g)$$

---

* Robert C. Juvinall, "Stress, Strain, and Strength," p. 147, McGraw-Hill Book Company, New York, 1967.
† Egor P. Popov, "Introduction to Mechanics of Solids," p. 487, Prentice-Hall, Inc., Englewood Cliffs, N.J., 1968.

Solving Eq. (3-24) gives

$$U = \frac{3}{5Gbh} \int_0^l \left(\frac{wl}{2} - wx \right)^2 dx = \frac{w^2 l^3}{20Gbh} \qquad (3\text{-}26)$$

3-8 THE THEOREM OF CASTIGLIANO

We have already learned that the deflection of a structure, or of any group of connected elements, can be obtained by treating each element as a spring and assembling these springs in the proper geometric configuration to obtain a single equivalent spring. Another approach to this problem is provided by *Castigliano's theorem*. The advantages of Castigliano's method are that it is nearly idiot-proof, that it is a "cookbook" approach, and that it is ideally suited for digital computation.

Castigliano's theorem states that *when forces operate on elastic systems, the displacement corresponding to any force may be found by obtaining the partial derivative of the total strain energy with respect to that force.* As in our studies of spring rates, the terms *force* and *displacement* should be broadly interpreted, since they apply equally to moments and to angular displacements. Mathematically, the theorem of Castigliano is

$$\delta_i = \frac{\partial U}{\partial F_i} \qquad (3\text{-}27)$$

where δ_i is the displacement of the point of application of the ith force F_i in the direction of F_i.

Sometimes the deflection of a structure is required at a point where no force or moment is acting. In this case we can place an imaginary force Q_i at that point, develop the expression for δ_i, and then set Q_i equal to zero. The remaining terms give the deflection at the point of application of the imaginary force Q_i in the direction in which it was imagined to be acting.

Equation (3-27) can also be used to determine the reactions in indeterminate structures. The deflection is zero at these reactions, and so we merely solve the equation

$$\frac{\partial U}{\partial R_j} = 0 \qquad (3\text{-}28)$$

to obtain the reaction force R_j. If there are several indeterminate reactions, Eq. (3-28) is written once for each to obtain a set of equations, which are then solved simultaneously.

Castigliano's theorem, of course, is valid only for the condition in which the displacement is proportional to the force which produced it.

EXAMPLE 3-4   Find the maximum deflection of a simply supported beam with a uniformly distributed load.

SOLUTION   The maximum deflection will occur at the center of the beam, and so we place an imaginary force Q acting downward at this point. The end reactions are

$$R_1 = R_2 = \frac{wl}{2} + \frac{Q}{2}$$

Between $x = 0$ and $x = l/2$ the moment is

$$M = \left(\frac{wl}{2} + \frac{Q}{2}\right)x - \frac{wx^2}{2}$$

The strain energy for the whole beam is twice as much as for one-half; neglecting direct shear, this is

$$U = 2 \int_0^{l/2} \frac{M^2 \, dx}{2EI}$$

Therefore the deflection at the center is

$$y_{max} = \frac{\partial U}{\partial Q} = 2 \int_0^{l/2} \frac{2M}{2EI} \frac{\partial M}{\partial Q} \, dx = \frac{2}{EI} \int_0^{l/2} \left(\frac{wlx}{2} + \frac{Qx}{2} - \frac{wx^2}{2}\right)\frac{x}{2} \, dx$$

Since Q is imaginary, we can now set it equal to zero. Integration yields

$$y_{max} = \frac{2}{EI} \left[\frac{wlx^3}{12} - \frac{wx^4}{16}\right]_0^{l/2} = \frac{5wl^4}{384EI} \qquad ////$$

EXAMPLE 3-5   Use Castigliano's method to determine the spring rate of a cantilevered crank having length l, arm r, and concentrated force F, as shown in Fig. 3-8. Neglect the transverse-shear energy.

SOLUTION   The arm BC acts as a cantilever with a support at B. The strain energy is

$$U_{BC} = \int_0^r \frac{M_{BC}^2 \, dz}{2EI_2} \qquad (1)$$

where $M_{BC} = Fz$.

The reaction at B, due to F, is a torque $T = Fr$, and a force F producing a bending moment in AB of $M_{AB} = Fx$. Thus the strain energy stored in AB is

$$U_{AB} = \frac{T^2 l}{2GJ} + \int_0^l \frac{M_{AB}^2 \, dx}{2EI_1} \qquad (2)$$

Therefore the strain energy for the entire crank is

$$U = \int_0^r \frac{M_{BC}^2 \, dz}{2EI_2} + \frac{T^2 l}{2GJ} + \int_0^l \frac{M_{AB}^2 \, dx}{2EI_1} \qquad (3)$$

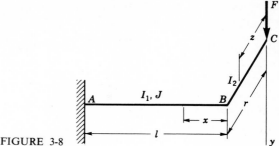

FIGURE 3-8

Taking the derivative of U with respect to F then gives

$$y = \frac{\partial U}{\partial F} = \frac{1}{EI_2} \int_0^r M_{BC} \frac{\partial M_{BC}}{\partial F} \, dz + \frac{Tl}{GJ} \frac{\partial T}{\partial F} + \frac{1}{EI_1} \int_0^l M_{AB} \frac{\partial M_{AB}}{\partial F} \, dx \qquad (4)$$

Now, since $M_{BC} = Fz$, $T = Fr$, and $M_{AB} = Fx$, we have

$$\frac{\partial M_{BC}}{\partial F} = z \qquad \frac{\partial T}{\partial F} = r \qquad \frac{\partial M_{AB}}{\partial F} = x$$

Substituting these values in Eq. (4) gives

$$y = \frac{1}{EI_2} \int_0^r Fz^2 \, dz + \frac{Fr^2l}{GJ} + \frac{1}{EI_1} \int_0^l Fx^2 \, dx = \frac{Fr^3}{3EI_2} + \frac{Fr^2l}{GJ} + \frac{Fl^3}{3EI_1}$$

Therefore the spring constant is

$$k = \frac{F}{y} = \frac{1}{(r^3/3EI_2) + (r^2l/GJ) + (l^3/3EI_1)} \qquad (5)$$

This result may have a surprising implication, depending upon your own intuition. So let us look into it in more detail.

A cantilever of length l loaded by a force F on the free end has a deflection of

$$y = \frac{Fl^3}{3EI} \qquad (6)$$

Therefore the spring rate is

$$k = \frac{F}{y} = \frac{3EI}{l^3} \qquad (7)$$

A bar of length l and polar moment of inertia J has, from Eq. (3-7), an angular deflection of

$$\theta = \frac{Tl}{GJ} \qquad (8)$$

Now, if this angular deflection is produced by a force F acting on the end of a lever r in long, then the torque is $T = Fr$ and the lever deflects a distance $r\theta$. For this reason, we can write Eq. (8) as

$$y = r\theta = \frac{Fr^2l}{GJ}$$

and hence

$$k = \frac{F}{y} = \frac{GJ}{r^2l} \qquad (9)$$

If we apply these results to Fig. 3-8, we find for BC

$$k_1 = \frac{3EI_2}{r^3} \qquad (10)$$

and for AB

$$k_2 = \frac{GJ}{r^2l} \qquad k_3 = \frac{3EI_1}{l^3} \qquad (11)$$

Note that these three relations appear in the denominator of Eq. (5), and hence that Eq. (5) can be written in the alternative form

$$k = \frac{1}{(1/k_1) + (1/k_2) + (1/k_3)} \qquad (12)$$

If you will now examine Eq. (3-3), you will see that these three springs are in series. Your own intuition should have told you that spring BC is in series with that of AB. But would you have guessed that the springs representing the torsion and bending of AB were in series with each other?   ////

3-9 DEFLECTION OF CURVED MEMBERS

Machine frames, springs, clips, fasteners, and the like, frequently occur as curved shapes. These members are easily analyzed for deflection by using Castigliano's theorem. Consider, for example, the curved frame of Fig. 3-9a. This frame is loaded by the force F and we wish the deflection in the x direction. If we cut the ring section at some position θ and show the force components on the cut section (Fig. 3-9b), then it can be seen that the force F causes bending, the component F_θ causes tension, and the component F_r causes direct shear. The total strain energy results from each of these effects and is

$$U = \int \frac{M^2\,ds}{2EI} + \int \frac{F_\theta^2\,ds}{2AE} + \frac{1}{2}\int \frac{F_r^2\,ds}{GA} \qquad (a)$$

where, by using the Juvinall approximation of $\frac{1}{2}$ in the third term, a nonrectangular cross section has been assumed (see Sec. 3-8). This is an application of

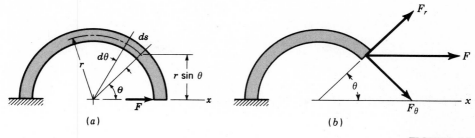

FIGURE 3-9

Eq. (3-24) and a factor of $\frac{3}{2}$ would be used for a rectangular section. According to Castigliano's theorem, the deflection produced by the force F is

$$\delta_x = \frac{\partial U}{\partial F} = \int_0^s \frac{M}{EI} \frac{\partial M}{\partial F} \, ds + \int_0^s \frac{F_\theta}{AE} \frac{\partial F_\theta}{\partial F} \, ds + \int_0^s \frac{F_r}{GA} \frac{\partial F_r}{\partial F} \, ds \qquad (b)$$

The factors for the first term in this equation are

$$M = Fr \sin \theta \qquad \frac{\partial M}{\partial F} = r \sin \theta \qquad ds = r \, d\theta$$

Using Fig. 3-9b we see that the factors for the second and third terms are

$$F_\theta = F \sin \theta \qquad \frac{\partial F_\theta}{\partial F} = \sin \theta$$

$$F_r = F \cos \theta \qquad \frac{\partial F_r}{\partial F} = \cos \theta$$

Substituting all these into Eq. (b) and factoring the result gives

$$U = \frac{Fr^3}{EI} \int_0^\pi \sin^2 \theta \, d\theta + \frac{Fr}{AE} \int_0^\pi \sin^2 \theta \, d\theta + \frac{Fr}{GA} \int_0^\pi \cos^2 \theta \, d\theta$$

$$= \frac{Fr^3}{2EI} + \frac{Fr}{2AE} + \frac{Fr}{2GA} \qquad (3\text{-}29)$$

This is the deflection of the free end of the frame in the direction of F. Because the radius is cubed in the first term, the second two terms will be negligible for large-radius frames.

3-10 THEORY OF COLUMNS

A short bar loaded in pure compression by a force P acting along the centroidal axis will shorten, in accordance with Hooke's law, until the stress reaches the elastic limit of the material. If P is increased still more, the material bulges, and is squeezed into a flat disk.

Now visualize a long thin straight bar, such as a yardstick, loaded in pure compression by another force P acting along the centroidal axis. As P is increased from zero, the member shortens according to Hooke's law, as before. However, if the member is sufficiently long, as P increases, a critical value will be reached designated P_{cr}, corresponding to a condition of unstable equilibrium. At this point any little crookedness of the member or slight movement of the load or support will cause the member to collapse by buckling.

If the compression member is long enough to fail by buckling, it is called a *column*; otherwise it is a simple compression member. Unfortunately, there is no line of demarcation which clearly distinguishes a column from a simple compression member. Then, too, a column failure can be a very dangerous failure because there is no warning that P_{cr} has been exceeded. In the case of a beam, an increase in the bending load causes an increase in the beam deflection, and the excessive deflection is a visible indication of the overload. But a column remains straight until the critical load is reached, after which there is sudden and total collapse. Depending upon the length, the actual stresses in a column at the instant of buckling may be quite low. For this reason the criterion of safety consists in a comparison of the actual load with the critical load.

The relationship between the critical load and the column material and geometry is developed with reference to Fig. 3-10a. We assume a bar of length l loaded by a force P acting along the centroidal axis on rounded or pinned ends. The figure shows that the bar is bent in the positive y direction. This requires a negative moment, and hence

$$M = -Py \qquad (a)$$

If the bar should happen to bend in the negative y direction, a positive moment would result, and so $M = -Py$, as before. Using Eq. (3-14), we write

$$\frac{d^2y}{dx^2} = -\frac{P}{EI}y \qquad (b)$$

or

$$\frac{d^2y}{dx^2} + \frac{P}{EI}y = 0 \qquad (3\text{-}30)$$

This resembles the well-known differential equation for simple harmonic motion. The solution is

$$y = A \sin \sqrt{\frac{P}{EI}}x + B \cos \sqrt{\frac{P}{EI}}x \qquad (c)$$

where A and B are constants of integration and must be determined from the boundary conditions of the problem.* We evaluate them using the conditions that $y = 0$ at $x = 0$ and at $x = l$. This gives $B = 0$, and

$$0 = A \sin \sqrt{\frac{P}{EI}}l \qquad (d)$$

---

* For a solution see Sec. 16-4.

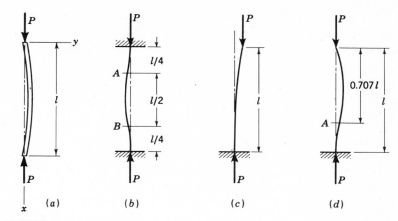

FIGURE 3-10
(a) Both ends rounded or pivoted; (b) both ends fixed; (c) one end free, one end fixed; (d) one end rounded, one end fixed.

The trivial solution of no buckling occurs with $A = 0$. However, if $A \neq 0$, then

$$\sin \sqrt{\frac{P}{EI}}\, l = 0 \qquad (e)$$

Equation (e) is satisfied by $\sqrt{P/EI}\, l = N\pi$, where N is an integer. Solving for $N = 1$ gives the critical load

$$P_{cr} = \frac{\pi^2 EI}{l^2} \qquad (3\text{-}31)$$

which is called the *Euler column formula*; it applies only to rounded-end columns. If we substitute these results back into Eq. (c), we get the equation of the deflection curve as

$$y = A \sin \frac{\pi x}{l} \qquad (3\text{-}32)$$

which indicates that the deflection curve is a half-wave sine. We are only interested in the minimum critical load, which occurs with $N = 1$. However, though it is not of any importance here, values of N greater than 1 result in deflection curves which cross the axis at points of inflection and are multiples of half-wave sines.

Using the relation $I = Ak^2$, where A is the area and k the radius of gyration enables us to rearrange Eq. (3-31) into the more convenient form

$$\frac{P_{cr}}{A} = \frac{\pi^2 E}{(l/k)^2} \qquad (3\text{-}33)$$

with l/k designated as the *slenderness ratio*. The solution to Eq. (3-33) is called the *critical unit load*. And though the unit load has the dimensions of stress, you are cautioned very particularly not to call it a stress! To do so might lead you into the error of comparing it with a strength, the yield strength, for example, and coming to the false conclusion that a margin of safety exists. Equation (3-33) shows that the critical unit load depends *only* upon the modulus of elasticity and the slenderness ratio. Thus a column obeying the Euler formula made of high-strength alloy steel is no better than one made of low-carbon steel, since E is the same for both.

The critical loads for columns with different end conditions can be obtained by solving the differential equation or by comparison. Figure 3-10*b* shows a column with both ends fixed. The inflection points are at A and B, a distance $l/4$ from the ends. The distance AB is the same curve as a rounded-end column. Substituting the length $l/2$ for l in Eq. (3-31), we obtain

$$P_{cr} = \frac{\pi^2 EI}{(l/2)^2} = \frac{4\pi^2 EI}{l^2} \qquad (3\text{-}34)$$

In Fig. 3-10*c* is shown a column with one end free and one end fixed. This curve is equivalent to half the curve for columns with rounded ends, so that if a length of $2l$ is substituted into Eq. (3-31), the critical load becomes

$$P_{cr} = \frac{\pi^2 EI}{(2l)^2} = \frac{\pi^2 EI}{4l^2} \qquad (3\text{-}35)$$

A column with one end fixed and one end rounded, as in Fig. 3-10*d*, occurs frequently. The inflection point is at A, a distance of $0.707l$ from the rounded end. Therefore

$$P_{cr} = \frac{\pi^2 EI}{(0.707l)^2} = \frac{2\pi^2 EI}{l^2} \qquad (3\text{-}36)$$

We can account for these various end conditions by writing the Euler equation in the two following forms:

$$P_{cr} = \frac{n\pi^2 EI}{l^2} \qquad \frac{P_{cr}}{A} = \frac{n\pi^2 E}{(l/k)^2} \qquad (3\text{-}37)$$

Table 3-1 END-CONDITION CONSTANTS FOR EULER COLUMNS [TO BE USED WITH EQ. (3-37)]

Column end conditions	End-condition constant n		
	Theoretical value	Conservative value	Recommended value*
Fixed-free	$\frac{1}{4}$	$\frac{1}{4}$	$\frac{1}{4}$
Rounded-rounded	1	1	1
Fixed-rounded	2	1	1.2
Fixed-fixed	4	1	1.2

* To be used only with liberal factors of safety when the column load is accurately known.

Here, the factor n is called the *end-condition constant*, and it may have any one of the theoretical values $\frac{1}{4}$, 1, 2, or 4, depending upon the manner in which the load is applied. In practice it is difficult, if not impossible, to fix the column ends so that the factors $n = 2$ or $n = 4$ would apply. Even if the ends are welded, some deflection will occur. Because of this, some designers never use a value of n greater than unity. However, if liberal factors of safety are employed, and if the column load is accurately known, then a value of n not exceeding 1.2 for both ends fixed, or for one end rounded and one end fixed, is not unreasonable, since it supposes only partial fixation. Of course, the value $n = \frac{1}{4}$ must always be used for a column having one end fixed and one end free. These recommendations are summarized in Table 3-1.

3-11 COLUMN DESIGN

We have previously noted the absence of any clear distinction between a simple compression member and a column. To illustrate the problem, Fig. 3-11a is a plot of the criteria of failure of both simple compression members and of Euler columns. If a member is short, it will fail by yielding; if it is long, it will fail by buckling. Consequently, the graph has the slenderness ratio l/k as the abscissa, and the unit load P/A as the ordinate. Then an ordinate through any desired slenderness ratio will either intersect the line AB and define a simple compression member, since failure would be by yielding, or the ordinate will intersect the line BD and define a column, because failure would occur by buckling. Unfortunately, this theory does not work out quite so beautifully in practice.

The results of a large number of experiments indicate that, within a broad region around point B (Fig. 3-11a), column failure begins before the unit load reaches a point represented by the graph ABD. Furthermore, the test points

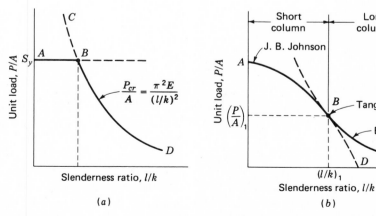

FIGURE 3-11

obtained from these experiments are scattered. Researchers have surmised that the failure of experiment to verify the theory in the vicinity of point B is explained by the fact that it is virtually impossible to construct an ideal column. Very small deviations can have an enormous effect upon the value of the critical load. Such factors as built-in stresses, initial crookedness, and very slight load eccentricities must all contribute to the scatter and to the deviation from theory.

Many different column formulas, most of them empirical, have been devised to overcome some of the disadvantages of the Euler equation. The *parabolic*, or *J. B. Johnson, formula* is widely used in the machine, automotive, aircraft, and structural-steel construction fields. This formula often appears in the form

$$\frac{P_{cr}}{A} = a - b\left(\frac{l}{k}\right)^2 \qquad (3\text{-}38)$$

where a and b are constants that are adjusted to cause the formula to fit experimental data. Figure 3-11b is a graph of two formulas, the Euler equation and the parabolic equation. Notice that curve ABD is the graph of the parabolic formula, while curve BC represents the Euler equation. In analyzing a column to determine the critical buckling load, only part AB of the parabolic graph and part BC of the Euler graph should be used.

The constants a and b in Eq. (3-38) are evaluated by deciding where the intercept A, in Fig. 3-11b, is to be located, and where the tangent point B is desired. Notice that the coordinates of point B are specified as $(P/A)_1$ and $(l/k)_1$.

One of the most widely used versions of the parabolic formula is obtained by making the intercept A correspond to the yield strength S_y of the material, and making the parabola tangent to the Euler curve at $(P/A)_1 = S_y/2$. Thus, the first constant in Eq. (3-38) is $a = S_y$. To get the second constant, substitute $S_y/2$ for P_{cr}/A and solve for $(l/k)_1$. Using Eq. (3-37) we get

$$\left(\frac{l}{k}\right)_1 = \sqrt{\frac{2\pi^2 nE}{S_y}} \qquad (3\text{-}39)$$

Then, substituting all this into Eq. (3-38) yields

$$\frac{S_y}{2} = S_y - b\frac{2\pi^2 nE}{S_y}$$

or

$$b = \left(\frac{S_y}{2\pi}\right)^2 \frac{1}{nE} \qquad (3\text{-}40)$$

which is to be used for the constant in

$$\frac{P_{cr}}{A} = S_y - b\left(\frac{l}{k}\right)^2 \qquad (3\text{-}41)$$

Of course, this equation should only be used for slenderness ratios up to $(l/k)_1$. Then the Euler equation is used when l/k is greater than $(l/k)_1$.

Equations (3-37) and (3-40) have been solved using the end conditions of

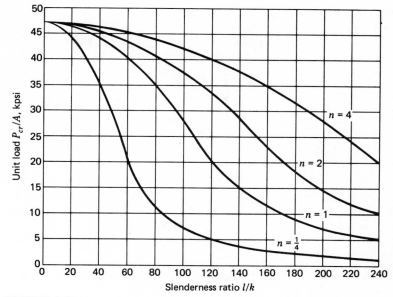

FIGURE 3-12

Design chart using parabolic equation and Euler equation for UNS G10150 cold-drawn steel for theoretical end conditions.

Table 3-1 for UNS G10150 cold-drawn steel, and the results are plotted in Fig. 3-12. This is called a design chart.

Sometimes, the parabolic equation is derived with the factor of safety included. For example, the original AISC column formula for ASTM No. A7 steel for both main and secondary members is

$$\frac{P}{A} = 17\,000 - 0.485\left(\frac{l}{k}\right)^2 \qquad (3\text{-}42)$$

which is valid only for l/k values less than 120. Since A7 steel has a tensile strength of 60 kpsi or more, this equation has a factor of safety built in. Note that this fact is also evident because it has been written in terms of P/A, instead of P_{cr}/A.

3-12 THE SECANT FORMULA

An eccentric column load is one in which the line of action of the column forces is not coincident with the centroidal axis of the cross section. The distance between the two axes is called the eccentricity e. The product of the force and the eccentricity produces an initial moment Pe. When this moment is introduced into the

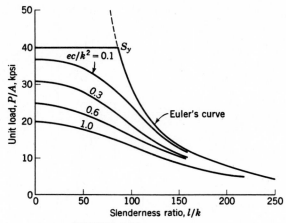

FIGURE 3-13
Comparison of the secant and Euler formulas.

analysis a rational formula can be deduced, valid for any slenderness ratio. The result is called the *secant formula*; it is usually expressed as

$$\frac{P}{A} = \frac{S_y}{1 + (ec/k^2)\sec[(l/k)\sqrt{P/4AE}]} \qquad (3\text{-}43)$$

In this equation c is the distance from the neutral plane of bending to the outer surface. The term ec/k^2 is called the eccentricity ratio. Figure 3-13 is a plot of Eq. (3-43) for various values of the eccentricity ratio and for a steel having a yield strength of 40 kpsi. Euler's equation is shown for comparison purposes.

Equation (3-43) is inconvenient to use for design purposes because the cross-sectional area A appears on both sides of the equation. Thus, once the column material has been selected, a chart after the fashion of Fig. 3-13 should be prepared for design purposes. Such charts should be saved for possible use in future design problems.

PROBLEMS

Sections 3-1 and 3-2

3-1   The stepped shaft shown in the figure is steel. Find the torsional spring rate.

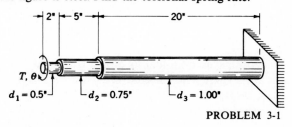

PROBLEM 3-1

3-2   This problem is intended to demonstrate that the spring rate depends upon how the force is applied. In *a* of the figure a force $F = 20$ lb is applied at the end of a 40-lb/in spring. This results in a deflection $y = 0.50$ in. Now suppose the same spring is attached to a rigid lever 10 in long as in *b* of the figure. Suppose, too, that the same force is applied at the end of the lever. This will produce a deflection z of the end of the lever. The spring rate is now $k' = F/z$. Compute k' and z. Let *a* be the distance from the pivot to the spring, and *b* the distance from the pivot to the force, and develop a general relation for springs on levers. Be sure to record the result here for reference purposes.

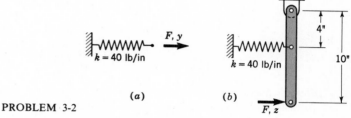

PROBLEM 3-2   (*a*)   (*b*)

3-3   Shown in the figure is a rigid lever 300 mm in length, pivoted at O, having a spring $k_A = 70$ kN/m attached at A, and a spring $k_B = 40$ kN/m attached at B. For this system a certain force F applied at C will produce a corresponding deflection x. We can then replace the entire system with an equivalent spring whose rate is $k = F/x$. Find k.

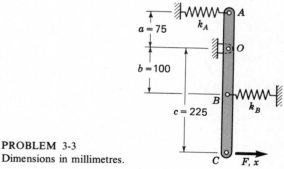

PROBLEM 3-3
Dimensions in millimetres.

3-4   The figure shows a geared system which we wish to assume has infinitely rigid teeth. Compute the torsional spring rate of the system. Explain how the velocity ratio of the gears enters into the calculation. Assume steel shafts.

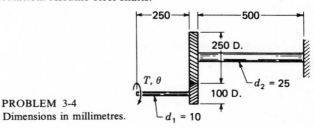

PROBLEM 3-4
Dimensions in millimetres.

3-5 The two gears shown in the figure have the respective tooth numbers N_1 and N_2. An input torque T_1 is applied to the end of a shaft, having a torsional spring constant k_1. This torque is resisted by an output torque T_2 at B, exerted at the end of the driven shaft, whose torsional stiffness is k_2.
   (a) Assume the shaft is fixed at B and find an expression for the torsional spring rate of the entire system based upon a deflection measured at A.
   (b) Assume the shaft is fixed at A and find an expression for the torsional spring rate corresponding to the deflection at B.

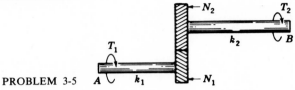

PROBLEM 3-5

3-6 Find the diameter of a solid round steel shaft which is to transmit $\frac{1}{20}$ hp at 1 rpm if the angular deflection is not to exceed 1° in a length of 30 diameters. Find the corresponding torsional stress.

3-7 Find the diameter of a solid round steel shaft to transmit 15 kW at 60 s^{-1} if the angular deflection is not to exceed 1° in a length of 30 diameters. Compute the corresponding torsional stress, and comment on the result.

Sections 3-3 to 3-5

3-8 Determine a set of cross-sectional dimensions for a steel straightedge 1 m long such that, when it is optimally supported, the deflection due to its own weight will be less than 12.5 μm.

3-9 A $2 \times 2 \times \frac{3}{8}$-in steel angle supports the load shown in the figure. Find the maximum deflection.

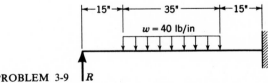

PROBLEM 3-9

3-10 A round steel shaft supports the loads shown in the figure. The shaft has a diameter of 30 mm and is supported by preloaded antifriction (ball) bearings that introduce some end constraint. Calculate the deflection at the center by assuming, first, that the ends are fixed as shown in the figure, and by assuming, second, that the shaft is simply supported at the ends. What is the ratio of these two deflections?

PROBLEM 3-10
Dimensions in millimetres.

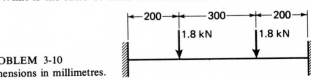

3-11 Select a standard steel angle with equal legs from Table A-8 to support the load in the figure such that the deflection will not exceed $\frac{1}{16}$ in.

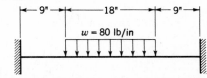

PROBLEM 3-11

3-12 Select a round steel bar to support the load shown in the figure such that the maximum deflection will not exceed 0.40 mm.

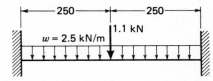

PROBLEM 3-12
Dimensions in millimetres.

3-13 A torsion spring, as shown in part *a* of the figure, can be treated as a cantilever for analysis purposes.

    (*a*) Determine the dimensions *b* and *h* of a rectangular steel section such that a force $F = 20$ N will produce a deflection of exactly 75 mm. Use $b = 10h$.

    (*b*) The yield strength of the steel used is 400 MPa. Will this yield strength be exceeded by the stress when the force is 20 N?

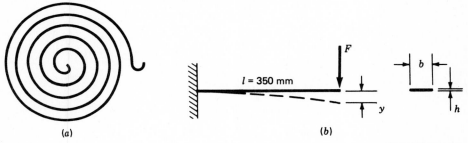

(*a*)   (*b*)

PROBLEM 3-13

3-14 The figure shows a cantilever steel spring of rectangular cross section. Determine the dimensions of the spring so that it has a scale of 140 lb/in. What is the maximum stress if the operating range is $\frac{1}{4}$ in?

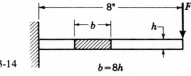

PROBLEM 3-14 $b = 8h$

3-15 The figure is a structural-engineering drawing of a beam spanning two columns. The beam is made up of two channel sections mounted back to back and spaced $\frac{1}{4}$ in apart, using short sections of $\frac{1}{4}$-in plates. Four short angles are used to secure the beam to the columns by riveting, bolting, or welding. Structural shapes are specified by stating their major dimension or dimensions, together with the weight per foot.

Thus the wide-flange beam, used here as a column, weighs 50 lb/ft.* Dimensions and properties of shapes frequently used in mechanical design are listed in Tables A-8 to A-11. Examination of Table A-11, for example, indicates that the 8-in back-to-back channels have a flange width of 2.527 in and a web thickness of 0.487 in.

In this problem two of these beams are used to support a machine whose total weight is W. The figure shows that this weight is to be transferred into the beams at four points, two on each beam. Find:

(*a*) The shear and moment diagrams for the beams.

(*b*) The weight W that can be supported by the two beams, allowing a maximum bending stress of 16 kpsi.

(*c*) The maximum deflection caused by this weight.

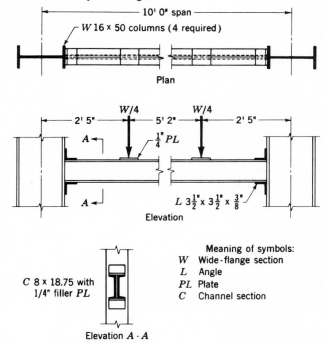

PROBLEM 3-15

Section 3-6

3-16 The figure shows the drawing of a countershaft and its loading diagram. Bearings A and B are self-aligning; find the maximum deflection.

---

* The dimensions and properties of standard rolled steel structural shapes may be found in "AISC Steel Construction Manual," American Institute of Steel Construction (AISC), New York.

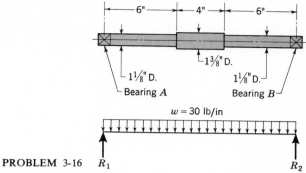

PROBLEM 3-16 R_1 R_2

3-17 Determine the maximum deflection of the shaft shown in the figure. The material is steel.

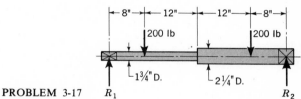

PROBLEM 3-17 R_1 R_2

Sections 3-7 to 3-9

3-18 Using the theorem of Castigliano, find the maximum deflection of the cantilever shown in the figure. Neglect direct shear.

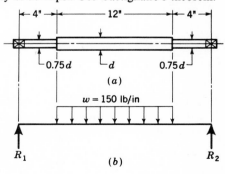

PROBLEM 3-18 I $2I$

3-19 In the figure, a is a drawing of a shaft and b is the loading diagram. Determine the diameter d so that the maximum deflection does not exceed 0.010 in. The material is an alloy steel with a modulus of elasticity of 30 Mpsi. Use Castigliano's theorem.

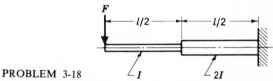

PROBLEM 3-19 R_1 (b) R_2

3-20 Find the deflection at the point of application of the force F to the cantilever in the figure and the deflection at the free end. Use Castigliano's theorem.

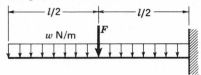

PROBLEM 3-20

3-21 The frame shown in the figure is composed of three structural aluminum angles welded at points B and C and bolted to a supporting structure at points A and D. The moments of inertia about the bending axes are $I_{AB} = I_{CD} = 112.5(10)^3$ mm^4 and $I_{BC} = 55.6(10)^3$ mm^4. Determine the magnitude of the load F in kilonewtons such that the maximum deflection of BC is not greater than 1.5 mm.

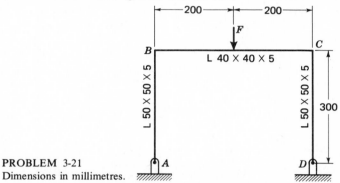

PROBLEM 3-21
Dimensions in millimetres.

3-22 The figure illustrates an X frame in the xz plane. The frame is made of two members of length $2l$ welded at the center of each at an X angle of θ. Each member has a moment of inertia I. The ends A, B, and D are simply supported. Find an expression for the downward deflection at C due to the force F.

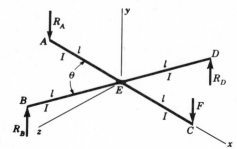

PROBLEM 3-22

3-23 The L frame shown in the figure has one end A twisted by the torque T and the other end C is simply supported by the reaction R. The rectangular and polar moments of inertia are I_1, J_1, and I_2, J_2 as shown, and both legs are made of the same material. Use Castigliano's theorem and find the angular rotation of the frame at A in the direction of the applied torque.

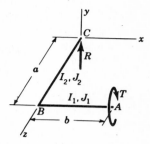

3-24 Using Castigliano's theorem, develop an expression for the deflection of end A of the bell-crank lever shown in the figure. B and C are pivoted to a fixed frame.

PROBLEM 3-24

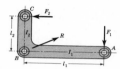

3-25 The figure shows a welded steel bracket loaded by a force $F = 5$ kN. Using the assumed loading diagram and Castigliano's method, find the maximum deflection of the end.

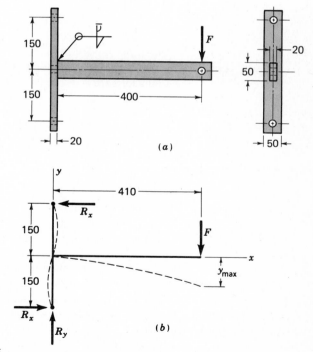

PROBLEM 3-25

Dimensions in millimetres. (a) Bracket; (b) assumed loading and deflection diagram.

3-26 The rectangular-shaped *C* frame shown in the figure is welded of three wide-flange beams. Find an expression for the deflection of the frame at the force *F* and in the direction of *F*.

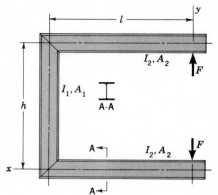

PROBLEM 3-26

Sections 3-10 to 3-12

3-27 Derive the parabolic column formula for UNS G10350 cold-drawn steel for a column with rounded ends such that the tangent point is located at $0.50S_y$. Plot a graph of P_{cr}/A vs. the slenderness ratio for l/k values up to 200.

3-28 The yield strength of cold-drawn UNS G10180 steel in metric units is 372 MPa. Use a factor of safety of 3 and derive a column equation for the allowable unit column load using the parabolic formula for both ends rounded. The tangent point is to be located at $0.50S_y$. Plot a design chart similar to Fig. 3-12 for slenderness ratios up to and including 200.

3-29 A column with both ends rounded is made of hot-rolled UNS G10150 steel with a 10×25-mm rectangular cross section. Find the buckling load in kilonewtons for the following column lengths: 85, 175, 400, and 600 mm.

3-30 A column with one end fixed and one end rounded is made of hot-rolled UNS G10100 steel. The member is a rectangular-section bar $\frac{1}{2} \times 1\frac{1}{2}$ in. Use the theoretical end-condition constant and find the buckling load for the following column lengths: 0.50, 2, and 4 ft.

3-31 Find the safe compressive load for a $4 \times 4 \times \frac{1}{2}$-in structural steel angle 5 ft long. Equation (3-42) applies.

3-32 A Euler column with one end fixed and one end free is to be made of an aluminum alloy. The cross-sectional area of the column is to be 600 mm² and it is to have a length of 2.5 m. Determine the column buckling load corresponding to the following shapes:

(*a*) A solid round bar.
(*b*) A round tube with a 50 mm OD.
(*c*) A 50-mm square tube.
(*d*) A square bar.

3-33 A steel tube having a $\frac{3}{16}$-in wall thickness is to be designed to safely support a column load of 3.60 kip. The column will have both ends rounded and will be made of UNS G10350 cold-drawn steel. Use the design chart of Prob. 3-27, a factor of safety of 4, and find the outside diameter to the nearest $\frac{1}{8}$ in for the following column lengths: 3 in, 15 in, and 45 in.

3-34 A steel tube having a 5-mm wall thickness is to be designed to safely support a column load of 15 kN. The column will have both ends rounded and will be made of UNS G10180 cold-drawn steel. Use the design chart of Prob. 3-28 and find an appropriate outside diameter to the nearest 2.5 mm for the following column lengths: 50 mm, 400 mm, and 1 m.

3-35 The hydraulic cylinder shown in the figure operates at a pressure of 3500 psi and has a 3-in cylinder bore. When the clevis mount, which is shown, is used, the piston rod is sized using l_{max} as the column length and unity for the end-condition constant. Determine appropriate safe diameters d for the piston rod using a medium carbon steel having a yield strength of 75 kpsi, and a factor of safety of 3, for column lengths of 96 in, 48 in, and 24 in.

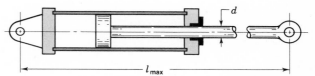

PROBLEM 3-35

3-36 Bars OA and AB in the figure are made of UNS G10100 hot-rolled steel and have a cross section of 1 by $\frac{1}{4}$ in as shown. Based on recommended values of end-condition constants, what weight W would cause a column failure?

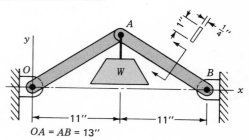

PROBLEM 3-36 $OA = AB = 13''$

4

STATISTICAL CONSIDERATIONS IN DESIGN

Almost any product that you can conceive of can be made with a high degree of perfection. It is not made in this state of perfection simply because of economic reasons. The more inspection you buy, the more design you buy, the greater your chances of putting out a zero-defect product. But there comes a point where, if you have gotten your product 98 percent defect-free, then if you spend a hundred times this much in design and in inspection you may only get it 99 percent zero-free, and you may spend a thousand times as much to make it really perfect. So it just isn't economically feasible to make every product perfect.

This is acceptable manufacturing philosophy to the manufacturer. It is totally unacceptable to the consumer who gets that one rare product that is defective and meets serious injury or death as a result of it.*

As indicated by this quotation, an uncertainty exists as to whether a product will actually perform satisfactorily. It is the purpose of this chapter to provide the tools needed by the design engineer to determine the numerical probability of success or of failure.

* By Craig Spangenberg, "You, Your Product, and the Law," *Mechanical Engineering*, vol. 90, no. 6, p. 19 June, 1968, by permission of the American Society of Mechanical Engineers.

4-1 PERMUTATIONS AND COMBINATIONS

In studying the likelihood that a particular machine assembly will be a success or a failure, we must first learn to count. That is, we must be able to count the number of ways it can succeed and the number of ways it can fail. When these counts are weighed against one another, it is possible to arrive at a figure which indicates the probability of success. In a group of ten bolts and ten nuts, say, it may be that a defective joint is obtained when one particular bolt is mated with one particular nut. So, to compute the probability of success, we must be able to count the number of ways that ten bolts and ten nuts can be mated.

Permutation refers to *order*, or *arrangement*. The permutations of the three letters *ABC* are

$$ABC \quad ACB \quad BAC \quad BCA \quad CAB \quad CBA$$

In some branches of engineering we are interested in finding the particular permutations. This is true of kinematic synthesis, for example, where we might be interested in looking at all the different ways of assembling four links to form a four-bar mechanism. In other engineering studies we are not so much interested in the details of each permutation as we are in the number of them. Thus we need a systematic way of counting them.

Suppose we wish to find all the permutations of n things using all of them at a time. We have n choices for the first thing; after this, there are $n - 1$ choices left for the second, $n - 2$ for the third, and so on. Thus the number of permutations is

$$P(n, n) = n(n - 1)(n - 2) \cdots (3)(2)(1) = n! \qquad (4\text{-}1)$$

where $P(n, n)$ means the number of permutations of n things taken n at a time. Thus the number of permutations of the three letters A, B, and C taken three at a time is

$$P(3, 3) = 3! = 3 \cdot 2 \cdot 1 = 6$$

Sometimes we need to permute n things r at a time. Thus all possible arrangements of the three letters *ABC* in pairs is

$$AB \quad AC \quad BA \quad BC \quad CA \quad CB$$

This is called the permutations of three things taken two at a time. Note that the first letter can be chosen in any one of three different ways; after this, the second letter can be chosen in any one of two different ways. Thus the number of permutations of three things taken two at a time is

$$P(3, 2) = 3 \cdot 2 = 6$$

Suppose we wish to find the number of permutations of eight things taken three at a time. The first thing can be selected eight different ways, the second seven different ways, and the third six different ways. Thus

$$P(8, 3) = 8 \cdot 7 \cdot 6$$

Similarly,

$$P(10, 4) = 10 \cdot 9 \cdot 8 \cdot 7 \qquad P(14, 3) = 14 \cdot 13 \cdot 12$$
$$P(4, 3) = 4 \cdot 3 \cdot 2 \qquad P(4, 4) = 4 \cdot 3 \cdot 2 \cdot 1$$

Inspection of these shows that there are r factors in the product, and hence we can write, in general,

$$P(n, r) = n(n - 1)(n - 2) \cdots (n - r + 1) \qquad (4\text{-}2)$$

A more convenient form, however, results from multiplying the numerator and denominator by $(n - r)!$. This gives

$$P(n, r) = \frac{n!}{(n - r)!} \qquad (4\text{-}3)$$

By defining $0! = 1$, Eq. (4-3) reduces to (4-1) for $r = n$.

Of course, for large values of n it is difficult to compute $n!$. In such cases Stirling's approximation should be used.* It is written

$$n! \sim n^n e^{-n} \sqrt{2\pi n} \qquad (4\text{-}4)$$

where the symbol $\sim$ means asymptotic to, and where $e = 2.718 \ldots$, the natural base of logarithms. This approximation becomes better as n becomes larger. Thus the error is 1 percent for $n = 10$ and 0.1 percent for $n = 100$.

Often, the things we have to order or arrange in engineering analysis are not all different. We may have a group of n things of which p of them are alike. If we permute them using all of them each time, some of the permutations will be the same because we cannot distinguish between the p's. If we permute the p things among themselves, we find there are $P(p, p) = p!$ arrangements possible, if they could be identified. Since they cannot be identified, the number of permutations of n things, of which p are alike, taken n at a time, is

$$P = \frac{n!}{p!} \qquad (4\text{-}5)$$

This formula is readily extended to complex groupings. For example, a collection of n things may contain one group of p things that are alike and another group of q things that are alike. The number of permutations is

$$P = \frac{n!}{p!q!} \qquad (4\text{-}6)$$

In mechanical engineering studies we frequently encounter the problem of arranging things on a circle instead of on a straight line; the placement of bolts on a round cylinder head is an example of this. In such a situation two arrangements

* See Mary L. Boas, "Mathematical Methods in the Physical Sciences," p. 409, John Wiley & Sons, Inc., New York, 1966.

are the same if one of them can be obtained by rotating the other. For this problem we count the permutations by placing the first thing anywhere. Then the remaining things can be permuted $(n - 1)!$ ways. Thus the number of permutations on a circle, of n things taken n at a time, is

$$P(n, n) = (n - 1)! \qquad (4\text{-}7)$$

It is especially important to remember that permutation refers to order, or arrangement. Thus the arrangement ABC is *different* from the arrangement CAB.

We now define the word *combination* as a group of things taken without any regard to the order. Thus ABC is the *same* combination as CAB or BCA. So our next problem is to learn how to count the number of combinations of n things taken r at a time. The expression $C(n, r)$ will be used to mean the number of combinations of n things taken r at a time.

Consider any one combination of three letters from the four letters $ABCD$, say, the combination ABC. The permutations of ABC are ABC, ACB, BAC, BCA, CAB, and CBA; that is, $P(3, 3) = 3!$. Thus, for each combination of r things, there are $r!$ permutations. Then for n combinations of r things there must be $r! \, C(n, r)$ permutations. This is the same as the number of permutations of n things taken r at a time, and hence

$$r! \, C(n, r) = P(n, r)$$

or
$$C(n, r) = \frac{P(n, r)}{r!} = \frac{n!}{r! \, (n - r)!} \qquad (4\text{-}8)$$

It is interesting to know that the number of combinations of n things taken r at a time is the same as the number of combinations of n things taken $n - r$ at a time. That is,

$$C(n, r) = C(n, n - r) \qquad (4\text{-}9)$$

You can easily prove this by substituting Eq. (4-8) for both sides of (4-9). Computations are often greatly simplified by using this relation.

4-2 PROBABILITY

Mathematical probability, meaning likelihood, prospect, or chance, is a number ranging between 0 and 1 which measures the expectancy that a named event will happen. A probability of 0 is given to an event if it is impossible for it to occur. If the event is certain to occur, the probability is 1. If there is an even chance that it will or will not occur, the probability is $\frac{1}{2}$.

As is the case with many concepts used in engineering, there is no rigorous mathematical definition of probability. As engineers we accept this because it furnishes us a means to get answers to engineering problems which we cannot obtain in any other way. Thus we define *numerical probability as a number p, between* 0 *and* 1, *which indicates the likelihood that a particular event E will occur, given that it can happen f ways out of n equally likely ways.* In this definition, the words "probability," "likelihood," and "equally likely" mean about the same thing, and so we find ourselves arguing around a circle. Nevertheless, when we say there is a 50 percent chance that it will snow today, we are using the concept of probability. It is not difficult, therefore, to accept probability as an intuitive concept. Thus, for the definition given above, we can write

$$p = P(E) = \frac{f}{n} \qquad (4\text{-}10)$$

where p is the probability that the event E will occur.

Suppose we toss a coin $n = 10$ times and that we get $f = 4$ heads. Then, using Eq. (4-10) for this experiment, we learn that the probability of heads is

$$p = P(\text{heads}) = \frac{4}{10}$$

But we know intuitively that if we were to toss the coin a very large number of times, it would come up heads about half the time. Thus a better definition of probability is

$$p = \lim_{n \to \infty} \frac{f}{n} \qquad (4\text{-}11)$$

It is also convenient to describe the probability that an event will not happen. If the event is E, then $\tilde{E}$ is called NOT E, and the probability that E will not happen is written

$$q = P(\tilde{E}) = 1 - P(E) \qquad (4\text{-}12)$$

EXAMPLE 4-1   Suppose a merchant accidentally mixes 2 used flashlight cells with his stock of 98 new cells. As a purchaser, what is the probability that you will get a used cell? A new cell?

SOLUTION   Call the event, getting a used cell, E. E can happen in $f = 2$ ways out of $n = 100$ equally likely ways. The probability p is

$$p = P(E) = \frac{f}{n} = \frac{2}{100}$$

The probability of *not* getting a used cell is $P(\tilde{E})$; there are $f = 98$ ways out of $n = 100$ ways, and so

$$q = P(\tilde{E}) = \frac{f}{n} = \frac{98}{100}$$

Note that $p + q = 1$, indicating the certainty of getting either a new or a used cell.

////

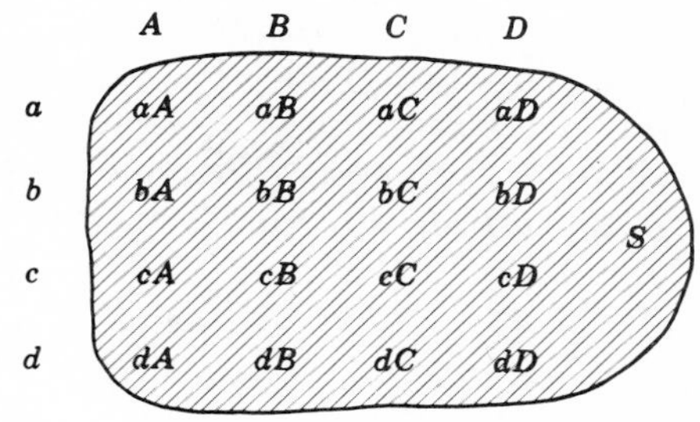

	A	B	C	D
a	aA	aB	aC	aD
b	bA	bB	bC	bD
c	cA	cB	cC	cD
d	dA	dB	dC	dD

S

FIGURE 4-1
Sample space S is the set of all possible outcomes.

In studying probability we also need to develop the idea of a *sample space*; this is a map, or set, or collection, of all possible outcomes. Suppose that you have four bearings A, B, C, and D which are to be assembled at random on shafts a, b, c, and d. The sample space S, shown in Fig. 4-1, indicates that there are 16 possible outcomes, designated aA, aB, aC, etc. Each outcome is called a *sample point*. If the outcomes are equally likely, a weight $w_i = \frac{1}{16}$ can be assigned to each point; note that w_i must be a positive number. Then, for any sample space,

$$\sum_{1}^{n} w_i = 1 \qquad (4\text{-}13)$$

An *event* E is any subset of the sample space. Thus, in Fig. 4-2, the event that bearings A, B, C, and D will be assembled with shafts a, b, c, and d, *in that order*, is represented as a subspace of the sample space. The event E consists of four outcomes; each outcome has a weight of $\frac{1}{16}$, and hence a probability of $\frac{1}{16}$. Thus the probability of E is $\frac{4}{16}$, or $\frac{1}{4}$, since it consists of four outcomes. The probability of an event E is therefore the sum of the weights of the sample points in E. We can also write this

$$P(E) = \frac{n(E)}{n(S)} \qquad (4\text{-}14)$$

where $n(E)$ and $n(S)$ are the number of sample points in E and S, respectively.

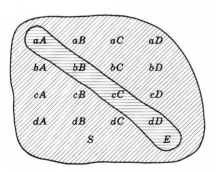

FIGURE 4-2
Event E is a subset of S.

The phrase *mutually exclusive* is used in probability studies when several events are under study. Two or more events are said to be mutually exclusive when the occurrence of any one of these events prevents the occurrence of the others. For example, occurrence of the event aA in Fig. 4-1 prevents the occurrence of aB, aC, aD, bA, cA, and dA. Thus aA and cA, say, are mutually exclusive.

4-3   PROBABILITY THEOREMS

In studying probability, it is often convenient to think of an event as consisting of several simpler events. In the preceding section we studied the problem of assembling bearings A, B, C, and D with shafts a, b, c, and d. The event aA of matching shaft a with bearing A is made up of the two events, getting shaft a from a, b, c, and d and getting bearing A from A, B, C, and D. Thus we can regard a, b, c, and d as the sample space for the first event, getting a shaft, and A, B, C, and D as the sample space for the second event, getting a bearing. For some problems the sample spaces may become so large as to be unmanageable, and so it is desirable to develop some theorems for handling these situations.

Now define $E_1 E_2$, called a *compound event*, such that $P(E_1 E_2)$ means the probability that *both E_1 and E_2* will occur. It is helpful to associate the word AND with the operation $P(E_1 E_2)$.

Also, let $E_1 + E_2$ be associated with the word OR such that $P(E_1 + E_2)$ means the probability that E_1 *or* E_2 *or both* will occur.

Returning to the shaft-and-bearing problem, suppose that a workman chooses a shaft and a bearing simultaneously. Let the event E_1 be, getting shaft a, and let the event E_2 be getting bearing A or bearing B. Then the problem can be mapped as shown in Fig. 4-3. By counting the weights we see that $P(E_1) = \frac{4}{16}$ and $P(E_2) = \frac{8}{16}$. However, there are only two points in $E_1 E_2$, and so $P(E_1 E_2) = \frac{2}{16}$. But $E_1 + E_2$ means either E_1 or E_2 or both, and we count 10 points; therefore $P(E_1 + E_2) = \frac{10}{16}$. This leads to the statement of the *addition theorem*, which is written

$$P(E_1 + E_2) = P(E_1) + P(E_2) - P(E_1 E_2) \qquad (4\text{-}15)$$

We observe that $P(E_1 E_2)$ must be subtracted because it is included once in E_1 and again in E_2.

If there are no points in $E_1 E_2$, then $P(E_1 E_2) = 0$, the simple events E_1 and E_2 are mutually exclusive, and Eq. (4-15) reduces to

$$P(E_1 + E_2) = P(E_1) + P(E_2) \qquad (4\text{-}16)$$

For three events, Eq. (4-15) becomes

$$P(E_1 + E_2 + E_3) = P(E_1) + P(E_2) + P(E_3)$$
$$- P(E_1 E_2) - P(E_1 E_3) - P(E_2 E_3) + P(E_1 E_2 E_3) \qquad (4\text{-}17)$$

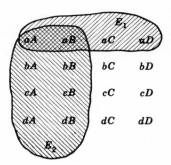

FIGURE 4-3

For more than three events the relation is rather lengthy. However, for i mutually exclusive events, the addition theorem is

$$P(E_1 + E_2 + E_3 + \cdots + E_i) = P(E_1) + P(E_2) + P(E_3) + \cdots + P(E_i) \qquad (4\text{-}18)$$

Next, define *conditional probability* $P(E_2 \,|\, E_1)$ to mean *the probability of* E_2, *given that* E_1 *has already occurred*. Now, referring to Fig. 4-3 again, we see that if E_1 has already occurred, then 12 points have been eliminated and the new sample space contains only the four points aA, aB, aC, and aD. E_2 now contains only the two points aA and aB. Consequently, $P(E_2 \,|\, E_1) = \frac{2}{4}$. We can also get this result using the original weights. Thus

$$P(E_2 \,|\, E_1) = \frac{\frac{2}{16}}{\frac{4}{16}} = \frac{2}{4}$$

or more generally,

$$P(E_2 \,|\, E_1) = \frac{P(E_1 E_2)}{P(E_1)} \qquad (4\text{-}19)$$

Equation (4-19) is called the *multiplication theorem;* it can also be written

$$P(E_1 E_2) = P(E_1)P(E_2 \,|\, E_1) \qquad (4\text{-}20)$$

For i events, Eq. (4-20) is

$$P(E_1 E_2 E_3 \cdots E_i) = P(E_1)P(E_2 \,|\, E_1)P(E_3 \,|\, E_1 E_2) \cdots P(E_i \,|\, E_1 E_2 E_3 \cdots E_{i-1}) \qquad (4\text{-}21)$$

It may be that the event E_1 has no effect on the probability that E_2 will or will not occur. In this case the events E_1 and E_2 are *independent*, and Eq. (4-21) becomes

$$P(E_1 E_2 E_3 \cdots E_i) = P(E_1)P(E_2)P(E_3) \cdots P(E_i) \qquad (4\text{-}22)$$

For example, the event E_1, choosing a shaft, is independent of the event E_2, choosing a bearing.

In applying probability to design, it often turns out that we are dealing with things that appear to be identical, but are not. These differences may not be evident until the parts are subjected to a close inspection. Thus two cold-drawn steel bars, nominally 1 in in diameter, when inspected might be 1.0005 and 1.0035 in in diameter, respectively. A quantity of connecting rods may contain some having a yield strength, say, of 84 kpsi and others having 92 kpsi. We also find that the successful performance of one element in a system sometimes depends upon the success of others. We could use particular machine elements in the examples and problems which follow, but our thinking will be clearer if abstract representation is employed. Thus we shall be using the letters of the alphabet, A, B, C, etc., to represent things such as bearings, bolts, shafts, rivets, and the like.

EXAMPLE 4-2   A shipment of 48 items, thoroughly mixed, consists of 6 A's, 12 B's, 18 C's, and 12 D's.

(a) If one is selected, what is the probability that it is either an A or a B? Neither a B nor a D?

(b) An item is selected, identified, then replaced, and another item selected. Find the probability that they are both B's; that the first is a B and the second an A; that neither of them are C's; that one is an A and the other a B; that the second is a D.

(c) Solve part b if the first item selected is not replaced.

SOLUTION   (a) Since only one is selected, the event A excludes B and the event B excludes A. Therefore

$$P(A + B) = P(A) + P(B) = \tfrac{6}{48} + \tfrac{12}{48} = \tfrac{3}{8} \qquad Ans.$$

Neither a B nor a D means NOT B AND NOT D, and is written $\tilde{B}\tilde{D}$. Since the set contains only A, B, C, and D, $\tilde{B}\tilde{D}$ means $A + C$. Therefore

$$P(\tilde{B}\tilde{D}) = P(A + C) = P(A) + P(C) = \tfrac{6}{48} + \tfrac{18}{48} = \tfrac{1}{2} \qquad Ans.$$

because A and C are mutually exclusive (there is only one selection). Another approach is to write

$$P(\tilde{B}\tilde{D}) = P(\tilde{B})P(\tilde{D} \,|\, \tilde{B}) \qquad (1)$$

Note that $\tilde{B} = A + C + D$ while $\tilde{D} = A + B + C$. Since there is only one selection, $\tilde{D}$ is *dependent* on $\tilde{B}$. Therefore, from Eq. (1),

$$P(\tilde{B}\tilde{D}) = \tfrac{36}{48}\left(\tfrac{36}{48} \cdot \tfrac{24}{36}\right) = \tfrac{3}{8} \qquad Ans.$$

(b) Here we designate the selections using the subscripts 1 and 2, respectively. Since the first selection is replaced, the selections are independent, and so

$$P(B_1 B_2) = P(B_1)P(B_2) = \tfrac{12}{48} \cdot \tfrac{12}{48} = \tfrac{1}{16} \qquad Ans.$$

Similarly,
$$P(B_1 A_2) = P(B_1)P(A_2) = \tfrac{12}{48} \cdot \tfrac{6}{48} = \tfrac{1}{32} \qquad Ans.$$

Next, since the successive events are independent,

$$P(\tilde{C}_1) = P(\tilde{C}_2) = 1 - P(C_1) = 1 - \tfrac{18}{48} = \tfrac{5}{8}$$

Then
$$P(\tilde{C}_1 \tilde{C}_2) = P(\tilde{C}_1)P(\tilde{C}_2) = \tfrac{5}{8} \cdot \tfrac{5}{8} = \tfrac{25}{64} \qquad Ans.$$

The probability that one is an A and the other is a B is written $P(A_1 B_2 + B_1 A_2)$, which means the first is an A AND the second is a B, OR the first is a B AND the second is an A. Expanding, we have

$$P(A_1 B_2 + B_1 A_2) = P(A_1)P(B_2) + P(B_1)P(A_2) = \tfrac{6}{48} \cdot \tfrac{12}{48} + \tfrac{12}{48} \cdot \tfrac{6}{48} = \tfrac{1}{16} \qquad Ans.$$

Note that $A_1 B_2$ *excludes* the possibility of $B_1 A_2$, and vice versa; that is, they are mutually exclusive.

Since the first selection is replaced, it cannot affect getting a D on the second selection. Therefore

$$P(D_2) = \tfrac{12}{48} = \tfrac{1}{4} \qquad Ans.$$

(c) Here the first selection is not replaced, and so the second event is *dependent* on the first. Therefore

$$P(B_1 B_2) = P(B_1)P(B_2 \mid B_1) = \tfrac{12}{48} \cdot \tfrac{11}{47} = \tfrac{11}{188} \qquad Ans.$$

Similarly,
$$P(B_1 A_2) = P(B_1)P(A_2 \mid B_1) = \tfrac{12}{48} \cdot \tfrac{6}{47} = \tfrac{3}{94} \qquad Ans.$$

The event $\tilde{C}_1 \tilde{C}_2$ follows the same approach. Thus

$$P(\tilde{C}_1 \tilde{C}_2) = P(\tilde{C}_1)P(\tilde{C}_2 \mid \tilde{C}_1) = \tfrac{30}{48} \cdot \tfrac{29}{47} = \tfrac{145}{376} \qquad Ans.$$

Getting an A AND a B means getting an A on the first selection AND a B on the second, OR getting a B on the first selection AND an A on the second. Therefore

$$P(A_1 B_2 + B_1 A_2) = P(A_1)P(B_2 \mid A_1) + P(B_1)P(A_2 \mid B_1)$$
$$= \tfrac{6}{48} \cdot \tfrac{12}{47} + \tfrac{12}{48} \cdot \tfrac{6}{47} = \tfrac{3}{47} \qquad Ans.$$

Note that $A_1 B_2$ and $B_1 A_2$ are mutually exclusive.

Finally, the probability that the second selection is a D is the probability of $D_1 D_2$ OR $\tilde{D}_1 D_2$. Thus

$$P(D_2) = P(D_1 D_2) + P(\tilde{D}_1 D_2) = P(D_1)P(D_2 \mid D_1) + P(\tilde{D}_1)P(D_2 \mid \tilde{D}_1)$$
$$= \tfrac{12}{48} \cdot \tfrac{11}{47} + \tfrac{36}{48} \cdot \tfrac{12}{47} = \tfrac{1}{4} \qquad Ans.$$

Note that the probability of D on the second selection is the same as the probability of D on the first, even though one item has been withdrawn. You can understand this by noting that if you remove all the items except one, without identifying them, the last has the same probability of being a D as the first would have. ////

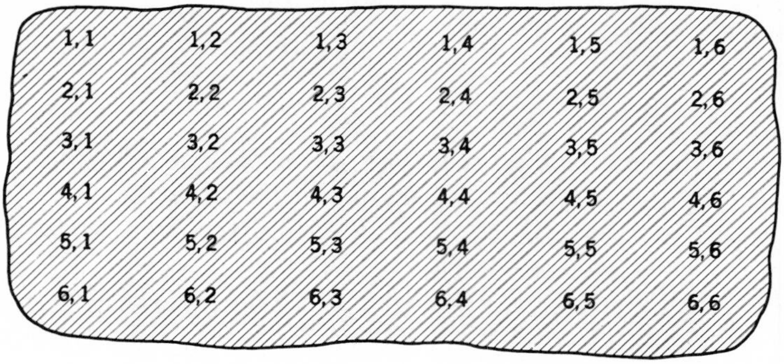

1,1
1,2
1,3
1,4
1,5
1,6
2,1
2,2
2,3
2,4
2,5
2,6
3,1
3,2
3,3
3,4
3,5
3,6
4,1
4,2
4,3
4,4
4,5
4,6
5,1
5,2
5,3
5,4
5,5
5,6
6,1
6,2
6,3
6,4
6,5
6,6

FIGURE 4-4

4-4 RANDOM VARIABLES

Consider a collection of 20 tensile-test specimens that have been machined from a like number of samples selected at random from a carload shipment of, say, UNS G10200 cold-drawn steel. It is reasonable to expect that there will be differences in the ultimate tensile strengths S of each of these test specimens. Such differences may occur because of differences in the sizes of the specimens, in the strength of the material itself, or both. Such an experiment is called a *random experiment*, because the specimens were selected at random. The strength S determined by this experiment is called a *random*, or a *stochastic, variable*. So a random variable is a variable quantity, such as strength, size, or weight, for example, whose value depends upon the outcome of a random experiment.

To extend the concept of a random variable, suppose we toss two dice and define a random variable x as the sum of the numbers which appear. Either die can display any number from 1 to 6, and so there are 36 points in the sample

p = f(x)
6/36
5/36
4/36
3/36
2/36
1/36
0
0 1 2 3 4 5 6 7 8 9 10 11 12 13
x

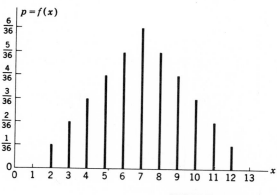

FIGURE 4-5
Frequency distribution.

space. Let us designate these points by the notation a, b, where a is the number on the first die and b the number on the second. Then Fig. 4-4 is the sample space for all possible outcomes. This figure shows that the random variable x has a specific value for each possible outcome. For the event 5, 4, $x = 5 + 4 = 9$. Values of x for each of the remaining events is computed in the same manner. It is useful to form a table showing the values of x and the corresponding values of the probability of x, called $p = f(x)$. This is easily done from Fig. 4-4 since there are 36 points, each outcome having a weight $w = \frac{1}{36}$. The results are shown in Table 4-1. Any table such as this, listing all possible values of a random variable, together with the corresponding probabilities, is called a *probability distribution*.

Note in Fig. 4-5 that the probabilities can be plotted in graphical form. Here it is clear that the probability is a function of x. This *probability function* $p = f(x)$ is often called the *frequency function*, or sometimes, the *probability density*.

Sometimes we are not so much interested in the probability of a particular value of x as we are in the probability that x is less than some particular value. For example, if the random variable is the strength of a machine element, we would be vitally interested in knowing the probability of the strength being less than some particular value, say, the design value. The probability that x is less than or equal to a certain value x_i can be obtained from the probability function $p = f(x)$ simply by summing the probabilities of all x's up to and including x_i. If we do this with Table 4-1, letting x_i equal 2, then 3, and so on, up to 12, we get Table 4-2, which is called a *cumulative probability distribution*. The function $F(x)$, in Table 4-2, is called a *cumulative probability function*. In terms of $f(x)$ it may be expressed mathematically in the general form

$$F(x_i) = \sum_{x_j \le x_i} f(x_j) \qquad (4\text{-}23)$$

where $F(x_i)$ is properly called the *distribution function*. The cumulative distribution may also be plotted as a graph (Fig. 4-6).

In the example of the two dice, the variable x is called a discrete random variable, because the variable x can assume only discrete values. In many cases,

Table 4-1 A PROBABILITY DISTRIBUTION

x	2	3	4	5	6.	7	8	9	10	11	12
$f(x)$	$\frac{1}{36}$	$\frac{2}{36}$	$\frac{3}{36}$	$\frac{4}{36}$	$\frac{5}{36}$	$\frac{6}{36}$	$\frac{5}{36}$	$\frac{4}{36}$	$\frac{3}{36}$	$\frac{2}{36}$	$\frac{1}{36}$

Table 4-2 A CUMULATIVE PROBABILITY DISTRIBUTION

x	2	3	4	5	6	7	8	9	10	11	12
$F(x)$	$\frac{1}{36}$	$\frac{3}{36}$	$\frac{6}{36}$	$\frac{10}{36}$	$\frac{15}{36}$	$\frac{21}{36}$	$\frac{26}{36}$	$\frac{30}{36}$	$\frac{33}{36}$	$\frac{35}{36}$	$\frac{36}{36}$

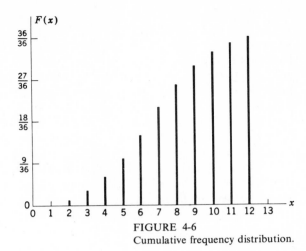

FIGURE 4-6
Cumulative frequency distribution.

however, we must employ the idea of a *continuous random variable*. A continuous random variable is one that can take on any value in a specified interval. In such a case the graphs corresponding to Figs. 4-5 and 4-6 would be constructed as continuous curves.

4-5 SAMPLE AND POPULATION

In studying the variations in the mechanical properties and characteristics of mechanical elements, we shall generally be dealing with a finite number of elements. The total number of elements, called the *population*, may in some cases be quite large. In such cases it is usually impractical to measure the characteristics of each member of the population, because this involves destructive testing in some cases, and so we select a small part of the group, called a *sample*, for these determinations. Thus the *population* is the entire group, and the *sample* is a part of the population.

The arithmetic mean of a sample, called the *sample mean*, consisting of N elements, is defined by the equation

$$\bar{x} = \frac{x_1 + x_2 + x_3 + \cdots + x_N}{N} = \frac{1}{N} \sum_1^N x_j \qquad (4\text{-}24)$$

In a similar manner a population consisting of N elements has a *population mean* defined by the equation

$$\mu = \frac{x_1 + x_2 + x_3 + \cdots + x_N}{N} = \frac{1}{N} \sum_1^N x_j \qquad (4\text{-}25)$$

The *mode* and the *median* are also used as measures of central value. The *mode* is the value that occurs most frequently. The *median* is the middle value if there are an odd number of cases; it is the mean of the two middle values if there are an even number.

Suppose now that we conduct an experiment to determine the tensile strength of, say, class 40 gray cast iron.* So we cast a tensile-test specimen from each of 61 ladles to get our sample. Our first problem is to learn how to classify the data that result from the testing procedure.

Since 42.5 is just as far from 42 as it is from 43, how should we round it to avoid cumulative rounding errors? The rule is, *round to the even integer preceding the 5*. Thus 42.5 is rounded off to 42, and 43.5 is rounded off to 44.

Our next problem is that of dividing the data into intervals. If the tensile strengths of the 61 specimens tested should happen to lie in the range of 30 to 60 kpsi, we might be tempted to count the number of specimens that failed in the interval of 30–35 kpsi, 35–40 kpsi, etc. But if a specimen failed at 35 kpsi, there would be a question as to how it should be assigned. We therefore define *class interval*, without ambiguity, as 30–34 kpsi, 35–39 kpsi, etc. The values 30 and 34 kpsi are called the *class limits*. But the *class boundaries* are actually 29.5, 34.5, 39.5, etc. Notice that there are no ambiguities here because of our rules for rounding. Thus 29.5 would be rounded to 30, and 34.5 would be rounded to 34.

The difference between the upper and lower boundaries is called the *class width;* in this example the width is 5 kpsi.

Using these ideas, Table 4-3 shows the classified data taken from an actual experiment. The *frequency* is the number of observations in each class interval. The *relative frequency* is the frequency divided by the number of specimens.

These data can be displayed graphically to illustrate the distribution as in Fig. 4-7. This chart is called a *histogram*. Notice that the vertical sides of the rectangles are the class boundaries and their heights correspond to the frequency. The *class mark*, or *class midpoint*, is at the center of the class interval. The *frequency polygon* is obtained by connecting these points with straight lines.

Once the data have been classified, a shorter method of computing the mean value can be used. Since we are dealing with strengths, let the midpoint of the class

---

* See Table A-20. ASTM No. 40 cast iron has a minimum tensile strength of 40 kpsi.

Table 4-3

Class intervals, kpsi	Frequency	Relative frequency, %
30–34	1	1.6
35–39	7	11.5
40–44	27	44.3
45–49	21	34.4
50–54	5	8.2
Total	61	100

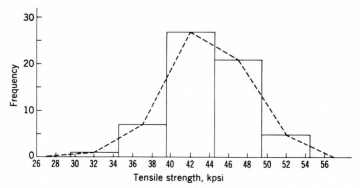

FIGURE 4-7
Histogram of 61 tensile tests on No. 40 gray cast iron.

intervals be designated $S_1 = 32$ kpsi, $S_2 = 37$ kpsi, etc. Then using f_1, f_2, etc., as the frequencies, the mean is

$$\bar{S} = \frac{\sum\limits_1^k f_i S_i}{\sum f_i} \qquad (4\text{-}26)$$

where k is the number of intervals. This equation is then solved in tabular form, as shown in Table 4-4, giving the sample mean $\bar{S} = 43.8$ kpsi.

Besides the average, it is useful to have another kind of measure which will tell us something about the spread, or dispersion, of the distribution. For any random variable x the deviation of the ith observation from the mean is $x_i - \bar{x}$. But since the sum of these is always zero, we square them, and define *sample variance* as

$$s^2 = \frac{(x_1 - \bar{x})^2 + (x_2 - \bar{x})^2 + \cdots + (x_N - \bar{x})^2}{N - 1} = \frac{\sum\limits_1^N (x_j - \bar{x})^2}{N - 1} \qquad (4\text{-}27)$$

Table 4-4 COMPUTATION OF THE SAMPLE MEAN FOR 61 TENSILE TESTS OF NO. 40 GRAY CAST IRON

Midpoint of interval, S_i	Frequency, f_i	$f_i S_i$
$S_1 = 32$	1	32
$S_2 = 37$	7	259
$S_3 = 42$	27	1 134
$S_4 = 47$	21	987
$S_5 = 52$	5	260
Total	61	2 672

$$\bar{S} = \frac{\sum\limits_1^k f_i S_i}{\sum f_i} = \frac{2\,672}{61} = 43.8 \text{ kpsi}$$

The *sample standard deviation*, defined as the square root of the variance, is

$$s = \sqrt{\frac{\sum\limits_{1}^{N}(x_j - \bar{x})^2}{N-1}} \qquad (4\text{-}28)$$

It should be observed that some authors define the variance and the standard deviation using N instead of $N-1$ in the denominator. For large values of N there is very little difference. For small values, the denominator $N-1$ actually gives a better estimate of the variance of the population from which the sample is taken.

Sometimes we shall be dealing with the standard deviation of the strength of an element. So you must be careful not to be confused by the notation. Note that we are using the *capital letter S* for *strength* and the *lower-case letter s* for *standard deviation*.

A more convenient form of Eq. (4-27) for grouped data is

$$s^2 = \frac{\sum f_i x_i^2 - \dfrac{(\sum f_i x_i)^2}{N}}{N-1} \qquad (4\text{-}29)$$

where the summation is carried out for all the intervals.*

The use of Eq. (4-29) is illustrated in Table 4-5 for the experiment on class 40 gray cast iron.

* For a development of this relation, see Wilfred J. Dixon and Frank J. Massey, Jr., "Introduction to Statistical Analysis," 2d ed., McGraw-Hill Book Company, New York, 1957.

Table 4-5 COMPUTATION OF VARIANCE AND STANDARD DEVIATION FOR 61 TENSILE TESTS OF NO. 40 GRAY CAST IRON

S_i, kpsi	f_i	$f_i S_i$	$f_i S_i^2$
32	1	32	1 024
37	7	259	9 583
42	27	1 134	47 628
47	21	987	46 389
52	5	260	13 520
Total	61	2 672	118 144

$$\bar{S} = \frac{\sum f_i S_i}{N} = \frac{2672}{61} = 43.8 \text{ kpsi}$$

$$s_S^2 = \frac{\sum f_i S_i^2 - \dfrac{(\sum f_i S_i)^2}{N}}{N-1} = \frac{118\,144 - (2672)^2/61}{60}$$

$$= 18.37 \text{ kpsi}^2$$

$$s_S = (18.37)^{1/2} = 4.29 \text{ kpsi}$$

In engineering an entire population is rarely available for experimentation since this often involves destructive testing. However, it is convenient to have a definition of *population variance* and *population standard deviation*. These are defined as follows:

$$\hat{\sigma}^2 = \frac{\sum\limits_{1}^{N} (x_j - \mu)^2}{N} \qquad (4\text{-}30)$$

$$\hat{\sigma} = \sqrt{\frac{\sum\limits_{1}^{N} (x_j - \mu)^2}{N}} \qquad (4\text{-}31)$$

The practice of using the Greek letter σ (sigma) for standard deviation is universal in mathematical statistics. But it is also widely used to represent normal stress, as we have seen in Chap. 2, and so we are faced with a serious problem in notation. Substitution of another symbol for stress is not satisfactory, and neither would such a symbol be satisfactory for standard deviation. To resolve this difficulty we shall continue to use the Greek letter σ for normal stress. To represent standard deviation we shall use the same Greek letter but with a caret over it, like this $\hat{\sigma}$. With this notation the standard deviation of the normal stress would be written $\hat{\sigma}_\sigma$.

4-6 THE NORMAL DISTRIBUTION

One of the most important of the many distributions which occur in the study of statistics is the *normal*, or *gaussian*, *distribution*. But authors differ on the reasons for using the word "normal." Dixon and Massey* state that it is called normal because of the particular manner in which the area below the distribution curve is arranged, and *not* as the opposite of abnormal. But Wadsworth and Bryant† state that the word normal is used because of historical attempts to establish this distribution as a basic law governing all continuous random variables.

The equation of the normal curve is

$$f(x) = \frac{1}{\hat{\sigma}\sqrt{2\pi}} e^{-(x-\mu)^2/2\hat{\sigma}^2} \qquad (4\text{-}32)$$

where $f(x)$ is the frequency function.‡ The total area under this curve, from $x = -\infty$ to $x = +\infty$, is one square unit. Therefore the area between any two points, say, from $x = a$ to $x = b$, is the proportion of cases which lie between the

---

* Dixon and Massey, *op. cit.*, p. 48.
† George P. Wadsworth and Joseph G. Bryan, "Probability and Random Variables," p. 104, McGraw-Hill Book Company, New York, 1960.
‡ For a derivation of this function see M. F. Spotts, "Mechanical Design Analysis," pp. 265–272, Prentice-Hall, Inc., Englewood Cliffs, N.J., 1964.

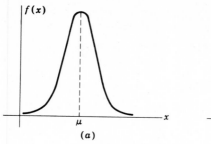
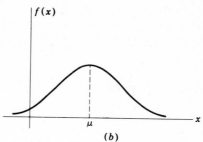

FIGURE 4-8
The shape of the normal distribution curve; (a) small $\hat{\sigma}$; (b) large $\hat{\sigma}$.

two points. A plot of Eq. (4-32) yields a bell-shaped curve having a maximum ordinate at $x = \mu$ and decreasing to zero at both plus and minus infinity. If $\hat{\sigma}$ is small, the curve is tall and thin; if $\hat{\sigma}$ is large, the curve is short and broad; see Fig. 4-8.

Equation (4-32) is difficult to solve for the ordinates and just as difficult to integrate for the areas, and so it is convenient for the engineer to have access to the results in tabular form. For this purpose we first express the variable x in so-called *standard units* as

$$z = \frac{x - \mu}{\hat{\sigma}} \qquad (4\text{-}33)$$

where now z is to be normally distributed with unit variance and a mean of zero. Equation (4-32) then becomes a *unit normal distribution* and is written

$$f(z) = \frac{1}{\sqrt{2\pi}} e^{-z^2/2} \qquad (4\text{-}34)$$

Values of $f(z)$ and its integral are tabulated in Tables A-13 and A-14. A few examples will serve to illustrate the use of these tables.

EXAMPLE 4-3　In a shipment of 250 connecting rods, the mean tensile strength is found to be 45 kpsi and the standard deviation 5 kpsi. Assume the strengths have a normal distribution.

　　(a) How many have a strength less than 40 kpsi?
　　(b) How many have a strength between 40 and 60 kpsi?

SOLUTION　(a) Since a specimen having a strength of 39.5 kpsi would have its strength rounded off to 40, we actually wish to find the number of rods having strengths less than 39.5 kpsi. In standard units, using Eq. (4-33), this becomes

$$z_{39.5} = \frac{S - \mu}{\hat{\sigma}} = \frac{39.5 - 45}{5} = -1.1$$

Now the area to the left of $z = -1.1$ is the same as the area to the right of $z = +1.1$ because the normal curve is symmetrical about the mean. Therefore, from Table A-14,

$$A = 0.5000 - 0.3643 = 0.1357$$

Thus 13.57 percent have a strength less than 40 kpsi, and so

$$N = (250)(0.1357) = 33.925$$

and rounding off, we find that 34 connecting rods are likely to have strengths less than 40 kpsi.

(b) Since 59.5 kpsi would be rounded off to 60, the corresponding standardized variable is

$$z_{59.5} = \frac{S - \mu}{\hat{\sigma}} = \frac{59.5 - 45}{5} = +2.9$$

From Table A-14 we find the area between $z = 0$ and $z = 2.9$ to be $A_{2.9} = 0.4981$. And from part a the area from $z = 0$ to $z = -1.1$ is $A_{1.1} = 0.3643$. Therefore

$$A = 0.4981 + 0.3643 = 0.8624$$

which is the proportion having strengths between 40 and 60 kpsi. So the number in this range is

$$N = (250)(0.8624) = 215.600$$

or rounding off, 216 connecting rods may be expected to be in the range of 40 to 60 kpsi. Since $216 + 34 = 250$, we note that all the rods are likely to have strengths less than 60 kpsi. ////

EXAMPLE 4-4 A group of shafts are to be machined to 1.001 in diameter with a tolerance of ± 0.001 in. If a sample of 200 shafts is taken and found to have a mean of 1.001 in, what must be the standard deviation in order to assure that 96 percent of the shafts are within acceptable dimensions? Assume the diameters are normally distributed.

SOLUTION   Since the shafts are distributed according to the normal curve, we need only assure that 2 percent of them do not exceed 1.0025 in, because this value rounds off to 1.002 in. The area under the positive half of the normal curve must correspond to 48 percent. By interpolation, we find from Table A-14, $z = 2.054$. From Eq. (4-33)

$$x - \mu = z\hat{\sigma}$$

or

$$1.0025 - 1.001 = 2.054\hat{\sigma}$$

and hence

$$\hat{\sigma} = 0.0007 \text{ in}$$

to four decimal places. ////

4-7   SAMPLING DISTRIBUTIONS

Suppose you have just received the first shipment of 10 000 bolts for an important fastener application and it is your task to accept or reject this shipment, depending upon their conformity to specifications. Perhaps these specifications relate to strength, and so you have decided to sample the population of 10 000 bolts and conduct hardness tests on this sample. How many specimens must you include in the sample, say, to be 95 percent certain that the bolts conform? Here and in the next several sections, this and other questions relating to sampling will be explored.

For purposes of investigating the sampling of a population let us devise a manageable situation consisting of a population of only six helical springs. For this population the spring rates in pounds-force per inch are

$$k_1 = 3 \qquad k_4 = 6$$
$$k_2 = 5 \qquad k_5 = 7$$
$$k_3 = 6 \qquad k_6 = 7$$

Before we begin sampling, let us compute the population mean μ and the population variance $\hat{\sigma}^2$. The calculations are shown in Table 4-6. Note that the variance is computed using N instead of $N - 1$ in the denominator; this follows from Eq. (4-30). The frequency distribution of this population, shown in Fig. 4-9, clearly does not resemble a normal distribution.

With this out of the way, let us learn something about sampling. First, we choose a sample size of two springs, and then obtain all possible samples from this population. For each sample we shall compute the sample mean $\bar{k}$ and the sample variance s^2. We shall permit a single spring to be chosen both times in a sample, noting that the chance of duplication in sampling a large population is quite small. With such a small population it is possible to diagram the entire sample space,

Table 4-6   POPULATION STATISTICS FOR SIX HELICAL SPRINGS

k	f	fk	fk^2
3	1	3	9
5	1	5	25
6	2	12	72
7	2	14	98
Total	6	34	204

$$\mu = \tfrac{34}{6} = 5\tfrac{2}{3} \text{ lb/in}$$
$$\hat{\sigma}^2 = \frac{\sum fk^2 - [(\sum fk)^2/N]}{N} = \frac{204 - [(34)^2/6]}{6} = \frac{17}{9}$$
$$\hat{\sigma} = (\tfrac{17}{9})^{1/2} = 1.374 \text{ lb/in}$$

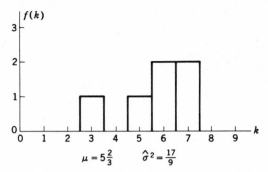

FIGURE 4-9
Frequency distribution of a population of
six springs.

$$\mu = 5\tfrac{2}{3} \qquad \hat{\sigma}^2 = \tfrac{17}{9}$$

and this is shown in Fig. 4-10. In this figure, the spring rates are the numbers separated by commas, the sample means are enclosed in parentheses, and the sample variances are enclosed in brackets.

Because we have used *all* possible samples of size 2, the results displayed in Fig. 4-10 are, in fact, two new populations which have been derived from the original population of six springs. Each of these populations, of course, has a frequency distribution, a mean, and a variance. Table 4-7 shows the distribution of the *sample means* $\bar{k}$, together with the *population mean of the sample means* $\mu_{\bar{k}}$ and the *population variance of the sample means* $\hat{\sigma}_{\bar{k}}^2$. Similarly, Table 4-8 is the *distribution of the sample variances* s_k^2 and the corresponding population mean of these variances. In this case there appears to be no particular value in computing the population variance of the sample variances.

Before drawing any conclusions from these results, which are now complete, it is worth emphasizing that Eq. (4-29) has been used in computing the sample

Table 4-7 POPULATION STATISTICS FOR THE SAMPLING DISTRIBUTION OF THE MEANS

$\bar{k}$	f	$f\bar{k}$	$f\bar{k}^2$
3	1	3	9
4	2	8	32
4.5	4	18	81
5	5	25	125
5.5	4	22	121
6	8	48	288
6.5	8	52	338
7	4	28	196
Total	36	204	1190

$$\mu_{\bar{k}} = \tfrac{204}{36} = 5\tfrac{2}{3} \text{ lb/in}$$

$$\hat{\sigma}_{\bar{k}}^2 = \frac{\sum f\bar{k}^2 - [(\sum f\bar{k})^2/N]}{N} = \frac{1190 - [(204)^2/36]}{36} = \frac{17}{18}$$

	k_1	k_2	k_3	k_4	k_5	k_6
k_1	3, 3 (3) [0]	3, 5 (4) [2]	3, 6 (4.5) [4.5]	3, 6 (4.5) [4.5]	3, 7 (5) [8]	3, 7 (5) [8]
k_2	5, 3 (4) [2]	5, 5 (5) [0]	5, 6 (5.5) [0.5]	5, 6 (5.5) [0.5]	5, 7 (6) [2]	5, 7 (6) [2]
k_3	6, 3 (4.5) [4.5]	6, 5 (5.5) [0.5]	6, 6 (6) [0]	6, 6 (6) [0]	6, 7 (6.5) [0.5]	6, 7 (6.5) [0.5]
k_4	6, 3 (4.5) [4.5]	6, 5 (5.5) [0.5]	6, 6 (6) [0]	6, 6 (6) [0]	6, 7 (6.5) [0.5]	6, 7 (6.5) [0.5]
k_5	7, 3 (5) [8]	7, 5 (6) [2]	7, 6 (6.5) [0.5]	7, 6 (6.5) [0.5]	7, 7 (7) [0]	7, 7 (7) [0]
k_6	7, 3 (5) [8]	7, 5 (6) [2]	7, 6 (6.5) [0.5]	7, 6 (6.5) [0.5]	7, 7 (7) [0]	7, 7 (7) [0]

FIGURE 4-10
Sample space for all samples of size 2 from a population of six springs.

Table 4-8 POPULATION STATISTICS FOR THE SAMPLING DISTRIBUTION OF THE VARIANCES

s_k^2	f	$f s_k^2$
0	10	0
0.5	12	6
2	6	12
4.5	4	18
8	4	32
Total	36	68

Population mean of $s_k^2 = \frac{68}{36} = \frac{17}{9}$

variance s^2 and the population variance $\hat{\sigma}^2$. In particular, use $N-1$ in the denominator when s^2 is desired and N when $\hat{\sigma}^2$ is desired. Of course, in computing s^2,
N is the size of the sample, and in computing $\hat{\sigma}^2$, N is the size of the population.

One of the conclusions to be observed is that the population mean μ of the
six springs is the same as the population mean $\mu_{\bar{x}}$ of the sampling distribution of
the means. Also note, from Tables 4-6 and 4-7, that $\hat{\sigma}^2 = \frac{17}{9}$, whereas $\hat{\sigma}_{\bar{x}}^2 = \frac{17}{18}$. In
other words, $\hat{\sigma}^2/\hat{\sigma}_{\bar{x}}^2 = 2$, the sample size N. This is a general and useful relation,
and it may be written

$$\hat{\sigma}_{\bar{x}}^2 = \frac{\hat{\sigma}^2}{N} \qquad (4\text{-}35)$$

where N is the sample size used to compute the sample mean $\bar{x}$. The meaning of
this, with $\hat{\sigma}_{\bar{x}}^2 = \hat{\sigma}^2/2$, is that a sample size of 2 has a better chance of being near the
population mean than a single selection.

Finally, we note from Tables 4-6 and 4-8 that the mean of the distribution of
$s_{\bar{x}}^2$ is the same as the variance $\hat{\sigma}^2$ of the population of springs. Thus, by using $N-1$
as the divisor for the sample variance and N for the population variance, we can
say that the two "usually," or "on the average," are equal.

In the example we have just concluded, the assumption was made that a
single spring could be duplicated in a sample. If the samples are not replaced, the
results noted above must be modified. The general relations for sampling without
replacement are

$$\hat{\sigma}_{\bar{x}}^2 = \frac{\hat{\sigma}^2}{N}\left(\frac{N_p - N}{N_p - 1}\right) \qquad (4\text{-}36)$$

$$\mu_s^2 = \frac{\hat{\sigma}^2 N}{N_p - 1} \qquad (4\text{-}37)$$

where N and N_p are the sample and population sizes, respectively.

Dixon and Massey* show that the sampling distribution of the means of a
population will be approximately normal if the sample size is large and if the
population has a finite variance. But in engineering we seldom encounter a population that does not have a finite variance. On the other hand, it may be expensive
to obtain a large sample size, and so this restriction can cause some trouble if the
population distribution is unknown. However, if the population is normal,
the sampling distribution of the mean will also be normal. And when we are
dealing with approximations, it is a fact that the sampling distribution of the mean
will be more normal than the distribution of the population. This you can discover
for yourself by plotting the frequency distribution of the sample means in
Table 4-7 and comparing the result with the frequency distribution of the population in Fig. 4-9. Do this.

* *Op. cit.*, p. 50.

4-8 POPULATION COMBINATIONS

Cases often arise in design in which we must make a statistical analysis of the result of combining two or more populations in some specified manner. For example, we may wish to assemble a population of cranks into a population of machines. The cranks all have a population of strengths S_i, and the machines subject them to a population of stresses σ_j. For safety, the strengths must be greater than the stresses, but since we are dealing with distributions, it may happen that some of the cranks will be subjected to stresses greater than their strengths. Thus we are interested in determining the reliability of these assemblies.

Another problem involving several populations occurs in the dimensioning of mating parts. We may wish, say, to specify a range of shaft diameters and another range of bearing sizes such that the shafts will assemble into the bearings with a specified clearance range. Thus we have a population of shaft diameters and a population of bearing diameters which result in a population of clearances. Our problem might be to specify the shaft and bearing sizes such that 99 percent of the assemblies have satisfactory clearances.

One of the very interesting problems arising in design is called "tolerance stacking." Suppose eight parts having mean lengths $\bar{x}_1, \bar{x}_2, \ldots, \bar{x}_8$ and tolerances of $\pm x_1, \pm x_2, \ldots, \pm x_8$ are to be assembled end to end into a space whose mean length and tolerances are $\bar{y}$ and $\pm y$, respectively. And suppose, furthermore, that we require that these parts fit with a clearance $\bar{c}$ within the limits $\pm c$. What is the probability of getting a satisfactory assembly when the elements are selected at random from each of the several populations?

The means of two or more populations may be either added or subtracted to obtain a resultant mean. Thus, for addition,

$$\mu = \mu_1 + \mu_2 \qquad (4\text{-}38)$$

and for subtraction,

$$\mu = \mu_1 - \mu_2 \qquad (4\text{-}39)$$

where in each case μ is the mean of the resulting distribution.

The standard deviations follow the Pythagorean theorem. Thus the standard deviation for both addition and subtraction is

$$\hat{\sigma} = \sqrt{\hat{\sigma}_1^2 + \hat{\sigma}_2^2} \qquad (4\text{-}40)$$

4-9 DIMENSIONING—DEFINITIONS AND STANDARDS

Though many decisions relating to the dimensioning of parts and assemblies can be made quite satisfactorily by the design draftsman, the engineer must be familiar with standard practices in dimensioning in order to retain control of the design for which he is directly responsible. Furthermore, the tolerancing of mating parts and assemblies has a significant effect on the cost and reliability of the finished product, and hence some of these decisions cannot be left to the judgment of technicians.

The following terms are used generally in dimensioning:

Nominal size The size we use in speaking of an element. For example, we may specify a $\frac{1}{2}$-in bolt or a $1\frac{1}{2}$-in pipe. Either the theoretical size or a measured size may be quite different. The bolt, say, may actually measure 0.492 in. And the theoretical size of a $1\frac{1}{2}$-in standard-weight pipe is 1.61 in for the outside diameter.

Basic size The exact theoretical size. Limiting variations in either the plus or minus direction begin from the basic dimension.

Limits The stated maximum and minimum permissible dimensions.

Tolerance The difference between the two limits.

Bilateral tolerance The variation in both directions from the basic dimension. That is, the basic size is between the two limits; for example, 1.005 ± 0.002 in.

Unilateral tolerance The basic dimension is taken as one of the limits, and variation is permitted in only one direction; for example,

$$1.005 \begin{array}{l} + 0.004 \\ - 0.000 \end{array} \text{in}$$

Natural tolerance A tolerance equal to plus and minus three standard deviations from the mean. For a normal distribution this assures that 99.73 percent of production is within the tolerance limits.

Clearance A general term which refers to the mating of cylindrical parts such as a bolt and a hole, or a journal and a bearing. The word clearance is used only when the internal member is smaller than the external member. The *diametral clearance* is the measured difference in the two diameters. The *radial clearance* is the difference in the two radii.

Interference The opposite of clearance for the mating of cylindrical parts in which the internal member is larger than the external member.

Allowance The minimum stated clearance or the maximum stated interference for mating parts.

The USASI (United States of America Standards Institute) has adopted a useful set of standards for the mating of cylindrical parts. A portion of these standards has been reproduced as Table A-15 for use in design. This table gives the shaft and hole tolerances for both clearance and interference fits ranging from very loose running fits to heavy force and shrink interference fits. These fits are tabulated for the *basic hole system* in which the lower limit of the hole dimension is made equal to the theoretical dimension and all other limits are taken from this starting point. If the holes are finished by reaming, the basic dimension should be the size of a standard reamer. If the holes are finished by grinding, any suitable dimension may be used as the basic size.

There is, of course, no law which requires the engineer to employ these standards. They do, however, furnish a good starting point for the development of other tolerances.

4-10 STATISTICAL ANALYSIS OF TOLERANCING

Variations in dimensions of parts in a production process may occur purely by chance as well as for specific reasons. For example, the operating temperature of a machine tool changes during start-up, and this may have an effect on the dimensions of the parts produced during the first 30 min of running time. When all such causes of variations in part dimensions have been eliminated, the production process is said to be "in statistical control." Under these conditions all variations in dimensions occur at random.

We will assume that production is in statistical control for our investigations in this section and that the variations in part dimensions have a normal distribution.

Problems involving the tolerancing of mating cylindrical parts occur frequently. If the two parts must mate with a clearance fit, the problem is to discover the percentage of rejected assemblies due either to a fit that is too tight or to one that is too loose. A similar problem occurs when the parts must mate with an interference fit. Both of these problems can be treated using the statistical approaches already developed. An example will serve to illustrate the notation and approach.

EXAMPLE 4-5   A shaft and hole are to mate with a class RC8 loose running fit. The nominal diameter is 1 in. What percentage of assemblies are likely to be rejected because the clearance is less than 0.006 in? What percentage is likely to be rejected for having clearances greater than 0.009 in? Assume natural tolerances.

SOLUTION   Using Table A-15, we find the hole and shaft dimensions shown in Fig. 4-11. Using c for clearance and the subscripts H and S for hole and shaft, respectively, we find the maximum and minimum clearances to be

$$c_{max} = d_{H,\,max} - d_{S,\,min} = 1.0035 - 0.9935 = 0.0100 \text{ in}$$

$$c_{min} = d_{H,\,min} - d_{S,\,max} = 1.0000 - 0.9955 = 0.0045 \text{ in}$$

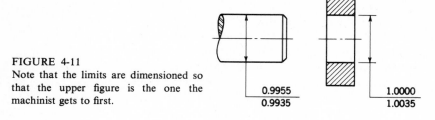

FIGURE 4-11
Note that the limits are dimensioned so that the upper figure is the one the machinist gets to first.

0.9955
0.9935

1.0000
1.0035

Therefore the mean clearance is

$$\mu_c = \frac{c_{max} + c_{min}}{2} = \frac{0.0100 + 0.0045}{2} = 0.007\,25 \text{ in}$$

The means of the hole and shaft diameters are easily found. They are

$$\bar{d}_H = 1.001\,75 \text{ in} \qquad \bar{d}_S = 0.9945 \text{ in}$$

Subtracting these from the maximums and minimums shown in Fig. 4-11 gives the tolerances, respectively, as

$$t_H = \pm 0.001\,75 \text{ in} \qquad t_S = \pm 0.0010 \text{ in}$$

Since these are assumed to be natural tolerances, the standard deviations are

$$\hat{\sigma}_H = \frac{0.001\,75}{3} = 0.000\,55 \text{ in} \qquad \hat{\sigma}_S = \frac{0.0010}{3} = 0.000\,33 \text{ in}$$

Then, using Eq. (4-40), the standard deviation of the clearance is

$$\hat{\sigma}_c = \sqrt{\hat{\sigma}_H^2 + \hat{\sigma}_S^2} = \sqrt{(0.000\,55)^2 + (0.000\,33)^2} = 0.000\,644 \text{ in}$$

The standardized variable is to correspond to a clearance not less than 0.006 in. Therefore, from Eq. (4-33), we have

$$z = \frac{c - \mu_c}{\hat{\sigma}_c} = \frac{0.006 - 0.007\,25}{0.000\,644} = -1.945$$

From Table A-14 find $A_z = 0.4741$. Thus the proportion of clearances less than 0.006 in is

$$P(c < 0.006) = 0.5000 - 0.4741 = 0.0259$$

or 2.59 percent.

For clearances greater than 0.009 in we have

$$z = \frac{c - \mu_c}{\hat{\sigma}_c} = \frac{0.009 - 0.007\,25}{0.000\,644} = 2.717$$

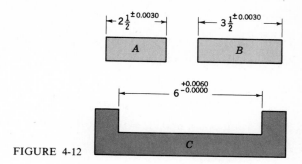

FIGURE 4-12

Then $A_z = 0.4967$, and hence

$$P(c > 0.009) = 0.5000 - 0.4967 = 0.0033$$

or 0.33 percent.   ////

Another problem which arises in dimensioning is illustrated in Fig. 4-12. Here parts A and B are to be placed end to end and then assembled into C. One of the questions that might be asked is: What percentage of the assemblies are likely to be rejected because of interference? The solution consists in summing the population of A's to the population of B's to obtain a resultant population of X's. Then the X's and the C's may be treated in exactly the same manner as the shaft-and-hole problem in Example 4-5. For the assembly $X = A + B$ we have, from Eqs. (4-39) and (4-40),

$$\mu_X = \mu_A + \mu_B \qquad (4\text{-}41)$$

$$\hat{\sigma}_X = \sqrt{\hat{\sigma}_A^2 + \hat{\sigma}_B^2} \qquad (4\text{-}42)$$

Since A and B could in themselves be subassemblies, the process described by these equations can be extended indefinitely.

Next, we combine Eqs. (4-39) and (4-40) to predict interference between X and C. The result is

$$z = \frac{\mu_C - \mu_X}{\sqrt{\hat{\sigma}_C^2 + \hat{\sigma}_X^2}} \qquad (4\text{-}43)$$

EXAMPLE 4-6   Suppose the standard deviations for the assembly of Fig. 4-12 are $\hat{\sigma}_A = 0.0010$ in, $\hat{\sigma}_B = 0.0010$ in, and $\hat{\sigma}_C = 0.0015$ in. Find the percentage of assemblies in which interference between $X = A + B$ and C is likely to occur.

SOLUTION   From Fig. 4-12 we find

$$\mu_A = 2.5000 \text{ in} \qquad \mu_B = 3.5000 \text{ in} \qquad \mu_C = 6.0030 \text{ in}$$

Then
$$\mu_X = \mu_A + \mu_B = 2.5000 + 3.5000 = 6.0000 \text{ in}$$

Also
$$\hat{\sigma}_X = \sqrt{\hat{\sigma}_A^2 + \hat{\sigma}_B^2} = \sqrt{(0.0010)^2 + (0.0010)^2} = 0.001\ 41 \text{ in}$$

Then, using Eq. (4-43),

$$z = \frac{\mu_C - \mu_X}{\sqrt{\hat{\sigma}_C^2 + \hat{\sigma}_X^2}} = \frac{6.0030 - 6.0000}{\sqrt{(0.0015)^2 + (0.00141)^2}} = 1.453$$

From Table A-14, find $A_z = 0.4269$, using interpolation. Therefore the probability of interference is

$$P = 0.5000 - 0.4269 = 0.0731$$

or 7.31 percent.

The problem can also be solved backward. Another problem would be that of selecting a dimension for μ_C, say, such that only 1 percent interference is obtained. In this case we use Table A-14 to find z corresponding to 1 percent and then solve Eq. (4-43) for μ_C.   ////

PROBLEMS

Section 4-1

4-1   A joint is to be assembled using one bolt and one nut, but there are five bolts and four nuts from which to choose. In how many ways can the joint be assembled?

4-2   A chairman and three committee members are to be appointed from a group of 15 people. How many ways can this be done?

4-3   In how many ways can five gears be placed in a line on a shaft? In how many ways can five bolts be arranged around a bolt circle?

4-4   How many numbers can be formed from the set 1, 2, 3, 4 if the digits are not repeated in any one number? How many numbers can be formed if repetition of digits is permitted?

4-5   A bolt circle has holes for ten $\frac{1}{2}$-in bolts of which eight are coarse-pitch and two are fine-pitch. If the fine-pitch bolts are to be opposite each other, how many ways can the bolts be inserted?

4-6   A planar four-bar linkage consists of four members, each of which can be connected to another by a pin joint (*revolute*), as shown in the figure. In the study of mechanisms such assemblages are called *kinematic chains*. If one of the links is fixed to a frame and another is used as a driver, or input member, the result is called a *four-bar mechanism*. Either of the remaining two links can then be used as output members, and their motion characteristics are dependent on all the link lengths. Suppose you are given four links, each having a different length and with lengths such that the four can be assembled. How many different four-bar mechanisms can be formed?

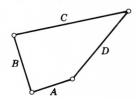

PROBLEM 4-6

Sections 4-2 and 4-3

4-7 A small company owns two drill presses, each of which is in use, respectively, 60 and 90 percent of the time. If a press is suddenly required for a rush job, what is the probability that one will be available?

4-8 A shelf holds 30 boxed items in storage with the labels missing, but it is known that these items consist of 6 A's, 10 B's, and 14 C's. Two boxes are removed, opened, and not replaced.
   (a) What is the probability that the first box contains an A and the second a B if the first box is selected and opened before the second is selected?
   (b) What is the probability that one box contains an A and the other a B if they are both selected and opened at the same time?
   (c) What is the probability that both boxes contain B's if one box is selected and opened at a time?

4-9 A package contains five loose bolts and nuts which are intended to be mating parts. One of the nuts, call it A, however, will not mate with one of the bolts, call it a, though nut A will mate with b, c, d, or e. If these are used to assemble a machine, what is the probability of obtaining a successful assembly if all the fasteners are used and if they are selected at random?

4-10 A tray holds 30 mixed items consisting of 6 A's, 10 B's, and 14 C's.
   (a) An item is selected, identified, and then replaced, and another one selected. What is the probability that neither one of them is a C? That the second is an A? That the first is a C and the second an A?
   (b) Solve part a if the first item selected is not replaced.

4-11 How many different ways can a 2- × 2-in color slide be loaded into a projector? If the slide is not marked for loading and the picture is not examined, what is the probability of loading the projector correctly in one trial? If the first trial was unsuccessful and the slide is loaded differently on a second trial, what is the probability of success in two trials?

Sections 4-4 and 4-5

4-12 The data shown in the table on page 160 represent the distribution of yield strengths from specimens of UNS G10180 round cold-drawn steel bars of diameters from $\frac{3}{4}$ to

$1\frac{1}{4}$ in. These data are from the "Metals Handbook," 8th ed., vol. 1, p. 76, 1961.*
Draw the corresponding histogram and find the mean, the variance, and the standard
deviation for this sample.

Class limits, kpsi	Frequency	Class limits, kpsi	Frequency
62.5–63.5	2	78.5–79.5	13
64.5–65.5	0	80.5–81.5	6
66.5–67.5	2	82.5–83.5	5
68.5–69.5	4	84.5–85.5	5
70.5–71.5	2	86.5–87.5	4
72.5–73.5	4	88.5–89.5	0
74.5–75.5	4	90.5–91.5	0
76.5–77.5	5	92.5–93.5	2

4-13 The yield strengths from 901 heats of 1035 steel, as rolled, are tabulated below,
together with the frequencies. The specimens had diameters ranging from 1 to 9 in.
Compute the mean and the standard deviation, and find the standard deviation as a
percent of the mean.

Class mark S_y, kpsi	Frequency f	Class mark S_y, kpsi	Frequency f
40	2	51	77
41	10	52	52
42	7	53	37
43	25	54	37
44	15	55	27
45	40	56	20
46	65	57	15
47	105	58	15
48	110	59	12
49	120	60	10
50	98	61	2

Sections 4-6 to 4-10

4-14 The commercial diametral tolerance of music wire up to 0.026 in in diameter is
± 0.0003 in. A 0.010-in-diameter music wire is used to manufacture coil springs in
which the wire is stressed in torsion. If the diametral tolerance is the same as the
natural tolerance, determine the standard deviation of the stress, assuming all other
factors are constant. The applied torque is 0.25 oz·in.

4-15 The standard manufacturing tolerance of a cold-drawn steel flat† up to $\frac{3}{4}$ in in width is
0.003 in. A $\frac{1}{8} \times \frac{5}{8}$ steel flat is subjected to a bending load such that the neutral plane is
parallel to the $\frac{5}{8}$-in dimension. Suppose both the thickness and width have natural

---

* Used with the permission of the publishers, The American Society for Metals, Metals Park,
Ohio.

† A *flat* is any steel bar having a rectangular cross section. Flats generally range in sizes from $\frac{1}{8}$ in
thick $\times \frac{3}{16}$ in wide up to 3 in thick $\times$ 12 in wide. Sizes $\frac{3}{16}$ in and thinner are called *strip*. The
overlap is intentional.

tolerances of $+0.000$ and -0.003 in. What is the standard deviation of the bending stress if the moment is 64 lb·in and all other factors are constant?

4-16 Based on the tolerances shown in Fig. 4-12 and the dimensions of A and B, find the mean dimension for C such that only 1 percent interference is obtained. Use the information from Example 4-6.

4-17 A shaft and hole have the dimensions 0.6230 ± 0.0010 and 0.6260 ± 0.0010 in with standard deviations of 0.0004 and 0.0005 in, respectively. Compute the allowance. What percentage of assemblies can be expected to have a smaller allowance?

4-18 A shaft and hole are to have natural tolerances of 0.002 and 0.003 in, respectively. Determine the mean clearance such that not more than 5 percent of the assemblies will have clearances less than 0.002 in. What percentage of these will have clearances greater than 0.007 in?

5

THE STRENGTH OF MECHANICAL ELEMENTS

One of the primary considerations in designing any machine or structure is that the strength must be sufficiently greater than the stress to assure both safety and reliability. To assure that mechanical parts do not fail in service, it is necessary to learn why they sometimes do fail. Then we shall be able to relate the stresses with the strengths to achieve safety.

5-1 SOME REMARKS ON STRENGTH

Ideally, in designing any machine element, the engineer should have at his disposal the results of a great many strength tests of the particular material chosen. These tests should have been made on specimens having the same heat treatment, surface finish, and size as the element he proposes to design; and the tests should be made under exactly the same loading conditions as the part will experience in service. This means that, if the part is to experience a bending load, it should be tested with a bending load. If it is to be subjected to combined bending and torsion, it should be tested under combined bending and torsion. If it is made of heat-treated UNS G10400 steel drawn at 900°F with a ground finish, the specimens tested should be of the same material prepared in the same manner. Such tests will provide very useful and precise information. They tell the engineer what factor of safety to use and what the reliability is for a given service life. Whenever such data are available for design purposes, the engineer can be assured that he is doing the best possible job of engineering.

The cost of gathering such extensive data prior to design is justified if failure of the part may endanger human life, or if the part is manufactured in sufficiently large quantities. Automobiles and refrigerators, for example, have very good reliabilities because the parts are made in such large quantities that they can be thoroughly tested in advance of manufacture. The cost of making these tests is very low when it is divided by the total number of parts manufactured.

You can now appreciate the following four design categories:

1 Failure of the part would endanger human life, or the part is made in extremely large quantities; consequently, an elaborate testing program is justified during design.
2 The part is made in large enough quantities so that a moderate series of tests is feasible.
3 The part is made in such small quantities that testing is not justified at all; or the design must be completed so rapidly that there is not enough time for testing.
4 The part has already been designed, manufactured, and tested and found to be unsatisfactory. Analysis is required to understand why the part is unsatisfactory and what to do to improve it.

It is with the last three categories that we shall be mostly concerned in this book. This means that the designer will usually have only published values of yield strength, ultimate strength, and percentage elongation, such as those listed in the Appendix, and little else. With this meager information the engineer is expected to design against static and dynamic loads, biaxial and triaxial stress states, high and low temperatures, and large and small parts! The data usually available for design have been obtained from the simple tension test, where the load was applied gradually and the strain given time to develop. Yet these same data must be used in designing parts with complicated dynamic loads applied thousands of times per minute. No wonder machine parts sometimes fail.

To sum up, *the fundamental problem of the designer is to use the simple tension-test data and relate them to the strength of the part, regardless of the stress state or the loading situation.* The balance of this chapter is devoted to the solution of this problem.

5-2 DUCTILITY AND HARDNESS

It is possible for two metals to have exactly the same strength and hardness, yet one of these metals may have a superior ability to absorb overloads, because of the

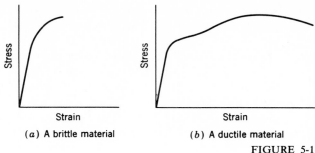

(a) A brittle material   (b) A ductile material

FIGURE 5-1

property called *ductility.* Two such materials having approximately the same strength and hardness are illustrated in Fig. 5-1. In (a), a brittle material is shown, and it is seen that it is one which has a small plastic deformation. On the other hand, a ductile material as in (b) will be capable of a relatively large plastic deformation before fracture. Ductility is measured by the *percentage elongation* which occurs in the material at fracture. The usual dividing line between ductility and brittleness is 5 percent elongation. A material having less than 5 percent elongation at fracture is said to be *brittle,* while one having more is said to be *ductile.*

The elongation of a material is usually measured over a 2-in gauge length. Since this is not a measure of the actual strain, another method of determining ductility is sometimes used. After the specimen has been fractured, measurements are made of the area of the cross section at the fracture. Ductility can then be expressed as the *percentage reduction in cross-sectional area.*

The characteristic of a ductile material which permits it to absorb large overloads is an additional safety factor in design. Ductility is also important because it is a measure of that property of a material which permits it to be cold-worked. Such operations as bending, drawing, heading, and stretch forming are metal-processing operations which require ductile materials.

When a material is to be selected to resist wear, erosion, or plastic deformation, hardness is generally the most important property. Several methods of hardness testing are available, depending upon which particular property is most desired. The four hardness numbers in greatest use are the Brinell, Rockwell, Vickers, and Knoop.

Most hardness-testing systems employ a standard load which is applied to a ball or pyramid in contact with the material to be tested. The hardness is then expressed as a function of the size of the resulting indentation. This means that hardness is an easy property to measure, because the test is nondestructive and test specimens are not required. Usually the test can be conducted directly on an actual machine element.

The fact that the Brinell hardness number H_B can be used to obtain a very good estimate of the tensile strength of steel is of particular value. The relation is

$$S_u = 500H_B \qquad (5\text{-}1)$$

where S_u is in psi. In SI units the corresponding relation is

$$S_u = 3.45H_B \qquad (5\text{-}2)$$

where S_u is in MPa.

Datsko* shows that the tensile strength is related to the hardness by means of the strain-hardening exponent and the applied load, and he has graphed these relations.

5-3 MECHANICAL PROPERTIES

Tables A-17 to A-24 list the properties of a wide variety of materials. For the student, these tables furnish a source of useful information for problem solving and design projects. For the practicing engineer, the tables can serve as a standard in the sense that they provide values which he has a right to expect with proper control of the production processes.

The new numbering system for metals and alloys, called the Unified Numbering System† (UNS), is used in this book where appropriate. The Society of Automotive Engineers (SAE) was the first to recognize the need, and adopted a system of numbering steels. Later the American Iron and Steel Institute (AISI) adopted a similar system. In this system a letter prefix designates the process by which the steel is made. For instance, A designates a basic open-hearth alloy steel, B is an acid-bessemer carbon steel, C is a basic open-hearth carbon steel, D is an acid open-hearth carbon steel, E is an electric-furnace steel, and BOF is for steels made by the relatively new basic oxygen furnace process. The first two numbers following the letter prefix indicate the composition, excluding the carbon content. The various compositions used are as follows:

10	plain carbon	46	nickel-molybdenum
11	free-cutting carbon steel with more sulfur or phosphorus	48	nickel-molybdenum
		50	chromium
13	manganese	51	chromium
23	nickel	52	chromium
25	nickel	61	chromium-vanadium
31	nickel-chromium	86	chromium-nickel-molybdenum
33	nickel-chromium	87	chromium-nickel-molybdenum
40	molybdenum	92	manganese-silicon
41	chromium-molybdenum	94	nickel-chromium-molybdenum
43	nickel-chromium-molybdenum		

---

* Joseph Datsko, "Material Properties and Manufacturing Processes," John Wiley & Sons, Inc., p. 38, New York, 1966.

† Available from Society of Automotive Engineers, Warrendale, Pa.

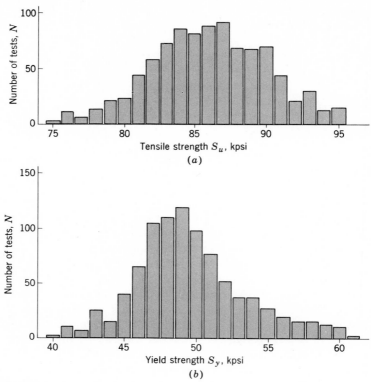

FIGURE 5-2
Distribution of tensile properties of hot-rolled UNS G10350 steel, as rolled. These tests were made from round bars varying in diameter from 1 to 9 in. (a) Tensile-strength distributions from 930 heats; $\overline{S}_u = 86.0$ kpsi, $s_{S_u} = 4.04$ kpsi. (b) Yield-strength distribution from 899 heats; $\overline{S}_y = 49.5$ kpsi, $s_{S_y} = 5.36$ kpsi. (*By permission,* "*Metals Handbook,*" *vol. 1, 8th ed., p. 64, American Society for Metals, Metals Park, Ohio, 1961.*)

The last two numbers following the prefix (three for the high-carbon steels in the 51- and 52-chromium groups) refer to the approximate carbon content. Thus AISI C1040 is a basic open-hearth carbon steel with a carbon content of 0.37 to 0.44 percent. Similarly, SAE 2330 is a nickel-alloy steel with 0.28 to 0.33 percent carbon.

Examination of Table A-17 shows that the UNS number for steels includes the former AISI and SAE numbers. Thus a UNS G10350 steel is the same as AISI 1035 and SAE 1035.

UNS numbers for other materials also include the original number. For example, a UNS A93004 wrought aluminum alloy is the same material as the Aluminum Association designation 3004 alloy.

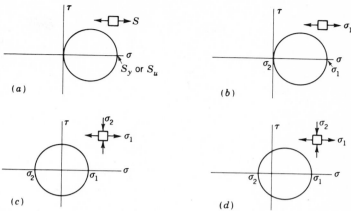

FIGURE 5-3
(a) Mohr's circle for the simple tension test; (b) Mohr's circle for pure tension; (c) Mohr's circle for pure torsion; (d) Mohr's circle for combined tension and compression.

Cast iron is not designated by composition but rather by the tensile-strength level. For example, a No. 30 cast iron is expected to have a tensile strength not less than 30 kpsi. No UNS numbers have as yet been assigned to the cast irons.

Published information regarding the statistical properties of materials is almost impossible to find. However, tests to discover these properties should probably be conducted in parallel with production anyway, because the results could vary widely, depending upon the amount of quality control exercised. Nevertheless, some information regarding the distribution of strengths from a large number of tests is quite useful as a standard against which one can compare his own distributions. Accordingly, Fig. 5-2 has been included. The information from this distribution will also be useful as student problem-solving material.

5-4 THE MAXIMUM-NORMAL-STRESS THEORY

This theory of failure is of importance for comparison purposes. Its predictions do not agree with experiment and, in fact, may even give results on the unsafe side.

The maximum-normal-stress theory states that failure occurs whenever the largest principal stress equals the yield strength, or the ultimate strength, of the material. In Chap. 2 it was found that the principal stresses can always be found when the stress state at a point is given. If we designate the largest of these three stresses as σ_1 such that $\sigma_1 > \sigma_2 > \sigma_3$, then the maximum-normal-stress theory states that failure occurs whenever $\sigma_1 = S_y$ or $\sigma_1 = S_u$, whichever is applicable. Here S_y is the yield strength and S_u the ultimate strength.

Figure 5-3a shows Mohr's circle for the simple tension test, and (b), (c), and (d) various other biaxial stress situations. The uniaxial stress state described in (b)

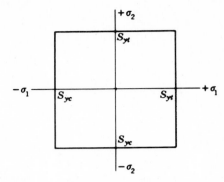

FIGURE 5-4
The maximum-normal-stress theory of failure for biaxial stresses. The tensile yield strength is S_{yt}, and the compressive yield strength S_{yc}.

corresponds exactly to the tension test, and so there is no problem in this case. Now the maximum-normal-stress theory states that only the largest principal stress predicts failure and that the others may be neglected. But Fig. 5-3c, which is the diagram for pure torsion, shows that $\tau = \sigma_1$, and consequently that failure would occur when the shear stress became equal to the strength in tension or compression. This prediction is not in agreement with experimental data.

Let us plot σ_1 and σ_2 on axes at right angles to each other so that we can better picture the implications of this theory of failure. This has been done in Fig. 5-4, with tension plotted above and to the right and compression below and to the left. We have assumed, in plotting this diagram, that the yield strengths are equal in tension and compression. Note that an actual test point is obtained only where the diagram crosses an axis. The theory states that failure occurs whenever a point whose coordinates are σ_1 and σ_2 falls on or outside the graph. We shall defer a correlation of test data with this theory until later. For the present it is sufficient to mention that points inside the figure and in the first and third quadrants are on the safe side, whereas points in the second and fourth quadrants may be on the unsafe side, according to this theory.

The maximum-normal-stress theory may be used to predict yielding or fracture, whichever is used as the criterion of failure. If σ_1 is largest in absolute value of the three principal stresses, then failure by yielding occurs whenever

$$\sigma_1 = S_{yt} \qquad \text{or} \qquad \sigma_1 = -S_{yc} \qquad (5\text{-}3)$$

Similarly, failure by fracture will occur whenever

$$\sigma_1 = S_{ut} \qquad \text{or} \qquad \sigma_1 = -S_{uc} \qquad (5\text{-}4)$$

It is also convenient to define safety using the factor of safety n. Thus, if yielding is the criterion of failure, the factor of safety n is obtained whenever

$$\sigma_1 = \frac{S_{yt}}{n} \qquad \text{or} \qquad \sigma_1 = -\frac{S_{yc}}{n} \qquad (5\text{-}5)$$

Similarly, for fracture,

$$\sigma_1 = \frac{S_{ut}}{n} \qquad \text{or} \qquad \sigma_1 = -\frac{S_{uc}}{n} \qquad (5\text{-}6)$$

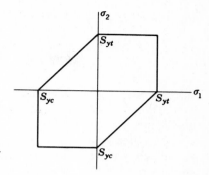

FIGURE 5-5
The maximum-shear-stress theory of failure.

5-5 THE MAXIMUM-SHEAR-STRESS THEORY

This is an easy theory to use, it is always on the safe side of test results, and it has been used in many design codes. It is used only to predict yielding, and hence it applies only to ductile materials.

The maximum-shear-stress theory states that yielding begins whenever the maximum shear stress in a mechanical element becomes equal to the maximum shear stress in a tension-test specimen when that specimen begins to yield. Thus, from Fig. 5-3a, yielding begins when $\tau_{max} = S_y/2$. For a general stress state three maximum shear stresses may be found (Fig. 2-6b). They are

$$\tau_{12} = \frac{\sigma_1 - \sigma_2}{2} \qquad \tau_{23} = \frac{\sigma_2 - \sigma_3}{2} \qquad \tau_{13} = \frac{\sigma_1 - \sigma_3}{2}$$

Call the largest of these τ_{max}. Then, according to the maximum-shear-stress theory, failure by yielding occurs whenever

$$\tau_{max} = \frac{S_y}{2} \qquad (5\text{-}7)$$

Note that this theory predicts that the yield strength in shear is half the yield strength in tension. That is,

$$S_{sy} = 0.50 S_y \qquad (5\text{-}8)$$

(Before reading further, return to Sec. 5-4 and discover for yourself the relation between S_{sy} and S_y as predicted by the maximum-normal-stress theory.)

It is also necessary to define safety using the maximum-shear-stress theory. This theory states that safety is obtained whenever

$$\tau_{max} = \frac{S_y}{2n} \qquad (5\text{-}9)$$

where n is the factor of safety.

A graph of the maximum-shear-stress theory for biaxial stresses is shown in Fig. 5-5. Note that the graph is the same as that for the maximum-normal-stress theory when both principal stresses have the same sign.

5-6 THE DISTORTION-ENERGY THEORY

This failure theory is also called the *shear-energy theory* and the *von Mises-Hencky theory*. It is only slightly more difficult to use than the maximum-shear-stress theory, and it is the best theory to use for ductile materials. Like the maximum-shear-stress theory, it is employed to define only the beginning of yield.

The distortion-energy theory originated because of the observation that ductile materials stressed hydrostatically (equal tension or compression) had yield strengths greatly in excess of the values given by the simple tension test. It was postulated that yielding was not a simple tensile or compressive phenomenon at all, but rather, that it was related somehow to the angular distortion of the stressed element. Now, one of the earlier theories of failure predicted that yielding would begin whenever the total strain energy stored in the stressed element became equal to the strain energy stored in an element of the tension-test specimen at the yield point. This theory, called the *maximum-strain-energy theory*, is no longer used, but it was a forerunner of the distortion-energy theory. It was argued, Why not take the total strain energy and subtract from it whatever energy is used only to produce a volume change? Then whatever energy is left will be that which produces angular distortion alone. Let us see how this works.

Figure 5-6a shows an element acted upon by stresses arranged so that $\sigma_1 > \sigma_2 > \sigma_3$. For a unit cube the work done in any one of the principal directions is

$$u_n = \frac{\sigma_n \epsilon_n}{2} \qquad (a)$$

where $n = 1$, 2, or 3. Therefore, from Eq. (2-22), the total strain energy is

$$u = u_1 + u_2 + u_3 = [1/(2E)][\sigma_1^2 + \sigma_2^2 + \sigma_3^2 - 2\mu(\sigma_1\sigma_2 + \sigma_2\sigma_3 + \sigma_3\sigma_1)] \qquad (b)$$

We next define a stress

$$\sigma_{av} = \frac{\sigma_1 + \sigma_2 + \sigma_3}{3} \qquad (c)$$

and apply this stress to each of the principal directions of a unit cube (Fig. 5-6b). The remaining stresses $\sigma_1 - \sigma_{av}$, $\sigma_2 - \sigma_{av}$, and $\sigma_3 - \sigma_{av}$, shown in Fig. 5-6c, will produce only angular distortion. Substituting σ_{av} for σ_1, σ_2, and σ_3 in Eq. (b) gives the amount of strain energy producing only volume change.

$$u_v = \frac{1}{2E}[3\sigma_{av}^2 - 2\mu(3)\sigma_{av}^2] = \frac{3\sigma_{av}^2}{2E}(1 - 2\mu) \qquad (d)$$

If we now substitute $\sigma_{av}^2 = [(\sigma_1 + \sigma_2 + \sigma_3)/3]^2$ into Eq. (d) and simplify the expression, we get

$$u_v = \frac{1 - 2\mu}{6E}(\sigma_1^2 + \sigma_2^2 + \sigma_3^2 + 2\sigma_1\sigma_2 + 2\sigma_2\sigma_3 + 2\sigma_3\sigma_1) \qquad (e)$$

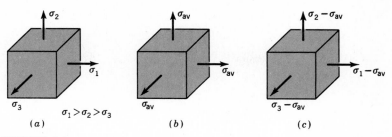

FIGURE 5-6
(a) Element with triaxial stresses; this element undergoes both volume change and angular distortion. (b) Element under hydrostatic tension undergoes only volume change. (c) Element has angular distortion without volume change.

Then the distortion energy is obtained by subtracting Eq. (e) from Eq. (b). This yields

$$u_d = u - u_v = \frac{1 + \mu}{3E} \left[\frac{(\sigma_1 - \sigma_2)^2 + (\sigma_2 - \sigma_3)^2 + (\sigma_3 - \sigma_1)^2}{2} \right] \qquad (5\text{-}10)$$

Note that the distortion energy is zero if $\sigma_1 = \sigma_2 = \sigma_3$.

For the simple tension test, $\sigma_1 = S_y$ and $\sigma_2 = \sigma_3 = 0$. Therefore the distortion energy is

$$u_d = \frac{1 + \mu}{3E} S_y^2 \qquad (5\text{-}11)$$

The criterion is obtained by equating Eqs. (5-10) and (5-11).

$$2S_y^2 = (\sigma_1 - \sigma_2)^2 + (\sigma_2 - \sigma_3)^2 + (\sigma_3 - \sigma_1)^2 \qquad (5\text{-}12)$$

which defines the beginning of yield for a triaxial stress state. When $\sigma_3 = 0$, the stress situation is biaxial and Eq. (5-12) reduces to

$$S_y^2 = \sigma_1^2 - \sigma_1 \sigma_2 + \sigma_2^2 \qquad (5\text{-}13)$$

For pure torsion $\sigma_2 = -\sigma_1$ and $\tau = \sigma_1$; consequently

$$S_{sy} = 0.577 S_y \qquad (5\text{-}14)$$

Comparison of Eq. (5-14) with (5-8) shows that the distortion-energy criterion predicts a shearing-yield strength appreciably higher than that predicted by the maximum-shear-stress theory. How does it compare with the yield strength in shear as predicted by the maximum-normal-stress theory?

For analysis and design purposes, it is convenient to define a *von Mises stress*, from Eq. (5-13), as

$$\sigma' = \sqrt{\sigma_1^2 - \sigma_1 \sigma_2 + \sigma_2^2} \qquad (5\text{-}15)$$

Then failure by yielding is predicted by the distortion-energy theory whenever

$$\sigma' = S_y \qquad \text{(5-16)}$$

Similarly, safety is predicted by

$$\sigma' = \frac{S_y}{n} \qquad \text{(5-17)}$$

where n is the factor of safety, as before.

EXAMPLE 6-1   A mechanical element is fabricated from a steel having a yield strength of 50 kpsi. The stress state at one point on the element is $\sigma_x = 20$ kpsi, $\sigma_y = -8$ kpsi, $\tau_{xy} = 12$ kpsi. Determine the factor of safety by the maximum-normal-stress theory, the maximum-shear-stress theory, and the distortion-energy theory.

SOLUTION   First, construct Mohr's circle and find the principal stresses and the maximum shear stresses. The results are

$$\sigma_1 = 24.4 \text{ kpsi} \qquad \sigma_2 = -12.4 \text{ kpsi} \qquad \sigma_3 = 0 \text{ kpsi}$$

Therefore

$$\tau_{max} = \frac{\sigma_1 - \sigma_2}{2} = \frac{24.4 - (-12.4)}{2} = 18.4 \text{ kpsi}$$

For the maximum-normal-stress theory use Eq. (5-5). The answer is

$$n = \frac{S_{yt}}{\sigma_1} = \frac{50}{24.4} = 2.05$$

For the maximum-shear-stress theory use Eq. (5-9). The answer is

$$n = \frac{S_{yt}}{2\tau_{max}} = \frac{50}{(2)(18.4)} = 1.36$$

For the distortion-energy theory compute the von Mises stress, first, using Eq. (5-15). The result is

$$\sigma' = \sqrt{\sigma_1^2 - \sigma_1 \sigma_2 + \sigma_2^2} = \sqrt{(24.4)^2 - (24.4)(-12.4) + (-12.4)^2} = 32.5 \text{ kpsi}$$

Then, using Eq. (5-17), the factor of safety is found to be

$$n = \frac{S_{yt}}{\sigma'} = \frac{50}{32.5} = 1.54 \qquad \text{////}$$

5-7 FAILURE OF DUCTILE MATERIALS WITH STATIC LOADS

It is now time to summarize the results of the three preceding sections and relate them to experimental results. Plotting all three failure theories on the σ_1, σ_2

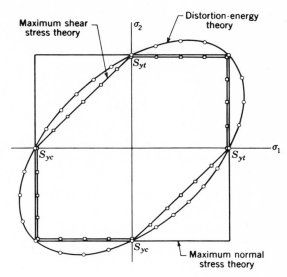

FIGURE 5-7
Comparison of three static-failure theories for ductile materials and biaxial stresses.

coordinate system gives the graph of Fig. 5-7. Well-documented experiments indicate that the distortion-energy theory predicts yielding with greatest accuracy in
all four quadrants. Thus, accepting the distortion-energy theory as the correct one
means that the maximum-shear-stress theory always gives results which are
conservative. And it means that the maximum-normal-stress theory gives conservative results *only* if the signs of the two principal stresses are alike. For pure
torsion the signs of the two principal stresses are opposite, and consequently, only
the distortion-energy theory or the maximum-shear-stress theory can be used.

The decision about which theory to employ must be made by the designer
after a consideration of the facts involved in the particular problem that is being
solved. A designer who is trying to discover why a part failed so that a better job of
redesigning it can be done, should certainly use the distortion-energy theory. If
one is solving to obtain dimensions which need not be held too closely, and if the
problem must be solved quickly, the maximum-shear-stress theory should be
used. But when the margin of safety must be held within close limits and the
designer is trying for the best possible job of designing without actually testing the
parts, the distortion-energy theory of failure should be employed.

5-8 FAILURE OF BRITTLE MATERIALS WITH STATIC LOADS

In selecting a failure theory for use with brittle materials we first note the following characteristics of most brittle materials:

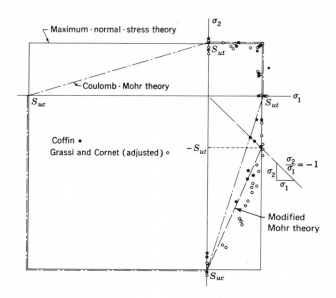

FIGURE 5-8
A plot of experimental data points from tests of gray cast iron subjected to biaxial stresses. The data were adjusted to correspond to $S_{ut} = 32$ kpsi and $S_{uc} = 105$ kpsi. Superimposed on the plot are graphs of the maximum-normal-stress theory, the Coulomb-Mohr theory, and the modified Mohr theory.

1 A graph of stress versus strain is a smooth continuous line to the failure point; failure occurs by fracture, and hence these materials do not have a yield strength.

2 The compressive strength is usually many times greater than the tensile strength.

3 The ultimate torsional strength S_{su}, that is, the modulus of rupture, is approximately the same as the tensile strength.

The maximum-normal-stress theory and the Coulomb-Mohr theory have both been used to predict fracture of brittle materials. The maximum-normal-stress theory has already been investigated. In using this theory the test points shown in Fig. 5-4 would be changed to the values S_{ut} and S_{uc}.

The *Coulomb-Mohr theory*, sometimes called the *internal-friction theory*, is based upon the results of two tests, the tensile test and the compression test. On the σ, τ coordinate system, plot both circles, one for S_{ut} and one for S_{uc}. Then the Coulomb-Mohr theory states that fracture occurs for any stress situation which produces a circle tangent to the envelope of the two test circles. If we arrange the principal stresses so that $\sigma_1 > \sigma_2 > \sigma_3$, then the largest circle will be formed by σ_1

and σ_3. These two stresses and the strengths are related by the equation

$$\frac{\sigma_1}{S_{ut}} + \frac{\sigma_3}{S_{uc}} = 1 \qquad (5\text{-}18)$$

which defines fracture by the Coulomb-Mohr theory. To define safety, the equation is written

$$\frac{\sigma_1}{S_{ut}} + \frac{\sigma_3}{S_{uc}} = \frac{1}{n} \qquad (5\text{-}19)$$

where n is the factor of safety. For both of these equations it is important to note that S_{uc} and σ_3 are negative quantities.

Fortunately, there is a rather large number of data points available which we can use to test the validity of these two theories.* This comparison has been made in Fig. 5-8. In the first quadrant, where σ_1 and σ_2 both have the same sense, we note that the two theories are identical, and hence that either theory can be used satisfactorily to predict fracture. It is in the fourth quadrant, where σ_1 and σ_2 have opposite senses, that the two theories differ. We note first that the Coulomb-Mohr theory is on the conservative side since all the data points fall outside.† Note also, in the figure, the line having the slope $\sigma_2/\sigma_1 = -1$. For pure torsion, $\sigma_2 = -\sigma_1$, and hence the intersection of this line with the graph of a failure theory yields the value of S_{su} as predicted by that theory. Note that its intersection with the maximum-normal-stress theory yields $S_{su} = S_{ut}$, which we have already noted is one of the characteristics of brittle materials. But the Coulomb-Mohr theory predicts a value of S_{su} somewhat less than S_{ut}.

The *modified Mohr theory*, shown in the fourth quadrant of Fig. 5-8, is not as conservative as the Coulomb-Mohr theory, but it does a better job of predicting failure. Burton Paul‡ proposed a similar theory, though slightly different, which we shall not include here. The modified Mohr theory is best applied by a graphical approach, as demonstrated in the following example.

EXAMPLE 5-2   A small, 6-mm-diameter pin was designed of ASTM No. 40 cast iron to take an axial compressive load of 3.5 kN combined with a torsional load of 9.8 N·m. Compute the factor of safety guarding against a static failure using each of the three theories for brittle materials.

SOLUTION   The axial compressive stress is

$$\sigma_x = \frac{F}{A} = \frac{4F}{\pi d^2} = -\frac{4(3.5)(10)^3}{\pi(6)^2} = -124 \text{ MPa}$$

* L. F. Coffin, The Flow and Fracture of a Brittle Material, *Trans. ASME*, vol. 72, *J. Appl. Mech.*, vol. 17, pp. 233–248, 1950; R. C. Grassi and I. Cornet, Fracture of Gray Cast Iron Tubes under Biaxial Stresses, *Trans. ASME*, vol. 71, *J. Appl. Mech.*, vol. 16, pp. 178–182, 1949.

† A conservative theory is perfectly satisfactory for design purposes where the object is to determine a set of dimensions such that the part will not fail, but it is completely useless in an analysis if the object is to learn why something failed.

‡ Burton Paul, A Modification of the Coulomb-Mohr Theory of Fracture, *Trans. ASME*, *J. Appl. Mech.*, ser. E, vol. 28, no. 2, pp. 259–268, June 1961.

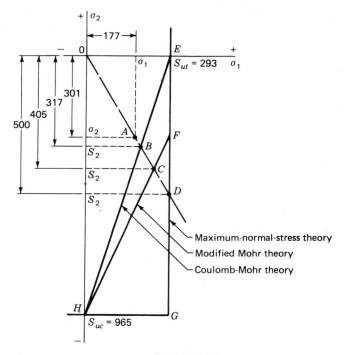

FIGURE 5-9
All stresses and strengths are in MPa.

The torsional shear stress is

$$\tau_{xy} = \frac{16T}{\pi d^3} = \frac{16(9.8)(10)^3}{\pi(6)^3} = 231 \text{ MPa}$$

When Mohr's circle diagram is constructed, the principal stresses are found to be $\sigma_1 = 177$ MPa, $\sigma_2 = -301$ MPa.

If we employ the typical values of the strengths instead of the minimums, then from Table A-20 we have $S_{ut} = 42.5$ kpsi and $S_{uc} = 140$ kpsi. Converting these to SI units gives $S_{ut} = 293$ MPa and $S_{uc} = 965$ MPa.

The next step is to plot a graph to scale corresponding to the fourth quadrant of Fig. 5-8, using the stress and strength magnitudes just found. This has been done in Fig. 5-9. Point A represents the coordinates σ_1, σ_2 of the actual stress state. If σ_1 and σ_2 increased in magnitude, but retained the same ratio to each other, then points B, C, and D would represent failure by each theory. Thus, if OA represents the stress state, then AB, AC, and AD represent the respective margins of safety (see Sec. 1-6 for definition of margin of safety). The corresponding factors of safety are equal to OB divided by OA, OC divided by OA, and OD divided by OA.

Another way to obtain the factor of safety is to project points B, C, and D to the σ_1 or σ_2 axis. The resulting intersections define corresponding strengths S_1 or S_2, if a mnemonic notation is used. Thus, in Fig. 5-9, we can read out the strength S_2 for each failure theory. The factor of safety for the Coulomb-Mohr theory is

$$n = \frac{S_2}{\sigma_2} = \frac{-317}{-301} = 1.05 \qquad Ans.$$

For the modified Mohr theory we get

$$n = \frac{-405}{-301} = 1.35 \qquad Ans.$$

And for the maximum-normal-stress theory we find

$$n = \frac{-500}{-301} = 1.6 \qquad Ans.$$

Of course the last result can also be obtained by dividing S_{ut} by σ_1 and S_{uc} by σ_2 and selecting the smaller of the two resulting values.   ////

5-9 FATIGUE

In obtaining the properties of materials relating to the stress-strain diagram, the load is applied gradually, giving sufficient time for the strain to develop. With the usual conditions, the specimen is tested to destruction so that the stresses are applied only once. These conditions are known as static conditions and are closely approximated in many structural and machine members.

The condition frequently arises, however, in which the stresses vary or fluctuate between values. For example, a particular fiber on the surface of a rotating shaft, subjected to the action of bending loads, undergoes both tension and compression for each revolution of the shaft. If the shaft is a part of an electric motor rotating at 1725 rpm, the fiber is stressed in tension and in compression 1725 times each minute. If, in addition, the shaft is also axially loaded (caused, for example, by a helical or worm gear), an axial component of stress is superimposed upon the bending component. This results in a stress, in any one fiber, which is still fluctuating but which is fluctuating between different values. These and other kinds of loads occurring in machine members produce stresses which are called repeated, alternating, or fluctuating stresses.

Machine members are often found to have failed under the action of repeated or fluctuating stresses, and yet the most careful analysis reveals that the actual maximum stresses were below the ultimate strength of the material and quite frequently even below the yield strength. The most distinguishing character-istic of these failures has been that the stresses have been repeated a very large number of times. Hence the failure is called a *fatigue* failure.

A fatigue failure begins with a small crack. The initial crack is so minute that it cannot be detected by the naked eye and is even quite difficult to locate in a Magnaflux or x-ray inspection. The crack will develop at a point of discontinuity

FIGURE 5-10
A fatigue failure of a $7\frac{1}{2}$-in-diameter forging at a press fit. The specimen is UNS G10450 steel, normalized and tempered, and has been subjected to rotating bending. (*Courtesy of The Timken Company.*)

in the material, such as a change in cross section, a keyway, or a hole. Less obvious points at which fatigue failures are likely to begin are inspection or stamp marks, internal cracks, or even irregularities caused by machining. Once a crack has developed, the stress-concentration effect becomes greater and the crack progresses more rapidly. As the stressed area decreases in size, the stress increases in magnitude until, finally, the remaining area fails suddenly. A fatigue failure, therefore, is characterized by two distinct areas of failure (Fig. 5-10). The first of these is that due to the progressive development of the crack, while the second is due to the sudden fracture. The zone of the sudden fracture is very similar in appearance to the fracture of a brittle material, such as cast iron, which has failed in tension.

When machine parts fail statically, they usually develop a very large deflection, because the stress has exceeded the yield strength, and the part is replaced before fracture actually occurs. Thus many static failures are visible ones and give warning in advance. But a fatigue failure gives no warning; it is sudden and total, and hence dangerous. It is a relatively simple matter to design against a static failure because our knowledge is quite complete. But fatigue is a much more complicated phenomenon, only partially understood, and the engineer seeking to rise to the top of the profession must acquire as much knowledge of the subject as possible. Anyone can double or triple factors of safety, because of a lack of knowledge of fatigue, and get a design that will not fail. But such designs will not compete in today's market, and neither will the engineers who produce them.

5-10 FATIGUE STRENGTH AND ENDURANCE LIMIT

To determine the strength of materials under the action of fatigue loads, specimens are subjected to repeated or varying forces of specified magnitudes

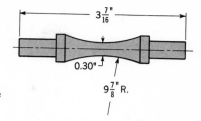

FIGURE 5-11
Test specimen for the R. R. Moore rotating-beam machine.

while the cycles or stress reversals are counted to destruction. The most widely used fatigue-testing device is the R. R. Moore high-speed rotating-beam machine. This machine subjects the specimen to pure bending (no transverse shear) by means of weights. The specimen, shown in Fig. 5-11, is very carefully machined and polished, with a final polishing in an axial direction to avoid circumferential scratches.

Other fatigue-testing machines are available for applying fluctuating or reversed axial stresses, torsional stresses, or combined stresses to the test specimens.

To establish the fatigue strength of a material, quite a number of tests are necessary because of the statistical nature of fatigue. For the rotating-beam test a constant bending load is applied, and the number of revolutions (stress reversals) of the beam required for failure is recorded. The first test is made at a stress which is somewhat under the ultimate strength of the material. The second test is made with a stress which is less than that used in the first. This process is continued, and the results plotted as an S-N diagram (Fig. 5-12). This chart may be plotted on semilog paper or on log-log paper. In the case of ferrous metals and alloys, the

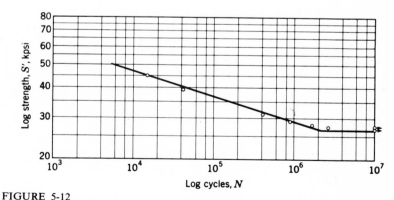

FIGURE 5-12
An S-N diagram for annealed UNS G10400 steel. The data observed and plotted are 45.0 kpsi at $15.1(10)^3$ cycles, 39.1 kpsi at $41.3(10)^3$ cycles, 30.9 kpsi at $408.3(10)^3$ cycles, 28.9 kpsi at $897.4(10)^3$ cycles, 28.05 kpsi at $1.696(10)^6$ cycles, 27.4 kpsi at $2.692(10)^6$ cycles, 27.25 kpsi at $10.012(10)^6$ cycles, and 27.0 kpsi at $10.089(10)^6$ cycles. The last two specimens did not fail.

graph becomes horizontal after the material has been stressed for a certain number of cycles. Plotting on log paper emphasizes the knee of the curve, which might not be apparent if the results were plotted by using cartesian coordinates.

The ordinate of the *S-N* diagram is called the *fatigue strength* S_f; a statement of this strength must always be accompanied by a statement of the number of cycles *N* to which it corresponds.

In the case of the steels, a knee occurs in the graph, and beyond this knee failure will not occur no matter how great the number of cycles. The strength corresponding to the knee is called the *endurance limit* S_e, or the *fatigue limit*. The graph of Fig. 5-12 never does become horizontal for nonferrous metals and alloys, and hence these materials do not have an endurance limit.

As noted previously, it is always good engineering to conduct a testing program on the materials to be employed in design and manufacture. This, in fact, is a requirement, not an option, in guarding against the possibility of a fatigue failure. *Because of this necessity for testing it would really be unnecessary for us to proceed any further in the study of fatigue failure except for one important reason: the desire to know why fatigue failures occur so that the most effective method or methods can be used to improve fatigue strength.* Thus our primary purpose in studying fatigue is to understand why failures occur so that we can guard against them in an optimum manner. For this reason, the analytical and design approaches presented in this book, or in any other book for that matter, do not yield absolutely precise results. The results should be taken as a guide, as something which indicates what is important and what is not important in designing against fatigue failure.

The methods of fatigue-failure analysis represent a combination of engineering and science. Often science fails to provide the answers which are needed. But the airplane must still be made to fly—safely. And the automobile must be manufactured with a reliability that will ensure a long and trouble-free life and at the same time produce profits for the stockholders of the industry. Fatigue is a similar situation. Science has not yet completely explained the actual mechanism of fatigue. But still the engineer must design things that will not fail. In a sense this is a classic example of the true meaning of engineering as contrasted with science. Engineers use science to solve their problems *if* the science is available. But available or not, the problem must be solved, and whatever form the solution takes under these conditions is called engineering.

One of the first problems to be solved is that of learning whether there is any general relationship between the endurance limit and the strengths obtained from the simple tension test. When a search using great quantities of data from the tension test and the rotating-beam test is made, it is found that there is indeed a relation between the results of these two tests. This relationship can be observed in Fig. 5-13. Because of the scatter, this relation cannot be plotted as a single smooth curve; instead, it would be necessary to use a wide paint brush to blacken out most of the data points. Close inspection of Fig. 5-13 shows that the endurance limit ranges from about 40 to 60 percent of the tensile strength for steels up to

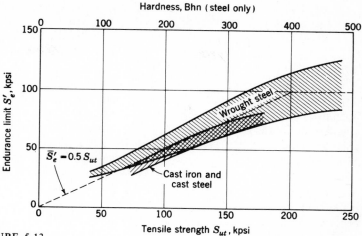

FIGURE 5-13

Observed relationship between tensile strength and endurance limit for wrought and cast steel and cast iron. Approximate values can be obtained from the Brinell hardness if the tensile strength is unknown. The bands containing the observations are called *scatter bands*. [*By permission from Charles Lipson and Robert C. Juvinall (eds.), "Application of Stress Analysis to Design and Metallurgy," The University of Michigan Summer Conference, Ann Arbor, Mich., 1961.*]

about $S_{ut} = 200$ kpsi. And the endurance limit appears to be about 100 kpsi for tensile strengths of 200 kpsi and over.

Now it is important to observe here that the dispersion of the endurance limit is *not* due to a dispersion, or spread, in the tensile strengths of the specimens tested at all. If we take, for example, a large number of rotating-beam specimens of a steel having $S_{ut} = 100$ kpsi exactly, the endurance limits of these specimens will range between 40 and 60 kpsi, with a mean of about 50 kpsi. It is for this reason that we decide on the following relation for predicting the *mean endurance limit of the rotating-beam specimens*:

$$\bar{S}'_e = 0.50 S_{ut} \qquad S_{ut} \le 200 \text{ kpsi}$$
$$\bar{S}'_e = 100 \text{ kpsi} \qquad S_{ut} > 200 \text{ kpsi} \qquad (5\text{-}20)$$

The corresponding rounded relationships in SI are

$$\bar{S}'_e = 0.50 S_{ut} \qquad S_{ut} \le 1400 \text{ MPa}$$
$$\bar{S}'_e = 700 \text{ MPa} \qquad S_{ut} > 1400 \text{ MPa} \qquad (5\text{-}21)$$

The prime mark on S'_e refers to the rotating-beam specimen itself because we wish to reserve the symbol S_e for the endurance limit of a particular machine element. Soon we shall learn that these two strengths may be quite different.

As shown in Fig. 5-13, the endurance limit for cast iron is somewhat less than for steel. We shall generally use the relation

$$\bar{S}'_e = 0.40 S_{ut} \qquad (5\text{-}22)$$

for cast iron. Note that this result differs only slightly from the values shown in Table A-20.

Processors of aluminum and magnesium alloys publish very complete tabulations of the properties of these materials, including the fatigue strengths. They ordinarily run from about 30 to 40 percent of the tensile strength, depending upon whether the material is cast or wrought. These materials do not have an endurance limit, and the fatigue strength is usually based on 10^8 or $5(10)^8$ cycles of stress reversal.

As indicated by the data displayed in Table 5-1, the standard deviations of the endurance limit vary from about 4 to 10 percent. Using reasonable care, inspection, and control techniques, it is probable that a standard deviation of 8 percent of the endurance limit can be held, and in this book, we shall use this figure in our reliability analyses.

The bar on the symbol $\bar{S}'_e$, to indicate the mean, is an inconvenient notation, which we shall now abandon. Instead we shall use S'_e for the endurance limit of the rotating-beam specimen corresponding to a stated reliability R. If the reliability is *not* stated, S'_e is taken as the endurance limit corresponding to $R = 0.50$, that is, the mean endurance limit.

Table 5-1 STANDARD DEVIATIONS OF ENDURANCE LIMIT*

Material†	Tensile strength		Endurance limit		Standard deviation	
UNS No.	MPa	kpsi	MPa	kpsi	kpsi	%
G43400 steel	965	140	489	71	3.5	4.9
	1310	190	586	85	6.7	7.8
	1580	230	620	90	5.3	5.9
	1790	260	668	97	6.3	6.5
G43500 steel	2070	300	689	100	4.4	4.4
R50001-series titanium alloy	1000	145	579	84	5.4	6.4
A97076 aluminum alloy	524	76	186	27	1.6	6.0
C63000 aluminum bronze	806	117	331	48	4.5	9.4
C17200 beryllium copper	1210	175	248	36	2.7	7.5

* Reported by F. B. Stulen, H. N. Cummings, and W. C. Schulte, Preventing Fatigue Failures, Part 5, *Machine Design*, vol. 33, p. 161, June 22, 1961.
† Alloys are heat-treated, hot-worked; specimens smooth, subjected to long-life rotating-beam tests.

5-11 FINITE-LIFE STRENGTH

In the past many designers have thoughtlessly designed all parts for an infinite life. This is inefficient; even a minor investigation will often reveal that a great many

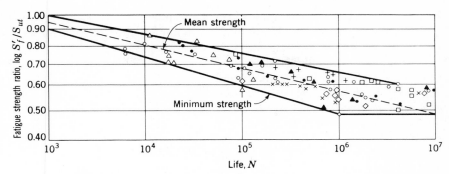

FIGURE 5-14
A compilation of a number of fatigue tests obtained by plotting the ratio of the fatigue strength to the tensile strength. All data from rotating-beam tests. [*By permission from Charles Lipson and Robert C. Juvinall (eds.), "Application of Stress Analysis to Design and Metallurgy," The University of Michigan Summer Conference, Ann Arbor, Mich., 1961.*]

parts are never cycled this long. Actually, it is a very simple matter to determine the fatigue strength S'_f corresponding to any finite life N. Figure 5-14 is a graph showing the results of a large number of fatigue tests of various materials and strengths in which the abscissa is the logarithm of the number of cycles N, as in the usual S-N diagram, but the ordinate is the logarithm of the ratio of the fatigue strength to the tensile strength. We can see from this graph that nearly all the strengths will lie on or above a line drawn from the endurance limit at 10^6 cycles to a point corresponding to $0.9S_{ut}$ at 10^3 cycles. And a dashed line, representing the mean fatigue strength, would begin at the endurance limit at 10^7 cycles and intersect the 1000-cycle line at $0.95S_{ut}$.

In this book anything which has a life of 1000 cycles or less will be treated as a problem in statics, that is, a problem in which the possibility of a fatigue failure is ruled out. We should note, however, that fatigue failures at $N < 1000$ cycles do occur in isolated instances. This is a specialized study and an interesting research topic which cannot be considered here.

The important fact to be observed about Fig. 5-14 is that, with the help of Eqs. (5-20) and (5-21), it furnishes us a means for estimating the fatigue strength S'_f corresponding to any number of cycles from 10^3 to infinity. Thus, using only published tensile-test data, such as that in Table A-17, one can construct a log S–log N chart and on it plot either the minimum-strength line or the mean-strength line. Such a chart will then enable us to find S'_f corresponding to any stated life N.

Now, this is important. Suppose you were asked to serve as an expert witness in a court of law in connection with a case involving fatigue failure. In view of the discussion in Sec. 5-10 on the statistical nature of fatigue, would you use the mean-strength line of Fig. 5-14 or the minimum-strength line to estimate the life N? Figure 5-14 is a plot for many different steels with differing heat treatments and processing methods. The minimum-strength line for the entire group might actually coincide with the mean-strength line for one particular steel. For this reason it might be dangerous to use anything but the minimum-strength line in Fig. 5-14 to define mean strength. Thus *it is recommended that a line on the* log *S–log N chart joining* $0.9S_{ut}$ *at* 10^3 *cycles and-*S'_e *at* 10^6 *cycles be used to define the mean fatigue strength* S'_f *corresponding to any life N.*

An easy way to obtain the fatigue strength S'_f corresponding to a given number of cycles N is to plot the S-N diagram on 2×3 log-log paper. The values are then easy to read off. A disadvantage of this approach is that the slope of the S-N line is so small that accurate results are difficult to obtain.

The availability of the scientific calculator has made an easier method feasible. Let us write the equation of the S-N line as

$$\log S'_f = -m \log N + b \qquad (5\text{-}23)$$

This line must intersect 10^6 cycles at S'_e and 10^3 cycles at $0.9S_{ut}$. When these are substituted into Eq. (5-23), the resulting equations can be solved for the constants m and b. The results are

$$m = \frac{1}{3} \log \frac{0.9S_{ut}}{S'_e} \qquad (5\text{-}24)$$

$$b = \log \frac{(0.9S_{ut})^2}{S'_e} \qquad (5\text{-}25)$$

When S_{ut} and S'_e are given, these two equations can be solved by using the pocket calculator. Then, if N is given and S'_f is to be found, we solve Eq. (5-23) in the form

$$S'_f = \frac{10^b}{N^m} \qquad 10^3 \le N \le 10^6 \qquad (5\text{-}26)$$

which is easily solved by using the calculator. Alternatively, if S_f is given and N is desired, then Eq. (5-23) yields

$$N = \frac{10^{b/m}}{S'^{1/m}_f} \qquad 10^3 \le N \le 10^6 \qquad (5\text{-}27)$$

Of course you should demonstrate for yourself that these relations are correct.

The terms S'_f and S'_e in Eqs. (5-24)–(5-27) designate, respectively, the fatigue strength and endurance limit of a rotating-beam specimen. If these equations are to be used for an actual machine or structural element, the terms should be replaced by S_f and S_e, as in the example which follows.

EXAMPLE 5-3   The endurance limit of a steel member is 112 MPa and the tensile strength is 385 MPa. What is the fatigue strength corresponding to a life of $70(10)^3$ cycles?

SOLUTION   Since $0.9S_{ut} = 0.9(385) = 346$ MPa, Eqs. (5-24) and (5-25) give, respectively,

$$m = \frac{1}{3} \log \frac{0.9S_{ut}}{S_e} = \frac{1}{3} \log \frac{346}{112} = 0.163$$

$$b = \log \frac{(0.9S_{ut})^2}{S_e} = \log \frac{(346)^2}{112} = 3.029$$

Then, from Eq. (5-26), we find the finite-life strength to be

$$S_f = \frac{10^b}{N^m} = \frac{10^{3.029}}{[70(10)^3]^{0.163}} = 173 \text{ MPa} \qquad \textit{Ans.}$$

$$////$$

5-12 CUMULATIVE FATIGUE DAMAGE

Instead of a single reversed stress σ for n cycles, suppose a part is subjected to σ_1 for n_1 cycles, σ_2 for n_2 cycles, etc. Under these conditions our problem is to estimate the fatigue life of a part subjected to these reversed stresses, or to estimate the factor of safety if the part has an infinite life. A search of the literature reveals that this problem has not been solved completely. Therefore the results obtained using either of the approaches presented here should be employed as guides to indicate how one might seek improvement. They should *never* be used to obtain absolute values unless your own experiments indicate the feasibility of doing so. An approach consistently in agreement with experiment has not yet been reported in the literature of the subject.

The theory which is in greatest use at the present time to explain cumulative fatigue damage is the *Palmgren-Miner cycle-ratio summation theory*, also called *Miner's rule*.* Mathematically, this theory is stated as

$$\frac{n_1}{N_1} + \frac{n_2}{N_2} + \cdots + \frac{n_i}{N_i} = C \qquad (5\text{-}28)$$

where n is the number of cycles of stress σ applied to the specimen, and N is the life corresponding to σ. The constant C is determined by experiment and is usually found in the range

$$0.7 \le C \le 2.2$$

Many authorities recommend using $C = 1$, and then Eq. (5-28) may be written

$$\sum \frac{n}{N} = 1 \qquad (5\text{-}29)$$

---

* A. Palmgren, Die Lebensdauer von Kugellagern, *ZVDI*, vol. 68, pp. 339–341, 1924; M. A. Miner, Cumulative Damage in Fatigue, *J. Appl. Mech.*, vol. 12, *Trans. ASME*, vol. 67, pp. A159–A164, 1945.

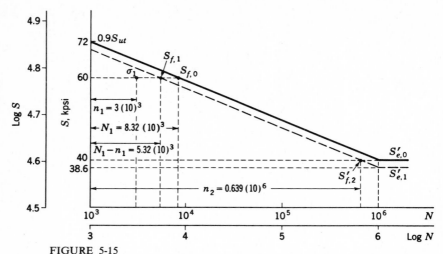

FIGURE 5-15
Use of Miner's rule to predict the endurance limit of a material that has been overstressed for a finite number of cycles.

     To illustrate the use of Miner's rule, let us choose a steel having the properties $S_{ut} = 80$ kpsi and $S'_{e,0} = 40$ kpsi, where we have used the designation $S'_{e,0}$ instead of the more usual S'_e, to indicate the endurance limit of the *virgin*, or *undamaged, material*. The log S–log N diagram for this material is shown in Fig. 5-15 by the heavy solid line. Now apply, say, a reversed stress $\sigma_1 = 60$ kpsi for $n_1 = 3000$ cycles. Since $\sigma_1 > S'_{e,0}$, the endurance limit will be damaged, and we wish to find the new endurance limit $S'_{e,1}$ of the damaged material using Miner's rule. The figure shows that the material has a life $N_1 = 8320$ cycles, and consequently, after the application of σ_1 for 3000 cycles, there are $N_1 - n_1 = 5320$ cycles of life remaining. This locates the finite-life strength $S_{f,1}$ of the damaged material, as shown on Fig. 5-15. To get a second point, we ask the question, With n_1 and N_1 given, how many cycles of stress $\sigma_2 = S'_{e,0}$ can be applied before the damaged material fails? This corresponds to n_2 cycles of stress reversal, and hence, from Eq. (5-29), we have

$$\frac{n_1}{N_1} + \frac{n_2}{N_2} = 1 \qquad (a)$$

or

$$n_2 = \left(1 - \frac{n_1}{N_1}\right) N_2 \qquad (b)$$

Then

$$n_2 = \left[1 - \frac{3(10)^3}{8.32(10)^3}\right](10)^6 = 0.639(10)^6 \text{ cycles}$$

This corresponds to the finite-life strength $S_{f,2}$ in Fig. 5-15. A line through $S_{f,1}$

and $S_{f,2}$ is the log S–log N diagram of the damaged material according to Miner's rule. The new endurance limit is $S_{e,1} = 38.6$ kpsi.

Though Miner's rule is quite generally used, it fails in two ways to agree with experiment. First, note that this theory states that the static strength S_{ut} is damaged, that is, decreased, because of the application of σ_1; see Fig. 5-15 at $N = 10^3$ cycles. Experiments fail to verify this prediction.

Miner's rule, as given by Eq. (5-29), does not account for the order in which the stresses are applied, and hence ignores any stresses less than $S'_{e,0}$. But it can be seen in Fig. 5-15 that a stress σ_3 in the range $S'_{e,1} < \sigma_3 < S'_{e,0}$ would cause damage if applied after the endurance limit had been damaged by the application of σ_1.

Manson's* approach overcomes both of the deficiencies noted for the Palmgren-Miner method; historically it is a much more recent approach, and it is just as easy to use. Except for a slight change, we shall use and recommend the Manson method in this book. Manson plotted the S–log N diagram instead of a log S–log N plot as is recommended here. Manson also resorted to experiment to find the point of convergence of the S–log N lines corresponding to the static strength, instead of arbitrarily selecting the intersection of $N = 10^3$ cycles with $S = 0.9 S_{ut}$ as is done here. Of course, it is always better to use experiment, but our purpose in this book has been to use the simple tension-test data to learn as much as possible about fatigue failure.

* S. S. Manson, A. J. Nachtigall, C. R. Ensign, and J. C. Freche, Further Investigation of a Relation for Cumulative Fatigue Damage in Bending, *Trans. ASME, J. Eng. Ind.*, ser. B, vol. 87, no. 1, pp. 25–35, Febuary 1965.

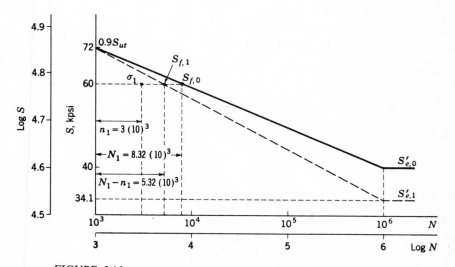

FIGURE 5-16
Use of Manson's method to predict the endurance limit of a material that has been overstressed for a finite number of cycles.

The method of Manson, as presented here, consists in having all log S–log N lines, that is, lines for both the damaged as well as the virgin material, converge to the same point, $0.9S_{ut}$ at 10^3 cycles. In addition, the log S–log N lines must be constructed in the same historical order in which the stresses occur.

The data from the preceding example are used for illustrative purposes. The results are shown in Fig. 5-16. Note that the strength $S_{f,1}$ corresponding to $N_1 - n_1 = 5.32(10)^3$ cycles is found in the same manner as before. Through this point and through $0.9S_{ut}$ at 10^3 cycles draw the heavy dashed line to meet $N = 10^6$ cycles and define the endurance limit $S'_{e,1}$ of the damaged material. In this case the new endurance limit is 34.1 kpsi, somewhat less than that found by Miner's method.

It is now easy to see from Fig. 5-16 that a reversed stress $\sigma = 36$ kpsi, say, would not harm the endurance limit of the virgin material, no matter how many cycles it might be applied. However, if $\sigma = 36$ kpsi should be applied *after* the material had been damaged by $\sigma_1 = 60$ kpsi, then additional damage would be done.

5-13   ENDURANCE-LIMIT MODIFYING FACTORS

The endurance limit S_e of a machine element may be considerably smaller than the endurance limit S'_e of the rotating-beam specimen. This difference may be explained by employing a variety of modifying factors, each of which accounts for a separate effect. Using this idea, we may write

$$S_e = k_a k_b k_c k_d k_e k_f S'_e \qquad (5\text{-}30)$$

where S_e = endurance limit of mechanical element
$\quad S'_e$ = endurance limit of rotating-beam specimen
$\quad k_a$ = surface factor
$\quad k_b$ = size factor
$\quad k_c$ = reliability factor
$\quad k_d$ = temperature factor
$\quad k_e$ = modifying factor for stress concentration
$\quad k_f$ = miscellaneous-effects factor

5-14   SURFACE FINISH

The surface of the rotating-beam specimen is highly polished, with a final polishing in the axial direction to smooth out any circumferential scratches. Obviously, most machine elements do not have such a high-quality finish. The modification factors k_a, shown in Fig. 5-17, depend upon the quality of the finish and upon the tensile strength. You should examine this chart carefully. It emphasizes the great importance of having a good surface finish when fatigue failure is a possibility.

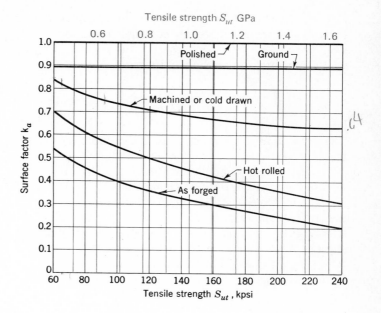

FIGURE 5-17
Surface-finish modification factors for steel. These are the k_a factors for use in Eq. (5-30).

The factors shown on this chart have been obtained by condensing large compilations of data from tests of wrought steels and are probably also valid for cast steels and the better grades of cast iron.

The surface factors for nonferrous materials, such as the aluminum alloys, should be taken as unity because the tabulated endurance limits of these materials include the surface-finish effect.

5-15 SIZE EFFECTS

The rotating-beam test gives the endurance limit for a specimen 0.30 in in diameter. When specimens of larger size are tested with completely reversed stresses in bending or in torsion, it is found that the endurance limit is 10 to 15 percent less for specimens up to 2 in (50 mm). For specimens larger than 2 in the endurance limit may be as much as 25 percent less, but individual tests should be run in such cases.

The triangular stress distribution, in the case of bending and torsion, is very similar to the stress distribution in a notched bar. That is, bending and torsion resemble stress concentration. It is probably this, combined with the fact that a large specimen will probably have more surface defects than a small one, which accounts for the reduction in the bending and torsional endurance limits due to size.

Therefore, for bending and torsion, k_b should be selected as follows:

$$k_b = \begin{cases} 1 & d \leq 0.30 \text{ in (7.6 mm)} \\ 0.85 & 0.30 < d \leq 2 \text{ in (50 mm)} \\ 0.75 & d > 2 \text{ in (50 mm)} \end{cases}$$

The dimension d corresponds to the section depth h for nonround sections in bending.

Surprisingly enough, the values of k_b listed above are also appropriate for reversed axial loads, but for completely different reasons. In the case of axial loading, the stress distribution is constant, and hence the stress concentration near the surface is absent. However, careful measurement of the endurance limits of both cast and wrought steels shows some decrease in the endurance limit for specimens taken near the core as compared with specimens taken near the surface. It therefore seems perfectly appropriate to use the values listed above for axial loads too. Then, for nonround sections, d is taken as the smallest cross-sectional dimension.

5-16 RELIABILITY*

In this section we shall outline an analytical approach which will enable you to design a mechanical element subjected to fatigue loads such that it will last for any desired life at any stated reliability. In many ways life and reliability may constitute a more effective method of measuring design efficiency than the use of a factor

---

* The following books and papers on reliability are recommended for additional study and information:

Jack R. Benjamin and C. Allin Cornell, "Probability, Statistics, and Decision for Civil Engineers," McGraw-Hill Book Company, New York, 1970

Charles Lipson and Narendra J. Sheth, "Statistical Design and Analysis of Engineering Experiments," McGraw-Hill Book Company, New York, 1973.

Edward B. Haugen and Paul H. Wirsching, Probabilistic Design, *Machine Design*, vol. 47, nos. 10–14, 1975.

Dimitri Kececioglu, Louie B. Chester, and Thomas M. Dodge, Combined Bending-Torsion Fatigue Reliability of AISI 4340 Steel Shafting with $K_t = 2.34$, *ASME* paper no. 74-WA/DE-12, 1974.

C. Mischke, A Method of Relating Factor of Safety and Reliability, *ASME* paper no. 69-WA/DE-6, 1969.

C. Mischke, Designing to a Reliability Specification, *SAE* paper no. 740643, 1974.

C. R. Mischke, A Rationale for Mechanical Design to a Reliability Specification, Implementing Mechanical Design to a Reliability Specification, and Organizing the Computer for Mechanical Design, *ASME* paper in The Design Engineering Technical Conference, New York, Oct. 5–9, 1974.

My Dao-Thien and M. Massoud, On the Probabilistic Distributions of Stress and Strength in Design Problems, *ASME* paper no. 74-WA/DE-7, 1974.

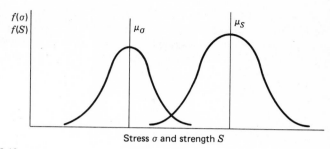

FIGURE 5-18
Graph of stress and strength distributions showing the mean stress μ_σ and the mean strength μ_S.

of safety because both life and reliability are easily measured. The approach presented here is a logical one but additional testing is needed before it can be recommended for general use. In particular, do not expect the method to yield absolute values. Its greatest use will be as a guide to reveal the most effective thing to do to improve life and reliability of actual parts.

In order to define the exact meaning of reliability let us suppose that we have a large group or population of mechanical parts. We can associate a certain strength S and a certain stress σ with each part. But since there are a large number of them, we have a population of strengths and a population of stresses. These two populations might be distributed somewhat as shown in Fig. 5-18. Using the notation of Sec. 4-8 we designate μ_σ and $\hat{\sigma}_\sigma$ as the mean and the standard deviation of the stress, and μ_S and $\hat{\sigma}_S$ as the mean and standard deviation of the strength. Although strength is generally greater than the stress, Fig. 5-18 shows that the forward tail of the stress distribution may overlap the rearward tail of the strength distribution and result in some failures. To determine the reliability, we combine both populations using Eqs. (4-39) and (4-40). The combined population then has a mean value and standard deviation of

$$\mu = \mu_S - \mu_\sigma \qquad \hat{\sigma} = \sqrt{\hat{\sigma}_S^2 + \hat{\sigma}_\sigma^2}$$

The corresponding standardized variable z_R is

$$z_R = \frac{\mu}{\hat{\sigma}} = \frac{\mu_S - \mu_\sigma}{\sqrt{\hat{\sigma}_S^2 + \hat{\sigma}_\sigma^2}} \qquad (5\text{-}31)$$

By entering this value of z_R in Table A-14, the area A_z under the normal distribution curve corresponding to the combined population can be found. Then the reliability R is

$$R = 0.5 + A_z \qquad (5\text{-}32)$$

Equation (5-32) enables us to determine the standardized variable z_R corresponding to any desired reliability. Thus, using Table A-14 we find $z_R = 1.288$ for 90 percent reliability ($R = 0.90$ and $A_z = 0.4000$).

An examination of Table 5-1 shows that the standard deviation of the endurance limit for steels is not likely to exceed 8 percent. In fact data presented by Haugen and Wirsching* also show standard deviations of less than 8 percent. This means that we can obtain the endurance limit corresponding to any specified reliability R merely by subtracting a number of standard deviations from the mean endurance limit. Thus, the reliability factor k_c is

$$k_c = 1 - 0.08z_R \qquad (5\text{-}33)$$

Table 5-2 shows the standardized variable z_R corresponding to the various reliabilities required in design together with the corresponding reliability factor k_c computed from Eq. (5-33).†

In using the approach suggested here, it is important to remember that the actual distribution of fatigue strengths can be better approximated by the Weibull distribution‡,§ than by the normal distribution. The normal distribution is preferred here because of the convenience of matching stress with strength.

In using Table 5-2 you should observe carefully the conditions applying to the use of Eqs. (5-20) and (5-21) to find S'_e. Briefly, these conditions require that the ultimate tensile strength S_{ut} be known with certainty, unless S'_e is found by some other method.

Students who are solving problems for practice purposes require a standard approach for selecting S_{ut} so that the "correct answer" will be the same for all members of the class. You may have noted in referring to Table A-17 that the strengths are estimated minimum values for hot-rolled and cold-drawn bars, and that these are typical values when the steels are heat-treated. Thus these values can be obtained provided one adheres to the specifications and takes sufficient care in

* *Op. cit.*, no. 12, May 15, 1975.
† I am very grateful to Professor Charles Mischke of Iowa State University for working out the higher values in this table and permitting me to include them here. J.E.S.
‡ Lipson and Sheth, *op. cit.*, p. 324
§ See Sec. 9-2.

Table 5-2 RELIABILITY FACTORS k_c CORRESPONDING TO AN 8 PERCENT STANDARD DEVIATION OF THE ENDURANCE LIMIT

Reliability R	Standardized variable z_R	Reliability factor k_c
0.50	0	1.000
0.90	1.288	0.897
0.95	1.645	0.868
0.99	2.326	0.814
0.999	3.091	0.753
0.999 9	3.719	0.702
0.999 99	4.265	0.659
0.999 999	4.753	0.620
0.999 999 9	5.199	0.584
0.999 999 99	5.612	0.551
0.999 999 999	5.997	0.520

inspection and in processing. In this book, therefore, we shall assume that quality-control procedures have been established which assure that the tensile strength S_{ut} is always equal to or greater than the values shown in Table A-17 when fatigue failure must be guarded against.

You are cautioned very particularly that these recommendations are for students only. Corresponding guidelines in the practice of engineering must be obtained from in-plant experience.

5-17 TEMPERATURE EFFECTS

The temperature factor k_d should be obtained from actual tests if possible when high-temperature operation is required. In such cases it may be necessary to apply k_d to both ends of the S-N diagram because the static strength may be reduced too. It is also desirable in such cases to check against the possibility of failure due to creep. For the temperature factor use, for steels,

$$k_d = \frac{620}{460 + T} \qquad (5\text{-}34)$$

when $T > 160$ F; otherwise use $k_d = 1$.

5-18 STRESS CONCENTRATION*

In the development of the basic stress relations it is assumed that cross sections remain constant and that there are no irregularities in the member. However, most mechanical parts do have holes, grooves, notches, or other kinds of discontinuities present. Any such discontinuity alters the stress distribution so that the basic stress relations no longer describe the stress state. These discontinuities are called *stress raisers*, and the regions in which they occur are called areas of stress concentration.

A *theoretical*, or *geometric*, *stress-concentration factor* K_t or K_{ts} is used to relate the actual maximum stress at the discontinuity to the nominal stress. Then the maximum stresses are given by the equations

$$\sigma_{max} = K_t \sigma_0 \qquad \tau_{max} = K_{ts} \tau_0 \qquad (5\text{-}35)$$

where σ_0 is the usual Mc/I or F/A type of normal stress and τ_0 is a Tc/J or F/A shear stress; these are mostly based on the net section areas. Table A-25 contains a number of charts from which most values of K_t, or K_{ts}, can be obtained. For others, use the index to find the particular mechanical element under consideration.

---

* The most authoritative reference on stress concentration to be found is R. E. Peterson, "Stress Concentration Factors," John Wiley & Sons, New York, 1974.

Stress concentration is a highly localized effect. The high stresses actually exist in only a very small region in the vicinity of the discontinuity. In the case of ductile materials the first load applied to the member will cause yielding at the discontinuity which relieves the stress concentration. Thus when the parts are made of ductile materials and the loads are static, it isn't necessary to use a stress-concentration factor at all.*

Stress concentration does have to be considered, however, when parts are made of brittle materials or when they are subject to fatigue loading. Even under these conditions, however, it turns out that some materials are not very sensitive to the existence of notches or discontinuities and hence that the full values of the theoretical stress concentration factors need not be used. For these materials it is convenient to use a reduced value of K_t. The resulting factor is defined by the equation

$$K_f = \frac{\text{endurance limit of notch-free specimens}}{\text{endurance limit of notched specimens}} \qquad (a)$$

This factor is usually called a *fatigue stress-concentration factor* although it is used for brittle materials under static loads too.

Now, in using K_f, it does not matter, algebraically, whether it is used as a factor for *increasing the stress* or whether it is used for *decreasing the strength*. This simply means that it may be placed on one side of the equation or the other. But a great many troublesome problems can be avoided if K_f is treated as a factor which reduces the strength of a member. Therefore we shall call K_f a *fatigue-strength reduction factor* and shall nearly always use it in this sense. This means that the modifying factor for stress concentration k_e of Eq. (5-30) and K_f have the relation

$$k_e = \frac{1}{K_f} \qquad (5\text{-}36)$$

Notch sensitivity q is defined by the equation

$$q = \frac{K_f - 1}{K_t - 1} \qquad (5\text{-}37)$$

where q is usually between zero and unity. Equation (5-37) shows that, if $q = 0$, $K_f = 1$, and the material has no sensitivity to notches at all. On the other hand, if $q = 1$, then $K_f = K_t$, and the material has full sensitivity. In analysis or design work one first determines K_t from the geometry of the part. Then, the material having been specified, q can be found, and the equation is solved for K_f.

$$K_f = 1 + q(K_t - 1) \qquad (5\text{-}38)$$

For steels and UNS A92024 aluminum alloys use Fig. 5-19 to find q when the parts are subjected either to rotating-beam action or to reversed axial loading. Use Fig. 5-20 for parts subjected to reversed shear.

---

* In a previous edition of this book, pp. 225–228 (2d ed.), it is shown that the yielding in the vicinity of a stress raiser can actually improve the static strength of the part.

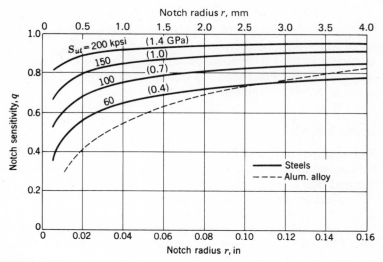

FIGURE 5-19
Notch-sensitivity charts for steels and UNS A92024-T wrought aluminum alloys subjected to reversed bending or reversed axial loads. For larger notch radii use the values of q corresponding to $r = 0.16$ in (4 mm). [*Reproduced by permission from George Sines and J. L. Waisman (eds.), "Metal Fatigue," pp. 296, 298, McGraw-Hill Book Company, New York, 1959.*]

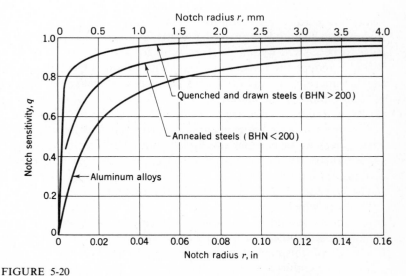

FIGURE 5-20
Notch-sensitivity curves for materials in reversed torsion. For larger notch radii use the values of q corresponding to $r = 0.16$ in (4 mm).

Both charts (Figs. 5-19 and 5-20) show that for large notch radii, and especially for high-strength materials, the sensitivity index approaches unity. This means that whenever there is any doubt, one can always make $K_f = K_t$ and err on the safe side. Also, if the notch radius is quite large—and it should be designed this way if at all possible—then q is not far from unity and the error of assuming K_f equal to K_t will be quite small.

Brittle Materials

The notch sensitivity of the cast irons is very low, varying from about zero to 0.20, depending upon the tensile strength. To be on the conservative side, it is recommended that a notch sensitivity $q = 0.20$ be used for all grades of cast iron.

Since brittle materials have no yield strength, the stress-concentration factor K_f must be applied to the static strength S_{ut} or S_{uc} as well as to the endurance limit as observed earlier. This means that both ends of the S-N diagram for cast iron would be lowered the same amount since K_f must be used to reduce the strength at each end.

5-19 MISCELLANEOUS EFFECTS

Though the factor k_f is intended to account for the reduction in endurance limit due to all other effects, it is really intended as a reminder that these must be accounted for, because actual values of k_f are not available.

Residual stresses may either improve the endurance limit or affect it adversely. Generally, if the residual stress in the surface of the part is compression, the endurance limit is improved. Fatigue failures appear to be tensile failures, or at least to be caused by tensile stress, and so anything which reduces tensile stress will also reduce the possibility of a fatigue failure. Operations such as shot peening, hammering, and cold rolling build compressive stresses into the surface of the part and improve the endurance limit significantly. Of course, the material must not be worked to exhaustion.

The endurance limits of parts which are made from rolled or drawn sheets or bars, as well as parts which are forged, may be affected by the so-called *directional characteristics* of the operation. Rolled or drawn parts, for example, have an endurance limit in the transverse direction which may be 10 to 20 percent less than the endurance limit in the longitudinal direction.

Parts which are case-hardened may fail at the surface or at the maximum core radius, depending upon the stress gradient. Figure 5-21 shows the typical triangular stress distribution of a bar under bending or torsion. Also plotted as a heavy line on this figure are the endurance limits S_e for the case and core. For this example the endurance limit of the core rules the design because the figure shows that the stress σ or τ, whichever applies, at the outer core radius, is appreciably larger than the core endurance limit.

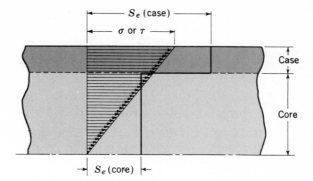

FIGURE 5-21
The failure of a case-hardened part in bending or torsion. In this example failure occurs in the core.

Of course, if stress concentration is also present, the stress gradient is much steeper, and hence failure in the core is unlikely.

Corrosion

It is to be expected that parts which operate in a corrosive atmosphere will have a lowered fatigue resistance. This is of course true, and it is due to the roughening or pitting of the surface by the corrosive material. But the problem is not so simple as the one of finding the endurance limit of a specimen which has been corroded. The reason for this is that the corrosion and the stressing occur at the same time, so that the resultant weakening is much greater than that produced by the two factors acting one at a time. In fact, the result of the stresses is to increase the amount of corrosion.

Plating

Metallic coatings, such as chromium plating, nickel plating, or cadmium plating, reduce the endurance limit by as much as 35 percent. In some cases the reduction by coatings has been so severe that it has been necessary to eliminate the plating process.

EXAMPLE 5-4   Corresponding to a reliability of 99 percent, find the endurance limit of a 1-in round UNS G10150 cold-drawn steel bar.

SOLUTION   From Table A-17, we find $S_{ut} = 56$ kpsi and $H_B = 111$.
From Fig. 5-17, $k_a = 0.84$.
Since the endurance limit means bending endurance limit unless specified differently, we find $k_b = 0.85$ for reversed bending, from Sec. 5-15.

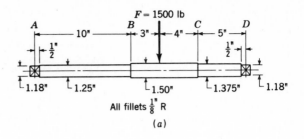

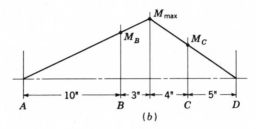

FIGURE 5-22
(a) Shaft drawing. The shaft rotates and the load is stationary. The material is UNS G10350 steel, drawn at 1000°F, with a machined finish. (b) Bending-moment diagram.

From Table 5-2, $k_c = 0.814$.
Since nothing is stated, we assume $k_d = k_e = k_f = 1$.
From Eq. (5-20)

$$S_e' = 0.50 S_{ut} = 0.50(56) = 28.0 \text{ kpsi}$$

Then, from Eq. (5-30)

$$S_e = (0.84)(0.85)(0.814)(28.0) = 16.3 \text{ kpsi} \qquad Ans.$$

////

EXAMPLE 5-5   Figure 5-22a shows a rotating shaft supported in ball bearings at A and D and loaded by the nonrotating force F. Using the methods of the preceding sections, estimate the life of the part.

SOLUTION   From Fig. 5-22b we learn that failure will probably occur at B rather than at C. Point B has a smaller cross section, a higher bending moment, and a higher stress-concentration factor. It is unlikely that failure would occur under the load F, even though the maximum moment occurs there, because there is no stress concentration and the section is larger.
We shall solve the problem by finding the strength at point B, since it will

probably be different at other points, and comparing this strength with the stress at point B.

From Table A-17 find that $S_{ut} = 103$ kpsi and $S_{yt} = 72$ kpsi. Therefore

$$S'_e = (0.5)(103) = 51.5 \text{ kpsi}$$

The surface and size factors are next found to be $k_a = 0.73$ and $k_b = 0.85$. The reliability is always taken as 50 percent unless a specific value is given; therefore $k_c = 1$. Also, $k_d = 1$, since nothing concerning temperature is stated. Then, using Fig. A-25-9, we calculate

$$\frac{D}{d} = \frac{1.5}{1.25} = 1.20 \qquad \frac{r}{d} = \frac{0.125}{1.25} = 0.10$$

and find $K_t = 1.60$. Then, entering Fig. 5-19 with $S_{ut} = 103$ kpsi, find $q = 0.82$. The fatigue-strength reduction factor is found to be

$$K_f = 1 + q(K_t - 1) = 1 + 0.82(1.60 - 1) = 1.49$$

Consequently, the modifying factor for stress concentration is $k_e = 1/K_f = 1/1.49 = 0.671$. The endurance limit at point B is therefore

$$S_e \text{ (at } B) = k_a k_b k_e S'_e = (0.73)(0.85)(0.671)(51.5) = 21.4 \text{ kpsi}$$

Now, to determine the stress at B: The bending moment is

$$M_B = 10 \frac{9F}{22} = (10) \frac{(9)(1500)}{22} = 6140 \text{ lb·in}$$

The section modulus is $I/c = \pi d^3/32 = \pi(1.25)^3/32 = 0.192 \text{ in}^3$. Therefore the stress is

$$\sigma = \frac{M}{I/c} = \frac{6140}{0.192} = 32(10)^3 \text{ psi}$$

Since this stress is greater than the endurance limit, the part will have only a finite life.

Following the procedure of Example 5-3, we first find $0.9S_{ut} = 0.9(103) = 92.7$ kpsi. Then

$$m = \frac{1}{3} \log \frac{0.9S_{ut}}{S_e} = \frac{1}{3} \log \frac{92.7}{21.4} = 0.212$$

$$b = \log \frac{(0.9S_{ut})^2}{S_e} = \log \frac{(92.7)^2}{21.4} = 2.604$$

Next, we compute $b/m = 2.604/0.212 = 12.28$, and $1/m = 1/0.212 = 4.717$. Then, from Eq. (5-27) the estimated life at 50 percent reliability is found to be

$$N = \frac{10^{b/m}}{S_f^{1/m}} = \frac{10^{12.28}}{32^{4.717}} = 151(10)^3 \text{ cycles} \qquad Ans.$$

////

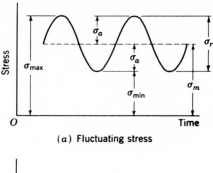

(a) Fluctuating stress

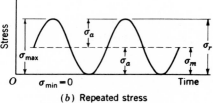

(b) Repeated stress

FIGURE 5-23
Types of fatigue stress.

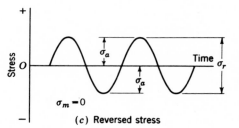

(c) Reversed stress

5-20 FLUCTUATING STRESSES

Quite frequently it is necessary to determine the strength of parts corresponding to stress situations other than complete reversals. Many times in design the stresses fluctuate without passing through zero. Figure 5-23 illustrates some of the various stress-time relationships which may occur. The components of stress with which we must deal, some of which are shown in Fig. 5-23a, are

σ_{min} = minimum stress σ_m = mean stress

σ_{max} = maximum stress σ_r = stress range

σ_a = stress amplitude σ_s = steady, or static, stress

The steady, or static, stress is *not* the same as the mean stress; in fact, it may have any value between σ_{min} and σ_{max}. The steady stress exists because of a fixed load or preload applied to the part, and it is usually independent of the varying portion of the load. A helical compression spring, for example, is always loaded into a space shorter than the free length of the spring. The stress created by this initial compression is called the steady, or static, component of the stress. It is not the same as the mean stress.

We shall have occasion to apply the subscripts of these components to shear stresses as well as normal stresses.

The following relations are evident from Fig. 5-23:

$$\sigma_m = \frac{\sigma_{max} + \sigma_{min}}{2} \qquad (5\text{-}39)$$

$$\sigma_a = \frac{\sigma_{max} - \sigma_{min}}{2} \qquad (5\text{-}40)$$

Although the stress components have been defined by using a sine stress-time relation, the exact shape of the curve does not appear to be of particular significance.

5-21 FATIGUE STRENGTH UNDER FLUCTUATING STRESSES

Now that we have defined the various components of stress associated with a part subjected to fluctuating stress, we want to vary both the mean stress and the stress amplitude, to learn something about the fatigue resistance of parts when subjected to such situations. Two methods of plotting the results of such tests are in general use and are both shown in Fig. 5-24.

The *modified Goodman diagram* of Fig. 5-24a has the mean stress plotted along the abscissa and all other components of stress plotted on the ordinate, with tension in the positive direction. The endurance limit, fatigue strength, or finite-life strength, whichever applies, is plotted on the ordinate above and below the origin. The mean-stress line is a 45° line from the origin to point A, representing the tensile strength of the part. The modified Goodman diagram consists of the lines constructed from point A to S_e (or S_f) above and below the origin. A better average through the points of failure would be obtained by constructing curved lines from point A, but straight lines are easier to construct, and so we shall accept the modified Goodman diagram as defining failure in this book. Note that the yield strength is also plotted on both axes, because yielding would be the criterion of failure if σ_{max} exceeded S_y.

Another fatigue diagram which is often used is that of Fig. 5-24b. Here the mean stress is also plotted on the abscissa, tension to the right and compression to the left. But the ordinate has only the stress amplitude σ_a plotted along it. Thus, with this diagram, only two of the stress components are used. The endurance limit, fatigue strength, or finite-life strength, whichever applies to the particular problem, is the limiting value of the stress amplitude, and so it is plotted on the

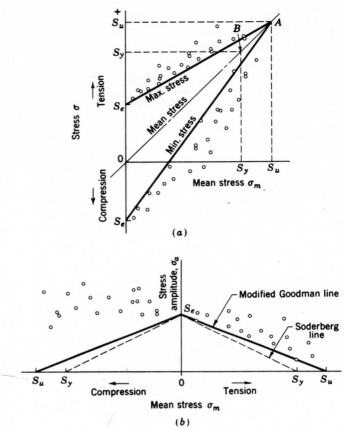

FIGURE 5-24
Two types of fatigue diagrams showing typical points of failure. (a) Modified
Goodman diagram; (b) diagram showing the modified Goodman line.

ordinate. A straight line from S_e to S_u on the abscissa is also the modified Goodman criterion of failure. Note that when the mean stress is tension, most of the
failure points fall above this line. On the compression side, however, the failure
points show that the magnitude of the mean stress has no effect. The Soderberg
line, drawn from S_e to S_y, has also been proposed as a criterion for design, because
yielding is also used to define failure. Note, however, that the modified Goodman
line errs on the safe side, and so the Soderberg line is even more conservative.

Now that we have used these two diagrams to learn how parts fail, we can
use them to construct a well-defined criterion of fatigue or static failure.

As noted, the modified Goodman diagram is particularly useful because it
contains all the stress components. In Fig. 5-25 the diagram has been redrawn to
show all these stress components and also the manner in which it will be used to

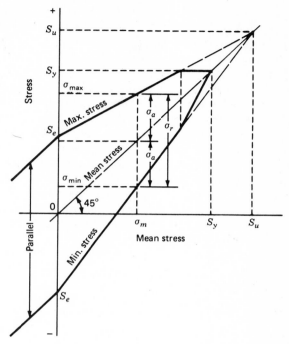

FIGURE 5-25
Modified Goodman diagram showing all the strengths and the limiting values of all stress components for a particular mean stress.

define failure. When the mean stress is compression, failure is defined by the two heavy parallel lines originating at $+S_e$ and $-S_e$ and drawn downward and to the left. When the mean stress is tension, failure is defined by the maximum-stress line or by the yield strength as indicated by the heavy outline to the right of the ordinate. The modified Goodman diagram is particularly useful for analysis when all the dimensions of the part are known and the stress components can be easily calculated. But it is rather difficult to use for design, that is, when the dimensions are unknown.

The fatigue diagram of Fig. 5-26* is the one we shall employ for design purposes, where the heavy lines will be taken as the criterion of failure. To explain,

* A justifiable confusion exists with respect to the correct terminology for the fatigue diagrams of Figs. 5-24 to 5-26. Figure 5-24a uses the axes originally employed by Goodman, but the diagram is a modification of his proposal. This modified Goodman diagram is modified again in Fig. 5-24b because a new set of axes have been employed. Still another modification is introduced in Fig. 5-25, where the yield strength has been added as a limiting factor and the diagram has been extended into the compressive-stress region. The fourth modification is obtained by converting this to the axes of Fig. 5-26. The original Goodman diagram, which is not shown, has not been employed for many years. For these reasons many people, quite correctly, simplify the whole situation and call Fig. 5-25 the *Goodman diagram*, and Fig. 5-26 the *modified Goodman diagram*.

note that the yield strength has been plotted on the mean-stress axis, both for tension and for compression, and also on the stress-amplitude axis. A line from S_y to S_{yc} defines failure by compressive yielding; from S_y to S_{yt}, failure by tensile yielding. We construct the modified Goodman line for tensile mean stress and a horizontal line from S_e to the left for compressive mean stress. The intersections of the two lines in each quadrant are the transition points between a failure by fatigue and a failure by yielding. The heavy outline therefore specifies when failure by either method will occur.

In design work the force amplitude and the mean force can usually be calculated or determined. Sometimes we are working with bending moments or torsional moments. In these cases one can usually calculate the mean moment and the amplitude of the moment. The stress amplitude and the mean stress are related to these through the dimensions which are to be found. The ratio σ_a/σ_m is the same as F_a/F_m or M_a/M_m. So a line from the origin through point A can be drawn and the limiting values of σ_a or σ_m found as the projections of this point on the two axes.

The simplicity of the approach described here is worth noting. The value of the fatigue strength plotted on the ordinate has previously been corrected for size, surface finish, reliability, stress concentration, and other effects. Consequently one need not worry about which stress components these factors should have been applied to.*

The modified Goodman criterion applies for the cast irons when the mean stress is tension.† The existence of a compressive mean stress has no effect on the endurance limit, however.

EXAMPLE 5-6   It is desired to determine the size of a UNS G10500 cold-drawn steel bar to withstand a tensile preload of 8 kip and a fluctuating tensile load varying from 0 to 16 kip. Owing to the design of the ends, the bar will have a geometric stress-concentration factor of 2.02 corresponding to a fillet whose radius is $\frac{3}{16}$ in. Determine a suitable diameter for an infinite life and a factor of safety of at least 2.0.

---

* More sophisticated approaches are available however. Some of these include a factor of safety, some are directed specifically to design, in contrast to analysis, and others are quite analytical. The following references are highly authoritative and are recommended:

W. N. Findley, J. J. Coleman, and B. C. Hanley, Theory for Combined Bending and Torsion Fatigue, *Proceedings of the International Conference on the Fatigue of Metals*, London, 1956.

Robert C. Juvinall, "Engineering Considerations of Stress, Strain, and Strength," pp. 268–314, McGraw-Hill Book Company, New York, 1967.

Robert E. Little, "Analysis of the Effect of Mean Stress on Fatigue Strength of Notched Steel Specimens," doctoral dissertation, publication IP-63Q, The University of Michigan, 1963.

L. D. Mitchell and D. T. Vaughan, A General Method for the Fatigue-Resistant Design of Mechanical Components, Part 1 Graphical, *ASME* paper no. 74-WA/DE-4, Part 2 Analytical, *ASME* paper no. 74WA/DE-5, 1974.

George Sines and J. L. Waisman (eds.), "Metal Fatigue," chap. 7, McGraw-Hill Book Company, New York, 1959.

† See "Metals Handbook," 8th ed., pp. 356 and 357, American Society for Metals, Metals Park, Ohio, 1961.

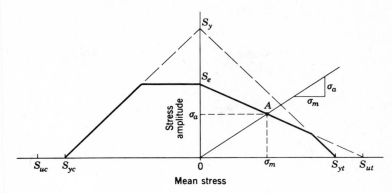

FIGURE 5-26
Fatigue diagram showing how to find the limiting values of σ_a and σ_m when the ratio of the two is given.

SOLUTION From Table A-17, $S_y = 84$ kpsi and $S_{ut} = 100$ kpsi. So $S'_e = 0.50S_{ut} = 0.50(100) = 50$ kpsi. We next find $k_a = 0.73$ from Fig. 5-17. For axial loads, $k_b = 0.85$. From Fig. 5-19, $q = 0.86$, and so

$$K_f = 1 + q(K_t - 1) = 1 + 0.86(2.02 - 1) = 1.87$$

so that $k_e = 1/K_f = 1/1.87 = 0.535$. These are all the necessary corrections, so that

$$S_e = (0.73)(0.85)(0.535)(50) = 16.6 \text{ kpsi}$$

Next we determine the stresses in terms of their dimensions. The static stress is

$$\sigma_s = \frac{F_s}{A} = \frac{8}{\pi d^2/4} = \frac{10.2}{d^2} \text{kpsi}$$

The stress range is

$$\sigma_r = \frac{F_r}{A} = \frac{16}{\pi d^2/4} = \frac{20.4}{d^2} \text{kpsi}$$

Then

$$\sigma_a = \frac{\sigma_r}{2} = (10.2/d^2) \text{ kpsi}$$

and in this case,

$$\sigma_m = \sigma_s + \sigma_a = (20.4/d^2) \text{ kpsi}$$

Therefore

$$\sigma_a/\sigma_m = 0.50$$

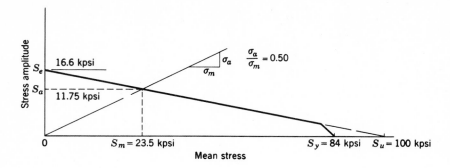

FIGURE 5-27
Corresponding to $\sigma_a/\sigma_m = 0.50$, S_a is the *strength amplitude* and S_m the *mean strength*.

To relate the stresses and strengths, we plot a fatigue diagram (Fig. 5-27). Note that only the tensile side is needed. Both axes are plotted to the same scale in this example. Sometimes, though, more accuracy can be obtained if they are plotted at different scales. If this is done, be careful in constructing the line from the yield strength to the transition point, as it will not make a 45° angle with the mean-stress axis. The intersection of the modified Goodman line with another line at a slope $\sigma_a/\sigma_m = 0.50$ defines *two* values of *strength*. Using a mnemonic notation for these, S_a is a strength corresponding to the stress σ_a, and S_m is a strength corresponding to the stress σ_m. For a factor of safety of 2.0 we have

$$\sigma_a \leq S_a/2.0$$

Consequently

$$10.2/d^2 \leq 11.75/2.0 \qquad \text{or} \qquad d \geq 1.32 \text{ in}$$

We therefore choose $d = 1\frac{3}{8}$ in to get it in fractional-inch or stock size.   ////

5-22 TORSIONAL FATIGUE STRENGTH

In Sec. 5-5 we learned that the maximum-shear-stress theory predicted the yield strength in shear to be

$$S_{sy} = 0.50S_y \qquad (a)$$

and that this relation gives conservative values. Thus Eq. (*a*) is useful for design, because it is easy to apply and to remember, but not for the analysis of failure. A more accurate prediction of failure, we learned, is given by the distortion-energy theory, which predicts the yield strength in shear to be

$$S_{sy} = 0.577S_y \qquad (b)$$

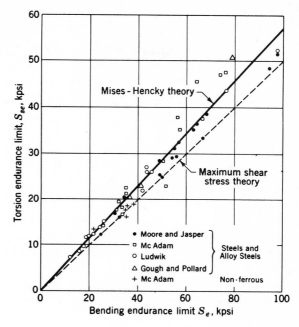

FIGURE 5-28
Relation between endurance limits in torsion and in bending. [*From Thomas J. Dolan, Stress Range, in Oscar J. Horger (ed.), "ASME Handbook-Metals Engineering— Design," sec. 6-2, p. 97, McGraw-Hill Book Company, New York, 1953. Reproduced by permission of the publishers.*]

Interestingly enough, as indicated by experiments whose results are shown in Fig. 5-28, these two theories are also useful in predicting the endurance limit in shear S_{se} when the bending endurance limit S_e is known. Thus the maximum-shear-stress theory conservatively predicts

$$S_{se} = 0.50S_e \qquad (5\text{-}41)$$

and the distortion-energy theory yields

$$S_{se} = 0.577S_e \qquad (5\text{-}42)$$

We shall employ only Eq. (5-42) in this book because, as shown in Fig. 5-28, it predicts failure more accurately.

Let us now consider the case in which there is a torsional stress amplitude τ_a and a torsional mean stress τ_m. Corresponding strengths are the torsional or shear endurance limit S_{se}, the yield strength in shear S_{sy}, and the torsional modulus of rupture S_{su}. Using these strengths, it should be possible to construct a torsional fatigue diagram corresponding to that of Fig. 5-26. When we do this, and also plot

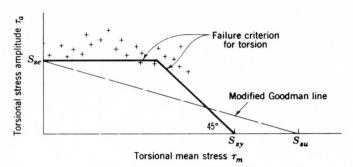

FIGURE 5-29
Fatigue diagram for combined alternating and mean torsional stress showing failure points.

a number of experimental observations of failures on it, we get the diagram of Fig. 5-29.

The interesting thing about Fig. 5-29 is that, up to a certain point, torsional mean stress has no effect on the torsional endurance limit. Thus it is not necessary to construct such a diagram for torsion at all! Instead, a fatigue failure is indicated if

$$\tau_a = S_{se} \qquad (5\text{-}43)$$

and a static failure if

$$\tau_{\max} = \tau_a + \tau_m = S_{sy} \qquad (5\text{-}44)$$

as indicated by the solid line of Fig. 5-29. Of course, these equations can be used for design too if a factor of safety is used.

5-23 FATIGUE FAILURE DUE TO COMBINED STRESSES

One of the most frequently encountered problems in design is that of a rotating shaft subjected to a constant torque and a stationary bending load. An element on the surface of the shaft has a torsional stress $\tau = Tc/J$ which is constant in magnitude and direction when referred to a mark made on the shaft surface. But, owing to the bending moment, the same element will have a normal stress $\sigma = \pm Mc/I$ varying from tension to compression and back again, as the shaft rotates. If the stresses on the element are analyzed using a Mohr's circle diagram, it will be found that the principal stresses do not maintain the same orientation, relative to a mark on the surface, as the shaft rotates.

The problem is even more complicated when it is realized that the normal stresses σ_x and σ_y as well as the shear stress τ_{xy}, in the general two-dimensional stress state, may have both mean and alternating components. In this book we shall present a method of using the distortion-energy theory applied to fatigue to solve this problem because all available experimental evidence shows it to be conservative and because the method employs the basic theory already developed in this chapter.*

To apply the theory, construct two stress elements, one for the mean stresses and one for the alternating stresses. Two Mohr's circles are then drawn, one for each element, and the principal mean stresses obtained from one circle and the principal alternating stresses obtained from the other. We can then define mean and alternating von Mises stresses as

$$\sigma'_m = \sqrt{\sigma_{1m}^2 - \sigma_{1m}\sigma_{2m} + \sigma_{2m}^2}$$
$$\sigma'_a = \sqrt{\sigma_{1a}^2 - \sigma_{1a}\sigma_{2a} + \sigma_{2a}^2} \qquad (5\text{-}45)$$

These two stress components may then be applied to a fatigue diagram like Fig. 5-26, using the modified Goodman criteria exactly as before.

Several words of caution are appropriate. Formulas like Eq. (5-45) can also be written by using the maximum- and minimum-stress components, but these will not produce the same results. Also, be sure to use the method of Sec. 5-22 when the normal stresses σ_x or σ_y are zero. While Eq. (5-45) could be used in such cases, it will not give the same results.

A simplification of Eq. (5-45) is easily developed which eliminates the necessity of the Mohr's-circle analysis when τ_{xy} is accompanied by a single normal stress, say σ_x. In this special case, the von Mises stresses are found to be

$$\sigma'_m = \sqrt{\sigma_{xm}^2 + 3\tau_{xym}^2}$$
$$\sigma'_a = \sqrt{\sigma_{xa}^2 + 3\tau_{xya}^2} \qquad (5\text{-}46)$$

EXAMPLE 5-7 A bar of steel has $S_{ut} = 700$ MPa, $S_y = 500$ MPa, and a fully corrected endurance limit $S_e = 200$ MPa. For each of the cases below find the factors of safety which guard against static and fatigue failures.

(a) $\tau_m = 140$ MPa
(b) $\tau_m = 140$ MPa, $\tau_a = 70$ MPa
(c) $\tau_{xym} = 100$ MPa, $\sigma_{xa} = 80$ MPa
(d) $\sigma_{xm} = 60$ MPa, $\sigma_{xa} = 80$ MPa, $\tau_{xym} = 70$ MPa, $\tau_{xya} = 35$ MPa

SOLUTION (a) The yield strength in shear is

$$S_{sy} = 0.577 S_y = 0.577(500) = 288 \text{ MPa}$$

---

* The following papers discuss this subject in much greater detail:
R. E. Little, Fatigue Stresses from Complex Loadings, *Machine Design*, Jan. 6, 1966, pp. 145–149.
W. R. Miller, K. Ohji, and J. Marin, Rotating Principal Stress Axes in High-Cycle Fatigue, *ASME* paper no. 66-WA/Met-9, 1966.
Dimitri Kececioglu et. al, *op. cit.*

Therefore

$$n(\text{static}) = S_{sy}/\tau_{\max} = 288/140 = 2.06 \qquad Ans.$$

Since τ_m is a steady stress, there is no fatigue.

(b) The maximum shear stress, from Fig. 5-23, is

$$\tau_{\max} = \tau_m + \tau_a = 140 + 70 = 210 \text{ MPa}$$

Therefore

$$n(\text{static}) = S_{sy}/\tau_{\max} = 288/210 = 1.37 \qquad Ans.$$

From Eq. (5-42) we find the endurance limit in shear to be

$$S_{se} = 0.577S_e = 0.577(200) = 115 \text{ MPa}$$

Based on Eq. (5-43) we therefore have

$$n(\text{fatigue}) = S_{se}/\tau_a = 115/70 = 1.64 \qquad Ans.$$

(c) The maximum von Mises stress occurs when the alternating component is summed with the mean component. Using Eq. (5-46), to bypass the use of a Mohr's circle, gives

$$\sigma'_{\max} = \sqrt{\sigma_{xa}^2 + 3\tau_{xym}^2} = \sqrt{(80)^2 + 3(100)^2} = 191 \text{ MPa}$$

Therefore the factor of safety guarding against a static failure is

$$n(\text{static}) = \frac{S_y}{\sigma'_{\max}} = \frac{500}{191} = 2.62 \qquad Ans.$$

The distortion-energy theory must also be used to find the possibility of a fatigue failure. Using Eq. (5-46) again we obtain

$$\sigma'_m = \sqrt{\sigma_{xm}^2 + 3\tau_{xym}^2} = \sqrt{3(100)^2} = 173 \text{ MPa}$$

$$\sigma'_a = \sigma_{xa} = 80 \text{ MPa}$$

Next we plot these two components on the stress axes of the fatigue diagram of Fig. 5-30. A line drawn from the origin to point A, determined by these two stress coordinates, intersects the modified Goodman line at B giving a mean strength $S_m = 270$ MPa as shown. Therefore the factor of safety guarding against fatigue failure is

$$n(\text{fatigue}) = S_m/\sigma'_m = 270/173 = 1.56 \qquad Ans.$$

(d) To determine the possibility of a static failure we first compute the maximum normal stresses and maximum shear stresses, assuming that eventually the maximums might occur simultaneously. This gives

$$\sigma_{x\,\max} = \sigma_{xm} + \sigma_{xa} = 60 + 80 = 140 \text{ MPa}$$

$$\tau_{xy\,\max} = \tau_{xym} + \tau_{xya} = 70 + 35 = 105 \text{ MPa}$$

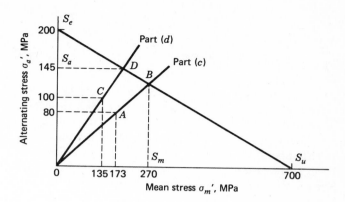

FIGURE 5-30

Bypassing the use of Mohr's circle again, we find the maximum von Mises stress to be

$$\sigma'_{max} = \sqrt{(140)^2 + 3(105)^2} = 229 \text{ MPa}$$

Then the factor of safety is found to be

$$n(\text{static}) = S_y/\sigma'_{max} = 500/229 = 2.18 \qquad Ans.$$

To determine the possibility of a fatigue failure we use Eq. (5-46) to get

$$\sigma'_m = \sqrt{\sigma^2_{xm} + 3\tau^2_{xym}} = \sqrt{(60)^2 + 3(70)^2} = 135 \text{ MPa}$$
$$\sigma'_a = \sqrt{\sigma^2_{xa} + 3\tau^2_{xya}} = \sqrt{(80)^2 + 3(35)^2} = 100 \text{ MPa}$$

These two stress components are plotted on Fig. 5-30 as before, giving point C. A line through the origin and point C intersects the modified Goodman line at D and yields the alternating strength as $S_a = 145$ MPa. Note that we could just as well have used the mean strength, as before. The factor of safety is

$$n(\text{fatigue}) = S_a/\sigma'_a = 145/100 = 1.45 \qquad Ans.$$

$////$

5-24 SURFACE STRENGTH

Our studies in this chapter thus far have dealt with the failure of a machine element by yielding, by fracture, and by fatigue. The endurance limit obtained by the rotating-beam test is frequently called the *flexural endurance limit* because it is a test of a rotating beam. In this section we shall study a property of *mating materials* called the *surface endurance limit*. The design engineer must frequently solve problems in which two machine elements mate with one another by rolling, sliding, or a combination of rolling and sliding contact. Obvious examples of such combinations are the mating teeth of a pair of gears, a cam and follower, a wheel and rail, or a chain and sprocket. A knowledge of the surface strength of materials is necessary if the designer is to create machines having a long and satisfactory life.

When two surfaces roll or roll and slide against one another with sufficient force, a pitting failure will occur after a certain number of cycles of operation. Authorities are not in complete agreement on the exact mechanism of the pitting;[*] although the subject is quite complicated, they do agree that the Hertz stresses, the number of cycles, the surface finish, the hardness, the degree of lubrication, and the temperature all influence the strength. In Sec. 2-20 it was learned that, when two surfaces are pressed together, a maximum shear stress is developed slightly below the contacting surface. It is postulated by some authorities that a surface fatigue failure is initiated by this maximum shear stress and then is propagated rapidly to the surface. The lubricant then enters the crack which is formed and, under pressure, eventually wedges the chip loose.

To determine the surface fatigue strength of mating materials, Buckingham[†] designed a simple machine for testing a pair of contacting rolling surfaces in connection with his investigation of the wear of gear teeth. Buckingham and, later, Talbourdet[‡] gathered large numbers of data from many tests so that considerable design information is now available. To make the results useful for designers, Buckingham defined a *load-stress factor*, also called a *wear factor*, which is derived from the Hertz equations. Equations (2-88) and (2-89), for contacting cylinders, are found to be

$$b = \sqrt{\frac{2F}{\pi l}\frac{(1-\mu_1^2)/E_1 + (1-\mu_2^2)/E_2}{(1/d_1)+(1/d_2)}} \tag{5-47}$$

$$p_{max} = -\frac{2F}{\pi b l} \tag{5-48}$$

where b = half-width of rectangular contact area
F = contact force
l = width of cylinders
μ = Poisson's ratio
E = modulus of elasticity
d = cylinder diameter

* See, for example, Stewart Way, Pitting Due to Rolling Contact, *Trans. ASME*, vol. 57, pp. A-49–A-58, 1935, and Charles Lipson and L. V. Colwell (eds.), "Handbook of Mechanical Wear," p. 95, University of Michigan Press, Ann Arbor, 1961.
† Earl Buckingham, "Analytical Mechanics of Gears," chap. 23, McGraw-Hill Book Company, New York, 1949.
‡ As reported by W. D. Cram, Experimental Load-Stress Factors, in Charles Lipson and L. V. Colwell (eds.), "Engineering Approach to Surface Damage," University of Michigan Summer Session, Ann Arbor, 1958.

On the average, $\mu = 0.30$ for engineering materials. Thus let $\mu = \mu_1 = \mu_2 = 0.30$. Also, it is more convenient to use the cylinder radius; so let $2r = d$. If we then designate the width of the cylinders as w instead of l and remove the square-root sign, Eq. (5-41) becomes

$$b^2 = 1.16 \frac{F}{w} \frac{(1/E_1) + (1/E_2)}{(1/r_1) + (1/r_2)} \qquad (5\text{-}49)$$

Next, define a new kind of endurance limit called *surface endurance limit*, using Eq. (5-48), as

$$S_{fe} = \frac{2F}{\pi b w} \qquad (5\text{-}50)$$

The *surface endurance limit*, therefore, is the contacting pressure which, after a large number of cycles, will cause failure of the surface. Such failures are often called *wear* because they occur after a very long time. They should not be confused with abrasive wear, however. By substituting the value of b from Eq. (5-49) into (5-50) and rearranging, we obtain

$$2.857 S_{fe}^2 \left(\frac{1}{E_1} + \frac{1}{E_2} \right) = \frac{F}{w} \left(\frac{1}{r_1} + \frac{1}{r_2} \right) \qquad (5\text{-}51)$$

The left side of this equation contains E_1, E_2, and S_{fe}, constants which come about because of the selection of a certain material for each element of the pair. We call this K_1, *Buckingham's load-stress factor*. Having selected the two materials, K_1 is computed from the equation

$$K_1 = 2.857 S_{fe}^2 \left(\frac{1}{E_1} + \frac{1}{E_2} \right) \qquad (5\text{-}52)$$

With K_1 known, we now write the design equation as

$$K_1 = \frac{F}{w} \left(\frac{1}{r_1} + \frac{1}{r_2} \right) \qquad (5\text{-}53)$$

which, if satisfied, defines a surface fatigue failure in 10^8 cycles of operation according to Talbourdet's experiments. Since we usually want to define safety instead of failure, we would write Eq. (5-53) in the form

$$\frac{K_1}{n} = \frac{F}{w} \left(\frac{1}{r_1} + \frac{1}{r_2} \right) \qquad (5\text{-}54)$$

where n is the factor of safety.

Values of the surface endurance limit for steels can be obtained from the equation

$$S_{fe} = \begin{cases} 0.4 H_B - 10 & \text{kpsi} \\ 2.76 H_B - 70 & \text{MPa} \end{cases} \qquad (5\text{-}55)$$

where H_B is the Brinell hardness number. If the two materials have different hardnesses the lesser value is generally, though not always, used. The results of this procedure agree with the values of the load-stress factors recommended by Buckingham.

PROBLEMS

Sections 5-1 to 5-7

5-1 A ductile steel has a yield strength of 40 kpsi. Find factors of safety corresponding to failure by the maximum-normal-stress theory, the maximum-shear-stress theory, and the distortion-energy theory, respectively, for each of the following stress states:

(a) $\sigma_x = 10$ kpsi, $\sigma_y = -4$ kpsi
(b) $\sigma_x = 10$ kpsi, $\tau_{xy} = 4$ kpsi cw
(c) $\sigma_x = -2$ kpsi, $\sigma_y = -8$ kpsi, $\tau_{xy} = 4$ kpsi ccw
(d) $\sigma_x = 10$ kpsi, $\sigma_y = 5$ kpsi, $\tau_{xy} = 1$ kpsi cw

5-2 A machine element is loaded so that $\sigma_1 = 20$ kpsi, $\sigma_2 = -15$ kpsi, and $\sigma_3 = 0$ kpsi; the material has a minimum yield strength in tension and compression of 60 kpsi. Find the factor of safety for each of the following failure theories:

(a) Maximum-normal-stress theory
(b) Maximum-shear-stress theory
(c) Distortion-energy theory

5-3 A machine part is statically loaded and has a yield strength of 350 MPa. For each stress state indicated below, find the factor of safety using each of the three static-failure theories:

(a) $\sigma_1 = 70$ MPa, $\sigma_2 = 70$ MPa
(b) $\sigma_1 = 70$ MPa, $\sigma_2 = 35$ MPa
(c) $\sigma_1 = 70$ MPa, $\sigma_2 = -70$ MPa
(d) $\sigma_1 = -70$ MPa, $\sigma_2 = 0$ MPa

5-4 Based on the use of UNS C27000 hard yellow brass rod as the material, find factors of safety for each of the three static-failure theories for the following stress states:

(a) $\sigma_x = 70$ MPa, $\sigma_y = 30$ MPa
(b) $\sigma_x = 70$ MPa, $\tau_{xy} = 30$ MPa cw
(c) $\sigma_x = -10$ MPa, $\sigma_y = -60$ MPa, $\tau_{xy} = 30$ MPa ccw
(d) $\sigma_x = 50$ MPa, $\sigma_y = 20$ MPa, $\tau_{xy} = 40$ MPa cw

5-5 A force F applied at D near the end of a 15-in lever shown in the figure results in certain stresses in the cantilevered bar $OABC$. The bar $(OABC)$ is made of UNS G10350 steel which is forged, machined, and heat-treated and tempered to 800°F. What force F would cause the cantilevered bar to yield?

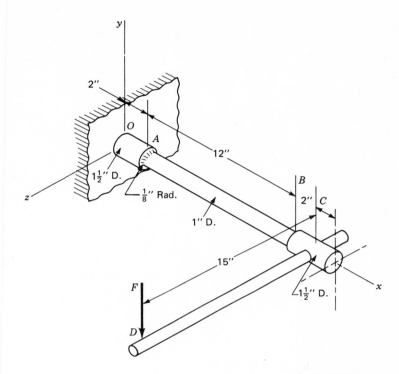

PROBLEM 5-5

5-6 The figure shows a round bar subjected to the vector moment $\mathbf{M} = 1.75\mathbf{i} + 1.10\mathbf{k}$ kN·m. The material is UNS A95056-H38 aluminum alloy. A stress element A located on top of the bar is oriented in the xz plane as shown. Using the stresses on this element determine the factor of safety guarding against a static failure by using the maximum-shear-stress theory and the distortion-energy theory.

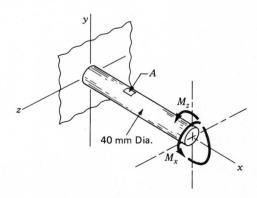

PROBLEM 5-6

5-7 A lever subjected to a downward static force of 400 lb is keyed to a 1-in round bar as
shown in the figure.

(*a*) Find the critical stresses in the round bar.

(*b*) The round bar is made of UNS G46200 steel, heat-treated and drawn to 800°F.
Based on static loading, find the factor of safety by using the distortion-energy
theory.

(*c*) As a check on (*b*) find the factor of safety using the maximum-shear-stress theory.
Should the result be greater or lesser than that obtained in (*b*)? Why?

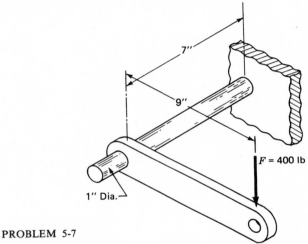

PROBLEM 5-7

5-8 A UNS A92024-T3 aluminum tube has a 3-in OD and a 0.049-in wall. It is subjected
to an internal pressure of 1200 psi. Find the factor of safety guarding against yielding
by using each of the three theories for ductile materials.

5-9 A thin-wall pressure vessel is made of UNS A93003-H14 aluminum-alloy tubing. The
vessel has an OD of 60 mm and a wall thickness of 1.50 mm. What internal pressure
would cause the material to yield?

5-10 A thick-walled cylinder is to have an inside diameter of 0.500 in, be made of UNS
G41400 cold-drawn steel, and it must resist an internal pressure of 5 kpsi based on a
factor of safety of at least 4. Specify a satisfactory outside diameter, basing your
decision on yielding as predicted by the maximum-shear-stress theory.

5-11 A 1½-in-diameter UNS G10350 cold-drawn steel shaft has a forged gear mounted on it
with a class FN 4 force fit. The gear hub is 3 in long and 2½ in in diameter. Based on a
mean fit, find the critical value of the von Mises stress.

5-12 A gun barrel is assembled by shrinking an outer barrel over an inner barrel so that the
maximum principal stress is 70 percent of the yield strength of the material. The
material of both members is steel, $S_y = 78$ kpsi, $E = 30$ Mpsi, and $\mu = 0.292$.
The nominal radii of the barrels are $\frac{3}{16}$, $\frac{3}{8}$, and $\frac{9}{16}$ in. Calculate the critical von Mises
stress for both members.

5-13 Suppose the gun of Prob. 5-12 is fired with an internal pressure of 40 kpsi. What are
the new values of the critical von Mises stress?

Section 5-8

5-14 Using typical values for the strengths of ASTM No. 40 cast iron, find the factors of safety corresponding to fracture by the maximum-normal-stress theory, the Coulomb-Mohr theory, and the modified Mohr theory, respectively, for each of the following stress states:

(*a*) $\sigma_x = 10$ kpsi, $\sigma_y = -4$ kpsi

(*b*) $\sigma_x = 10$ kpsi, $\tau_{xy} = 4$ kpsi cw

(*c*) $\sigma_x = -2$ kpsi, $\sigma_y = -8$ kpsi, $\tau_{xy} = 4$ kpsi ccw

(*d*) $\sigma_x = 10$ kpsi, $\sigma_y = -30$ kpsi, $\tau_{xy} = 10$ kpsi cw

5-15 Tests on a particular melt of ASTM No. 20 cast iron gave $S_{ut} = 150$ MPa and $S_{uc} = 600$ MPa. Find the factor of safety for each of the failure theories diagrammed in Fig. 5-8 for the following stress states:

(*a*) $\sigma_x = 50$ MPa, $\tau_{xy} = 30$ MPa cw

(*b*) $\sigma_x = -80$ MPa, $\sigma_y = -40$ MPa, $\tau_{xy} = 20$ MPa ccw

(*c*) $\sigma_x = 40$ MPa, $\sigma_y = 30$ MPa, $\tau_{xy} = 10$ MPa ccw

(*d*) $\sigma_x = 30$ MPa, $\sigma_y = -60$ MPa, $\tau_{xy} = 30$ MPa cw

5-16 Due to a heavy shrink fit, a hollow ASTM No. 30 cast-iron member, having a 1-in-diameter hole and a $1\frac{1}{2}$-in OD, is subjected to an external pressure of 38 kpsi. Using typical values of the strength, find the margin of safety *m*.

Sections 5-9 to 5-11

5-17 What is the fatigue strength of a rotating-beam specimen made of UNS G10180 hot-rolled steel corresponding to a life of $250(10)^3$ cycles of stress reversal? What would be the life of the specimen if the alternating stress were 40 kpsi?

5-18 Derive Eqs. (5-24) to (5-27).

Sections 5-13 to 5-19

5-19 A $\frac{3}{16}$-in drill rod was heat-treated and ground and the measured hardness was found to be $H_B = 490$. What is the endurance limit?

5-20 Find the endurance limit of a 1-in bar of UNS G10350 steel, heat-treated and tempered to 1000°F, if the bar has a machined finish.

5-21 Find the endurance limit corresponding to a reliability of 99.9 percent for a UNS G43400 steel bar, heat-treated and tempered to 1000°F, if the bar is about $1\frac{1}{2}$ in in diameter and has no points of stress concentration. The material is ground and polished.

5-22 Two connecting rods having approximately $\frac{3}{4}$ in diameters are made as forgings. They are to have reliabilities of 99.99 percent. One is made of the very best and most expensive steel available at the time, a UNS G43400 steel, heat-treated and tempered to 600°F. The other is made of an ordinary and easily obtainable medium carbon steel, a UNS G10400 steel, heat-treated and tempered to 1000°F. Find both endurance limits. Is there any advantage in using the high-priced steel? Why?

5-23 A portion of a machine member is shown in the figure. It is loaded by completely reversed axial forces F which are uniformly distributed across the width. The material is a UNS G10180 cold-drawn steel flat. For 90 percent reliability and infinite life determine the maximum force F that can be applied.

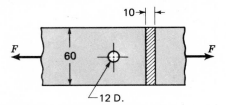

PROBLEM 5-23
Dimensions in millimetres.

5-24 The figure is an idealized representation of a machine member subjected to the action of an alternating force F which places the member in completely reversed bending. The material is UNS G10500 steel, heat-treated and tempered to 600°F, with a ground finish. Based on 50 percent reliability, infinite life, and no margin of safety, determine the maximum value of the alternating force F which can probably be applied.

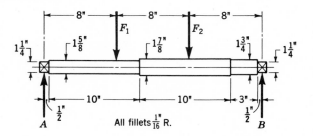

PROBLEM 5-24
Dimensions in millimetres.

5-25 The shaft shown in the figure rotates at 1720 rpm and is to have a life of 3 min at 50 percent reliability. The steel used has the following properties: $S_{ut} = 89$ kpsi, $E = 30$ Mpsi, 22.5 percent elongation in 2 in, $H_B = 178$. The shaft is finished by grinding. It is simply supported in antifriction bearings at A and B and is loaded by the static forces $F_1 = 2.0$ kip and $F_2 = 3.0$ kip. Progressing from A to B, the stress-concentration factors for the shoulders in bending are $K_t = 2.00, 1.94, 2.08,$ and 2.02. Find the factor of safety guarding against failure.

PROBLEM 5-25

5-26 The bar shown in the figure is of UNS A92017-T4 wrought-aluminum alloy, and it has a fatigue strength of 18 kpsi at $5(10)^8$ cycles for reversed axial loading. Find the factor of safety if the axial load shown is completely reversed.

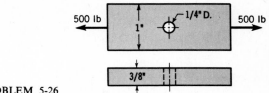

PROBLEM 5-26

5-27 The bar shown in the figure is machined from a UNS G10350 cold-drawn steel flat. The axial load shown is completely reversed. Find the factor of safety.

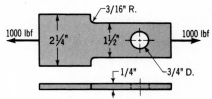

PROBLEM 5-27

5-28 The figure is a drawing of a rotating shaft loaded in completely reversed bending by the 400-lb force. The shaft is to be made of UNS G41400 steel, heat-treated to a hardness of $H_B = 376$, with a ground finish in critical places. Based on an infinite life, find the factor of safety.

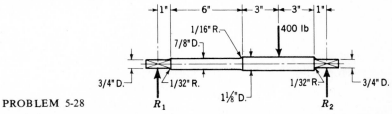

PROBLEM 5-28

5-29 The rotating shaft shown in the figure is machined from a 50-mm bar of cold-drawn UNS G10350 steel. The shaft is designed for an infinite life and a reliability of 99.99 percent. What factor of safety guards against a fatigue failure if the force F is 3.0 kN?

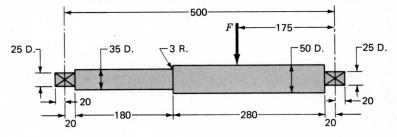

PROBLEM 5-29
Dimensions in millimetres.

5-30 The connecting rod shown in the figure is machined from a 10-mm-thick bar of UNS G10350 cold-drawn steel. The bar is axially loaded by completely reversed forces acting on pins through the two drilled holes. Theoretical stress-concentration factors for the fillets may be obtained from Fig. A-25-5, and for the pin-loaded holes from Fig. A-25-12. Based on infinite life, 99 percent reliability, and a factor of safety of 2, determine the maximum safe reversed axial load that can be employed.

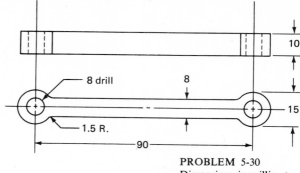

PROBLEM 5-30
Dimensions in millimetres.

5-31 The shaft shown in the figure has bearing reactions R_1 and R_2, rotates at 1150 rpm, and supports the 10-kip bending force. The specifications call for a ductile steel having $S_{ut} = 120$ kpsi and $S_y = 90$ kpsi. The shaft is to be machined and is to have a life of $80(10)^3$ cycles corresponding to 90 percent reliability. Find the safe diameter d based on a factor of safety of 1.60.

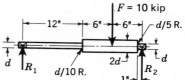

PROBLEM 5-31

5-32 The figure shows a rotating shaft loaded by two bending forces having the bearing reactions R_1 and R_2. Point A is a shaft shoulder which is required for positioning the left-hand bearing. The grinding-relief groove at B is 2.5 mm deep (see Table A-25-14). The surface AB is ground, but the groove is machined. The material of the shaft is UNS G43400 steel, heat-treated and tempered to 1000°F under conditions such that the ultimate tensile strength is 1.30 GPa. Determine the factor of safety corresponding to a life of $0.35(10)^6$ revolutions of the shaft.

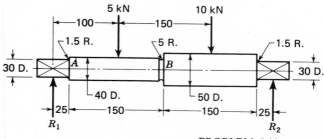

PROBLEM 5-32
Dimensions in millimetres.

Sections 5-20 and 5-21

5-33 A mechanical part is made of steel with the properties $S_u = 600$ MPa, $S_y = 480$ MPa, and $S_e = 200$ MPa. Determine the factor of safety for the following stress states:

(*a*) A bending stress alternating between 40 and 100 MPa.

(*b*) A bending stress alternating between 0 and 200 MPa.

(*c*) A pure axial compressive stress which fluctuates between 0 and 200 MPa.

5-34 The figure shows a formed round-wire cantilever spring subjected to a varying force. A hardness test made on 25 springs gave a minimum hardness of 380 Bhn. It is apparent from the mounting details that there is no stress concentration. A visual inspection of the springs indicates that the surface finish corresponds closely to a hot-rolled finish. Based on a 50 percent reliability, find the number of cycles of load application likely to cause failure.

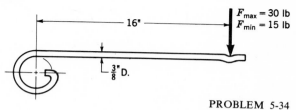

$F_{max} = 30$ lb
$F_{min} = 15$ lb

16"

$\frac{3}{8}$" D.

PROBLEM 5-34

5-35 The figure is a drawing of a 12-gauge (0.1094-in) $\times$ $\frac{3}{4}$-in latching spring. The spring is assembled by deflecting it 0.075 in initially, and then deflecting it an additional 0.15 in during each latching operation. The material is machined high-carbon steel heat-treated to 490 Bhn..The stress concentration at the bend is 1.70, corresponding to a fillet radius of $\frac{1}{8}$ in and uncorrected for notch sensitivity.

(*a*) Find the maximum and minimum latching forces F.

(*b*) Will the spring fail in fatigue? Why?

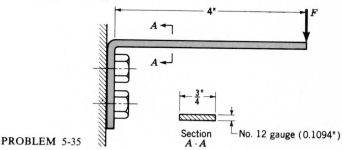

4"

F

A

A

$\frac{3}{4}$"

Section
A·A

No. 12 gauge (0.1094")

PROBLEM 5-35

5-36 The figure shows a short, rectangular link-rod which is loaded by the forces F acting upon the pins at each end. The forces F vary so as to produce axial tension ranging from 70 to 30 kN. A UNS G41300 steel has been selected for the link which is to be heat-treated and tempered to 1000°F after which all surfaces will be finished by grinding. Determine the dimension t to the nearest millimetre, basing the design on an infinite life at 99.999 percent reliability and a factor of safety of 1.35.

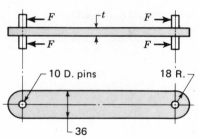

PROBLEM 5-36
Dimensions in millimetres.

5-37 The figure shows the free-body diagram of a connecting rod portion having stress concentration at two places. The forces F fluctuate between a tension of 4 kip and a compression of 16 kip. The material of the rod is cold-drawn UNS G10180 steel. Neglecting column action, find the factors of safety which guard against both static and fatigue failure, based on infinite life and $R = 0.50$.

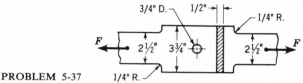

PROBLEM 5-37

5-38 The figure shows two views of a flat steel spring loaded in bending by the force F. The spring is assembled so as to produce a preload $F_{min} = 0.90$ kN. The force then varies from this minimum to a maximum of 3.0 kN. The spring is forged of a 95-point carbon steel and has the following properties after a suitable heat treatment: $S_u = 1400$ MPa, $S_y = 950$ MPa, $H_B = 399$, 12 percent elongation in 50 mm. Find the thickness t if $K_t = 2.50$ and a factor of safety of 1.90 is to be used.

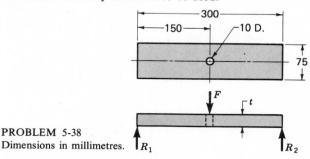

PROBLEM 5-38
Dimensions in millimetres.

Section 5-22

5-39 A rotating steel shaft has the following properties: $S_u = 90$ kpsi, $S_y = 70$ kpsi, $S_e = 30$ kpsi, $S_{su} = 67$ kpsi, $S_{sy} = 40$ kpsi, and $S_{se} = 17$ kpsi. This shaft is subjected to a steady torsional stress of 9 kpsi. In addition, due to torsional vibration, the shaft is subjected to completely reversed torsional stress of magnitude 6 kpsi. Find the factor of safety for a fatigue failure and a static failure.

5-40 A 20-mm-diameter shaft is made of cold-drawn UNS G10350 steel and has a 6-mm-diameter hole drilled transversely through it. Determine the factor of safety guarding against both fatigue and static failures for the following loads:

(a) The shaft is subject to a torque which fluctuates between 0 and 90 N·m.

(b) The shaft is subject to a completely reversed torque of 40 N·m.

(c) The shaft is loaded by a steady torque of 50 N·m together with an alternating component of 35 N·m.

Section 5-23

5-41 A bar of steel has the properties $S_e = 40$ kpsi, $S_y = 60$ kpsi, and $S_u = 80$ kpsi. For each of the cases below find the factor of safety guarding against a static failure and either the factor of safety guarding against a fatigue failure or the expected life of the part.

(a) A steady torsional stress of 15 kpsi and an alternating bending stress of 25 kpsi.

(b) A steady torsional stress of 20 kpsi and an alternating torsional stress of 10 kpsi.

(c) A steady torsional stress of 15 kpsi, an alternating torsional stress of 10 kpsi, and an alternating bending stress of 12 kpsi.

(d) An alternating torsional stress of 30 kpsi.

(e) An alternating torsional stress of 15 kpsi and a steady tensile stress of 15 kpsi.

5-42 A spherical pressure vessel 24 in in diameter is made of cold-drawn UNS G10180 steel No. 10 gauge (0.1345 in). The vessel is to withstand an infinite number of pressure fluctuations from 0 to p_{max}.

(a) What maximum pressure will cause static yielding?

(b) What maximum pressure will eventually cause a fatigue failure? In any case, the joints and connections are adequately reinforced and do not weaken the vessel.

5-43 The figure illustrates a steel shaft supported in bearings at R_1 and R_2 and loaded by completely reversed bending forces and by static (nonmoving) torsion. The shaft is of UNS G10500 steel, heat-treated and tempered to 600°F, and has a ground finish. Find the factor of safety based on the possibility of a fatigue failure for an infinite life at 90 percent reliability.

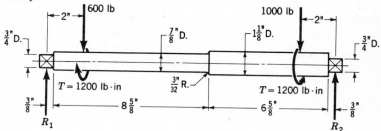

PROBLEM 5-43

5-44 The figure shows a stationary torsion-bar spring loaded statically by the forces $F = 35$ N and by a torque T which varies from 0 to 8 N·m. The material is UNS G61500 steel heat-treated and tempered to 1000°F. The ends of the spring are ground up to the shoulders. The 2.5-m body of the spring has a hot-rolled surface finish. The geometric stress-concentration factors at the shoulders are 1.68 for bending and 1.42 for torsion. Determine a suitable diameter d to the nearest millimetre, using a factor of safety of at least 1.80.

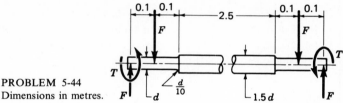

PROBLEM 5-44
Dimensions in metres.

5-45 A rotating shaft is loaded in static torsion and reversed bending by the stresses $\sigma_x = 10$ kpsi and $\tau_{xy} = 6$ kpsi referred to a mark on the shaft surface. Find the magnitude of the principal stresses and the direction of σ_1 from the x axis as the shaft makes one-half a revolution. Use $15°$ increments of shaft angle.

PART TWO

Design of
Mechanical Elements

THE DESIGN OF SCREWS, FASTENERS, AND CONNECTIONS

This book presupposes a knowledge of the elementary methods of fastening. Typical methods of fastening or joining parts include the use of such items as bolts, nuts, cap screws, setscrews, rivets, spring retainers, locking devices, and keys. Studies in engineering graphics and in metal processes often include instruction on various joining methods, and the curiosity of any person interested in engineering naturally results in the acquisition of a good background knowledge of fastening methods. Consequently, the purpose of this chapter is not to describe the various fasteners or tabulate available sizes, but rather to select and specify suitable ones in the design of machines and devices.

The subject is one of the most interesting in the entire field of mechanical design. The number of new inventions in the fastener field over any period you might care to mention has been tremendous. There is an overwhelming variety of fasteners available for the designer's selection. Another thing: Did you know that a good bolt material should be strong and tough, but a good nut material should be soft and ductile? Or did you know that there are certain applications where you should tighten the bolt as tightly as possible and, if it does not fail by twisting in two during tightening, there is a very good possibility that the bolt never will fail? In the material to follow you will discover the why of these questions. You will learn why a nut or bolt loosens and what you must do to keep it tight. Methods of joining parts are extremely important in the engineering of a quality design, and it is necessary to have a thorough understanding of the performance of fasteners and joints under all conditions of use and design.

Jumbo jets such as Boeing's 747 and Lockheed's 1011 require as many as 2.5 million fasteners, some of which cost several dollars apiece. The 747, for example, needs about 70,000 titanium fasteners, costing about $150,000 in all; 400,000 other close-tolerance fasteners, costing about $250,000; and 30,000 squeeze rivets priced at 50 cents each, installed. To keep costs down, Boeing, Lockheed, and their subcontractors constantly review new fastener designs, installation techniques, and tooling. Cost-saving designs and tooling will find a ready market, which will grow in value as jumbo jets proliferate.*

6-1 THREAD STANDARDS AND DEFINITIONS

The terminology of screw threads, illustrated in Fig. 6-1, is explained as follows:

The *pitch* is the distance between adjacent thread forms measured parallel to the thread axis. The pitch is the reciprocal of the number of thread forms per inch N.

The *major diameter* is the largest diameter of a screw thread.

The *minor diameter* is the smallest diameter of a screw thread.

The lead l, not shown, is the distance the nut moves parallel to the screw axis when the nut is given one turn. For a single thread, as in Fig. 6-1, the lead is the same as the pitch.

A *multiple-threaded* product is one having two or more threads cut beside each other (imagine two or more strings wound side by side around a pencil). Standardized products such as screws, bolts, nuts, etc., all have single threads. A

* *Product Engineering*, vol. 41, no. 8, p. 9, Apr. 13, 1970.

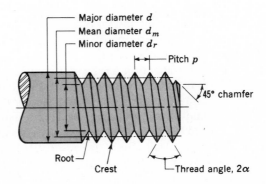

FIGURE 6-1
Terminology of screw threads.

double-threaded screw has a lead equal to twice the pitch; a *triple-threaded* screw has a lead equal to three times the pitch, etc.

All threads are made according to the *right-hand rule* unless otherwise noted.

Figure 6-2 shows the thread geometry for the three English thread standards in most general use. Figure 6-2a represents the *American National (Unified)* thread standard which has been approved in this country and in Great Britain for use on all standard threaded products. The thread angle is 60° and the crests of the thread may be either flat or rounded.

The International Standardization Organisation (ISO) has been attempting to standardize screw thread systems of the world. The ISO metric threads have the same 60° thread angle as the unified system and they also may be manufactured with either flat or rounded crests.

Tables 6-1 and 6-2 will be useful in specifying and designing threaded parts. Note that the thread size is specified by giving the pitch p for metric sizes and by giving the number of threads per inch N for the unified sizes. The screw sizes in Table 6-2 under $\frac{1}{4}$ in in diameter are numbered or gauge sizes. The second column in Table 6-2 shows that a No. 8 screw has a nominal diameter of 0.1640 in.

A great many tensile tests of threaded rods have shown that an unthreaded rod having a diameter equal to the mean of the pitch and minor diameters will

Table 6-1   DIAMETERS AND AREAS OF REGULAR-PITCH METRIC THREADS*

Nominal major diameter, d mm	Pitch p mm	Threads per inch N (approx)	Tensile stress area, A_t mm^2	Minor diameter area, A_r mm^2
3	0.5	51	5.06	4.51
4	0.7	36	8.83	7.81
5	0.8	32	14.26	12.77
6	1	26	20.23	18.03
8	1.25	20	36.79	33.07
10	1.5	17	58.27	52.64
12	1.75	14	84.66	76.74
16	2	13	157.27	144.90
20	2.5	10	245.75	226.40
24	3	8	353.87	326.02
30	3.5	7	562.60	521.57
36	4	6	819.50	762.84

* The minor diameter, used to obtain A_r, was found from the equation $d_r = d - 1.280\,655p$. The pitch diameter used was found from $d_p = d - 0.640\,327p$. The mean of the pitch and minor diameters was used to compute the tensile-stress areas. The threads per inch is presented only as a reference; do not use these approximate values in computations.

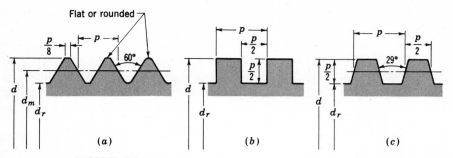

FIGURE 6-2
(*a*) American National or Unified thread; (*b*) square thread; (*c*) Acme thread.

Table 6-2 DIAMETERS AND AREAS OF UNIFIED SCREW THREADS UNC AND UNF

Size designation	Nominal major diameter, in	Coarse series—UNC			Fine series—UNF		
		Threads per inch, N	Tensile-stress area A_t, in^2	Minor-diameter area A_r, in^2	Threads per inch, N	Tensile-stress area A_t, in^2	Minor-diameter area A_r, in^2
0	0.0600				80	0.001 80	0.001 51
1	0.0730	64	0.002 63	0.002 18	72	0.002 78	0.002 37
2	0.0860	56	0.003 70	0.003 10	64	0.003 94	0.003 39
3	0.0990	48	0.004 87	0.004 06	56	0.005 23	0.004 51
4	0.1120	40	0.006 04	0.004 96	48	0.006 61	0.005 66
5	0.1250	40	0.007 96	0.006 72	44	0.008 80	0.007 16
6	0.1380	32	0.009 09	0.007 45	40	0.010 15	0.008 74
8	0.1640	32	0.0140	0.011 96	36	0.014 74	0.012 85
10	0.1900	24	0.017 5	0.014 50	32	0.020 0	0.017 5
12	0.2160	24	0.024 2	0.020 6	28	0.025 8	0.022 6
$\frac{1}{4}$	0.2500	20	0.031 8	0.026 9	28	0.036 4	0.032 6
$\frac{5}{16}$	0.3125	18	0.052 4	0.045 4	24	0.058 0	0.052 4
$\frac{3}{8}$	0.3750	16	0.077 5	0.067 8	24	0.087 8	0.080 9
$\frac{7}{16}$	0.4375	14	0.106 3	0.093 3	20	0.118 7	0.109 0
$\frac{1}{2}$	0.5000	13	0.141 9	0.125 7	20	0.159 9	0.148 6
$\frac{9}{16}$	0.5625	12	0.182	0.162	18	0.203	0.189
$\frac{5}{8}$	0.6250	11	0.226	0.202	18	0.256	0.240
$\frac{3}{4}$	0.7500	10	0.334	0.302	16	0.373	0.351
$\frac{7}{8}$	0.8750	9	0.462	0.419	14	0.509	0.480
1	1.0000	8	0.606	0.551	12	0.663	0.625
$1\frac{1}{4}$	1.2500	7	0.969	0.890	12	1.073	1.024
$1\frac{1}{2}$	1.5000	6	1.405	1.294	12	1.315	1.260

have the same tensile strength as the threaded rod. The area of this unthreaded rod is called the tensile-stress area A_t of the threaded rod; values of A_t are listed in both tables.

Unified threads are specified by stating the nominal diameter, the number of threads per inch, and the thread series, like this:

$$\tfrac{5}{8}\text{''-18UNF}$$

Note the use of double tick marks to mean inches.

Metric threads are specified by writing the diameter and pitch in millimetres, in that order. Thus

$$\text{M12} \times 1.75$$

is a thread having a nominal major diameter of 12 mm and a pitch of 1.75 mm. Note the letter M, which precedes the diameter, is the clue to the metric designation.

Square and Acme threads are used on screws when power is to be transmitted. Since each application is a special one, there is really no need for a standard relating the diameter to the number of threads per inch.

Modifications are frequently made to both Acme and square threads. For instance, the square thread is sometimes modified by cutting the space between the teeth so as to have an included thread angle of 10–15°. This is not difficult, since these threads are usually cut with a single-point tool anyhow; the modification retains most of the high efficiency inherent in square threads and makes the cutting simpler. Acme threads are sometimes modified to a stub form by making the tooth shorter. This results in a larger minor diameter, and consequently a stronger screw.

6-2   THE MECHANICS OF POWER SCREWS

A power screw is a device used in machinery to change angular motion into linear motion and, usually, to transmit power. Familiar applications include the lead screws of lathes and the screws for vises, presses, and jacks.

A schematic representation of the application of power screws to a power-driven press is shown in Fig. 6-3. In use, a torque T is applied to the ends of the screws through a set of gears, thus driving the head of the press downward against the load.

In Fig. 6-4 a square-threaded power screw with single thread having a mean diameter d_m, a pitch p, and a helix angle ψ is loaded by the axial compressive force F. We wish to find an expression for the torque required to raise this load, and another expression for the torque required to lower the load.

First, imagine that a single thread of the screw is unrolled or developed (Fig. 6-5) for exactly one turn. Then one edge of the thread will form the hypotenuse of a right triangle whose base is the circumference of the mean-thread-

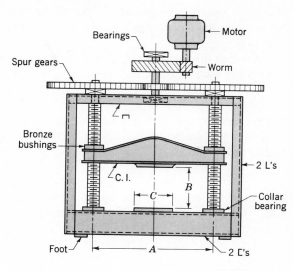

FIGURE 6-3

diameter circle and whose height is the lead. The angle ψ, in Figs. 6-4 and 6-5, is the helix angle of the thread. We represent the summation of all the unit axial forces acting upon the normal thread area by F. To raise the load, a force P acts to the right (Fig. 6-5a), and to lower the load, P acts to the left (Fig. 6-5b). The friction force is the product of the coefficient of friction μ with the normal force N,

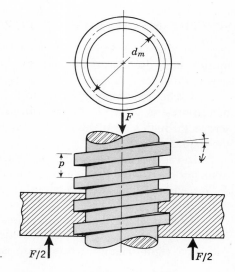

FIGURE 6-4
A power screw.

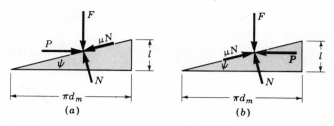

FIGURE 6-5
Force diagrams. (a) Lifting the load; (b) lowering the load.

and acts to oppose the motion. The system is in equilibrium under the action of these forces, and hence, for raising the load, we have

$$\sum F_H = P - N \sin \psi - \mu N \cos \psi = 0$$
$$\sum F_V = F + \mu N \sin \psi - N \cos \psi = 0$$

(a)

In a similar manner, for lowering the load, we have

$$\sum F_H = -P - N \sin \psi + \mu N \cos \psi = 0$$
$$\sum F_V = F - \mu N \sin \psi - N \cos \psi = 0$$

(b)

Since we are not interested in the normal force N, we eliminate it from each of these sets of equations and solve the result for P. For raising the load this gives

$$P = \frac{F(\sin \psi + \mu \cos \psi)}{\cos \psi - \mu \sin \psi}$$

(c)

and for lowering the load,

$$P = \frac{F(\mu \cos \psi - \sin \psi)}{\cos \psi + \mu \sin \psi}$$

(d)

Next, divide numerator and denominator of these equations by $\cos \psi$ and use the relation $\tan \psi = l/\pi d_m$ (Fig. 6-5). We then have, respectively,

$$P = \frac{F[(l/\pi d_m) + \mu]}{1 - (\mu l/\pi d_m)}$$

(e)

$$P = \frac{F[\mu - (l/\pi d_m)]}{1 + (\mu l/\pi d_m)}$$

(f)

Finally, noting that the torque is the product of the force P and the mean radius $d_m/2$, we can write

$$T = \frac{F d_m}{2} \left(\frac{l + \pi \mu d_m}{\pi d_m - \mu l} \right)$$

$(6\text{-}1)$

where T is the torque required for two purposes: to overcome thread friction and to raise the load.

The torque required to lower the load, from Eq. (f), is found to be

$$T = \frac{Fd_m}{2}\left(\frac{\pi\mu d_m - l}{\pi d_m + \mu l}\right) \qquad (6\text{-}2)$$

This is the torque required to overcome a part of the friction in lowering the load. It may turn out, in specific instances where the lead is large or the friction is low, that the load will lower itself by causing the screw to spin without any external effort. In such cases, the torque T from Eq. (6-2) will be negative or zero. When a positive torque is obtained from this equation, the screw is said to be *self-locking*. Thus the condition for self-locking is

$$\pi\mu d_m > l$$

Now divide both sides of this inequality by πd_m. Recognizing that $l/\pi d_m = \tan \psi$, we get

$$\mu > \tan \psi \qquad (6\text{-}3)$$

This relation states that self-locking is obtained whenever the coefficient of thread friction is equal to or greater than the tangent of the thread helix angle.

An expression for efficiency is also useful in the evaluation of power screws. If we let $\mu = 0$ in Eq. (6-1), we obtain

$$T_0 = \frac{Fl}{2\pi} \qquad (h)$$

which, since thread friction has been eliminated, is the torque required only to raise the load. The efficiency is therefore

$$e = \frac{T_0}{T} = \frac{Fl}{2\pi T} \qquad (6\text{-}4)$$

The preceding equations have been developed for square threads where the normal thread loads are parallel to the axis of the screw. In the case of Acme or Unified threads, the normal thread load is inclined to the axis because of the thread angle 2α and the helix angle ψ. Since helix angles are small, this inclination can be neglected and only the effect of the thread angle (Fig. 6-6a) considered. The effect of the angle α is to increase the frictional force by the wedging action of the threads. Therefore the frictional terms in Eq. (6-1) must be divided by $\cos \alpha$. For raising the load, or for tightening a screw or bolt, this yields

$$T = \frac{Fd_m}{2}\left(\frac{l + \pi\mu d_m \sec \alpha}{\pi d_m - \mu l \sec \alpha}\right) \qquad (6\text{-}5)$$

In using Eq. (6-5), remember that it is an approximation because the effect of the helix angle has been neglected.

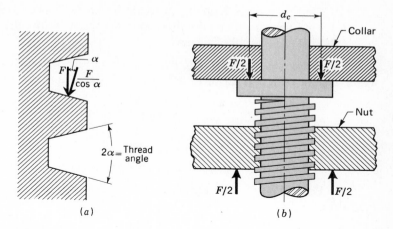

FIGURE 6-6
(a) Normal thread force is increased because of angle α; (b) thrust collar has frictional diameter d_c.

For power screws the Acme thread is not so efficient as the square thread because of the additional friction due to the wedging action, but it is often preferred because it is easier to machine and permits the use of a split nut, which can be adjusted to take up for wear.

Usually a third component of torque must be applied in power-screw applications. When the screw is loaded axially, a thrust or collar bearing must be employed between the rotating and stationary members in order to take out the axial component. Figure 6-6b shows a typical thrust collar in which the load is assumed to be concentrated at the mean collar diameter d_c. If μ_c is the coefficient of collar friction, the torque required is

$$T_c = \frac{F \mu_c d_c}{2} \qquad (6\text{-}6)$$

For large collars the torque should probably be computed in a manner similar to that employed for disk clutches.*

EXAMPLE 6-1   A power screw has 6 square threads per inch, double threads, and a major diameter of 1 in and is to be used in an application similar to that of Fig. 6-3. The given data include $\mu = \mu_c = 0.08$, $d_c = 1.25$ in, and $F = 1500$ lb per screw.

(a) Find the pitch, thread depth, thread width, mean diameter, minor diameter, and lead.

* See Sec. 14-5.

(b) Find the torque required to rotate the screw "against" the load.

(c) Find the torque required to rotate the screw "with" the load.

(d) Find the overall efficiency.

SOLUTION   (a) Since $N = 6$, $p = \frac{1}{6}$ in. From Fig. 6-2b the thread depth and width are the same and equal to half the pitch, or $\frac{1}{12}$ in. Also,

$$d_m = d - \frac{p}{2} = 1 - \frac{1}{12} = 0.9167 \text{ in} \qquad Ans.$$

$$d_r = d - p = 1 - \frac{1}{6} = 0.8333 \text{ in} \qquad Ans.$$

$$l = np = (2)(\tfrac{1}{6}) = 0.333 \text{ in} \qquad Ans.$$

(b) Using Eqs. (6-1) and (6-6), the torque required to turn the screw against the load is

$$T = \frac{Fd_m}{2}\left(\frac{l + \pi\mu d_m}{\pi d_m - \mu l}\right) + \frac{F\mu_c d_c}{2}$$

$$= \frac{(1500)(0.9167)}{2}\left[\frac{0.333 + \pi(0.08)(0.9167)}{\pi(0.9167) - (0.08)(0.333)}\right]$$

$$+ \frac{(1500)(0.08)(1.25)}{2}$$

$$= 136 + 75 = 211 \text{ lb·in} \qquad Ans.$$

(c) The torque required to lower the load, that is, to rotate the screw with the load, is obtained using Eqs. (6-2) and (6-6). Thus

$$T = \frac{Fd_m}{2}\left(\frac{\pi\mu d_m - l}{\pi d_m + \mu l}\right) + \frac{F\mu_c d_c}{2}$$

$$= \frac{(1500)(0.9167)}{2}\left[\frac{\pi(0.08)(0.9167) - 0.333}{\pi(0.9167) + (0.08)(0.333)}\right]$$

$$+ \frac{(1500)(0.08)(1.25)}{2}$$

$$= -24.4 + 75 \approx 50 \text{ lb·in} \qquad Ans.$$

The minus sign in the first term indicates that the screw alone is not self-locking and would rotate due to the action of the load except for the fact that collar friction is present and must be overcome too. Thus the torque required to rotate the screw "with" the load is less than is necessary to overcome collar friction alone.

(d) The overall efficiency is

$$e = \frac{Fl}{2\pi T} = \frac{(1500)(0.333)}{(2\pi)(211)} = 0.377 \qquad Ans.$$

////

6-3   THREAD STRESSES

In Fig. 6-6b a force F is transmitted through a square-threaded screw into a nut. We are interested in finding the stresses in the nut threads and in the screw threads which might cause these threads to fail, say, by yielding.

If we assume that the load is uniformly distributed over the nut height h and that the screw threads would fail by shearing off on the minor diameter, then the average screw-thread shear stress is

$$\tau = \frac{2F}{\pi d_r h} \qquad (6\text{-}7)$$

The threads on the nut would shear off on the major diameter, and so the average nut-thread shear stress is

$$\tau = \frac{2F}{\pi dh} \qquad (6\text{-}8)$$

We note very particularly that these are *average stresses* because we have assumed that the threads share the load equally. There are many cases, as we shall discover later, where this assumption is grossly in error. In view of this, rather large factors of safety, $n > 2$, should be used when Eqs. (6-7) and (6-8) are used for design purposes.

The bearing stress in the threads is

$$\sigma = \frac{-4pF}{\pi h(d^2 - d_r^2)} \qquad (6\text{-}9)$$

and this is an average stress too, because the force is assumed to be uniformly distributed over the face of the threads. Actually, there may be some bending of the thread, and so a high factor of safety should be employed in this case too.

Similar thread-stress formulas can easily be developed for other thread forms.

6-4   THREADED FASTENERS

Fasteners are named according to how they are intended to be used rather than how they are actually employed in specific instances. If this basic fact is remembered, it will not be difficult to distinguish between a *screw* and a *bolt*.

If a product is designed so that its primary purpose is assembly into a tapped hole, it is a *screw*. Thus a screw is tightened by exerting torque on the *head*.

If a product is designed so that it is intended to be used with a nut, it is a *bolt*. A bolt is tightened by exerting torque on the *nut*.

A *stud* resembles a threaded rod; one end assembles into a tapped hole, the other end receives a nut.

It is the intent, rather than the actual use, which determines the name of a

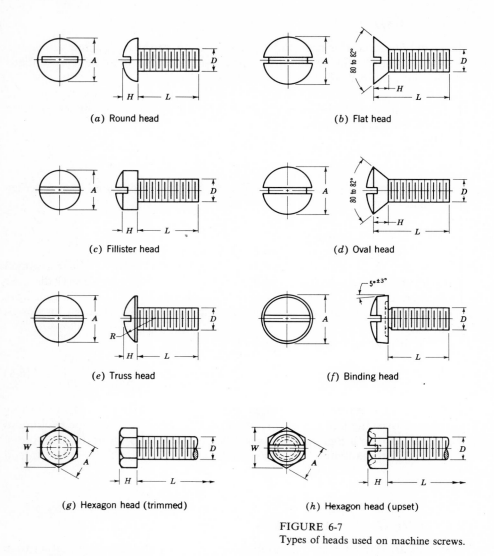

(a) Round head

(b) Flat head

(c) Fillister head

(d) Oval head

(e) Truss head

(f) Binding head

(g) Hexagon head (trimmed)

(h) Hexagon head (upset)

FIGURE 6-7
Types of heads used on machine screws.

product. Thus it may be desirable on various occasions to use a drill through two sheets of steel, say, and join them using a machine screw and nut.

Space does not permit a complete tabulation of the dimensions of a large variety of threaded products, but Tables A-26 to A-29 show some of the sizes of bolts, screws, and nuts. Figure 6-7 also shows a variety of head forms available on standard *machine screws*. A machine screw is a small screw; though shown in Table A-26 in sizes up to $\frac{3}{4}$ in, it is usually available only in sizes of $\frac{3}{8}$ in and under.

Cap screws are employed in sizes from $\frac{1}{4}$ in up to and including $1\frac{1}{2}$ in, as shown in Table A-27. Figures 6-8 and 6-9a show four of the head forms most

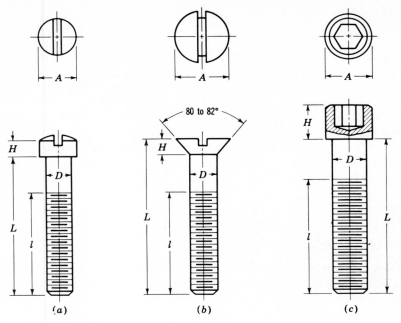

FIGURE 6-8
Cap-screw heads: (a) fillister head; (b) flat head; (c) hexagon socket head.

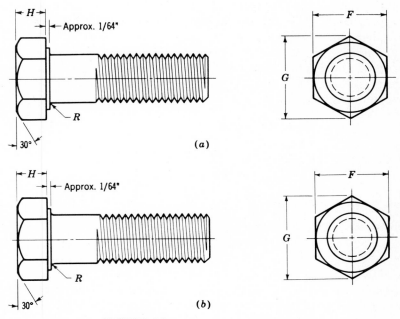

FIGURE 6-9
(a) Hexagon-head cap screw; (b) hexagon-head bolt (finished).

commonly used with cap screws. The heads in Fig. 6-8 require no wrench clearance. Metric cap screws are made in sizes from 3 mm up to 48 mm in diameter. They are available in fine pitch as well as the regular pitch series.

There is no difference between a hexagon-head cap screw and a *finished* hexagon-head bolt, as an examination of Fig. 6-9a and b and Tables A-27 and A-28 will show. Of course, bolts are also made in *semifinished* and in *unfinished* forms which may be dimensionally different. The following characteristics should be noted from the figure: the chamfer on the head and on the body, the length of threads $2D + \frac{1}{4}$ in, the washer face under the head, and the fillet radius.

6-5 PRELOADING OF BOLTS

When a connection is desired which can be disassembled without destructive methods and which is strong enough to resist both external tensile loads and shear loads, or a combination of these, then the simple bolted joint using hardened washers is a good solution. Such a joint is illustrated in Fig. 6-10, in which the bolt has first been tightened to produce an initial tensile preload F_i, after which the external tensile load P and the external shear load F_s are applied. The effect of the preload is to place the parts in compression for better resistance to the external tensile load and to create friction between the parts to resist the shear load. The shear load does not affect the final bolt tension, and we shall neglect this load for the time being in order to study the effect of the external tensile load on the compression of the parts and the resultant bolt tension.

The *spring constant*, or *stiffness constant*, of an elastic member such as a bolt, as we learned in Chap. 3, is the ratio of the force applied to the member to the deflection produced by that force. The deflection of a bar in simple tension or compression was found to be

$$\delta = \frac{Fl}{AE} \qquad (a)$$

where δ = deflection
F = force
A = area
E = modulus of elasticity

Therefore the stiffness constant is

$$k = \frac{F}{\delta} = \frac{AE}{l} \qquad (b)$$

In finding the stiffness of a bolt, A is the area based on the nominal or major diameter because the effect of the threads is neglected. The grip l is the total thickness of the parts which have been fastened together. Note that this is somewhat less than the length of the bolt.

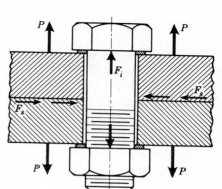

·FIGURE 6-10
A bolted connection.

There may be more than two members included in the grip of the bolt. These act like compressive springs in series, and hence the total spring rate of the *members* is

$$\frac{1}{k_m} = \frac{1}{k_1} + \frac{1}{k_2} + \frac{1}{k_3} + \cdots + \frac{1}{k_i} \qquad (c)$$

If one of the members is a soft gasket, its stiffness relative to the other members is usually so small that for all practical purposes the others can be neglected and only the gasket stiffness used.

If there is no gasket, the stiffness of the members is rather difficult to obtain, except by experimentation, because the compression spreads out between the bolt head and nut and hence the area is not uniform. In many cases the geometry is such that this area can be determined. When it cannot, a safe approach might be to use a hollow cylinder for the members having a hole the same size as the bolt and an outside diameter three times the bolt diameter. If we use this assumption, and also assume that all materials included in the grip are the same, then the stiffness of the members from Eq. (b) is

$$k_m = \frac{2\pi d^2 E}{l} \qquad (6\text{-}10)$$

Also, the stiffness of the bolt, from Eq. (b), is

$$k_b = \frac{\pi d^2 E}{4l} \qquad (6\text{-}11)$$

In Eqs. (6-10) and (6-11) d is the nominal bolt diameter. Thus, if the bolt and members have the same modulus of elasticity, the members are eight times as stiff as the bolt with this assumption.

Let us now visualize a tension-loaded bolted connection. We use the following nomenclature:

P = total external load on bolted assembly

F_i = preload on bolt due to tightening and in existence
before P is applied

P_b = portion of P taken by bolt

P_m = portion of P taken by members

F_b = resultant bolt load

F_m = resultant load on members

When the external load P is applied to the preloaded assembly, there is a change in the deformation of the bolt and also in the deformation of the connected members. The bolt, initially in tension, gets longer. This *increase* in deformation of the bolt is

$$\Delta\delta_b = \frac{P_b}{k_b} \qquad (d)$$

The connected members have initial compression due to the preload. When the external load is applied, this compression will *decrease*. The decrease in deformation of the members is

$$\Delta\delta_m = \frac{P_m}{k_m} \qquad (e)$$

On the assumption that the members have not separated, the increase in deformation of the bolt must equal the decrease in deformation of the members, and consequently

$$\frac{P_b}{k_b} = \frac{P_m}{k_m} \qquad (f)$$

Since $P = P_b + P_m$, we have

$$P_b = \frac{k_b P}{k_b + k_m} \qquad (g)$$

Therefore the resultant load on the bolt is

$$F_b = P_b + F_i = \frac{k_b P}{k_b + k_m} + F_i \qquad (6\text{-}12)$$

In the same manner, the resultant compression of the connected members is found to be

$$F_m = \frac{k_m P}{k_b + k_m} - F_i \qquad (6\text{-}13)$$

Equations (6-12) and (6-13) hold only as long as some of the initial compression remains in the members. If the external force is large enough to remove this compression completely, the members will separate and the entire load will be carried by the bolt.

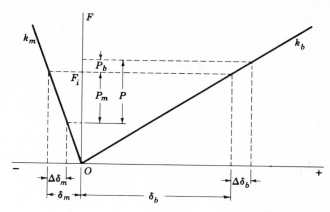

FIGURE 6-11

Figure 6-11 is a plot of the force-deflection characteristics and shows what is happening. The line k_m is the stiffness of the members; any force, such as the preload F_i, will cause a compressive deformation δ_m in the members. The same force will cause a tensile deformation δ_b in the bolt. When an external load is applied, δ_m is reduced by the amount $\Delta\delta_m$ and δ_b is increased by the same amount $\Delta\delta_b = \Delta\delta_m$. Thus the load on the bolt increases and the load in the members decreases.

The following example is used to illustrate the meaning of Eqs. (6-12) and (6-13). In Fig. 6-12a the fish scale with the 150-lb weight is analogous to a bolt tightened to a tensile preload of 150 lb. Then in Fig. 6-12b a block is forced in position as shown and the 150-lb weight removed and replaced with a 20-lb weight. Figure 6-12b now represents a bolted assembly having a bolt preload of 150 lb and an external load of 20 lb. Adding the 20-lb weight does not increase the tension in shank A, which represents the bolt. If the tension were greater than 150 lb, the scale would read more than 150 lb and the block would fall out. This is an extreme example, since the stiffness constant of the block k_m is a great deal more than that of the scale k_b, but it does illustrate the advantages to be gained by proper preloading. The following example is more realistic.

EXAMPLE 6-2   In Fig. 6-10, let $k_m = 8k_b$. If the preload is $F_i = 1000$ lb and the external load is $P = 1100$ lb, what is the resultant tension in the bolt and the compression in the members?

SOLUTION   From Eq. (6-12) the resultant bolt tension is

$$F_b = \frac{k_b P}{k_b + k_m} + F_i = \frac{k_b(1100)}{k_b + 8k_b} + 1000 = 1122 \text{ lb} \qquad Ans.$$

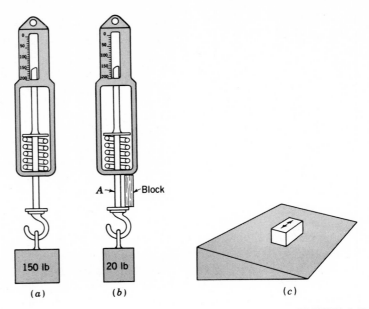

A→ ‹-Block

150 lb   20 lb

(a)   (b)   (c)

FIGURE 6-12

The resultant compression of the members, from Eq. (6-13), is

$$F_m = \frac{k_m P}{k_b + k_m} - F_i = \frac{8k_b(1100)}{k_b + 8k_b} - 1000 = -22 \text{ lb} \qquad Ans.$$

This example shows that the proportion (11 percent) of the load taken by the bolt is small and that it depends on the relative stiffness. The members are still in compression, and hence there is no separation of the parts even though the external load is greater than the preload.   ////

The importance of preloading of bolts cannot be overestimated. A high preload improves both the fatigue resistance of a bolted connection and the locking effect. To see why this is true, imagine an external tensile load which varies from 0 to P. If the bolts are preloaded, only about 10 percent of this load will cause a fluctuating bolt stress. Thus we will be operating with a very small σ_a/σ_m slope on the modified Goodman fatigue diagram.

To see why preloading of bolts improves the locking effect, visualize (Fig. 6-12c) an inclined plane representing a screw thread, and on this plane we place a block, which is analogous to the nut. Now if the block is wiggled back and forth, it will eventually work its way down the plane. This is analogous to the loosening of a nut. To retain a tight nut, the resultant bolt tension should vary as little as possible; in other words, σ_a should be very small compared with σ_m.

6-6 TORQUE REQUIREMENTS

Having learned that a high preload is very desirable in important bolted connections, we must next consider means of assuring that the preload is actually developed when the parts are assembled.

If the overall length of the bolt can actually be measured with a micrometer when it is assembled, the bolt elongation due to the preload F_i can be computed using the formula $\delta = F_i l/AE$. Then simply tighten the nut until the bolt elongates through the distance δ. This assures that the desired preload has been attained.

The elongation of a screw cannot usually be measured because the threaded end may be in a blind hole. It is also impractical in many cases to measure bolt elongation. In such cases the wrench torque required to develop the specified preload must be estimated. Then torque wrenching, pneumatic-impact wrenching, or the turn-of-the-nut method may be used.

The torque wrench has a built-in dial which indicates the proper torque.

With impact wrenching the air pressure is adjusted so that the wrench stalls when the proper torque is obtained, or in some wrenches, the air automatically shuts off at the desired torque.

The turn-of-the-nut method requires that we first define the meaning of snug-tight. The *snug-tight* condition is the tightness attained by a few impacts of an impact wrench, or the full effort of a man using an ordinary wrench. Having attained the snug-tight condition, all additional turning develops useful tension in the bolt. The turn-of-the-nut method requires that one compute the fractional number of turns necessary to develop the required preload from the snug-tight condition. For example, for heavy hexagon structural bolts the turn-of-the-nut specification states that the nut should be turned a minimum of 180° from the snug-tight condition under optimum conditions. Note that this is also about the correct rotation for the wheel nuts of a passenger car.

Although the coefficients of friction may vary widely, we can obtain a good estimate of the torque required to produce a given preload by combining Eqs. (6-5) and (6-6).

$$T = \frac{F_i d_m}{2}\left(\frac{l + \pi\mu d_m \sec \alpha}{\pi d_m - \mu l \sec \alpha}\right) + \frac{F_i \mu_c d_c}{2} \qquad (a)$$

Since $\tan \psi = l/\pi d_m$, we divide the numerator and denominator of the first term by πd_m and get

$$T = \frac{F_i d_m}{2}\left(\frac{\tan \psi + \mu \sec \alpha}{1 - \mu \tan \psi \sec \alpha}\right) + \frac{F_i \mu_c d_c}{2} \qquad (b)$$

Examination of Table A-29 shows that the diameter of the washer face of a hex nut is the same as the width across flats and equal to $1\frac{1}{2}$ times the nominal size. Therefore the mean collar diameter is $d_c = (d + 1.5d)/2 = 1.25d$. Equation (b) can now be arranged to give

$$T = \left[\left(\frac{d_m}{2d}\right)\left(\frac{\tan \psi + \mu \sec \alpha}{1 - \mu \tan \psi \sec \alpha}\right) + 0.625\mu_c\right] F_i d \qquad (c)$$

We now define a *torque coefficient* K as the term in brackets, and so

$$K = \left(\frac{d_m}{2d}\right)\left(\frac{\tan\psi + \mu\sec\alpha}{1 - \mu\tan\psi\sec\alpha}\right) + 0.625\mu_c \qquad (6\text{-}14)$$

Equation (c) now can be written

$$T = KF_id \qquad (6\text{-}15)$$

The coefficients of thread and collar friction for bolts, screws, and nuts range from about 0.12 to 0.20, depending upon thread finish, accuracy, and degree of lubrication. On the average, both μ and μ_c are about 0.15. The interesting thing about Eq. (6-14) is that $K \approx 0.20$ for $\mu = \mu_c = 0.15$, no matter what size bolts are

Table 6-3 THE DISTRIBUTION OF PRELOAD FOR 20 TESTS OF UNLUBRICATED BOLTS, SIZE $\frac{1}{2}$ in–20 UNF, TORQUED TO 800 lb·in

F_i, kip	f	fF_i	fF_i^2
5.3	1	5.3	28.1
6.2	1	6.2	38.5
6.3	1	6.3	39.7
6.6	1	6.6	43.5
6.8	1	6.8	46.2
6.9	1	6.9	47.6
7.4	1	7.4	54.8
7.6	3	22.8	173.4
7.8	1	7.8	60.8
8.0	2	16.0	128.0
8.4	1	8.4	70.6
8.5	2	17.0	144.5
8.8	1	8.8	77.4
9.0	1	9.0	81.0
9.1	1	9.1	82.8
9.6	1	9.6	92.2
Total	20	154.0	1209.1

$$\bar{F}_i = \frac{\sum fF_i}{N} = \frac{154.0}{20} = 7.7 \text{ kip}$$

$$s_F^2 = \frac{\sum fF_i^2 - \dfrac{(\sum fF_i)^2}{N}}{N-1}$$

$$= \frac{1209.1 - \dfrac{(154)^2}{20}}{19} = 1.226$$

$$s_F = \sqrt{1.226} = 1.107 \text{ kip}$$

Table 6-4 THE DISTRIBUTION OF PRELOAD FOR 10 TESTS OF LUBRICATED BOLTS, SIZE $\frac{1}{2}$ in–20 UNF, TORQUED TO 800 lb·in

F_i, kip	f	fF_i	fF_i^2
6.8	1	6.8	46.3
7.3	2	14.6	106.9
7.4	2	14.8	109.7
7.6	1	7.6	57.7
7.7	1	7.7	59.3
7.8	1	7.8	60.8
8.4	1	8.4	70.5
9.1	1	9.1	82.8
Total	10	76.8	594.0

$$\bar{F}_i = \frac{\sum fF_i}{N} = \frac{76.8}{10} = 7.68 \text{ kip}$$

$$s_F^2 = \frac{\sum fF_i^2 - \dfrac{(\sum fF_i)^2}{N}}{N-1}$$

$$= \frac{594.0 - \dfrac{(76.8)^2}{10}}{9} = 0.464$$

$$s_F = \sqrt{0.464} = 0.681 \text{ kip}$$

employed and no matter whether the threads are coarse or fine.* Thus Eq. (6-15) is more convenient as

$$T = 0.20F_i d \qquad (6\text{-}16)$$

In this form it is very simple to compute the wrench torque T needed to create a desired preload F_i when the size d of the fastener is known.

Blake and Kurtz have published results of numerous tests of the torquing of bolts.† By subjecting their data to a statistical analysis we can learn something about the distribution of the torque coefficients and the resulting preload. Blake and Kurtz determined the preload in quantities of unlubricated and lubricated bolts of size $\frac{1}{2}$ in–20 UNF when torqued to 800 lb·in. The statistical analyses of these two groups of bolts are displayed in Tables 6-3 and 6-4.

We first note that both groups have about the same mean preload, 7700 lb. The unlubricated bolts have a standard deviation of 1100 lb, which is about 15 percent of the mean. The lubricated bolts have a standard deviation of 680 lb, or about 9 percent of the mean, a substantial reduction. These deviations are quite large, though, and emphasize the necessity for quality-control procedures throughout the entire manufacturing and assembly process to assure uniformity.

The means obtained from the two samples are nearly identical, 7700 lb; using Eq. (6-15), we find, for both samples, $K = 0.208$, which is close to the recommended value.

6-7 BOLT STRENGTH AND PRELOAD

Table 6-5 lists the specifications most used for threaded fasteners. SAE grades 1 and 2 should only be used for unimportant or non-load-carrying connections; the carbon content of these is only about 10 to 20 points and so they are too ductile for loaded connections.

Table 6-6 shows how bolt grades are identified.

The terms proof load and proof strength appear frequently in fastener literature. The *proof load* of a bolt is the maximum tensile load a bolt can withstand without incurring a permanent set. The *proof strength* is the stress corresponding to the proof load based on the tensile-stress area (Tables 6-1 and 6-2). Thus proof strength is roughly equivalent to yield strength.

The recommendations included here for preload apply only to ungasketed joints using high-quality bolt materials, such as SAE 3 or better. Such bolt materials will have a stress-strain diagram in which there is no clearly defined yield point and a curve which progresses smoothly upward until fracture. Under these

---

* See Table 7-3, p. 246, in the original (1963) edition of this book for a complete listing of these values.

† J. C. Blake and H. J. Kurtz, The Uncertainties of Measuring Fastener Preload, *Machine Design*, vol. 37, pp. 128–131, Sept. 30, 1965.

conditions, *if the loads are static*, the *minimum* preload should be 90 percent of the proof load.

The recommendations for preload when fatigue is present are so important that we shall discuss this problem separately in the next section.

There are two sound reasons for the 90 percent recommendation for statically loaded joints composed of high-quality bolt materials. Because the stress-strain diagram proceeds smoothly to fracture, the bolt will retain its load-carrying capacity no matter how high the pre-tension. That is, there is no loss of capacity due to gross plastic yielding. The second reason for the 90 percent recommendation is that the torsional stress disappears after tightening. A joint subjected to slight movements will cause flattening of high spots, paint, or dirt and will relieve the torsional friction. Thus, if such a bolt does not fail during tightening, there is a very good chance that it never will fail! In some cases it may be desirable to turn the nut backward an eighth of a turn or so after tightening to relieve this torsion, but not enough to decrease the bolt tension.

EXAMPLE 6-3   Calculate the tightening torque for an SAE grade 6, $\frac{1}{2}''$–13UNC bolt. Compute the tensile stress and the torsional stress, and show on a Mohr's

Table 6-5 SPECIFICATIONS FOR BOLTS, CAP SCREWS, AND STUDS. MULTIPLY THE STRENGTH IN kpsi BY 6.89 TO GET THE STRENGTH IN MPa

SAE* grade	ASTM† grade	Metric‡ grade	Nominal diameter, in	Proof strength§ kpsi	Tensile strength¶ kpsi	Maximum hardness Bhn	Material
1	A307	4.6	$\frac{1}{4}$ to $1\frac{1}{2}$	33	55	207	Low-carbon steel
2		5.6	$\frac{1}{4}$ to $\frac{1}{2}$	55	69	241	Low-carbon steel
			Over $\frac{1}{2}$ to $\frac{3}{4}$	52	64	241	
			Over $\frac{3}{4}$ to $1\frac{1}{2}$	28	55	207	
3		6.8	$\frac{1}{4}$ to $\frac{1}{2}$	85	110	269	Medium-carbon steel
			Over $\frac{1}{2}$ to $\frac{5}{8}$	80	100	269	
5	A449	8.8	$\frac{1}{4}$ to $\frac{3}{4}$	85	120	302	Medium-carbon steel, heat treated
			Over $\frac{3}{4}$ to 1	78	115	302	
			Over 1 to $1\frac{1}{2}$	74	105	285	
7			$\frac{1}{4}$ to $1\frac{1}{2}$	105	133	321	Medium-carbon, alloy steel, heat treated
8	A354	10.9	$\frac{1}{4}$ to $1\frac{1}{2}$	120	150	352	Medium-carbon, alloy steel, heat treated

* Society of Automotive Engineers.
† American Society of Testing Materials.
‡ Cat. No. 2007, p. 177, Metric and Multistandard Components Corp., Elmsford, N.Y.
§, ¶ Minimum values.

circle diagram the reduction in principal stress due to the disappearance of the torsional stress.

SOLUTION For grade 6, the proof strength is $S_P = 110$ kpsi. Also, from Table 6-2, the tensile stress area is $A_t = 0.1419$ in². Therefore, the recommended preload is

$$F_i = 0.90S_P A_t = 0.90(110)(0.1419) = 14.05 \text{ kip}$$

With $K = 0.20$, the tightening torque is

$$T = 0.20F_i d = 0.20(14.05)(10)^3(0.500) = 1405 \text{ lb·in} \qquad Ans.$$

The tensile stress is 90 percent of the proof strength, and so $\sigma_x = 0.90(110) = 99$ kpsi.

The torque applied to the nut is used up in three ways. About 50 percent of it is used to overcome the friction between the bearing face of the nut and the member. About 40 percent of the applied torque is used to overcome thread friction, and the balance produces the bolt tension. The last two items are the only contributions to the torsion in the screw. From Table 6-2, $A_r = 0.1257$ in², and so the root diameter is $d_r = \sqrt{4A_r/\pi} = \sqrt{4(0.1257/\pi} = 0.400$ in. The torsion in the screw is $T = 0.50(1405) = 702$ lb·in, and so

$$\tau_{xy} = \frac{16T}{\pi d_r^3} = \frac{16(702)}{\pi(0.400)^3} = 56.0(10)^3 \text{ psi}$$

Drawing Mohr's circle corresponding to $\sigma_x = 99(10)^3$ psi and $\tau_{xy} = 56.0(10)^3$ psi gives the solid circle of Fig. 6-13. If the principal stresses are calculated, they will be found to be $\sigma_1 = 124.2(10)^3$ psi and $\sigma_2 = -25.2(10)^3$ psi.

For fracture, the maximum-normal-stress theory applies. Thus, during tightening, the factor of safety is

$$n = \frac{S_u}{\sigma_1} = \frac{140}{124.2} = 1.12$$

Table 6-6 IDENTIFICATION OF SAE BOLT GRADES; HEAD MARKINGS

Grades 0, 1, and 2, no marking

Grade 3: 2 radial dashes 180° apart

Grade 5: 3 radial dashes 120° apart

Grade 6: 4 radial dashes 90° apart

Grade 7: 5 radial dashes 72° apart

Grade 8: 6 radial dashes 60° apart

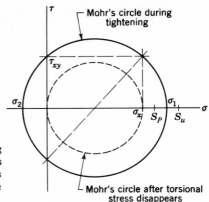

FIGURE 6-13
Mohr's circle diagram for a bolt during tightening. Tightening of the nut produces shear stress τ_{xy}. The bolt tensile stress is σ_x. The proof strength is S_P, and the tensile strength S_u.

or a 12 percent margin of safety. (Note that this is less than one standard deviation for unlubricated bolts!)

After the torsion disappears, the principal stress is σ_x and the factor of safety will be

$$n = \frac{S_u}{\sigma_x} = \frac{140}{99} = 1.41$$

You can now appreciate the statement: If a bolt does not fail during tightening, there is a good reason to believe that it never will fail. ////

In the design of a tension-loaded bolted joint, factor of safety n is used in a somewhat different way than in Example 6-3. In Example 6-2 it was learned that an external load about equal to the preload would remove most of the compression in the members. Therefore, to get a first approximation in the determination of bolt size, define factor of safety as

$$n = \frac{F_i}{P} \qquad (6\text{-}17)$$

Then the bolt size is determined such that its preload is larger than the external tensile load by the amount of the factor of safety. When high-quality bolt materials are used, when the members are stiff, and when the loads are static, this is about as far as we need go. If fatigue loads are present, a detailed analysis of the resulting design should be made using the approach of Sec. 6-9.

6-8 SELECTION OF THE NUT

Imagine three annular rings, analogous to square threads on a screw, cut on a male member, and three corresponding grooves with clearances as shown in

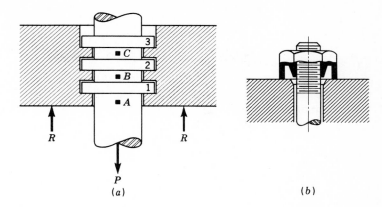

FIGURE 6-14
(a) Schematic diagram showing how first nut thread takes entire load. (b) Typical
method of distributing the thread load.

Fig. 6-14a on a female member or nut. Now apply a tension load P to the screw
and let the nut react to this load as shown in the figure. If the load is assumed to be
uniformly divided between the three threads, stress elements at A, B, and C on the
screw will have tensile loads of $F_A = P$, $F_B = 2P/3$, and $F_C = P/3$. Corresponding
stress elements in the nut, not shown, will have compressive loads $F_A = -P$,
$F_B = -2P/3$, and $F_C = -P/3$. Now, with the screw in tension, the screw gets
longer, and so threads 1, 2, and 3 will tend to move apart. However, the nut is in
compression, and so the nut threads will tend to move closer together. But these
actions prevent sharing of the loads as assumed in the beginning. We therefore
conclude that the load will not be shared at all and that, instead, the first thread
takes the entire force.

This tendency may be partially corrected by proportioning the nut so as to
cause more deformation to exist at the bottom. Figure 6-14b shows a nut design in
which material has been removed from the lower portion of the nut in order to
equalize the stress distribution.

In practice, conditions are not quite as severe as pictured, since yielding of
the threads in the nut will permit the other threads to transfer some of the load.
However, since such a tendency is present, it must be guarded against and knowledge of it used in selecting the nut.

Another factor which acts to reduce the tendency of the bottom thread to
take the entire load is that the wedging action of the threads tends to spread, or
dilate, the nut.

These conditions point to the fact that, when preloading is desired, careful
attention should be given to the nut material. Selecting a soft nut ensures plastic
yielding, which will enable the nut threads to divide the load more evenly.

Nuts are tested by determining their stripping strength. The test is made by
threading a nut on a hardened-steel mandrel and pulling it through the nut. The

strength is the load divided by the mean thread area. Common nuts have a stripping strength of approximately 90 kpsi.

It is interesting to know that three full threads are all that are required to develop the full bolt strength.

Another factor which must be considered in the design of bolted joints is the maintenance of the initial preload. This load may be relaxed by yielding of the clamped material, by extrusion of paint or plating from the contact surfaces, or by a compression of rough places. Extra contact area may be provided by hardened washers. This is especially necessary if the bolted parts are relatively soft and the bolt head or nut does not provide sufficient bearing area.

6-9   FATIGUE LOADING

Tension-loaded bolted joints subjected to fatigue action can be analyzed directly by the methods of Chap. 5. Table 6-7 lists average fatigue-strength reduction factors for the fillet under the bolt head, and also at the beginning of the threads on the bolt shank. Note that these are already corrected for notch sensitivity. Designers should be aware that situations may arise in which it would be advisable to investigate these factors more closely, since they are only average values. In fact Peterson* observes that the distribution of typical bolt failures is about 15 percent under the head, 20 percent at the end of the thread, and 65 percent in the thread at the nut face.

In using Table 6-7, it is usually safe to assume that the fasteners have rolled threads unless specific information is available. Also, in computing endurance limits, use a machined finish if nothing has been stated.

Tension-loaded connections subject to fatigue have to be designed very carefully. Very seldom can a preload equal to 90 percent of the proof load be used. In fact such a high bolt preload will usually lead directly to a fatigue failure. This means that the designer has a special responsibility to be sure that his recommendations are carried out during assembly.

Most of the time the type of fatigue loading encountered in the analysis of bolted joints is one in which the externally applied load fluctuates between zero and some maximum force P. This would be the situation in a pressure cylinder, for

---

* R. E. Peterson, "Stress Concentration Factors," p. 253, John Wiley & Sons, N.Y., 1964.

Table 6-7   FATIGUE-STRENGTH REDUCTION FACTORS K_f FOR THREADED ELEMENTS

SAE grade	Metric grade	Rolled threads	Cut threads	Fillet
0 to 2	3.6 to 5.8	2.2	2.8	2.1
3 to 8	6.6 to 10.9	3.0	3.8	2.3

example, where a pressure either exists or does not exist. In order to determine the mean and alternating bolt stresses for such a situation, let us employ the notation of Sec. 6-5. Then $F_{max} = F_b$ and $F_{min} = F_i$. Therefore, the alternating component of bolt stress is, from Eq. (6-12),

$$\sigma_a = \frac{F_b - F_i}{2A_t} = \frac{k_b}{k_b + k_m}\frac{P}{2A_t} \qquad (6\text{-}18)$$

Then, since the mean stress is equal to the alternating component plus the minimum stress, we have

$$\sigma_m = \sigma_a + \frac{F_i}{A_t} = \frac{k_b}{k_b + k_m}\frac{P}{2A_t} + \frac{F_i}{A_t} \qquad (6\text{-}19)$$

These two stress components are entered into the modified Goodman fatigue diagram, using the methods of Chap. 5, to determine the safeness of the design as will be shown in the examples to follow.

Suppose the design is not very safe; what can we do? The second term of Eq. (6-19) is the predominant term, because it contains the preload, and we can reduce its magnitude by reducing the preload. However, this is not the best solution because a reduced preload might lead to loosening of the nut.

To see what else can be done, take the first factor of Eq. (6-18) and divide the numerator and denominator by k_b. This gives

$$\frac{k_b}{k_b + k_m} = \frac{1}{1 + (k_m/k_b)} \qquad (a)$$

This equation shows that if we can design the connection so that the stiffness ratio k_m/k_b is very large, then the term $k_b/(k_b + k_m)$ will be small and so will σ_a. In most cases, this is the best way of designing a safe connection. Note that it can be done either by making k_m large or by making k_b small.

Of course another way of making σ_a small is to use more bolts, since this reduces the load P that must be carried by each bolt.

In the case of fully gasketed joints, the gasket is subject to the full compressive load between the members. Thus its stiffness would predominate and so Eq. (c) of Sec. 6-5 should include the gasket stiffness as one of the terms.

When bolted joints are fully gasketed, the bolt preload must be adjusted to provide a gasket pressure within the range recommended by the gasket manufacturers.

The correct gasket pressure for a confined gasket (Fig. 6-13) is obtained by matching the gasket size and shape to the dimensions of the retaining groove. Generally, a confined gasket is assumed to have no effect on the stiffness of the members.

EXAMPLE 6-4 Based on 50 percent reliability, find the endurance limit of the bolt shown in Fig. 6-15.

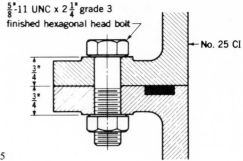

$\frac{5}{8}$"-11 UNC x $2\frac{1}{4}$" grade 3
finished hexagonal head bolt

No. 25 CI

$\frac{3}{4}$"

$\frac{3}{4}$"

FIGURE 6-15

SOLUTION   From Table 6-5 we find $S_u = 100$ kpsi. Therefore

$$S'_e = 0.50S_u = 0.50(100) = 50 \text{ kpsi}$$

Corresponding to a machined finish, the surface factor is $k_a = 0.73$. The size factor is $k_b = 0.85$. From Table 6-7 we select $K_f = 3.0$ for rolled threads, because it is larger than K_f at the fillet. Then $k_e = 1/K_f = 1/3.0 = 0.333$. The endurance limit of the bolt is now found to be

$$S_e = k_a k_b k_e S'_e = 0.73(0.85)(0.333)(50) = 10.3 \text{ kpsi} \qquad Ans.$$

////

EXAMPLE 6-5   A section of a tension-loaded connection employing a confined gasket is shown in Fig. 6-15. This connection has been tentatively designed to resist a load which fluctuates between 0 and 12 kip per bolt. Find the factor of safety of the connection based on a bolt preload of 10 kip and the endurance limit found in Example 6-4.

SOLUTION   From Table 6-5 we find $S_P = 80$ kpsi. Also, from Table 6-2, $A_t = 0.226$ in². Since F_i is given as 10 kip, we are only using

$$\frac{10}{0.226(80)} = 0.55$$

or 55 percent of the proof strength in tightening.

Using the assumptions of Sec. 6-5 and Eqs. (6-10) and (6-11), we find the stiffness constants to be

$$k_b = \frac{\pi d^2 E}{4l} = \frac{\pi(0.625)^2(30)(10)^6}{4(1.5)} = 6.13(10)^6 \text{ lb/in}$$

$$k_m = \frac{2\pi d^2 E}{l} = \frac{2\pi(0.625)^2(12)(10)^6}{1.5} = 19.6(10)^6 \text{ lb/in}$$

where the grip is $l = 1\frac{1}{2}$ in and $E = 12$ Mpsi for No. 25 cast iron.

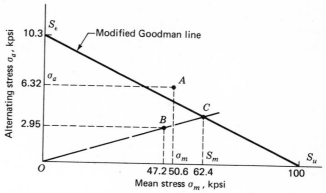

FIGURE 6-16
Fatigue diagram. Note that the two stress axes are not drawn to the same scale.

Next, we compute the mean and alternating stresses in the bolt from Eqs. (6-18) and (6-19). The results are

$$\sigma_a = \frac{k_b}{k_b + k_m}\frac{P}{2A_t} = \frac{6.13}{6.13 + 19.6}\frac{12}{2(0.226)} = 6.32 \text{ kpsi}$$

$$\sigma_m = \frac{k_b}{k_b + k_m}\frac{P}{2A_t} + \frac{F_i}{A_t} = \frac{6.13}{6.13 + 19.6}\frac{12}{2(0.226)} + \frac{10}{0.226}$$

$$= 6.32 + 44.25 = 50.6 \text{ kpsi}$$

These two stress components are plotted on the fatigue diagram of Fig. 6-16. They result in point A, outside the modified Goodman line. Thus there is no factor of safety and a fatigue failure can be expected according to this criteria.   ////

EXAMPLE 6-6 By making the members in Fig. 6-15 of steel, instead of cast iron, the stiffness should be larger than before, because the modulus of elasticity of steel is greater. With this change, what is the factor of safety?

SOLUTION   Since E for steel is 30 Mpsi, the new value of k_m is

$$k_m = (30/12)(19.6)(10)^6 = 49.0(10)^6 \text{ lb/in}$$

which is $2\frac{1}{2}$ times as much as the previous value. Equations (6-18) and (6-19) now give

$$\sigma_a = \frac{k_b}{k_b + k_m}\frac{P}{2A_t} = \frac{6.13}{6.13 + 49.0}\frac{12}{2(0.226)} = 2.95 \text{ kpsi}$$

$$\sigma_m = \frac{k_b}{k_b + k_m}\frac{P}{2A_t} + \frac{F_i}{A_t} = \frac{6.13}{6.13 + 49.0}\frac{12}{2(0.226)} + \frac{10}{0.226}$$

$$= 2.95 + 44.25 = 47.2 \text{ kpsi}$$

When these two stresses are plotted in Fig. 6-16 they locate point B inside the

Goodman line. Hence this is a safe connection. A line drawn through the origin O and point B locates C on the Goodman line. A projection from C to the σ_m axis gives $S_m = 62.4$ kpsi. Consequently, the factor of safety is

$$n = \frac{S_m}{\sigma_m} = \frac{62.4}{47.2} = 1.32$$

which is a substantial improvement.

It is worth observing from Fig. 6-16 that the change in material did not change the mean stress very much, but it did reduce σ_a by over 50 percent. About the same change could have been obtained by using two bolts instead of one. Then $P = 6$ kip would have been the only change made in Eqs. (6-18) and (6-19). This would also reduce σ_a by 50 percent.   ////

6-10   BOLTED AND RIVETED JOINTS LOADED IN SHEAR*

Riveted and bolted joints loaded in shear are treated exactly alike in design and analysis.

In Fig. 6-17a is shown a riveted connection loaded in shear. Let us now study the various means by which this connection might fail.

Figure 6-17b shows a failure by bending of the rivet or of the riveted members. The bending moment is approximately $M = Ft/2$, where F is the shearing force and t is the grip of the rivet, that is, the total thickness of the connected parts. The bending stress in the members or in the rivet is, neglecting stress concentration,

$$\sigma = \frac{M}{I/c} \qquad (6\text{-}20)$$

where I/c is the section modulus for the weakest member or for the rivet or rivets, depending upon which stress is to be found. The calculation of the bending stress in this manner is an assumption, because we do not know exactly how the load is distributed to the rivet nor the relative deformations of the rivet and the members. Although this equation can be used to determine the bending stress, it is seldom used in design; instead its effect is compensated for by an increase in the factor of safety.

In Fig. 6-17c failure of the rivet by pure shear is shown; the stress in the rivet is

$$\tau = \frac{F}{A} \qquad (6\text{-}21)$$

---

* The design of bolted and riveted connections for boilers, bridges, buildings, and other structures in which danger to human life is involved is strictly governed by various construction codes. When designing these structures the engineer should refer to the "American Institute of Steel Construction Handbook," the American Railway Engineering Association specifications, or the Boiler Construction Code of the American Society of Mechanical Engineers.

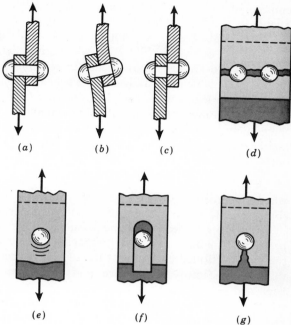

FIGURE 6-17
Modes of failure in shear loading of a bolted or riveted connection. (*a*) Shear loading; (*b*) bending of rivet; (*c*) shear of rivet; (*d*) tensile failure of members; (*e*) bearing of rivet on members or bearing of members on rivet; (*f*) shear tearout; (*g*) shear tearout.

where A is the cross-sectional area of all the rivets in the group. It may be noted that it is standard practice in structural design to use the nominal diameter of the rivet rather than the diameter of the hole, even though a hot-driven rivet expands and nearly fills up the hole.

Rupture of one of the connected members or plates by pure tension is illustrated in Fig. 6-17*d*. The tensile stress is

$$\sigma = \frac{F}{A} \qquad (6\text{-}22)$$

where A is the net area of the plate, that is, reduced by an amount equal to the area of all the rivet holes. For brittle materials and static loads and for either ductile or brittle materials loaded in fatigue, the stress-concentration effects must be included. It is true that the use of a bolt with an initial preload and, sometimes, a rivet will place the area around the hole in compression and thus tend to nullify the effects of stress concentration, but unless definite steps are taken to assure that the preload does not relax, it is on the conservative side to design as if the full stress-concentration effect were present. The stress-concentration effects are not considered in structural design because the loads are static and the materials ductile.

In calculating the area for Eq. (6-22) the designer should, of course, use the combination of rivet or bolt holes which gives the smallest area.

Figure 6-17e illustrates a failure by crushing of the rivet or plate. Calculation of this stress, which is usually called a *bearing stress*, is complicated by the distribution of the load on the cylindrical surface of the rivet. The exact values of the forces acting upon the rivet are unknown, and so it is customary to assume that the components of these forces are uniformly distributed over the projected contact area of the rivet. This gives for the stress

$$\sigma = \frac{F}{A} \qquad (6\text{-}23)$$

where the projected area for a single rivet is $A = td$. Here, t is the thickness of the thinnest plate and d is the rivet or bolt diameter.

Shearing, or tearing, of the margin is shown in Fig. 6-17f and g. In structural practice this failure is avoided by spacing the rivet at least $1\frac{1}{2}$ diameters away from the margin. Bolted connections usually are spaced an even greater distance than this for satisfactory appearance, and hence this type of failure may usually be neglected.

In structural design it is customary to select in advance the number of rivets and their diameters and spacing. The strength is then determined for each method of failure. If the calculated strength is not satisfactory, a change is made in the diameter, spacing, or number of rivets used, to bring the strength in line with expected loading conditions. It is not usual, in structural practice, to consider the combined effects of the various methods of failure.

6-11   CENTROIDS OF BOLT GROUPS

In Fig. 6-18 let A_1 through A_5 be the respective cross-sectional areas of a group of five bolts. These bolts need not be of the same diameter. In order to determine the shear forces which act upon each bolt, it is necessary to know the location of the centroid of the bolt group. Using statics, we learn that the centroid G is located by the coordinates

$$
\bar{x} = \frac{A_1 x_1 + A_2 x_2 + A_3 x_3 + A_4 x_4 + A_5 x_5}{A_1 + A_2 + A_3 + A_4 + A_5} = \frac{\sum\limits_{1}^{n} A_i x_i}{\sum\limits_{1}^{n} A_i}
$$

$$
\bar{y} = \frac{A_1 y_1 + A_2 y_2 + A_3 y_3 + A_4 y_4 + A_5 y_5}{A_1 + A_2 + A_3 + A_4 + A_5} = \frac{\sum\limits_{1}^{n} A_i y_i}{\sum\limits_{1}^{n} A_i}
$$

$$(6\text{-}24)$$

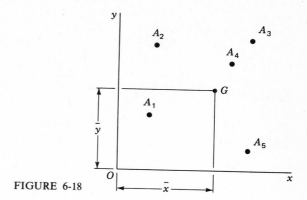

FIGURE 6-18

where x_i and y_i are the distances to the respective bolt centers. In many instances these centroidal distances can be located by symmetry.

6-12 SHEAR OF BOLTS AND RIVETS DUE TO ECCENTRIC LOADING

An example of eccentric loading of fasteners is shown in Fig. 6-19. This is a portion of a machine frame containing a beam A subjected to the action of a bending load. In this case, the beam is fastened to vertical members at the ends with bolts. You will recognize the schematic representation in Fig. 6-19b as an indeterminate beam with both ends fixed and with the moment reaction M and the shear reaction V at the ends.

For convenience, the centers of the bolts at one end of the beam are drawn to a larger scale in Fig. 6-19c. Point O represents the centroid of the group, and it is assumed in this example that all the bolts are of the same diameter. The total load taken by each bolt will be calculated in three steps. In the first step the shear V is divided equally among the bolts so that each bolt takes $F' = V/n$, where n refers to the number of bolts in the group, and the force F' is called the *direct load*, or *primary shear*.

It is noted that an equal distribution of the direct load to the bolts assumes an absolutely rigid member. The arrangement of the bolts or the shape and size of the members sometimes justify the use of another assumption as to the division of the load. The direct loads F' are shown as vectors on the loading diagram (Fig. 6-19c).

The *moment load*, or *secondary shear*, is the additional load on each bolt due to the moment M. If r_A, r_B, r_C, etc., are the radial distances from the centroid to the center of each bolt, the moment and moment load are related as follows:

$$M = F''_A r_A + F''_B r_B + F''_C r_C + \cdots \qquad (a)$$

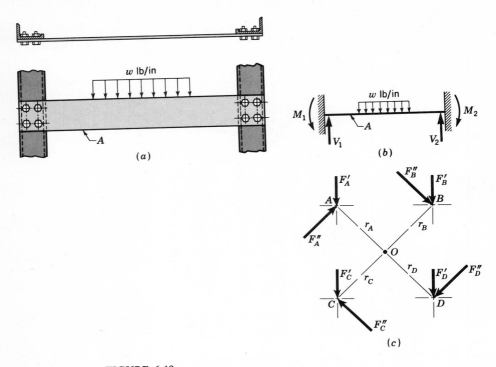

FIGURE 6-19
(a) Beam bolted at both ends with distributed load; (b) free-body diagram of beam; (c) enlarged view of bolt group showing primary and secondary shear forces.

where F'' is the moment load. The force taken by each bolt depends upon its radius; that is, the bolt farthest from the center of gravity takes the greatest load while the nearest bolt takes the smallest. We can therefore write

$$\frac{F''_A}{r_A} = \frac{F''_B}{r_B} = \frac{F''_C}{r_C} \qquad (b)$$

Solving Eqs. (a) and (b) simultaneously, we obtain

$$F''_n = \frac{Mr_n}{r_A^2 + r_B^2 + r_C^2 + \cdots} \qquad (6\text{-}25)$$

where the subscript n refers to the particular bolt whose load is to be found. These moment loads are also shown as vectors on the loading diagram.

In the third step the direct and moment loads are added vectorially to obtain the resultant load on each bolt. Since all the bolts or rivets are usually the same size, only that bolt having the maximum load need be considered. When the maximum load is found, the strength may be determined, using the various methods already described.

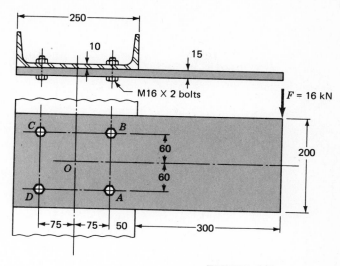

FIGURE 6-20
Dimensions in millimetres.

EXAMPLE 6-5 Shown in Fig. 6-20 is a 15- by 200-mm rectangular steel bar cantilevered to a 250-mm steel channel using four bolts. Based on the external load of 16 kN, find:

(a) The resultant load on each bolt.
(b) The maximum bolt shear stress.
(c) The maximum bearing stress.
(d) The critical bending stress in the bar.

SOLUTION (a) Point O, the centroid of the bolt group in Fig. 6-20, is found by symmetry. If a free-body diagram of the beam were constructed, the shear reaction V would pass through O and the moment reaction M would be about O. These reactions are

$$V = 16 \text{ kN} \qquad M = 16(425) = 6800 \text{ N·m}$$

In Fig. 6-21, the bolt group has been drawn to a larger scale and the reactions are shown. The distance from the centroid to the center of each bolt is

$$r = \sqrt{(60)^2 + (75)^2} = 96.0 \text{ mm}$$

The primary shear load per bolt is

$$F' = \frac{V}{n} = \frac{16}{4} = 4 \text{ kN}$$

Since the secondary shear forces are equal, Eq. (6-25) becomes

$$F'' = \frac{Mr}{4r^2} = \frac{M}{4r} = \frac{6800}{4(96.0)} = 17.7 \text{ kN}$$

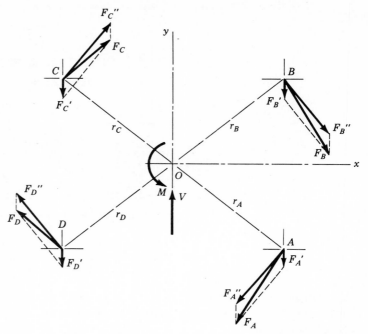

FIGURE 6-21

The primary and secondary shear forces are plotted to scale in Fig. 6-21 and the resultants obtained by using the parallelogram rule. The magnitudes are found by measurement (or analysis) to be

$$F_A = F_B = 21.0 \text{ kN} \qquad Ans.$$

$$F_C = F_D = 13.8 \text{ kN} \qquad Ans.$$

(b) Bolts A and B are critical because they carry the largest shear load. The bolt will tend to shear across its major diameter. Therefore the shear-stress area is $A_s = \pi d^2/4 = \pi(16)^2/4 = 201 \text{ mm}^2$. So the shear stress is

$$\tau = \frac{F}{A_s} = \frac{21.0(10)^3}{201} = 104 \text{ MPa} \qquad Ans.$$

(c) The channel is thinner than the bar so the largest bearing stress is due to the bolt pressing against the channel web. The bearing area is $A_b = td = 10(16) = 160 \text{ mm}^2$. So the bearing stress is

$$\sigma = \frac{F}{A_b} = -\frac{21.0(10)^3}{160} = -131 \text{ MPa} \qquad Ans.$$

(*d*) The critical bending stress in the bar is assumed to occur in a section parallel to the *y* axis and through bolts *A* and *B*. At this section the bending moment is

$$M = 16(300 + 50) = 5600 \text{ N·m}$$

The moment of inertia through this section is obtained by the use of the transfer formula, as follows:

$$I = I_{\text{bar}} - 2(I_{\text{holes}} + d^2 A)$$

$$= \frac{15(200)^3}{12} - 2\left[\frac{15(16)^3}{12} + (60)^2(15)(16)\right] = 8.26(10)^6 \text{ mm}^4$$

Then

$$\sigma = \frac{Mc}{I} = \frac{5600(100)}{8.26(10)^6}(10)^3 = 67.8 \text{ MPa} \qquad Ans.$$

$$////$$

6-13 KEYS, PINS, AND RETAINERS

Keys, pins, and retainers are normally used to secure elements such as gears or pulleys to shafts so that torque can be transferred between them. A pin may serve the double purpose of transferring torque as well as preventing relative axial motion between the mating parts. Relative axial motion can also be prevented by using a press or shrink fit, setscrews, or retainers or cotters. Figures 6-22 and 6-23

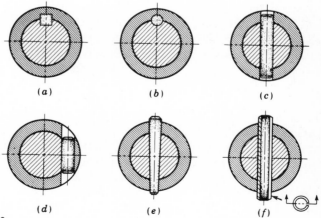

FIGURE 6-22
(*a*) Square key; (*b*) round key; (*c*)–(*d*) round pins; (*e*) taper pin; (*f*) split tubular spring pin. The pins in (*e*) and (*f*) are shown longer than necessary to illustrate the chamfer on the ends; but their lengths should not exceed the hub diameters, to avoid injuries which might be caused by projections on rotating parts.

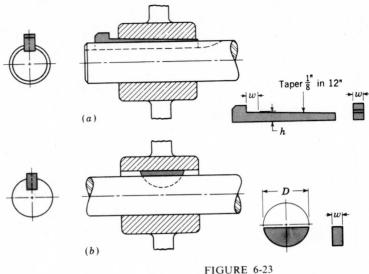

Taper $\frac{1}{8}$" in 12"

(a)

(b)

FIGURE 6-23
(a) Gib-head key; (b) Woodruff key.

illustrate a variety of these devices and methods of employing them. Notice that the gib-head key is tapered so that, when firmly driven, it also acts to prevent relative axial motion. The head makes removal possible without access to the other end, but this projection is hazardous with rotating parts. The Woodruff key is of general usefulness, and especially so when a wheel is to be positioned against a shaft shoulder, since the key slot need not be cut into the shoulder stress-concentration region. Standard sizes and numbering systems for keys and pins may be found in any of the mechanical engineering or machinery handbooks.

The usual practice is to choose a key whose size is one-fourth the shaft diameter. The length of the key is then adjusted according to the hub length and the strength required. It is sometimes necessary to use two keys to obtain the required strength.

In determining the strength of a key, the assumption may be made that the forces are uniformly distributed throughout the key length. This assumption is probably not true, since the torsional stiffness of the shaft will usually be less than that for the hub, causing large forces at one end of the key and small forces at the other end. This distribution may be still more complicated by the stiffening effect of the arms or web at the middle of the hub.

Geometric stress-concentration factors for keyways, when the shaft is in-bending, are given by Peterson* as 1.79 for an end-milled keyway (Fig. 6-24) and

* *Op. cit.*, pp. 245–247.

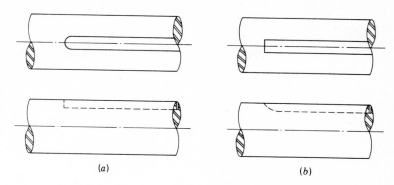

FIGURE 6-24
Keyways: (*a*) end-milled; (*b*) sled-runner.

1.38 for the sled-runner keyway; $K_t = 3$ should be used when the shafts are subject to combined bending and torsion. Charts A-25-10, A-25-11, and A-25-13 to A-25-15 should be used for shafts containing grooves or holes.

A force distribution having been assumed, it is customary to base the strength of the key on failure by crushing or by shearing. This is illustrated in the following example.

EXAMPLE 6-6   A UNS G 10350 steel shaft, heat-treated to a yield strength of 75 kpsi, has a diameter of $1\frac{7}{16}$ in. The shaft rotates at 600 rpm and transmits 40 hp through a gear. Select an appropriate key for the gear.

SOLUTION   A $\frac{3}{8}$-in square key is selected, UNS G10200 cold-drawn steel being used. The design will be based on a yield strength of 65 kpsi. A factor of safety of 2.80 will be employed in the absence of exact information about the nature of the load.

The torque is obtained from the horsepower equation

$$T = \frac{63\,000 \text{ hp}}{n} = \frac{(63\,000)(40)}{600} = 4200 \text{ lb·in}$$

Referring to Fig. 6-25, the force F at the surface of the shaft is

$$F = \frac{T}{r} = \frac{4200}{0.719} = 5850 \text{ lb}$$

By the distortion-energy theory, the shear strength is

$$S_{sy} = 0.577S_y = (0.577)(65) = 37.5 \text{ kpsi}$$

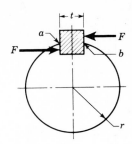

FIGURE 6-25

Failure by shear across the area ab will create a stress of $\tau = F/tl$. Substituting the strength divided by the factor of safety for τ gives

$$\frac{S_{sy}}{n} = \frac{F}{tl} \qquad \text{or} \qquad \frac{37.5(10)^3}{2.80} = \frac{5850}{0.375l}$$

or $l = 1.16$ in. To resist crushing, the area of one-half the face of the key is used:

$$\frac{S_y}{n} = \frac{F}{tl/2} \qquad \text{or} \qquad \frac{65(10)^3}{2.80} = \frac{5850}{0.375l/2}$$

and $l = 1.34$ in. The hub length of a gear is usually greater than the shaft diameter, for stability. If the key, in this example, is made equal in length to the hub, it would therefore have ample strength, since it would probably be $1\frac{7}{16}$ in or longer.   ////

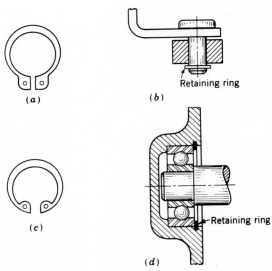

Retaining ring

(a)

(b)

(c)

Retaining ring

(d)

FIGURE 6-26
Typical uses for retaining rings. (a) External ring; (c) internal ring.

A retaining ring is frequently used instead of a shaft shoulder to axially position a component on a shaft or in a housing bore. As shown in Fig. 6-26, a groove is cut in the shaft or bore to receive the spring retainer. The tapered design of both the internal and external rings assures uniform pressure against the bottom of the groove. For sizes, dimensions, and ratings, the manufacturer's catalogs should be consulted.

Involute splines are the best means of transferring large amounts of torque. Since these are very closely related to gear teeth, selection methods will be presented in Chap. 11.

PROBLEMS

Sections 6-1 to 6-3

6-1 Show that for zero collar friction, the efficiency of a square-thread screw is given by the equation

$$e = \tan \psi \, \frac{1 - \mu \tan \psi}{\tan \psi + \mu}$$

Plot a curve of the efficiency for helix angles up to 45°. Use $\mu = 0.08$.

6-2 A 20-mm power screw has single square threads with a pitch of 4 mm and is used in a power-operated press. Each screw is subjected to a load of 5 kN. The coefficients of friction are 0.075 for the threads and 0.095 for the collar friction. The frictional diameter of the collar is 30 mm.

(a) Find the thread depth, the thread width, the mean and root diameters, and the lead.

(b) Find the torque required to "lower" and to "raise" the load.

(c) Find the overall efficiency.

6-3 The C clamp shown in the figure uses a $\frac{3}{8}$-in screw with 12 square threads per inch. The frictional coefficients are 0.15 for the threads and the collar. The collar has a

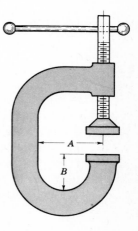

PROBLEM 6-3

frictional diameter of $\frac{5}{8}$ in. The handle is cold-drawn UNS G10100 steel. The capacity of the clamp is to be 150 lb.

(a) What torque is required to tighten the clamp to full capacity?

(b) Specify the length and diameter of the handle such that it will bend with a permanent set when the rated capacity of the clamp is exceeded. Use 3 lb as the handle force.

6-4 Find the horsepower required to drive a $1\frac{1}{2}$-in power screw having four square threads per inch. The threads are double and the load is 2.40 kip. The nut is to move at a velocity of 8 fpm. The frictional coefficients are 0.10 for the threads and collar. The frictional diameter of the collar is 3 in.

6-5 A single square-thread power screw is to raise a load of 70 kN. The screw has a major diameter of 36 mm and a pitch of 6 mm. The frictional coefficients are 0.13 for the threads and 0.10 for the collar. If the collar frictional diameter is 90 mm and the screw turns at a speed of 1 s^{-1}, find:

(a) The power input to the screw.

(b) The combined efficiency of the screw and collar.

Sections 6-4 to 6-7

6-6 The figure shows a gasketed joint to be used in a pressure vessel. The dimensions are $A = 4$ in, $B = 6$ in, $C = \frac{3}{4}$ in, $D = \frac{1}{8}$ in, and $E = 8$ in. The gasket manufacturers recommend a pressure of at least 2200 psi to obtain a proper leakproof seal.

(a) Determine the number of $\frac{5}{8}$"-18UNF grade 3 cap screws necessary to obtain a satisfactory seal.

(b) Screws should be spaced about 5 or more diameters for good wrench clearance. What spacing results from your result in (a)?

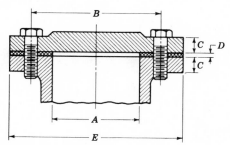

PROBLEMS 6-6 and 6-7

6-7 The dimensions of the gasketed pressure-vessel joint shown in the figure are: $A = 120$ mm, $B = 200$ mm, $C = 24$ mm, $D = 4$ mm, and $E = 250$ mm. A leakproof seal can be obtained if the average gasket pressure is at least 15 MPa.

(a) Determine the number of 12 mm, 16 mm, and 20 mm grade 8.8 metric cap screws necessary to develop the required gasket pressure.

(b) If the screw spacing is less than five screw diameters, the wrench clearance will be tight. And if the spacing is over ten diameters, the seal pressure may not be uniform from screw to screw. Based on these criteria, which result of (a) is the optimum?

6-8 The figure illustrates the connection of a cylinder head to a pressure vessel using 10 bolts and a confined gasket. The dimensions in millimetres are: $A = 100$, $B = 200$, $C = 300$, $D = 20$, and $E = 25$. The cylinder is used to store gas at a static pressure of 6 MPa. Metric grade 6.8 bolts are to be used with a factor of safety of at least 3. What size bolts should be used for this application?

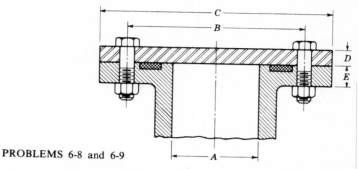

PROBLEMS 6-8 and 6-9

6-9 The figure illustrates the connection of a cylinder head which is to be bolted using 24 coarse-thread SAE grade 3 bolts. The dimensions are: $A = 20$ in, $B = 24$ in, $C = 26.5$ in, $D = \frac{3}{4}$ in, and $E = 1$ in. The cylinder will be used to store liquid at a static pressure of 100 psi. Use a factor of safety of 3 and determine the size of bolts to be used.

6-10 The upside-down A-frame shown in the figure is to be bolted to steel beams on the ceiling of a machine room using metric grade 8.8 bolts. This frame is to support the 40-kN radial load as illustrated. The total bolt grip is 48 mm, which includes the thickness of the steel beam, the A-frame feet, and the steel washers used.
   (a) What tightening torque should be used?
   (b) What portion of the external load is taken by the bolts? By the members?

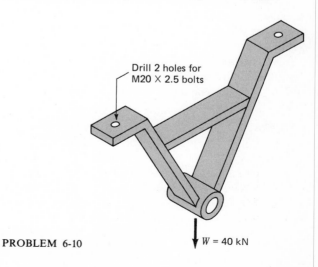

Drill 2 holes for
M20 X 2.5 bolts

PROBLEM 6-10

$W = 40$ kN

6-11 The figure shows a welded steel bracket which is to support a force F as shown. The bracket is to be bolted to a smooth vertical face, not shown, by means of four SAE grade 5 $\frac{3}{8}$″-16UNC bolts, two on centerline (abbreviated CL) A and the other two on CL B. One way of analyzing such a connection would be to assume that the bolts on CL A carry the entire moment load and those on CL B carry the entire shear load.

(*a*) Compute the external shear load carried by the bolts at B.

(*b*) Compute the external tensile load carried by the bolts at A.

(*c*) What is the factor of safety of the connection based on the A bolts?

(*d*) What is the factor of safety for the connection based on shear of the B bolts?

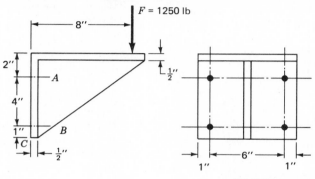

PROBLEM 6-11

Section 6-9

6-12 In the figure for Prob. 6-8, let $A = 0.9$ m, $B = 1$ m, $C = 1.10$ m, $D = 20$ mm, and $E = 25$ mm. The cylinder is made of ASTM No. 40 cast iron ($E = 120$ GPa), and the head of low-carbon steel. There are 36 M10 × 1.5 bolts of metric grade 10.9 steel. These bolts are to be tightened so that the preload is 60 percent of the proof load. During use, the cylinder pressure will vary between 0 and 550 kPa. Find the factor of safety that guards against a fatigue failure based on 90 percent reliability.

6-13 The section of the gasketed joint shown in the figure is loaded by a tensile force P which fluctuates between 0 and 6 kip. The bolts have been carefully preloaded to $F_i = 25$ kip per bolt. The members have $E = 16$ Mpsi.

(*a*) Find the endurance limit of the bolts based on 90 percent reliability.

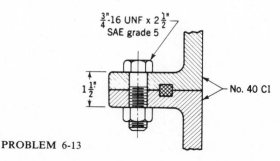

PROBLEM 6-13

(b) Find the stiffness constants k_m and k_b using the assumptions of Sec. 6-5.

(c) Find the factor of safety of the connection based on the possibility of a fatigue failure.

6-14 The figure shows a fluid-pressure linear actuator, or hydraulic cylinder, in which $D = 100$ mm, $t = 10$ mm, $L = 300$ mm, and $w = 20$ mm. Both brackets as well as the cylinder are of steel. The actuator has been designed for a working pressure of 4 MPa. Five M12 × 1.75 metric grade 6.8 bolts are used, tightened to 50 percent of the proof load.

(a) Find the stiffnesses of the bolts and members, assuming that the entire cylinder is compressed uniformly and that the end brackets are perfectly rigid.

(b) Find the mean and alternating stresses in the bolts.

(c) Find the endurance limit of the bolts based on 50 percent reliability.

(d) What factor of safety guards against a fatigue failure? A static failure?

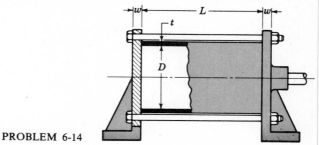

PROBLEM 6-14

Sections 6-10 to 6-12

6-15 The connections shown are made of UNS G 10180 cold-drawn steel and SAE grade 3 bolts. In each case find the static load F that would cause failure.

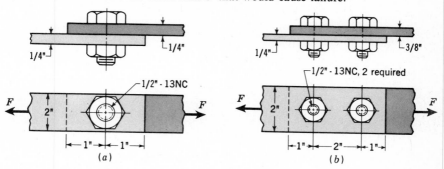

PROBLEM 6-15

6-16 The figure shows two cold-drawn bars of UNS G10180 steel bolted together to form a lap joint. Determine the factor of safety of the connection.

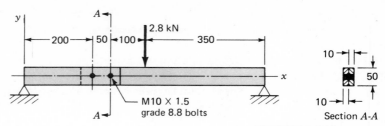

PROBLEM 6-16
Dimensions in millimetres.

6-17 A cold-drawn UNS G41300 steel bar is to be fastened using three M12 × 1.75 grade 8.8 bolts to the 150-mm channel shown in the figure. What maximum force F can be applied to this cantilever if the factor of safety is to be at least 2.8? Do not consider the channel.

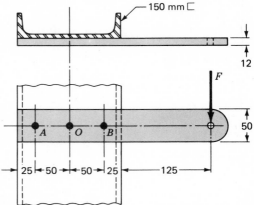

PROBLEM 6-17
Dimensions in millimetres.

6-18 Find the total shear load on each of the three bolts for the connection shown in the figure and compute the significant bolt shear stress and bearing stress. Compute the moment of inertia of the 8-mm plate on a section through the three bolt holes and find the maximum bending stress in the plate.

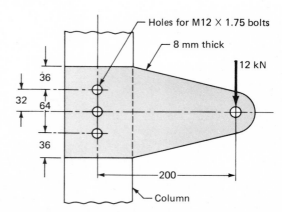

PROBLEM 6-18
Dimensions in millimetres.

6-19 A $\frac{3}{8}$ × 2″ UNS G10180 cold-drawn steel cantilever bar supports a static load of 300 lb
as shown in the figure. The bar is secured to the support using two $\frac{1}{2}$″-13 UNC grade 3
bolts as shown. Find the factor of safety for the following modes of failure: shear of
bolt, bearing on bolt, bearing on member, and strength of member.

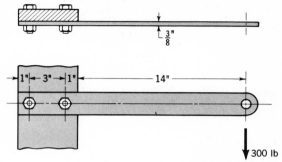

300 lb

PROBLEM 6-19

6-20 The figure shows a welded fitting which has been tentatively designed to be bolted to a
channel so as to transfer the 2500-lb load into the channel. The channel is made of
hot-rolled low-carbon steel with $S_y = 46$ kpsi; the two fitting plates are of hot-rolled
stock, $S_y = 45.5$ kpsi. The fitting is to be bolted using six standard SAE grade 2 bolts.
Check the strength of the design by calculating the factor of safety for all possible
modes of failure.

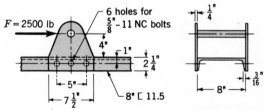

PROBLEM 6-20

7

WELDED, BRAZED, AND BONDED JOINTS

Processes such as welding, brazing, soldering, cementing, and gluing are used extensively in manufacturing today. Whenever parts have to be assembled or fabricated it is probable that one of these processes should be considered in preliminary design work.

One of the difficulties encountered by the design engineer in dealing with this subject is that it has not benefited by the rigorous treatment that so many other processes, materials, and mechanical elements have had. It is not clear why this should be so. It may be that the geometry does not lend itself well to mathematical treatment. Of course this means that an added element of uncertainty has been introduced and that this must be compensated for by the use of larger factors of safety in design. The fact that so many safe and reliable structures and devices utilizing these processes are in use today attests to the fact that engineers have been successful in overcoming these handicaps.

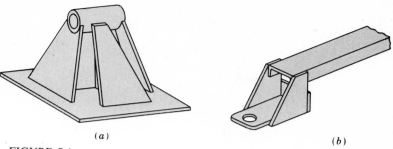

(a)   (b)

FIGURE 7-1
Examples of weldments. (a) A bearing base; can be made of relatively thin parts,
yet provides a good stiffness in both directions. (b) Another base detail; made from
a hot-rolled channel section and an assortment of cut plates. (*From "Procedure
Handbook of Arc Welding," 11th ed., pp. 5-219 and 5-226, Lincoln Electric Company,
Cleveland, 1957. Reproduced by permission of the publishers.*)

7-1 WELDING

Weldments are usually fabricated by clamping, jigging, or holding a collection of
hot-rolled low- or medium-carbon steel shapes, cut to particular configurations,
while the several parts are welded together. Two examples are shown in Fig. 7-1.
An ingenious designer, familiar with various hot-rolled shapes and methods of
cutting them, should be able to design strong and lightweight weldments that can
be easily and quickly welded with simple holding fixtures.

Figures 7-2 to 7-4* illustrate the type of welds used most frequently by
designers. For general machine elements most welds are fillet welds, though butt
welds are used a great deal in designing pressure vessels. Of course, the parts to be
joined must be arranged so that there is sufficient clearance for the welding
operation. If unusual joints are required because of insufficient clearance or because of the section shape, the design may be a poor one and the designer should
retrace his steps and endeavor to synthesize another solution.

Since heat is used in the welding operation, there is a possibility of metallurgical changes in the parent metal in the vicinity of the weld. Also, residual stresses
may be introduced because of clamping or holding or, sometimes, because of the
order of welding. Usually these residual stresses are not severe enough to cause
concern; in some cases a light heat treatment after welding has been found helpful
in relieving them. When the parts to be welded are thick, a preheating will also be
of benefit. If the reliability of the component is to be quite high, a testing program
should be established to learn what changes or additions to the operations are
necessary and to assure the best quality.

---

* From "Designers Guide for Welded Construction," The Lincoln Electric Co., Cleveland, by
permission. This is a six-page folder containing a summary of all welding symbols and other
useful information available to designers and students on request.

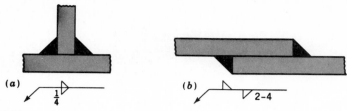

FIGURE 7-2
Fillet welds. (*a*) The fraction indicates the leg size; the arrow should point to only one weld when both sides are the same. (*b*) The symbol indicates that the welds are intermittent and staggered 2-in on 4-in centers.

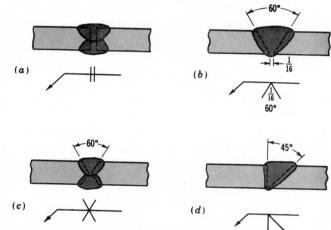

FIGURE 7-3
Butt or groove welds (*a*) Square butt-welded on both sides; (*b*) single V with 60° bevel and root opening of $\frac{1}{16}$ in; (*c*) double V; (*d*) single bevel.

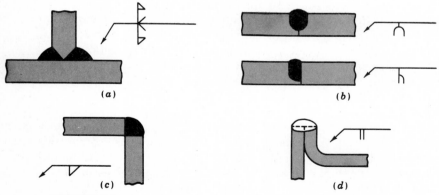

FIGURE 7-4
Special groove welds. (*a*) T joint for thick plates. (*b*) U and J welds for thick plates. (*c*) Corner weld may also have a bead weld on inside for greater strength, but should not be used for heavy loads. (*d*) Edge weld for sheet metal and light loads.

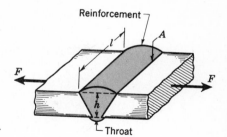

FIGURE 7-5
A typical butt joint.

7-2 BUTT AND FILLET WELDS

Figure 7-5 shows a single V-groove weld loaded by the tensile force F. For either tension or compression loading, the average normal stress is

$$\sigma = \frac{F}{hl} \qquad (7\text{-}1)$$

where h is the weld throat and l is the length of the weld, as shown in the figure. Note that the value of h does not include the reinforcement. The reinforcement is desirable in order to compensate for flaws, but it varies somewhat and does produce stress concentration at point A in the figure. If fatigue loads exist, it is good practice to grind or machine off the reinforcement.

The average stress in a butt weld due to shear loading is

$$\tau = \frac{F}{hl} \qquad (7\text{-}2)$$

The stress distribution in fillet welds has been investigated by photoelastic procedures, but attempts to solve the problem by the methods of the theory of elasticity have not been very successful. A model of the transverse fillet weld of Fig. 7-6 is easily constructed for photoelastic purposes and has the advantage of a balanced loading condition. Norris constructed such a model and reported the

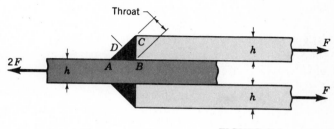

FIGURE 7-6
A transverse fillet weld.

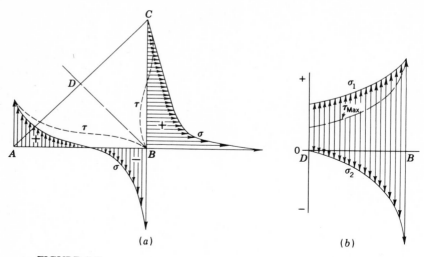

FIGURE 7-7
Stress distributions in fillet welds. (a) Stress distribution on the legs as reported by Norris; (b) distribution of principal stresses and maximum shear stress as reported by Salakian.

stress distribution along the sides AB and BC of the weld.* An approximate graph of the results he obtained is shown as Fig. 7-7a. Note that stress concentration exists at A and B on the horizontal leg and at B on the vertical leg. Norris states that he could not determine the stresses at A and B with any certainty.

Salakian† presents data for the stress distribution across the throat of a fillet weld (Fig. 7-7b). This graph is of particular interest because designers and stress analysts usually assume that failure will occur in the throat in determining the strength of a weld. Again, the figure shows stress concentration at point B.

Note that Fig. 7-7a applies either to the weld metal or to the parent members. Figure 7-7b, of course, gives the stress distribution only in the weld. No satisfactory analytical method of finding the stresses of these two figures is available. Consequently we shall utilize the methods developed earlier in the book, that is, by sizing the cross sections or areas so as to result in satisfactory average or nominal stresses when the loads are applied and, when the joints are subjected to fatigue loads, by applying fatigue-strength reduction factors K_f to the strength of the weld metal or to the strength of the parent members, depending on which set of calculations are being made.

* C. H. Norris, Photoelastic Investigation of Stress Distribution in Transverse Fillet Welds, *Welding J.*, vol. 24, p. 557s, 1945.
† A. G. Salakian and G. E. Claussen, Stress Distribution in Fillet Welds: A Review of the Literature, *Welding J.*, vol. 16, pp. 1–24, May 1937.

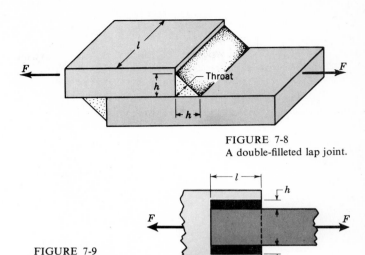

FIGURE 7-8
A double-filleted lap joint.

FIGURE 7-9
A parallel fillet weld.

The tension-loaded lap joint of Fig. 7-8 has a throat area of $0.707hl$ per weld. The method most often used for this problem is to assume that the throat section is in shear. The average stress is then

$$\tau = \frac{F}{1.414hl} \qquad (7\text{-}3)$$

Note particularly that the words "average stress" mean that we have assumed that this stress is uniformly distributed over the entire area. Furthermore, since this is the stress used to size the weld, the use of Eq. (7-3) implies that all normal stresses on the throat are assumed to be zero. This is far from the truth, as evidenced by the experimental results of Fig. 7-7. However, if the equation is used with the maximum stresses permitted by various construction codes, it turns out that the resulting welds are perfectly safe. Nevertheless, economies could be obtained if a rational approach were available.

In the case of parallel fillet welds (Fig. 7-9), the assumption of a shear stress along the throat is more realistic. Since there are two welds, the throat area for both is $A = (2)(0.707hl) = 1.414hl$. The average shear stress is therefore $\tau = F/1.414hl$, the same as in Eq. (7-3). It is quite probable that the stress distribution along the length of the welds is *not* uniform.

7-3 TORSION IN WELDED JOINTS

Figure 7-10 illustrates a cantilever of length l welded to a column by two fillet welds. The reaction at the support of a cantilever always consists of a shear force

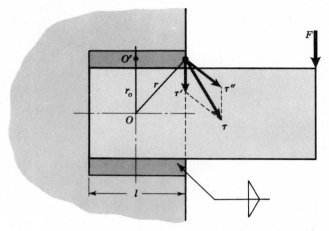

FIGURE 7-10
A moment connection.

V and a moment M. The shear force produces a *primary shear* in the welds of magnitude

$$\tau' = \frac{V}{A} \qquad (7\text{-}4)$$

where A is the throat area of all the welds.

The moment at the support produces *secondary shear* or *torsion* of the welds, and this stress is

$$\tau'' = \frac{Mr}{J} \qquad (7\text{-}5)$$

where r is the distance from the centroid of the weld group to the point in the weld of interest, and J is the polar moment of inertia of the weld group about the centroid of the group. When the size of the welds are known, these equations can be solved and the results combined to obtain the maximum shear stress.

The reverse procedure is that in which the allowable shear stress is given and we wish to find the weld size. The usual procedure would be to estimate the weld size, compute J and A, and then find and combine τ' and τ''. If the resulting maximum stress were too large, a larger weld size would be estimated and the procedure repeated. After a few such trials, a satisfactory result would be obtained.

A much more useful approach to the problem is to treat each fillet weld as a line. The resulting polar moment of inertia is then equivalent to a *unit polar moment of inertia*. The advantage of treating the weld as a line is that the unit polar moment of inertia is the same regardless of the weld size. Since the throat width of a fillet weld is $0.707h$, the relationship between the unit polar moment of inertia and the polar moment of inertia of a fillet weld is

$$J = 0.707hJ_u \qquad (7\text{-}6)$$

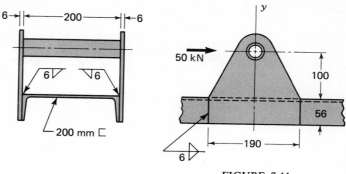

FIGURE 7-11
Dimensions in millimetres.

in which J_u is found by conventional methods for an area having unit width. The transfer formula for unit polar moment of inertia must be employed when the welds occur in groups, as in Fig. 7-10. Table 7-1 lists the throat areas and unit polar moments of inertia for the most common fillet welds encountered. The example that follows is illustrative of the calculations normally made.

EXAMPLE 7-1 A 50-kN load is transferred from a welded fitting into a 200-mm steel channel as illustrated in Fig. 7-11. Compute the maximum stress in the weld.

SOLUTION   As shown by the figure, each plate is welded to the channel using three 6-mm fillet welds. We shall divide the load in half and consider only a single plate in the analysis to follow. Two of the three welds are 56 mm long, and the length of the third is 190 mm. Using Table 7-1, we first locate the centroid of the weld group.

$$\bar{y} = \frac{b^2}{2b + d} = \frac{(56)^2}{2(56) + 190} = 10.4 \text{ mm}$$

$$\bar{x} = d/2 = 190/2 = 95 \text{ mm}$$

These dimensions are shown on the free-body diagram of Fig. 7-12a. Note that these dimensions also locate the origin O of the xy reference system. The reaction moment, a torque, is

$$M = 25(110.4) = 2760 \text{ N·m}$$

per plate.

Next, we refer to Table 7-1 and find the unit polar moment of inertia to be

$$J_u = \frac{8b^3 + 6bd^2 + d^3}{12} - \frac{b^4}{2b + d}$$

$$= \frac{8(56)^3 + 6(56)(190)^2 + (190)^3}{12} - \frac{(56)^4}{2(56) + 190}$$

$$= 1.67(10)^6 \text{ mm}^3$$

Table 7-1 TORSIONAL PROPERTIES OF FILLET WELDS; G IS THE CENTROID OF THE WELD GROUP; h IS THE WELD SIZE

Weld	Throat area	Location of G	Unit polar moment of inertia
	$A = 0.707hd$	$\bar{x} = 0$ $\bar{y} = d/2$	$J_u = d^3/12$
	$A = 1.414hd$	$\bar{x} = b/2$ $\bar{y} = d/2$	$J_u = \dfrac{d(3b^2 + d^2)}{6}$
	$A = 0.707h(b + d)$	$\bar{x} = \dfrac{b^2}{2(b + d)}$ $\bar{y} = \dfrac{d^2}{2(b + d)}$	$J_u = \dfrac{(b + d)^4 - 6b^2d^2}{12(b + d)}$
	$A = 0.707h(2b + d)$	$\bar{x} = \dfrac{b^2}{2b + d}$ $\bar{y} = d/2$	$J_u = \dfrac{8b^3 + 6bd^2 + d^3}{12} - \dfrac{b^4}{2b + d}$
	$A = 1.414h(b + d)$	$\bar{x} = b/2$ $\bar{y} = d/2$	$J_u = \dfrac{(b + d)^3}{6}$
	$A = 1.414\pi hr$		$J_u = 2\pi r^3$

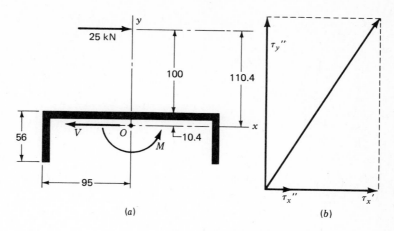

FIGURE 7-12
Dimensions in millimetres.

Then, from Eq. (7-6),

$$J = 0.707hJ_u = 0.707(6)(1.67)(10)^6 = 7.07(10)^6 \text{ mm}^4$$

Using Table 7-1 again, we find the throat area for the welds on one plate to be

$$A = 0.707h(2b + d) = 0.707(6)[2(56) + 190] = 1280 \text{ mm}^2$$

The primary shear stress is

$$\tau'_x = \frac{V}{A} = \frac{25(10)^3}{1280} = 19.5 \text{ MPa}$$

We can find the secondary shear stress in components parallel to x and y. The y component is

$$\tau''_y = \frac{Mr_x}{J} = \frac{2760(10)^3(95)}{7.07(10)^6} = 37.1 \text{ MPa}$$

The x component is

$$\tau''_x = \frac{Mr_y}{J} = \frac{2760(10)^3(10.4)}{7.07(10)^6} = 4.06 \text{ MPa}$$

These stress components should be combined to yield the maximum stresses, which occur at corners A and B. Thus, for A (Fig. 7-12b) we have

$$\tau = \sqrt{\tau_y^2 + \tau_x^2} = \sqrt{(37.1)^2 + (19.5 + 4.06)^2} = 43.9 \text{ MPa} \qquad Ans.$$

////

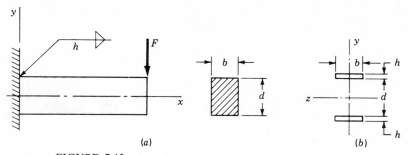

FIGURE 7-13
A rectangular cantilever welded to a support at the top and bottom edges.

7-4 BENDING IN WELDED JOINTS

Figure 7-13a shows a cantilever welded to a support by fillet welds at top and bottom. A free-body diagram of the beam would show a shear-force reaction V and a moment reaction M. The shear force produces a primary shear in the welds of magnitude

$$\tau' = V/A \qquad (a)$$

where A is the total throat area.

The moment M produces a normal bending stress σ in the welds. Though not rigorous, it is customary in the stress analysis of welds to assume that this stress acts normal to the throat area. By treating the two welds in Fig. 7-13b as lines, we find the unit moment of inertia to be

$$I_u = \frac{bd^2}{2} \qquad (b)$$

Then the moment of inertia based on the weld throat is

$$I = 0.707h\frac{bd^2}{2} \qquad (c)$$

The normal stress is now found to be

$$\sigma = \frac{Mc}{I} = \frac{M(d/2)}{0.707bd^2h/2} = \frac{1.414M}{bdh} \qquad (d)$$

The moment of inertia in Eq. (d) is based on the distance d between the two welds. If the moment of inertia is found by treating the two welds as rectangles instead, the distance between the weld centroids would be $(d + h)$. This would produce a slightly larger moment of inertia and result in a smaller value of the stress σ. Thus the method of treating welds as lines produces more conservative results. Perhaps the added safety is appropriate in view of the stress distributions of Fig. 7-7.

Table 7-2 BENDING PROPERTIES OF FILLET WELDS: THE UNIT MOMENT OF INERTIA I_u IS TAKEN ABOUT A HORIZONTAL AXIS THROUGH THE CENTROID OF THE WELD GROUP G; THE WELD SIZE IS GIVEN BY h

Weld	Throat area	Location of G	Unit moment of inertia
	$A = 0.707hd$	$\bar{x} = 0$ $\quad \bar{y} = d/2$	$I_u = \dfrac{d^3}{12}$
	$A = 1.414hd$	$\bar{x} = b/2$ $\quad \bar{y} = d/2$	$I_u = \dfrac{d^3}{6}$
	$A = 1.414hb$	$\bar{x} = b/2$ $\quad \bar{y} = d/2$	$I_u = \dfrac{bd^2}{2}$
	$A = 0.707h(2b + d)$	$\bar{x} = \dfrac{b^2}{2b + d}$ $\quad \bar{y} = d/2$	$I_u = \dfrac{d^2}{12}(6b + d)$
	$A = 0.707h(b + 2d)$	$\bar{x} = b/2$ $\quad \bar{y} = \dfrac{d^2}{b + 2d}$	$I_u = \dfrac{2d^3}{3} - 2d^2\bar{y} + (b + 2d)\bar{y}^2$
	$A = 1.414h(b + d)$	$\bar{x} = b/2$ $\quad \bar{y} = d/2$	$I_u = \dfrac{d^2}{6}(3b + d)$

(continued)

Table 7-2 BENDING PROPERTIES OF FILLET WELDS: THE UNIT MOMENT OF INERTIA I_u IS TAKEN ABOUT A HORIZONTAL AXIS THROUGH THE CENTROID OF THE WELD GROUP G; THE WELD SIZE IS GIVEN BY h

Weld	Throat area	Location of G	Unit moment of inertia
	$A = 0.707h(b + 2d)$	$\bar{x} = b/2$ $$\bar{y} = \frac{d^2}{b + 2d}$$	$$I_u = \frac{2d^3}{3} - 2d^2\bar{y} + (b + 2d)\bar{y}^2$$
	$A = 1.414h(b + d)$	$\bar{x} = b/2$ $\bar{y} = d/2$	$$I_u = \frac{d^2}{6}(3b + d)$$
	$A = 1.414\pi hr$		$I_u = \pi r^3$

Once the stress components σ and τ have been found for welds subjected to bending, they may be combined by using a Mohr's circle diagram to find the principal stresses or the maximum shear stress. Then an appropriate failure theory is applied to determine the likelihood of failure or safety. Because of the greater uncertainties in the analysis of weld stresses, the more conservative maximum shear stress theory is generally preferred.

Table 7-2 lists the bending properties most likely to be encountered in the analysis of welded beams.

7-5 THE STRENGTH OF WELDED JOINTS

The matching of the electrode properties with those of the parent metal is usually not so important as speed, operator appeal, and the appearance of the completed joint. The properties of electrodes vary considerably, but Table 7-3 lists the minimum properties for some electrode classes.

It is preferable, in designing welded components, to select a steel that will result in a fast, economical weld even though this may require a sacrifice of other qualities such as machinability. Under the proper conditions all steels can be

Table 7-3 MINIMUM WELD-METAL PROPERTIES

AWS electrode number*	Tensile strength, kpsi	Yield strength, kpsi	Percent elongation
E60xx	62	50	17–25
E70xx	70	57	22
E80xx	80	67	19
E90xx	90	77	14–17
E100xx	100	87	13–16
E120xx	120	107	14

* The American Welding Society (AWS) specification code numbering system for electrodes. This system uses an E prefix to a four- or five-digit numbering system in which the first two or three digits designate the approximate tensile strength. The last digit indicates variables in the welding technique, such as current supply. The next to the last digit indicates the welding position, as, for example, flat, or vertical, or overhead. The complete set of specifications may be obtained from the AWS upon request.

welded, but best results will be obtained if steels having a UNS specification between G10140 and G10230 are chosen. All these steels have a tensile strength in the hot-rolled condition in the range of 60 to 70 kpsi.

The designer can choose factors of safety or permissible working stresses with more confidence if he or she is aware of the values of those used by others. One of the best standards to use is the American Institute of Steel Construction (AISC) code for building construction.* The permissible stresses are now based on the yield strength of the material instead of the ultimate strength, and the code permits the use of a variety of ASTM structural steels having yield strengths varying from 33 to 50 kpsi. Provided that the loading is the same, the code permits the same stress in the weld metal as in the parent metal. For these ASTM steels, $S_y = 0.5S_u$. Table 7-4 lists the formulas specified by the code for calculating these permissible stresses for various loading conditions. The factors of safety implied by this code are easily calculated. For tension, $n = 1/0.60 = 1.67$. For shear, $n = 0.577/0.40 = 1.44$ if we accept the distortion-energy theory as the criterion of failure.

* For a copy write the AISC, New York.

Table 7-4 STRESSES PERMITTED BY THE AISC CODE FOR WELD METAL

Type of loading	Type of weld	Permissible stress
Tension	Butt	$0.60S_y$
Bearing	Butt	$0.90S_y$
Bending	Butt	$0.60S_y$–$0.66S_y$
Simple compression	Butt	$0.60S_y$
Shear	Butt or fillet	$0.40S_y$

288 DESIGN OF MECHANICAL ELEMENTS

Table 7-5 FATIGUE-STRENGTH REDUCTION FACTORS

Type of weld	K_f
Reinforced butt weld	1.2
Toe of transverse fillet weld	1.5
End of parallel fillet weld	2.7
T-butt joint with sharp corners	2.0

The AISC code, as well as the AWS code, for bridges includes permissible stresses when fatigue loading is present. The designer will have no difficulty in using these codes, but their empirical nature tends to obscure the fact that they have been established by means of the same knowledge of fatigue failure already discussed in Chap. 5. Of course, for structures covered by these codes the actual stresses *cannot* exceed the permissible stresses; otherwise the designer is legally liable. But in general, codes tend to conceal the actual margin of safety involved.

The fatigue-strength reduction factors listed in Table 7-5, as proposed by Jennings,* are suggested for use. These factors should be used for the parent metal as well as for the weld metal.

7-6 RESISTANCE WELDING

The heating and consequent welding which occurs when an electrical current is passed through several parts which are pressed together is called resistance welding. *Spot-welding* and *seam-welding* are the forms of resistance welding most often used. The advantages of resistance welding over other forms are the speed, the accurate regulation of time and heat, the uniformity of the weld and the mechanical properties which result, the elimination of filler rods or fluxes, and the fact that the process is easy to automate.

The spot- and seam-welding processes are illustrated schematically in Fig. 7-14. Seam-welding is actually a series of overlapping spot-welds since the current is applied in pulses as the work moves between the rotating electrodes.

Failure of a resistance weld is either by shearing of the weld or by tearing of the metal around the weld. Because of the tearing, it is good practice to avoid loading a resistance welded joint in tension. Thus, for the most part, design so that the spot or seam is loaded in pure shear. The shear stress is then simply the load divided by the area of the spot. Because of the fact that the thinner sheet of the pair being welded may tear, the strength of spot-welds is often specified by stating the load per spot based on the thickness of the thinnest sheet. Such strengths are best obtained by experiment.

* C. H. Jennings, Welding Design, *Trans. ASME*, vol. 58, pp. 497–509, 1936.

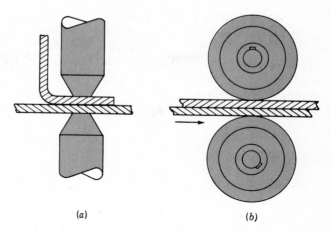

FIGURE 7-14
(a) Spot welding; (b) seam welding.

Somewhat larger factors of safety should be used when parts are fastened by spot-welding, rather than by bolts or rivets, to account for the metallurgical changes in the materials due to the welding.

7-7 BONDED JOINTS

When two parts or materials are connected together by a third material unlike the base materials, the process is called *bonding*. Thus *brazing, soldering,* and *cementing* or *gluing* are means of bonding parts together.

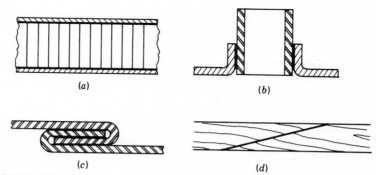

FIGURE 7-15
(a) Airplane wing section fabricated by bonding aluminum honeycomb to the skins, using resin bonding under heat and pressure; (b) tubing joined to sheet-metal section by brazing metal; (c) sheet-metal parts joined by soldering; (d) wood parts joined by gluing.

The connections between parts which are to be bonded should be designed so that the bonding material takes only a pure shear load. Since the strength of the bonding material may be considerably less than that of the parts to be joined, enough contact area must be provided in the joint to obtain an adequate margin of safety. Figure 7-15 shows some examples of bonded joints which represent good design practice. The properties of the bonding agents should be obtained directly from the manufacturers.

PROBLEMS

Sections 7-1 to 7-3

7-1 to 7-3   The permissible shear stress for the welds shown is 20 kpsi in English units and 140 MPa in SI units. For each case, find the load F that would cause such a stress.

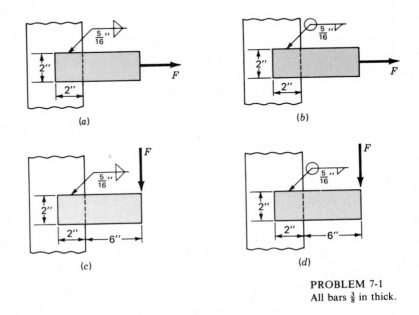

PROBLEM 7-1
All bars $\frac{3}{8}$ in thick.

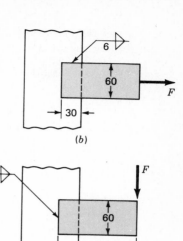

(a)   *(b)*

(c)   *(d)*

PROBLEM 7-2
Dimensions in millimetres; all bars 10 mm thick.

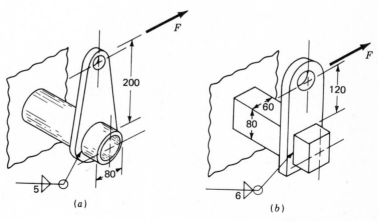

(a)   *(b)*

PROBLEM 7-3
Dimensions in millimetres.

7-4   For each weldment shown, find the torque T that can be applied if the permissible weld shear stress is 20 kpsi.

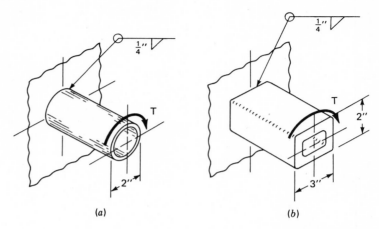

(a)   (b)

PROBLEM 7-4

7-5, 7-6   A collection of beam cross sections is illustrated. These are obtained by welding various structural-steel shapes together. Based on a permissible weld shear stress of 140 MPa in SI or 20 kpsi in English units, find the allowable direct shear load that each beam can carry. HINT: See Example 2-9.

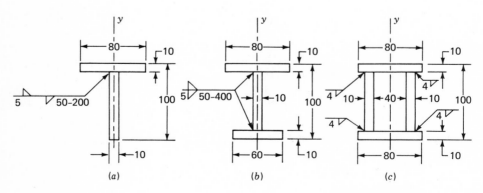

(a)   (b)   (c)

PROBLEM 7-5
Dimensions in millimetres.

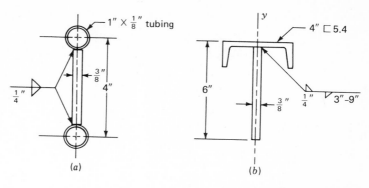

PROBLEM 7-6

Section 7-4

7-7 to 7-9 The beams shown in the figures are welded to fixed supports or to plates as shown. In each case find the maximum combined shear stress in the weld metal.

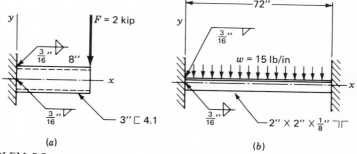

PROBLEM 7-7
(a) Beam is a 4.1-lb structural-steel channel; (b) beam is two structural angles, back-to-back, with same welds at each end.

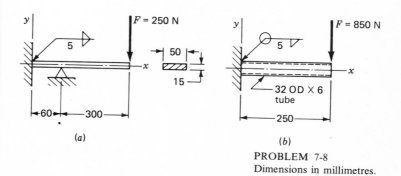

PROBLEM 7-8
Dimensions in millimetres.

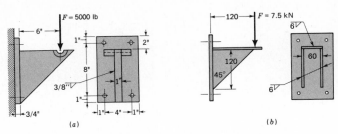

PROBLEM 7-9
(a) Dimensions in inches; (b) dimensions in millimetres.

8

MECHANICAL SPRINGS

Mechanical springs are used in machines to exert force, to provide flexibility, and to store or absorb energy. In general, springs may be classified as either wire springs or flat springs, although there are variations within these divisions. Wire springs include helical springs of round, square, or special-section wire and are made to resist tensile, compressive, or torsional loads. Under flat springs are included the cantilever and elliptical types, the wound motor- or clock-type power springs, and the flat spring washers, usually called Belleville springs.

8-1 STRESSES IN HELICAL SPRINGS

Figure 8-1a shows a round-wire helical compression spring loaded by the axial force F. We designate D as the *mean spring diameter* and d as the *wire diameter*. Now imagine that the spring is cut at some point (Fig. 8-1b), a portion of it removed, and the effect of the removed portion replaced by the internal forces. Then, as shown in the figure, the cut portion would exert a direct shear force F and a torsion T on the remaining part of the spring.

To visualize the torsion, picture a coiled garden hose. Now pull one end of

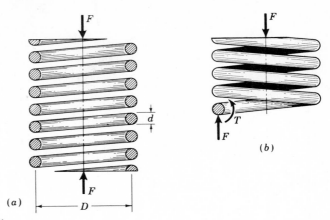

FIGURE 8-1
(a) Axially loaded helical spring; (b) free-body diagram showing that the wire is subjected to a direct shear and a torsional shear.

the hose in a straight line perpendicular to the plane of the coil. As each turn of hose is pulled off the coil, the hose twists or turns about its own axis. The flexing of a helical spring creates a torsion in the wire in a similar manner.

Using superposition, the maximum stress in the wire may be computed using the equation

$$\tau_{max} = \pm \frac{Tr}{J} + \frac{F}{A} \qquad (a)$$

where the term Tr/J is the torsion formula of Chap. 2. Replacing the terms by $T = FD/2$, $r = d/2$, $J = \pi d^4/32$, and $A = \pi d^2/4$ gives

$$\tau = \frac{8FD}{\pi d^3} + \frac{4F}{\pi d^2} \qquad (b)$$

In this equation the subscript indicating maximum shear stress has been omitted as unnecessary. The positive signs of Eq. (a) have been retained, and hence Eq. (b) gives the shear stress at the inside fiber of the spring.

Now define *spring index*

$$C = \frac{D}{d} \qquad (8\text{-}1)$$

as a measure of coil curvature. With this relation, Eq. (b) can be arranged to give

$$\tau = \frac{8FD}{\pi d^3}\left(1 + \frac{0.5}{C}\right) \qquad (c)$$

Or designating

$$K_s = 1 + \frac{0.5}{C} \qquad (8\text{-}2)$$

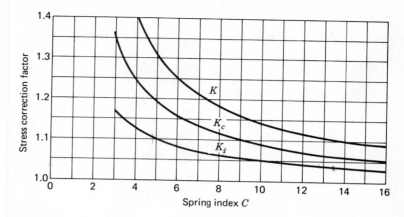

FIGURE 8-2
Values of the stress correction factors for round helical extension or compression springs.

then

$$\tau = K_s \frac{8FD}{\pi d^3} \qquad (8\text{-}3)$$

where K_s is called a *shear-stress multiplication factor*. This factor can be obtained from Fig. 8-2 for the usual values of C. For most springs, C will range from about 6 to 12. Equation (8-3) is quite general and applies for both static and dynamic loads. It gives the maximum shear stress in the wire, and this stress occurs at the inner fiber of the spring.

Many writers present the stress equation as

$$\tau = K \frac{8FD}{\pi d^3} \qquad (8\text{-}4)$$

where K is called the *Wahl correction factor*.* This factor includes the direct shear, together with another effect due to curvature. As shown in Fig. 8-3, curvature of the wire increases the stress on the inside of the spring but decreases it only slightly on the outside. The value of K may be obtained from the equation

$$K = \frac{4C - 1}{4C - 4} + \frac{0.615}{C} \qquad (8\text{-}5)$$

or from Fig. 8-2.

By defining $K = K_c K_s$, where K_c is the effect of curvature alone, we have

$$K_c = \frac{K}{K_s} \qquad (8\text{-}6)$$

---

* See A. M. Wahl, "Mechanical Springs," 2d ed., McGraw-Hill Book Company, New York, 1963. This book is the standard reference on springs.

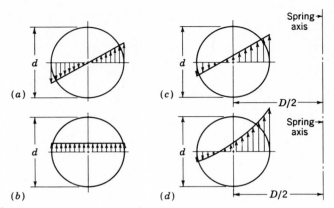

FIGURE 8-3
Superposition of stresses in a helical spring. (a) Pure torsional stress; (b) direct-shear stress; (c) resultant of direct- and torsional-shear stresses; (d) resultant of direct-, torsional-, and curvature-shear stresses.

Investigation reveals that curvature shear stress is highly localized on the inside of the spring. Springs subjected only to static loads will yield at the inside fiber and relieve this stress. Thus, for static loads, the curvature stress can be neglected and Eq. (8-3) used. For fatigue loads, K_c is used as a *fatigue-strength reduction factor*; therefore Eq. (8-3) gives the correct stress when fatigue is a factor too. Thus we shall not generally make use of Eq. (8-4) in this book. Values of K_c should be obtained by use of the equations if C is small; other values can be found directly from Fig. 8-2.

The use of square or rectangular wire is not recommended for springs unless space limitations make it necessary. Special-wire shapes are not made in large quantities, as are those of round wire; they have not had the benefit of refining development, and hence may not be as strong as springs made from round wire. When space is severely limited, the use of nested round-wire springs should always be considered. They may have an economical advantage over the special-section springs, as well as a strength advantage.

8-2 DEFLECTION OF HELICAL SPRINGS

To obtain the equation for the deflection of a helical spring, we shall consider an element of wire formed by two adjacent cross sections. Figure 8-4 shows such an element, of length dx, cut from wire of diameter d. Let us consider a line ab on the surface of the wire which is parallel to the spring axis. After deformation it will

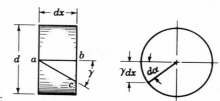

FIGURE 8-4
Cross-sectional element of a helical spring.

rotate through the angle γ and occupy the new position ac. From Eq. (2-15), which is the expression of Hooke's law for torsion, we have

$$\gamma = \frac{\tau}{G} = \frac{8FD}{\pi d^3 G} \qquad (a)$$

where the value of τ is obtained from Eq. (8-4), unity being used for the value of the Wahl correction factor.* The distance bc is $\gamma\, dx$, and the angle $d\alpha$, through which one section rotates with respect to the other, is

$$d\alpha = \frac{\gamma\, dx}{d/2} = \frac{2\gamma\, dx}{d} \qquad (b)$$

If the number of active coils is denoted by N, the total length of the wire is $\pi D N$. Upon substituting γ from Eq. (a) into Eq. (b) and integrating, the angular deflection of one end of the wire with respect to the other is

$$\alpha = \int_0^{\pi D N} \frac{2\gamma}{d}\, dx = \int_0^{\pi D N} \frac{16FD}{\pi d^4 G}\, dx = \frac{16FD^2 N}{d^4 G} \qquad (c)$$

The load F has a moment arm of $D/2$, and so the deflection is

$$y = \alpha \frac{D}{2} = \frac{8FD^3 N}{d^4 G} \qquad (8\text{-}7)$$

The deflection can also be obtained by using strain-energy methods. From Eq. (3-20) the strain energy for torsion is

$$U = \frac{T^2 l}{2GJ} \qquad (d)$$

Substituting $T = FD/2$, $l = \pi D N$, and $J = \pi d^4/32$ gives

$$U = \frac{4F^2 D^3 N}{d^4 G} \qquad (e)$$

---

* Wahl quotes the results of tests to show that K can be made unity for calculating deflections with very accurate results. *Ibid.*, p. 29.

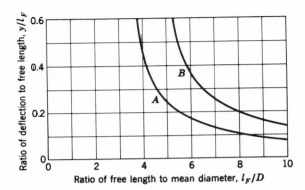

FIGURE 8-5
Curves show when buckling of compression coil springs may occur. Both curves are for springs having squared and ground ends. For curve A one end of the spring is compressed against a flat surface, the other against a rounded surface. For curve B both ends of the spring are compressed against flat and parallel surfaces.

and so the deflection is

$$y = \frac{\partial U}{\partial F} = \frac{8FD^3N}{d^4G} \qquad (f)$$

To find the spring constant, use Eq. (3-2), and substitute the value of y from Eq. (8-7). This gives

$$k = \frac{d^4G}{8D^3N} \qquad (8\text{-}8)$$

The equations presented in this section are valid for both compression and extension springs. Long coil springs having a free length more than four times the mean diameter may fail by buckling. This condition may be corrected by mounting the spring over a round bar or in a tube. Figure 8-5 will be helpful in deciding whether a compression spring is likely to buckle.

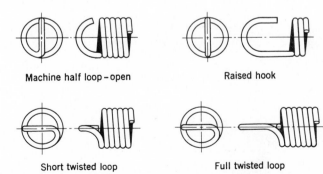

Machine half loop – open   Raised hook

Short twisted loop   Full twisted loop

FIGURE 8-6
Types of ends used on extension springs. (*Courtesy of Associated Spring Corporation.*)

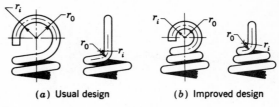

(a) Usual design   (b) Improved design

FIGURE 8-7
Ends for extension springs.

8-3 EXTENSION SPRINGS

Extension springs necessarily must have some means of transferring the load from the support to the body of the spring. Although this can be done with a threaded plug or a swivel hook, both of these add to the cost of the finished product, and so one of the methods shown in Fig. 8-6 is usually employed. In designing a spring with a hook end, the stress-concentration effect must be considered.

In Fig. 8-7a is shown a much used method of designing the end. Stress concentration due to the sharp bend makes it impossible to design the hook as strong as the body. Tests show that the stress-concentration factor is given approximately by

$$K = \frac{r_o}{r_i} \qquad (8\text{-}9)$$

which holds for bending stress and occurs when the hook is offset, and for torsional stress. Figure 8-7b shows an improved design due to a reduced coil diameter, not to elimination of stress concentration. The reduced coil diameter results in a lower stress because of the shorter moment arm.

Initial Tension

Close-wound springs are frequently made so that a load must be applied in order to separate the coils, one from another. The separating load is called the initial tension. Spring manufacturers prefer some initial tension in close-wound springs in order to hold the free length more accurately. However, the designer should specify the amount desired. The stress due to initial tension may be obtained from the equations in Sec. 8-1.

8-4 COMPRESSION SPRINGS

The type of end should be specified as follows: (1) plain ends; (2) plain ends, ground; (3) squared ends; or (4) squared and ground ends (Fig. 8-8). The type of end used results in dead or inactive turns at each end of the spring, and these must

(a) Plain end, right hand

(c) Squared and ground end, left hand

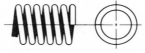

(b) Squared or closed end, right hand

(d) Plain end, ground, left hand

FIGURE 8-8
Ends for compression springs.

be subtracted from the total number of turns to obtain the number of active turns. There is no hard and fast rule, but the following specifications, when subtracted from the total number of turns, will give the approximate number of active coils:

Plain ends—subtract one-half turn
Plain and ground ends—subtract one turn
Squared ends—subtract one turn
Squared and ground ends—subtract two turns

It is customary in the design of springs to neglect the effects of eccentricity of loading due to the end turns. It is also customary to neglect the effect of residual ·stresses caused by heat treatment or overstressing. Instead, these two effects are usually accounted for by an increase in the factor of safety. It is the usual practice, in the manufacture of compression springs, to close them to their solid height; this induces a residual stress opposite in direction to the working stress and has the effect of improving the strength of the spring.

8-5 SPRING MATERIALS

Springs are manufactured either by hot- or cold-working processes, depending upon the size of the material, the spring index, and the properties desired. In general, prehardened wire should not be used if $D/d < 4$ or if $d > \frac{1}{4}$ in. Winding of the spring induces residual stresses through bending, but these are normal to the direction of the torsional working stresses in a coil spring. Quite frequently, in spring manufacture, they are relieved, after winding, by a mild thermal treatment.

A great variety of spring materials are available to the designer, including plain carbon steels, alloy steels, and corrosion-resisting steels, as well as nonferrous materials such as phosphor bronze, spring brass, beryllium copper, and various nickel alloys. Descriptions of the most commonly used steels will be found in Table 8-1. The UNS steels listed in the Appendix should be used in designing hot-worked, heavy-coil springs, as well as flat springs, leaf springs, and torsion bars.

Spring materials may be compared by an examination of their tensile strengths; these vary tremendously with wire size and to a lesser extent with the material and processing. In the past it has been customary to tabulate these

Table 8-1   HIGH-CARBON AND ALLOY SPRING STEELS*

Name of material	Similar specifications	Description
Music wire, 0.80–0.95C	UNS G10850 AISI 1085 ASTM A228-51	This is the best, toughest, and most widely used of all spring materials for small springs. It has the highest tensile strength and can withstand higher stresses under repeated loading than any other spring material. Available in diameters 0.12 to 3 mm (0.005 to 0.125 in). Do not use above 120°C (250°F) or at subzero temperatures.
Oil-tempered wire, 0.60–0.70C	UNS G10650 AISI 1065 ASTM 229-41	This general-purpose spring steel is used for many types of coil springs where the cost of music wire is prohibitive and in sizes larger than available in music wire. Not for shock or impact loading. Available in diameters 3 to 12 mm (0.125 to 0.5000 in), but larger and smaller sizes may be obtained. Not for use above 180°C (350°F) or at subzero temperatures.
Hard-drawn wire, 0.60–0.70C	UNS G10660 AISI 1066 ASTM A227-47	This is the cheapest general-purpose spring steel and should be used only where life, accuracy, and deflection are not too important. Available in diameters 0.8 to 12 mm (0.031 to 0.500 in). Not for use above 120°C (250°F) or at subzero temperatures.
Chrome vanadium	UNS G61500 AISI 6150 ASTM 231-41	This is the most popular alloy spring steel for conditions involving higher stresses than can be used with the high-carbon steels and for use where fatigue and long endurance are needed. Also good for shock and impact loads. Widely used for aircraft-engine valve springs and for temperatures to 220°C (425°F). Available in annealed or pretempered sizes 0.8 to 12 mm (0.031 to 0.500 in) in diameter.
Chrome silicon	UNS G92540 AISI 9254	This fairly new alloy is an excellent material for highly stressed springs requiring long life and subjected to shock loading. Rockwell hardnesses of C50 to C53 are quite common, and the material may be used up to 250°C (475°F). Available from 0.8 to 12 mm (0.031 to 0.500 in) in diameter.

* By permission from Harold C. R. Carlson, Selection and Application of Spring Materials, *Mech. Eng.*, vol. 78, pp. 331–334, 1956.

strengths for various wire sizes and materials.* But the availability of the scientific electronic calculator now makes such a tabulation unnecessary. The reason for this is that a log-log plot of the tensile strengths versus wire diameters is a straight line. The equation of this line can be written in terms of the ordinary logarithms of the strengths and wire diameters. This equation can then be solved to give

$$S_{ut} = \frac{A}{d^m} \qquad (8\text{-}10)$$

where A is a constant related to a strength intercept, and m is the slope of the line on the log-log plot. Of course such an equation is only valid for a limited range of wire sizes. Table 8-2 gives values of m and the constant A for both English and SI units for the materials listed in Table 8-1.

Although the torsional yield strength is needed to design springs, surprisingly, very little information on this property is available. Using an approximate relationship between yield strength and ultimate strength in tension,

$$S_y = 0.75 S_{ut} \qquad (8\text{-}11)$$

and then applying the distortion-energy theory gives

$$S_{sy} = 0.577 S_y \qquad (8\text{-}12)$$

and provides us with a means of estimating the torsional yield strength S_{sy}. But this method should not be used if experimental data are available; if used, a generous factor of safety should be employed, especially for extension springs, because of the uncertainty involved.

Variations in the wire diameter and in the coil diameter of the spring have an effect on the stress as well as on the spring scale. Large tolerances will result in

---

* See, for example, the second edition of this book: Joseph E. Shigley, "Mechanical Engineering Design," 2d ed., p. 362, McGraw-Hill Book Company, New York, 1972.

Table 8-2 CONSTANTS FOR USE IN EQ. (8-10) TO ESTIMATE THE TENSILE STRENGTH OF SELECTED SPRING STEELS

Material	Size range, in	Size range, mm	Exponent, m	Constant, A kpsi	Constant, A MPa
Music wire[a]	0.004–0.250	0.10–6.5	0.146	196	2170
Oil-tempered wire[b]	0.020–0.500	0.50–12	0.186	149	1880
Hard-drawn wire[c]	0.028–0.500	0.70–12	0.192	136	1750
Chrome vanadium[d]	0.032–0.437	0.80–12	0.167	169	2000
Chrome silicon[e]	0.063–0.375	1.6–10	0.112	202	2000

[a] Surface is smooth, free from defects, and with a bright lustrous finish.
[b] Has a slight heat-treating scale which must be removed before plating.
[c] Surface is smooth and bright, with no visible marks.
[d] Aircraft-quality tempered wire; can also be obtained annealed.
[e] Tempered to Rockwell C49 but may also be obtained untempered.

more economical springs, and so the defining of tolerances is an important phase of spring design. The commercial tolerance on wire diameter is usually not more than plus or minus 1.5 percent of the diameter. The tolerance on coil diameters varies from about 5 percent for springs having an index $D/d = 4$ up to more than 25 percent for D/d values of 16 or more. These tolerances correspond roughly to three standard deviations.

8-6 FATIGUE LOADING

Springs are made to be used, and consequently they are almost always subject to fatigue loading. In many instances the number of cycles of required life may be small, say, several thousand for a padlock spring or a toggle-switch spring. But the valve spring of an automotive engine must sustain millions of cycles of operation without failure; so they must be designed for infinite life.

In the case of shafts and many other machine members, fatigue loading in the form of completely reversed stresses is quite ordinary. Helical springs, on the other hand, are never used as both compression and extension springs. In fact, they are usually assembled with a preload so that the working load is additional. Thus the stress-time diagram of Fig. 5-23a expresses the usual condition for helical springs. The worst condition, then, would occur when there is no preload, that is, when $\tau_{min} = 0$.

In analyzing springs for the cause of a fatigue failure or in designing springs to resist fatigue, it is proper to apply the shear-stress multiplication factor K_s both to the mean stress τ_m and to the stress amplitude τ_a. The reason for this is that K_s is not really a stress-concentration factor at all, as indicated in Sec. 8-1, but merely a convenient means of calculating the shear stress at the inside of the coil. Now, we define

$$F_a = \frac{F_{max} - F_{min}}{2} \qquad (8\text{-}13)$$

and

$$F_m = \frac{F_{max} + F_{min}}{2} \qquad (8\text{-}14)$$

where the subscripts have the same meaning as those of Fig. 5-23a when applied to the axial spring force F. Then the stress components are

$$\tau_a = K_s \frac{8F_a D}{\pi d^3} \qquad (8\text{-}15)$$

$$\tau_m = K_s \frac{8F_m D}{\pi d^3} \qquad (8\text{-}16)$$

As indicated in Sec. 5-18, recent investigation indicates that notch sensitivities are higher than they were formerly thought to be. Furthermore, most of the knowledge gained refers to reversed bending rather than to alternating torsion. This, coupled with the fact that the sensitivity of high-hardness steels in Fig. 5-19 approaches unity, is a strong argument in favor of making the fatigue-strength reduction factor, for steels, equal to the full value of the Wahl curvature correction factor. In the past some authorities have used a reduced value, but this does not now seem proper. We have already seen (Sec. 5-22) that a torsional failure will occur whenever

$$\tau_a = S_{se} \qquad (8\text{-}17)$$

or whenever $$\tau_{max} = \tau_a + \tau_m = S_{sy} \qquad (8\text{-}18)$$

Consequently, these two equations will be the basis of our design to resist fatigue failure.

The best and most recent data on the torsional endurance limits of spring steels are those reported by Zimmerli.* He discovered the surprising fact that size, material, and tensile strength have no effect on the endurance limits (infinite life only) of spring steels in sizes under $\frac{3}{8}$ in (10 mm). We have already observed that endurance limits tend to level out at high tensile strengths (Fig. 5-13), but the reason for this is not clear. Zimmerli suggests that it may be because the original surfaces are alike or because plastic flow during testing makes them the same.

Interpreted in terms of the nomenclature of this book, Zimmerli's results are

$$S'_{se} = 45.0 \text{ kpsi (310 MPa)} \qquad \text{for unpeened springs}$$

$$S'_{se} = 67.5 \text{ kpsi (465 MPa)} \qquad \text{for peened springs}$$

These results are valid for music wire, carbon valve-spring wire, chromium-vanadium valve-spring wire, and chromium-silicon valve-spring wire. They are corrected for surface finish and size, but not for reliability, temperature, or stress concentration.

It is necessary to approximate the *S-N* diagram when springs are to be designed for finite life. One point on this diagram is provided by the endurance limits quoted above. The other point is more difficult to obtain, because no published data exist for the value of the modulus of rupture S_{su} (ultimate torsional strength) for spring steels. Until published data become available, it is suggested that the following relation be used:

$$S_{su} = 0.60 S_u \qquad (8\text{-}19)$$

EXAMPLE 8-1   A No. 13 W & M gauge (0.091-in) music-wire compression spring has an outside diameter of $\frac{9}{16}$ in, a free length of $3\frac{1}{8}$ in, 21 active coils, and squared and ground ends. The spring is to be assembled with a preload of 10 lb and will operate to a maximum load of 50 lb during use. Determine the factor of

---

* F. P. Zimmerli, Human Failures in Spring Applications, *The Mainspring*, publication of the Associated Spring Corporation, Bristol, Conn., no. 17, August–September, 1957.

safety, guarding against a fatigue failure based on a life of 50 000 cycles and 99 percent reliability.

SOLUTION The mean diameter is $D = 0.5625 - 0.091 = 0.4715$ in. Then $C = D/d = 0.4715/0.091 = 5.19$, and from Fig. 8-2, $K_s = 1.097$. Using Eqs. (8-13) and (8-14), find

$$F_a = \frac{F_{max} - F_{min}}{2} = \frac{50 - 10}{2} = 20 \text{ lb}$$

$$F_m = \frac{F_{max} + F_{min}}{2} = \frac{50 + 10}{2} = 30 \text{ lb}$$

Then, employing Eqs. (8-15) and (8-16), we have for the stresses

$$\tau_a = K_s \frac{8 F_a D}{\pi d^3} = 1.097 \frac{(8)(20)(0.4715)}{\pi (0.091)^3} = 34.9 (10)^3 \text{ psi}$$

$$\tau_m = K_s \frac{8 F_m D}{\pi d^3} = 1.097 \frac{(8)(30)(0.4715)}{\pi (0.091)^3} = 52.4 (10)^3 \text{ psi}$$

The endurance limit is $S'_{se} = 45$ kpsi, but this must be corrected for reliability and stress concentration and then for life. Using Table 5-2, find $k_c = 0.814$. Next, from Fig. 8-2, find $K = 1.30$, and so the curvature factor is $K_c = K/K_s = 1.30/1.097 = 1.185$. The notch sensitivity of spring steels is very close to unity because they have such a high ultimate strength, and so $K_f = K_c$. Therefore the modifying factor for stress concentration (Sec. 5-18) is $k_e = 1/K_f = 1/1.185 = 0.844$. Therefore

$$S_{se} = k_c k_e S'_{se} = 0.814(0.844)(45) = 30.8 \text{ kpsi}$$

From Table 8-2 we find $A = 196$ and $m = 0.146$. Therefore, from Eq. (8-10)

$$S_{ut} = \frac{A}{d^m} = \frac{196}{(0.091)^{0.146}} = 278 \text{ kpsi}$$

Therefore $S_{su} = 0.60 S_u = 0.60(278) = 167$ kpsi, and so $0.90 S_{su} = 0.90(167) = 150$ kpsi. Using Eqs. (5-24) to (5-26) to get the finite-life strength gives, successively,

$$m = \tfrac{1}{3} \log \frac{0.9 S_{su}}{S_{se}} = \tfrac{1}{3} \log \frac{150}{30.8} = 0.229$$

$$b = \log \frac{(0.9 S_{su})^2}{S_{se}} = \log \frac{(150)^2}{30.8} = 2.864$$

$$S_{sf} = \frac{10^b}{N^m} = \frac{10^{2.864}}{[50(10)^3]^{0.229}} = 61.4 \text{ kpsi}$$

Therefore the factor of safety guarding against a fatigue failure at $50(10)^3$ cycles is

$$n = \frac{S_{sf}}{\tau_a} = \frac{61.4}{34.9} = 1.76 \qquad ////$$

8-7 HELICAL TORSION SPRINGS

The torsion springs illustrated in Fig. 8-9 are used in door hinges and automobile starters and, in fact, for any application where torque is required. They are wound in the same manner as extension or compression springs but have the ends shaped to transmit torque.

A torsion spring is subjected to the action of a bending moment $M = Fr$, producing a normal stress in the wire. Note that this is in contrast to a compression or extension helical spring, in which the load produces a torsional stress in the wire. This means that the residual stresses built in during winding are in the same direction as the working stresses which occur during use. These residual stresses are useful in making the spring stronger by opposing the working stress, *provided* that the load is always applied so as to cause the spring to wind up. Because the residual stress opposes the working stress, torsional springs can be designed to operate at stress levels which equal or even exceed the yield strength of the wire.

The bending stress can be obtained by using curved-beam theory as explained in Sec. 2-18. It is convenient to write the expression in the form

$$\sigma = K\frac{Mc}{I} \qquad (a)$$

where K is a stress-concentration factor and, in this case, is treated as such, rather than as a strength-reduction factor. The value of K depends upon the shape of the wire and upon whether or not the stress is desired on the inner fiber of the coil or on the outer fiber. Wahl has analytically determined the following values for K for round wire:

$$K_i = \frac{4C^2 - C - 1}{4C(C-1)} \qquad K_o = \frac{4C^2 + C - 1}{4C(C+1)} \qquad (8\text{-}20)$$

where C is the spring index and the subscripts i and o refer to the inner and outer fibers, respectively. When the bending moment $M = Fr$ and the section modulus $I/c = \pi d^3/32$ are substituted in Eq. (a), we obtain

$$\sigma = K\frac{32Fr}{\pi d^3} \qquad (8\text{-}21)$$

which gives the bending stress for a round-wire torsion spring.

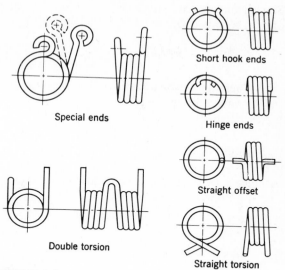

Short hook ends

Hinge ends

Straight offset

Special ends

Double torsion

Straight torsion

FIGURE 8-9
Torsion springs. (*Courtesy of Associated Spring Corporation.*)

Deflection

The strain energy in bending is, from Eq. (3-22),

$$U = \int \frac{M^2 \, dx}{2EI} \qquad (b)$$

For the torsion spring, $M = Fr$, and integration must be accomplished over the length of the wire. The force F will deflect through the distance $r\theta$, where θ is the total angular deflection of the spring. Applying Castigliano's theorem,

$$r\theta = \frac{\partial U}{\partial F} = \int_0^{\pi DN} \frac{\partial}{\partial F}\left(\frac{F^2 r^2 \, dx}{2EI}\right) = \int_0^{\pi DN} \frac{Fr^2 \, dx}{EI} \qquad (c)$$

Substituting $I = \pi d^4/64$ for round wire and solving Eq. (c) for θ gives

$$\theta = \frac{64FrDN}{d^4 E} \qquad (8\text{-}22)$$

where θ is the angular deflection of the spring in radians. The spring rate is therefore

$$k = \frac{Fr}{\theta} = \frac{d^4 E}{64DN} \qquad (8\text{-}23)$$

The spring rate may also be expressed as the torque required to wind up the spring one turn. This is obtained by multiplying Eq. (8-23) by 2π. Thus

$$k' = \frac{d^4 E}{10.2 D N} \qquad (8\text{-}24)$$

These deflection equations have been developed without taking into account the curvature of the wire. Actual tests show that the constant 10.2 should be increased slightly. Thus the equation

$$k' = \frac{d^4 E}{10.8 D N} \qquad (8\text{-}25)$$

will give better results. Corresponding corrections may be made to Eqs. (8-22) and (8-23) if desired.

8-8   BELLEVILLE SPRINGS

The inset of Fig. 8-10 shows a coned-disk spring, commonly called a *Belleville spring.* Although the mathematical treatment is beyond the purposes of this book, the reader should at least be familiar with the remarkable characteristics of these springs.

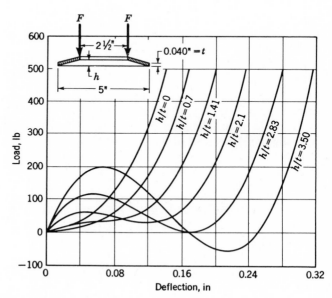

FIGURE 8-10
Load-deflection curves for Belleville springs. (*Courtesy of Associated Spring Corporation.*)

Aside from the obvious advantage of occupying a small space, a variation of the h/t ratio will produce a wide variety of load-deflection curve shapes, as illustrated in Fig. 8-10. For example, using an h/t ratio of 2.83 or larger gives an S curve which might be useful for snap-acting mechanisms. A reduction of the ratio to a value between 1.41 and 2.1 causes the central portion of the curve to become horizontal, which means that the load is constant over a considerable deflection range.

A higher load for a given deflection may be obtained by nesting, that is, by stacking the springs in parallel. On the other hand, stacking in series provides a larger deflection for the same load, but in this case there is danger of instability.

8-9 MISCELLANEOUS SPRINGS

The extension spring shown in Fig. 8-11 is made of slightly curved strip steel, not flat, so that the force required to uncoil it remains constant; thus it is called a *constant-force* spring. This is equivalent to a zero spring rate. Such springs can also be manufactured having either a positive or a negative spring rate.

A *volute spring* is a wide, thin strip, or "flat," of steel wound on the flat so that the coils fit inside one another. Since the coils do not stack, the solid height of the spring is the width of the strip. A variable-spring scale, in a compression volute spring, is obtained by permitting the coils to contact the support. Thus, as the deflection increases, the number of active coils decreases. The volute spring, shown in Fig. 8-12a, has another important advantage which cannot be obtained with round-wire springs: if the coils are wound so as to contact or slide on one another during action, the sliding friction will serve to damp out vibrations or other unwanted transient disturbances.

A *conical spring*, as the name implies, is a coil spring wound in the shape of a cone. Most conical springs are compression springs and are wound with round

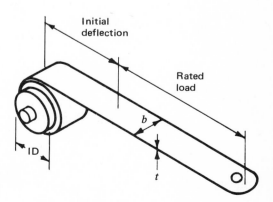

FIGURE 8-11
Constant-force spring. (*Courtesy of Vulcan Spring & Mfg. Co., Huntingdon Valley, Pa.*)

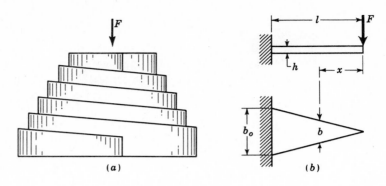

FIGURE 8-12
(a) A volute spring; (b) a flat triangular spring.

wire. But a volute spring is a conical spring too. Probably the principal advantage of this type of spring is that it can be wound so that the solid height is only a single wire diameter.

Flat stock is used for a great variety of springs, such as clock springs, power springs, torsion springs, cantilever springs, and hair springs; frequently it is specially shaped to create certain spring actions for fuse clips, relay springs, spring washers, snap rings, and retainers.

In designing many springs of flat stock or strip material it is often economical and of value to proportion the material so as to obtain a constant stress throughout the spring material. A uniform-section cantilever spring has a stress

$$\sigma = \frac{M}{I/c} = \frac{Fx}{I/c} \qquad (a)$$

which is proportional to the distance x if I/c is a constant. But there is no reason why I/c need be a constant. For example, one might design such a spring as that shown in Fig. 8-12b, in which the thickness h is constant but the width b is permitted to vary. Since, for a rectangular section, $I/c = bh^2/6$, we have, from Eq. (a),

$$\frac{bh^2}{6} = \frac{Fx}{\sigma}$$

or

$$b = \frac{6Fx}{h^2\sigma} \qquad (b)$$

Since b is linearly related to x, the width b_σ at the base of the spring is

$$b_\sigma = \frac{6Fl}{h^2\sigma} \qquad (8\text{-}26)$$

But the deflection of this triangular flat spring is more difficult to obtain, because the moment of inertia is now a variable. Probably the quickest solution could be obtained by using singularity functions, or the method of graphical integration.

The methods of stress and deflection analysis illustrated in previous sections of this chapter have served to illustrate that springs may be analyzed and designed by using the fundamentals discussed in the earlier chapters of this book. This is also true for most of the miscellaneous springs mentioned in this section, and you should now experience no difficulty in reading and understanding the literature of such springs.

8-10   CRITICAL FREQUENCY OF HELICAL SPRINGS

Coil springs are frequently used in applications imposing a very rapid reciprocating motion upon the coils, as, for example, in automotive valve springs. In these cases, the designer must be certain that the physical dimensions of the spring are not such as to create a natural vibratory frequency close to the frequency of the applied force. Such a condition would mean that the spring would resonate at the same frequency as the applied motion. Since helical springs are relatively free of damping forces, the internal stresses at resonance would be high.

Wahl has shown that the critical frequency of a helical spring is

$$f = \frac{m}{2} \sqrt{\frac{kg}{W}} \qquad (8\text{-}27)$$

where the fundamental frequency is found for $m = 1$, the second harmonic for $m = 2$, and so on, and where k is the spring constant as defined by Eq. (8-8). The frequency f is given in cycles per second. The weight of the spring is

$$W = AL\rho = \frac{\pi d^2}{4}(\pi DN)(\rho) = \frac{\pi^2 d^2 DN\rho}{4} \qquad (8\text{-}28)$$

where ρ is the wire density and the other quantities are as previously defined.

The fundamental critical frequency should be from fifteen to twenty times the frequency of the force in order to avoid resonance with the harmonics. If the frequency is not high enough, the spring should be redesigned to increase k or decrease W.

8-11   ENERGY-STORAGE CAPACITY

Quite frequently, in the selection and design of springs, the capacity of a spring to store energy is of major importance. Sometimes the designer is interested in absorbing shock and impact loads; at other times he is simply interested in storing the maximum energy in the smallest space. Equations (3-23) for strain energy can be particularly useful to the designer in choosing a particular form of spring. These equations are, or may be written,

$$u = \frac{\sigma^2}{2E} \qquad u = \frac{\tau^2}{2G} \qquad (8\text{-}29)$$

where u is the strain energy per unit volume. Of course, the particular equation to be used depends upon whether the spring is stressed axially, that is, in tension or compression, or whether it is stressed in shear. Maier* prefers to divide springs into two classes, which he calls E springs or G springs, depending upon which formula is applicable. Since the stress is usually not uniform, a form coefficient C_F is defined as follows:

$$u = C_F \frac{\sigma^2}{2E} \qquad u = C_F \frac{\tau^2}{2G} \qquad (8\text{-}30)$$

where $C_F = 1$, a maximum value, if the stress is uniformly distributed, meaning that the material is used most efficiently. For most springs the stress is not uniformly distributed, and so C_F will be less than unity. Thus the value of the form coefficient is a measure of the spring's capacity to store energy.

To calculate the form coefficient for a helical extension or compression spring, we write

$$u = \frac{U}{v} = \frac{Fy}{2v} \qquad (a)$$

where F = force
$\quad\quad y$ = deflection
$\quad\quad v$ = volume of active wire

* Karl W. Maier, Springs That Store Energy Best, *Prod. Eng.*, vol. 29, no. 45, p. 71, Nov. 10, 1958.

Table 8-3 FORM COEFFICIENTS—A MEASURE OF THE CAPACITY OF SPRINGS TO STORE ENERGY

Name of spring	Type	C_F
Tension bar	E	1.0
Clock spring	E	0.33
Torsion spring	E	0.25
Belleville washers	E	0.05–0.20
Cantilever beam	E	0.11
Torsion tube	G	About 0.90
Torsion bar	G	0.50
Compression spring	G	About 0.35

Since $y = 8FD^3N/d^4G$, $\tau = 8FDK/\pi d^3$, and $v = lA = (\pi DN)(\pi d^2/4)$, we have, from Eq. (a),

$$u = \frac{1}{2K^2}\left(\frac{\tau^2}{2G}\right) \tag{b}$$

And so $C_F = \frac{1}{2}K^2$. Note that, for $K = 1.20$, $C_F = 0.35$.

For a torsion bar, we use the relation

$$u = \frac{U}{v} = \frac{T\theta}{2v} \tag{c}$$

where θ is the angle of twist. Here $\tau = 16T/\pi d^3$, $\theta = 32Tl/\pi d^4G$, and $v = \pi d^2l/4$. The strain energy per unit volume is

$$u = \frac{1}{2}\frac{\tau^2}{2G} \tag{d}$$

and so $C_F = 0.50$.

Table 8-3 contains a list of form coefficients computed by Maier, which should be useful in selecting springs for energy-storage purposes.

PROBLEMS

Sections 8-1 to 8-4

8-1   A helical compression spring is made of No. 18 (0.047-in) wire having a torsional yield strength of 108 kpsi. It has an outside diameter of $\frac{1}{2}$ in and has 14 active coils.
(a) Find the maximum static load corresponding to the yield point of the material.
(b) What deflection would be caused by the load in (a)?
(c) Calculate the scale of the spring.
(d) If the spring has one dead turn at each end, what is the solid height?
(e) What should be the length of the spring so that when it is compressed solid the stress will not exceed the yield point?

8-2   The same as Prob. 8-1 except that the wire is No. 13 (0.092-in), the outside spring diameter is $\frac{3}{4}$ in, and the torsional strength is 90 kpsi.

8-3   A helical tension spring is made of 1.2-mm wire having a torsional yield strength of 740 MPa. The spring has an OD of 12 mm, 36 active coils, and hook ends, as in Fig. 8-7a. The spring is prestressed to 75 MPa during winding, which keeps it closed solid until an external load of sufficient magnitude is applied. When wound, the distance between hook ends is 70 mm.
(a) What is the spring preload?
(b) What load would cause yielding?
(c) What is the spring rate?
(d) What would be the distance between the hook ends if the spring were extended until the stress just reached the yield strength?

8-4   The compression helical spring shown in the figure is made of spring steel wire having a torsional yield strength of 640 MPa.
(*a*) Compute the spring rate.
(*b*) What force is required to close the spring to its solid height?
(*c*) After the spring has been closed to its solid height once, and the compressive force removed, will it spring back to its original free length?

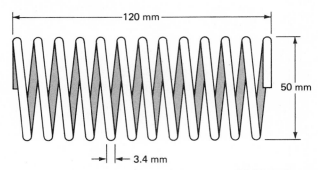

PROBLEM 8-4

8-5   Two steel compression coil springs are to be nested. The outer spring has an inside diameter of $1\frac{1}{2}$ in, a wire diameter of 0.120 in, and 10 active coils. The inner spring has an outside diameter of 1.25 in, a wire diameter of 0.091 in, and 13 active coils.
(*a*) Compute the spring rate of each spring.
(*b*) What force is required to deflect the nested spring assembly a distance of 1 in? (Both have the same free length.)
(*c*) Which spring will be stressed the most? Calculate this stress using the result of (*b*).

8-6   A compression coil spring of 3.4-mm music wire, having an outside diameter of 22 mm, has 8 active coils. Determine the stress and deflection caused by a static load of 270 N.

Section 8-5

8-7   A compression coil spring has 18 active coils, an outside diameter of $\frac{9}{32}$ in, and plain ends; it is made of No. 15 gauge (0.035-in) music wire.
(*a*) What should be the free length of the spring such that no permanent deformation will occur when it is compressed solid?
(*b*) What force is necessary to compress the spring to its solid length?

8-8   A helical compression spring uses No. 14 gauge (0.080-in) oil-tempered wire and has six active coils and squared ends. If the spring has an outside diameter of $\frac{1}{2}$ in, what should be the free length of the spring in order that, when it is compressed solid, the stress will not exceed 90 percent of the yield strength?

8-9   Design a compression coil spring of music wire having squared ends. The spring is to be assembled with a preload of 10 N and exert a force of 50 N when it is compressed an additional 140 mm. Find the wire diameter to the nearest 0.2 mm and the coil diameter to the nearest millimetre using $C = 12$, but do not use a larger wire diameter than is necessary. What is the free length and the solid height? The force corresponding to the solid height should be more than 50 N, say about 60 N.

8-10 Design a compression coil spring of hard-drawn wire having plain ends. When the spring is compressed a distance of 2.25 in it is to exert a force of 18 lb, but the force corresponding to the solid height should be a little more, say 24 lb, for safety. The spring index should be about 10 to avoid a large coil diameter. Use the smallest even-numbered W & M gauge wire, as tabulated in Table A-24, consistent with safety. Specify the number of coils, the free length, and the OD of the spring.

Section 8-6

8-11 A compression spring has 18 active coils, a free length of $1\frac{1}{4}$ in, and an outside diameter of $\frac{9}{32}$ in; it is made of No. 15 gauge (0.035-in) music wire and has plain ends.
   (a) Compute the spring rate, the solid height, and the stress in the spring when it is compressed to the solid height.
   (b) The spring operates with a minimum force of 2 lb and a maximum force of 4.5 lb. Compute the factor of safety guarding against a fatigue failure based on 50 percent reliability.

8-12 A helical compression spring is made of $\frac{1}{4}$-in-diameter steel wire and has an outside diameter of $2\frac{1}{4}$ in with squared and ground ends and 12 total coils. The length of the spring is such that, when it is compressed solid, the torsional stress is 120 kpsi.
   (a) Determine the spring rate.
   (b) Determine the free length of the spring.
   (c) This is a shot-peened spring; based on 90 percent reliability and infinite life, determine whether a fatigue failure can be expected if the spring is cycled between the loads $F_{min} = 50$ lb and $F_{max} = 250$ lb. Show computations to verify your decision.

8-13 An extension spring is made of 0.60-mm music wire and has an outside diameter of 4.8 mm. The spring is wound with a pre-tension of 1.10 N and the load fluctuates between this value and 6.8 N. Since the spring might fail statically or in fatigue, find the factor of safety for both types of failure.

8-14 A compression coil spring is made of 2-mm music wire and has an outside diameter of 12.5 mm. The maximum and minimum values of the fatigue load to which the spring is subjected are 90 and 45 N, respectively. For infinite life and 90 percent reliability, find the factor of safety.

8-15 A compression coil spring is wound from $\frac{1}{2}$-in-diameter bar stock over a mandrel which is 5 in in diameter. Assume that springback results in a spring having an ID 10 percent larger than the mandrel diameter. The spring is wound with 12 total coils and has squared and ground ends. After heat treatment the Brinell hardness is found to be $H_B = 380$. The free length of the spring is 20 in. The spring is assembled into a machine by compressing it to a length of 18 in. When the machine runs, the spring is compressed an additional 10 in so that the maximum load corresponds to a spring length of 8 in and the minimum load to a length of 18 in.

(a) Would this spring develop a permanent set if compressed solid? Why?

(b) What is the spring rate?

(c) Is the spring likely to buckle?

(d) Based on 50 percent reliability and infinite life, will the spring fail by fatigue? If not, what is the factor of safety guarding against a fatigue failure?

Section 8-7

8-16 The figure shows a finger exerciser used by law-enforcement officials and hand-gun enthusiasts to strengthen their grip. It is formed by winding cold-drawn steel wire around a mandrel so as to obtain $2\frac{1}{2}$ turns when the grip is in the closed position. After winding, the wire is cut so as to leave the two legs as handles. The plastic handles are then molded on, the grip squeezed together, and a wire clip placed around the legs to obtain initial "tension" and to space the handles for the best gripping position. The clip is formed like a figure eight to prevent it from coming off. The wire material is hard-drawn 0.60 carbon having a yield strength of 150 kpsi. The stress in the wire when the grip is closed should not, of course, exceed this figure. Based on the stress not exceeding 150 kpsi, find the configuration of the exerciser before the clip is assembled, and the hand force necessary to close the grip.

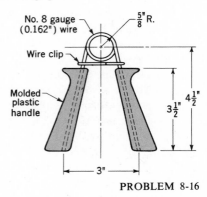

PROBLEM 8-16

9

ANTIFRICTION BEARINGS

The term *antifriction bearing* is used to describe that class of bearing in which the main load is transferred through elements in rolling contact rather than in sliding contact. In a rolling bearing the starting friction is about twice the running friction but still it is negligible in comparison to the starting friction of a sleeve bearing. Load, speed, and the operating viscosity of the lubricant do affect the frictional characteristics of a rolling bearing. It is probably a mistake to describe a rolling bearing as "antifriction," but the term is used generally throughout the industry.

From the mechanical designer's standpoint, the study of antifriction bearings differs in several respects when compared with the study of other topics. The specialist in antifriction-bearing design is confronted with the problem of designing a group of elements which compose a rolling bearing; these elements must be designed to fit into a space whose dimensions are specified; they must be designed to receive a load having certain characteristics; and, finally, these elements must be designed to have a satisfactory life when operated under the specified conditions. Bearing specialists must therefore consider such matters as fatigue loading, friction, heat, corrosion resistance, kinematic problems, material properties, lubrication, machining tolerances, assembly, use, and cost. From a consideration of all these factors, bearing specialists arrive at a compromise which, in their judgment, is a good solution to the problem as stated.

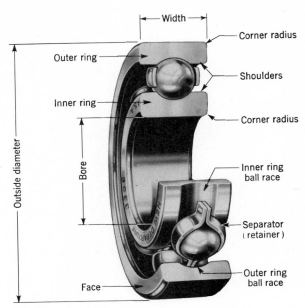

FIGURE 9-1
Nomenclature of a ball bearing. (*Courtesy of New-Departure-Hyatt Division, General Motors Corporation.*)

9-1 BEARING TYPES

Bearings are manufactured to take pure radial loads, pure thrust loads, or a combination of these two. The nomenclature of a *ball bearing* is illustrated in Fig. 9-1, which also shows the four essential parts of a bearing. These are the outer ring, the inner ring, the balls or rolling elements, and the separator. In low-priced bearings the separator is sometimes omitted, but it has the important function of separating the elements so that rubbing contact will not occur. Some of the various types of standardized bearings which are manufactured are shown in Fig. 9-2.

The single-row deep-groove bearing will take radial load as well as some thrust load. The balls are inserted into the grooves by moving the inner ring to an eccentric position. The balls are separated after loading, and the separator is then assembled.

The use of a filling notch (Fig. 9-2b) in the inner and outer rings enables a greater number of balls to be inserted, thus increasing the load capacity. The thrust capacity is decreased, however, because of the bumping of the balls against the edge of the notch when thrust loads are present.

The angular-contact bearing (Fig. 9-2c) provides a greater thrust capacity. All these bearings may be obtained with shields on one or both sides. The shields are not a complete closure but do offer a measure of protection against dirt.

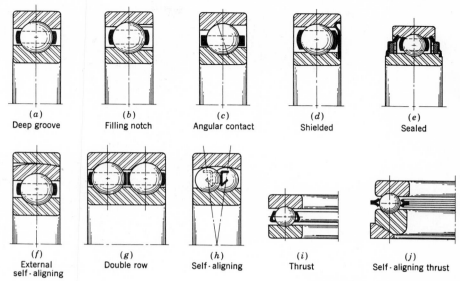

(a)
Deep groove

(b)
Filling notch

(c)
Angular contact

(d)
Shielded

(e)
Sealed

(f)
External
self - aligning

(g)
Double row

(h)
Self - aligning

(i)
Thrust

(j)
Self - aligning thrust

FIGURE 9-2
Various types of ball bearings.

A variety of bearings are manufactured with seals on one or both sides. When the seals are on both sides, the bearings are lubricated at the factory. Although a sealed bearing is supposed to be lubricated for life, a method of relubrication is sometimes provided.

Single-row bearings will withstand a small amount of shaft misalignment or deflection, but where this is severe, self-aligning bearings may be used.

Double-row bearings are made in a variety of types and sizes to carry heavier radial and thrust loads. Sometimes two single-row bearings are used together for the same reason, although a double-row bearing will generally require fewer parts and occupy less space.

The one-way ball thrust bearings (Fig. 9-2i) are made in many types and sizes.

Some of the large variety of standard roller bearings available are illustrated in Fig. 9-3. Straight roller bearings (Fig. 9-3a) will carry a greater load than ball bearings of the same size because of the greater contact area. However, they have the disadvantage of requiring almost perfect geometry of the raceways and rollers. A slight misalignment will cause the rollers to skew and get out of line. For this reason, the retainer must be heavy. Straight roller bearings will not, of course, take thrust loads.

Helical rollers are made by winding rectangular material into rollers, after which they are hardened and ground. Because of the inherent flexibility, they will take considerable misalignment. If necessary, the shaft and housing can be used

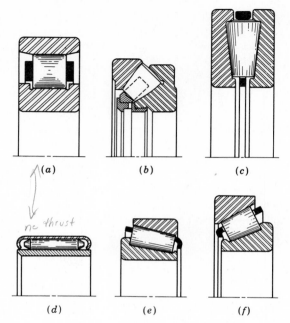

FIGURE 9-3
Types of roller bearings: (*a*) straight roller; (*b*) spherical roller thrust; (*c*) tapered roller thrust; (*d*) needle; (*e*) tapered roller; (*f*) steep-angle tapered roller. (*Courtesy of The Timken Company.*)

for raceways instead of separate inner and outer races. This is especially important if radial space is limited.

The spherical-roller thrust bearing (Fig. 9-3*b*) is useful where heavy loads and misalignment occur. The spherical elements have the advantage of increasing their contact area as the load is increased.

Needle bearings (Fig. 9-3*d*) are very useful where radial space is limited. They have a high load capacity when separators are used, but may be obtained without separators. They are furnished both with and without races.

Tapered roller bearings (Fig. 9-3*e*,*f*) combine the advantages of ball and straight roller bearings since they can take either radial or thrust loads or any combination of the two, and in addition, they have the high load-carrying capacity of straight roller bearings. The tapered roller bearing is designed so that all elements in the roller surface and the raceways intersect at a common point on the bearing axis.

The bearings described here represent only a small fraction of the many available for selection. Many special-purpose bearings are manufactured, and bearings are also made for particular classes of machinery, for example, instrument bearings. One of these, called a ball bushing, has balls that are recirculated. The advantage of the ball bushing is that it permits rotation, or sliding linear motion, or both.

9-2 BEARING LIFE*

When the ball or roller of an antifriction bearing rolls into the loading zone, Hertzian stresses occur on the inner ring, the rolling element, and the outer ring. Because the curvature of the contacting elements is different in the axial direction than it is in the radial direction, the formulas for these stresses are much more complicated than the Hertzian equations presented in Sec. 2-20.† If a bearing is clean and properly lubricated, is mounted and sealed against the entrance of dust or dirt, is maintained in this condition, and is operated at reasonable temperatures, then metal fatigue will be the only cause of failure. Since this implies many millions of stress applications, the term "bearing life" is in very general use.

The *life* of an *individual bearing* is defined as the total number of revolutions, or the number of hours at a given constant speed, of bearing operation required for the failure criteria to develop. Under ideal conditions the fatigue failure will consist of a spalling of the load-carrying surfaces. The Anti-Friction Bearing Manufacturers Association (AFBMA) standard states that the failure criterion is the first evidence of fatigue. It is noted, however, that the *useful* life is often used as the definition of fatigue life. The failure criterion used by The Timken Company laboratories‡ is the spalling or pitting of an area of 0.01 in². But Timken observes that the useful life may extend considerably beyond this point.

Rating life is a term sanctioned by the AFBMA and used by most bearing manufacturers. The rating life of a group of apparently identical ball or roller bearings is defined as the number of revolutions, or hours at some given constant speed, that 90 percent of a group of bearings will complete or exceed before the failure criterion develops. The terms *minimum life* and L_{10} *life* are also used to mean rating life.

The terms "average life" and "median life" are both used quite generally in discussing the longevity of bearings. Both terms are intended to have the same meaning. When groups consisting of large numbers of bearings are tested to failure, the median lives of the groups are averaged. Thus these terms are really intended to mean the average median life. In this book we shall use the term *median life* to mean the average of these medians.

In testing groups of bearings, the objective is to determine the median life

---

* For additional information see "AFBMA Standards," Anti-Friction Bearing Manufacturers Association, New York, 1972.
† These equations are not required here. For a complete presentation, see Hudson T. Morton, "Anti-Friction Bearings," 2d ed., pp. 223–236, Hudson T. Morton, Ann Arbor, Mich., 1965.
‡ *Timken Engineering Journal*, vol. 1, p. 22, The Timken Company, 1972.

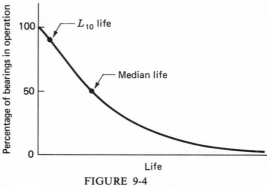

FIGURE 9-4
Typical curve of bearing life expectancy.

and the L_{10} or rated life. When many groups of bearings are tested, it is found that the median life is somewhere between 4 and 5 times the L_{10} life. The graph of Fig. 9-4 shows approximately how the failures are distributed. This curve is only approximate; it must not be used for analytical or prediction purposes.

The importance of knowing the probable survival of a group of bearings can be examined by the use of Eq. (4-22). Assume that the probability of any single bearing failure is independent of the others in the same machine. If the machine is assembled with a total of N bearings, each having the same reliability R, then the reliability of the group must be

$$R_N = (R)^N$$

from Eq. (4-22). Suppose we have a gear-reduction unit consisting of six bearings, all loaded so that the L_{10} lives are equal. Since the reliability of each bearing is 90 percent, the reliability of all the bearings in the assembly is

$$R_6 = (0.90)^6 = 0.531$$

This points up the need to select bearings having reliabilities greater than 90 percent.

The distribution of bearing failures can be best approximated by the Weibull distribution.* This distribution is widely used in engineering and is particularly useful in the study of fatigue failure. Lipson† shows how most distributions may be characterized by two or more parameters. In the case of the normal distribution the mean value μ describes the quality while the standard deviation $\hat{\sigma}$

---

* An excellent discussion and development of the Weibull distribution is contained in Charles Lipson and Narendra J. Sheth, "Statistical Design and Analysis of Engineering Experiments," pp. 36–44, 84–87, 111–113, McGraw-Hill Book Company, New York, 1973.
† *Op. cit.*, p. 84.

describes the uniformity of the distribution. In the case of the Weibull distribution the corresponding parameters are θ, a characteristic value, and b, the Weibull exponent. One form of the Weibull function may be written as

$$R = \exp\left[-\left(\frac{t}{\theta}\right)^{b}\right] \qquad (9\text{-}1)$$

where R = reliability
 t = time
 θ = design life
 b = Weibull exponent

For our purposes, the equation is more useful in the form

$$R = \exp\left[-\left(\frac{L}{mL_{10}}\right)^{b}\right] \qquad (9\text{-}2)$$

where R = reliability corresponding to life L
 L_{10} = rating life $(R = 0.90)$
 m = scale constant

The scale constant and the Weibull exponent can be found if two points on the life expectancy curve are known. As pointed out by Mischke,* the resulting equation is approximate for all other points on the curve. To be conservative we shall assume, as does Mischke, that the median life is 5 times as long as the rating life.

To find the two Weibull parameters, first substitute $R = 0.90$ and $L = L_{10}$ in Eq. (9-2). This gives

$$0.90 = \exp\left[-\left(\frac{L_{10}}{mL_{10}}\right)^{b}\right] = \exp\left[-\frac{1}{m^{b}}\right] \qquad (a)$$

Taking the natural logarithm of both sides gives

$$-0.105\,361 = -\frac{1}{m^{b}}$$

or

$$m^{b} = 9.491\,178 \qquad (b)$$

Now write Eq. (9-2) again, using $R = 0.50$ corresponding to $L = 5L_{10}$. This gives

$$0.50 = \exp\left[-\left(\frac{5L_{10}}{mL_{10}}\right)^{b}\right] = \exp\left[-\left(\frac{5}{m}\right)^{b}\right] \qquad (c)$$

Taking the natural logarithm of both sides of this equation yields

$$-0.693\,147 = -\frac{5^{b}}{m^{b}} \qquad (d)$$

* Charles Mischke, Bearing Reliability and Capacity, *Machine Design*, vol. 37, no. 22, pp. 139–140, Sept. 30, 1965.

Now substitute the value of m^b in Eq. (b) into (d) and solve for the exponent b. This process is as follows:

$$-0.693\,147 = -\frac{5^b}{9.491\,178}$$

$$5^b = (9.491\,179)(0.693\,146)$$

$$b \log 5 = \log 9.491\,178 + \log 0.693\,147$$

$$b = 1.17$$

Thus, from Eq. (b),

$$m = (9.491\,178)^{1/1.17} = 6.84$$

And so Eq. (9-2) can finally be written as

$$R = \exp\left[-\left(\frac{L}{6.84L_{10}}\right)^{1.17}\right] \qquad (9\text{-}3)$$

EXAMPLE 9-1*   A certain application requires a bearing to last for 1800 h with a reliability of 99 percent. What should be the rated life of the bearing selected for this application?

SOLUTION   Substitute into Eq. (9-3) as follows:

$$0.99 = \exp\left[-\left(\frac{1800}{6.84L_{10}}\right)^{1.17}\right]$$

Now take the natural logarithm of both sides and simplify the result. This gives

$$-0.010\,050 = -\frac{(1800)^{1.17}}{(6.84)^{1.17}(L_{10})^{1.17}} = -\frac{678.7}{(L_{10})^{1.17}}$$

$$L_{10} = \left(\frac{678.7}{0.010\,050}\right)^{1/1.17} = 13.4(10)^3 \text{ h} \qquad Ans.$$

////

9-3 BEARING LOAD

Experiments show that two groups of identical bearings tested under different loads F_1 and F_2 will have respective lives L_1 and L_2 according to the relation

$$\frac{L_1}{L_2} = \left(\frac{F_2}{F_1}\right)^a \qquad (9\text{-}4)$$

where $a = 3$ for ball bearings
$a = \frac{10}{3}$ for roller bearings

---

* See also Eugene Shube, Ball-Bearing Survival, *Machine Design*, vol. 34, no. 17, pp. 158–161, July 19, 1962.

The AFBMA has established a standard load rating for bearings in which speed is not a consideration. This rating is called the basic load rating. The *basic load rating C* is defined as *the constant radial load which a group of apparently identical bearings can endure for a rating life of one million revolutions of the inner ring* (stationary load and stationary outer ring). The rating life of one million revolutions is a base value selected for ease of computation. The corresponding load rating is so high that plastic deformation of the contacting surfaces would occur were it actually applied. Consequently the basic load rating is purely a reference figure; such a large load would probably never be applied.

Other names in common use for the basic load rating are *dynamic load rating, basic dynamic capacity,* and *specific dynamic capacity.*

Using Eq. (9-4), the life of a bearing subjected to any other load F will be

$$L = \left(\frac{C}{F}\right)^a \qquad (9\text{-}5)$$

However, the equation is more useful in the form

$$C = FL^{1/a} \qquad (9\text{-}6)$$

For example, if we desire a life of 27 million revolutions for a roller bearing, then the basic load rating must be

$$C = F(27)^{3/10} = 2.69F$$

or 2.69 times the actual radial load.

It is customary practice with bearing manufacturers to specify the rated radial bearing load corresponding to a certain speed in rpm and a certain L_{10} life in hours. For example, the *Timken Engineering Journal* tabulates the load ratings at 3000 h of L_{10} life at 500 rpm. By adopting the subscripts D to refer to design or required values, and R as the catalog or rated values, then Eq. (9-6) can be rewritten as

$$C_R = F\left[\left(\frac{L_D}{L_R}\right)\left(\frac{n_D}{n_R}\right)\right]^{1/a} \qquad (9\text{-}7)$$

where C_R is the basic load rating corresponding to L_R hours of L_{10} life at the speed n_R rpm. The force F is the actual radial bearing load; it is to be carried for L_D hours of L_{10} life at a speed of n_D rpm.

EXAMPLE 9-2 A roller bearing is to be selected to withstand a radial load of 4 kN and have an L_{10} life of 1200 h at a speed of 600 rpm. What load rating would you look for in searching the *Timken Engineering Journal*?

SOLUTION The quantities for use in Eq. (9-7) are $F = 4$ kN, $L_D = 1200$ h, $L_R = 3000$ h, $n_D = 600$ rpm, $n_R = 500$ rpm, and $a = \frac{10}{3}$. Therefore use

$$C_R = 4\left[\left(\frac{1200}{3000}\right)\left(\frac{600}{500}\right)\right]^{3/10} = 3.21 \text{ kN}$$

The Timken ratings are tabulated in English units and in dekanewtons (see Table A-1). Therefore the basic load rating used to enter the catalog is 321 daN.　　////

It is also possible to develop a relation to find the catalog rating corresponding to any desired reliability. To find this relation, note that the reciprocal of Eq. (9-3) is

$$\frac{1}{R} = \exp\left[\frac{L}{6.84 L_{10}}\right]^{1.17} \qquad (a)$$

where L is the desired life corresponding to the reliability R. Taking the natural logarithm of both sides gives

$$\ln\frac{1}{R} = \left(\frac{L}{6.84}\right)^{1.17} \frac{1}{(L_{10})^{1.17}}$$

Now, solve this expression for L_{10}. The result is

$$L_{10} = \frac{L}{6.84} \frac{1}{[\ln(1/R)]^{1/1.17}} \qquad (9\text{-}8)$$

Equation (9-8) gives the rating life corresponding to any desired life L at the reliability R. Incorporating this into Eq. (9-7) yields

$$C_R = F\left[\left(\frac{L_D}{L_R}\right)\left(\frac{n_D}{n_R}\right)\left(\frac{1}{6.84}\right)\right]^{1/a} \frac{1}{[\ln(1/R)]^{1/1.17a}} \qquad (9\text{-}9)$$

EXAMPLE 9-3　What load rating would be used if the application in Example 9-2 is to have a reliability of 99 percent?

SOLUTION　The terms are identical with those of Example 9-2 and in addition $R = 0.99$. Equation (9-9) yields

$$C_R = 4\left[\left(\frac{1200}{3000}\right)\left(\frac{600}{500}\right)\left(\frac{\cdot 1}{6.84}\right)\right]^{3/10} \frac{1}{[\ln(1/0.99)]^{1/(1.17)(10/3)}} = 5.86 \text{ kN}$$

Therefore, we enter the catalog with $C_R = 586$ daN.　　////

9-4　SELECTION OF BALL AND STRAIGHT ROLLER BEARINGS

Except for pure thrust bearings, as in Fig. 9-2i, ball bearings are usually operated with some combination of radial and thrust load. Since catalog ratings are based only on radial load, it is convenient to define an *equivalent radial load* F_e that will have the same effect on bearing life as do the applied loads. The AFBMA equation for equivalent radial load for ball bearings is the maximum of the two values

$$F_e = VF_r \qquad (9\text{-}10)$$

$$F_e = XVF_r + YF_a \qquad (9\text{-}11)$$

where F_e = equivalent radial load
   F_r = applied radial load
   F_a = applied thrust load
   V = a rotation factor
   X = a radial factor
   Y = a thrust factor

In using these equations the rotation factor V is to correct for the various rotating-ring conditions. For a rotating inner ring, $V = 1$. For a rotating outer ring, $V = 1.2$. The factor of 1.2 for outer-ring rotation is simply an acknowledgment that the fatigue life is reduced under these conditions. Self-aligning bearings are an exception; they have $V = 1$ for rotation of either ring.

The X and Y factors in Eq. (9-11) depend upon the geometry of the bearing, including the number of balls and ball diameter. When a theoretical derivation of the X and Y factors is made, it is found that the resulting curves can be approximated by pairs of straight lines. Thus there are two values of X and Y listed in Table 9-1. The set of values giving the largest equivalent load should always be used.

The AFBMA has established standard boundary dimensions for bearings which define the bearing bore, the outside diameter, the width, and the fillet sizes on the shaft and housing shoulders. The basic plan covers all ball and straight roller bearings in the metric sizes. The plan is quite flexible in that, for a given bore, there is an assortment of widths and outside diameters. Furthermore, the outside diameters selected are such that, for a particular outside diameter, one can usually find a variety of bearings having different bores and widths.

This basic AFBMA plan is illustrated in Fig. 9-5. The bearings are identified by a two-digit number called the *dimension-series code*. The first number in the code is from the *width series* 0, 1, 2, 3, 4, 5, and 6. The second number is from the *diameter series* (outside) 8, 9, 0, 1, 2, 3, and 4. Figure 9-5 shows the variety of bearings which may be obtained with a particular bore. Since the dimension-series code does not reveal the dimensions directly, it is necessary to resort to tabulations. The 02- and 03-series bearings are the most widely used, and the dimensions

Table 9-1   EQUIVALENT RADIAL-LOAD FACTORS

Bearing type	X_1	Y_1	X_2	Y_2
Radial-contact ball bearings	1	0	0.5	1.4
Angular-contact ball bearings with shallow angle	1	1.25	0.45	1.2
Angular-contact ball bearings with steep angle	1	0.75	0.4	0.75
Double-row and duplex ball bearings (type *DB* or *DF*)	1	0.75	0.63	1.25

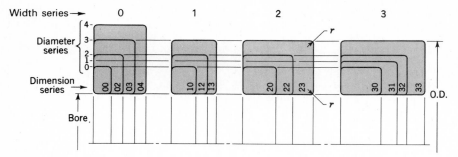

FIGURE 9-5

The basic AFBMA plan for boundary dimensions. These apply to ball bearings, straight roller bearings, and spherical roller bearings, but not to tapered roller bearings or to inch-series ball bearings. The contour of the corner is not specified; it may be rounded or chamfered, but it must be small enough to clear the fillet radius specified in the standards.

of some of these are tabulated in Tables 9-2 and 9-3. Shaft and housing shoulder diameters listed in the tables (Fig. 9-6) should be used whenever possible to secure adequate support for the bearing and to resist the maximum thrust loads. Table 9-4 lists the dimensions and load ratings of some straight roller bearings.

Table 9-2 DIMENSIONS AND BASIC LOAD RATINGS FOR THE 02-SERIES BALL BEARINGS

Bore, mm	OD, mm	Width, mm	Fillet radius, mm	Shoulder diameter, mm		Load rating, kN
				d_S	d_H	
10	30	9	0.6	12.5	27	3.58
12	32	10	0.6	14.5	28	5.21
15	35	11	0.6	17.5	31	5.87
17	40	12	0.6	19.5	34	7.34
20	47	14	1.0	25	41	9.43
25	52	15	1.0	30	47	10.8
30	62	16	1.0	35	55	14.9
35	72	17	1.0	41	65	19.8
40	80	18	1.0	46	72	22.5
45	85	19	1.0	52	77	25.1
50	90	20	1.0	56	82	26.9
55	100	21	1.5	63	90	33.2
60	110	22	1.5	70	99	40.3
65	120	23	1.5	74	109	44.1
70	125	24	1.5	79	114	47.6
75	130	25	1.5	86	119	50.7
80	140	26	2.0	93	127	55.6
85	150	28	2.0	99	136	64.1
90	160	30	2.0	104	146	73.9
95	170	32	2.0	110	156	83.7

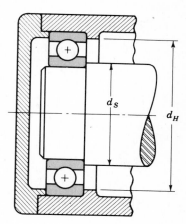

FIGURE 9-6
Shaft and housing shoulder diameters d_S and d_H should be adequate to assure good bearing support.

To assist the designer in the selection of bearings, most of the manufacturers' handbooks contain data on bearing life for many classes of machinery, as well as information on load-application factors. Such information has been accumulated the hard way, that is, by experience, and the beginning designer should utilize this

Table 9-3 DIMENSIONS AND BASIC LOAD RATINGS FOR THE 03-SERIES BALL BEARINGS

Bore, mm	OD, mm	Width, mm	Fillet radius, mm	Shoulder diameter, mm		Load rating, kN
				d_S	d_H	
10	35	11	0.6	12.5	31	6.23
12	37	12	1.0	16	32	7.48
15	42	13	1.0	19	37	8.72
17	47	14	1.0	21	41	10.37
20	52	15	1.0	25	45	12.24
25	62	17	1.0	31	55	16.2
30	72	19	1.0	37	65	21.6
35	80	21	1.5	43	70	25.6
40	90	23	1.5	49	80	31.4
45	100	25	1.5	54	89	40.5
50	110	27	2.0	62	97	47.6
55	120	29	2.0	70	106	55.2
60	130	31	2.0	75	116	62.7
65	140	33	2.0	81	125	71.2
70	150	35	2.0	87	134	80.1
75	160	37	2.0	93	144	87.2
80	170	39	2.0	99	153	94.8
85	180	41	2.5	106	161	101.9
90	190	43	2.5	111	170	110.8
95	200	45	2.5	117	179	117.9

Table 9-4 DIMENSIONS AND BASIC LOAD RATINGS FOR STRAIGHT ROLLER BEARINGS

Bore, mm	02-series			03-series		
	OD, mm	Width, mm	Load, kN	OD, mm	Width, mm	Load, kN
25	52	15	10.9	62	17	23.1
30	62	16	18.0	72	19	30.3
35	72	17	26.0	80	21	39.2
40	80	18	34.0	90	23	46.3
45	85	19	35.6	100	25	63.6
50	90	20	36.9	110	27	75.7
55	100	21	45.4	120	29	92.6
60	110	22	55.6	130	31	103.0
65	120	23	65.0	140	33	116.0
70	125	24	65.8	150	35	136.0
75	130	25	80.1	160	37	162.0
80	140	26	87.2	170	39	163.0
85	150	28	99.7	180	41	196.0
90	160	30	126.0	190	43	211.0
95	170	32	140.0	200	45	240.0
100	180	34	154.0	215	47	274.0
110	200	38	205.0	240	50	352.0
120	215	40	220.0	260	55	416.0
130	230	40	239.0	280	58	489.0
140	250	42	280.0	300	62	538.0

information until he gains enough experience to know when deviations are possible. Table 9-5 contains recommendations on bearing life for some classes of machinery. The load-application factors in Table 9-6 serve the same purpose as factors of safety; use them to increase the equivalent load before selecting a bearing.

Table 9-5 BEARING-LIFE RECOMMENDATIONS FOR VARIOUS CLASSES OF MACHINERY

Type of application	Life, kh
Instruments and apparatus for infrequent use	Up to 0.5
Aircraft engines	0.5–2
Machines for short or intermittent operation where service interruption is of minor importance	4–8
Machines for intermittent service where reliable operation is of great importance	8–14
Machines for 8-h service which are not always fully utilized	14–20
Machines for 8-h service which are fully utilized	20–30
Machines for continuous 24-h service	50–60
Machines for continuous 24-h service where reliability is of extreme importance	100–200

Table 9-6 LOAD-APPLICATION FACTORS

Type of application	Load factor
Precision gearing	1.0–1.1
Commercial gearing	1.1–1.3
Applications with poor bearing seals	1.2
Machinery with no impact	1.0–1.2
Machinery with light impact	1.2–1.5
Machinery with moderate impact	1.5–3.0

9-5 SELECTION OF TAPERED ROLLER BEARINGS

The nomenclature for a tapered roller bearing differs in some respects from that of ball and straight roller bearings. The inner ring is called the cone, and the outer ring is called the cup, as shown in Fig. 9-7. It can also be seen that a tapered roller bearing is separable in that the cup can be removed from the cone-and-roller assembly.

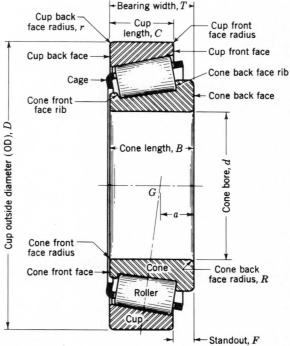

FIGURE 9-7
Nomenclature of a tapered roller bearing. Point G is the effective load center; use this point to calculate the radial bearing load. (*Courtesy of The Timken Company.*)

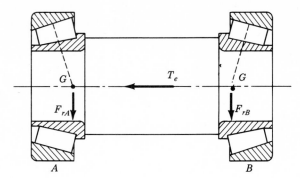

FIGURE 9-8

Schematic drawing showing a pair of tapered roller bearings assembled on a single shaft with indirect mounting. The radial bearing forces are F_{rA} and F_{rB}. T_e is the external thrust load.

A tapered roller bearing can carry both radial and thrust (axial) loads or any combination of the two. However, even when an external thrust load is not present, the radial load will induce a thrust reaction within the bearing because of the taper. To avoid separation of the races and rollers, this thrust must be resisted by an equal and opposite force. One way of generating this force is to always use at least two tapered roller bearings on a shaft. These can be mounted with the backs facing each other, called indirect mounting, or with the fronts facing each other, called direct mounting.

The thrust component F_a produced by a pure radial load F_r is specified by Timken as

$$F_a = \frac{0.47F_r}{K} \qquad (9\text{-}12)$$

where K is the ratio of the radial rating of the bearing to the thrust rating. The constant 0.47 is derived from a summation of the thrust components from the individual rollers supporting the load. The value of K is approximately 1.5 for radial bearings and 0.75 for steep-angle bearings. These values may be used for a preliminary bearing selection after which the exact values may be obtained from the *Timken Engineering Journal* in order to verify the selection.

Figure 9-8 shows a typical bearing mounting subjected to an external thrust load T_e. The radial reactions F_{rA} and F_{rB} are computed by taking moments about the effective load centers G. The distance a (Fig. 9-7) is obtained from the catalog rating sheets (*Timken Engineering Journal*). The equivalent radial loads are computed using an equation similar to Eq. (9-11) except that a rotation factor is not used with tapered roller bearings. We shall use subscripts A and B to designate each of the two bearings in Fig. 9-8. The equivalent radial load on bearing A is

$$F_{eA} = 0.4F_{rA} + K_A\left(\frac{0.47F_{rB}}{K_B} + T_e\right) \qquad (9\text{-}13)$$

For bearing B, we have

$$F_{eB} = 0.4F_{rB} + K_B\left(\frac{0.47F_{rA}}{K_A} - T_e\right) \qquad (9\text{-}14)$$

If the actual radial load on either bearing should happen to be larger than the corresponding value of F_e, then use the actual radial load instead of F_e for that bearing.

Figure 9-9 is a reproduction of a portion of a typical catalog page from the *Timken Engineering Journal*.

EXAMPLE 9-4   The gear-reduction unit shown in Fig. 9-10 is arranged to rotate the cup, while the cone is stationary. Bearing A takes the thrust load of 250 lb and, in addition, has a radial load of 875 lb. Bearing B is subjected to a pure radial load of 625 lb. The speed is 150 rpm. The desired L_{10} life is 90 kh. The desired shaft diameters are $1\frac{3}{8}$ in at A and $1\frac{1}{4}$ in at B. Select suitable tapered roller bearings using an application factor of unity.

SOLUTION   Since B carries only radial load, the thrust on A is augmented by the induced thrust due to B. Equation (9-13) applies. Using a trial value of 1.5 for K, we obtain

$$F_{eA} = 0.4F_{rA} + K_A\left(\frac{0.47F_{rB}}{K_B} + T_e\right) = 0.4(875) + 1.5\left[\frac{0.47(625)}{1.5} + 250\right]$$

$$= 1020 \text{ lb}$$

Thus $F_{eA} > F_{rA}$ and so we use 1020 lb as the equivalent radial load to select bearing A. We next use Eq. (9-7) to obtain the L_{10} rating. Using $L_R = 3$ kh and $n_R = 500$ rpm, we get

$$C_R = F\left[\left(\frac{L_D}{L_R}\right)\left(\frac{n_D}{n_R}\right)\right]^{1/a} = 1020\left[\left(\frac{90}{3}\right)\left(\frac{150}{500}\right)\right]^{3/10} = 1970 \text{ lb}$$

Using this figure and a bore of $1\frac{3}{8}$ in we enter the catalog sheets (Fig. 9-9 is typical) and select an LM48548 cone and an LM49510 cup. This bearing has an L_{10} rating of 2140 lb and $K = 1.55$. Since we assumed 1.5 for K the difference is small and we need not recalculate F_{eA}.

For bearing B, Eq. (9-14) applies. Thus

$$F_{eB} = 0.4F_{rB} + K_B\left(\frac{0.47F_{rA}}{K_A} - T_e\right) = 0.4(625) + 1.5\left[\frac{0.47(875)}{1.55} - 250\right]$$

$$= 273 \text{ lb}$$

Note that the actual value of K_A was used, but K_B was assumed to be 1.5 as before.

SINGLE-ROW STRAIGHT BORE—TS

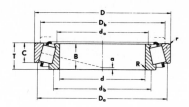

bore	outside diameter	width	rating at 500 RPM for 3000 hours L10		fac-tor	eff. load center	part numbers		cone				cup			
			one row radial	thrust					max. shaft fillet radius	width	backing shoulder diameters		max. housing fillet radius	width	backing shoulder	
			lb	lb			cone	cup								
d	D	T	lb	lb	K	a	cone	cup	R	B	d_b	d_a	r	C	D_b	D_a
1.2500	2.3125	0.5781	1280	1040	1.23	−0.05	08125	08231	0.04	0.5937	1.48	1.42	0.04	0.4219	2.05	2.17
1.2500	2.3280	0.6250	1580	1110	1.42	−0.12	▲LM67048	LM67010	Spec.	0.6600	1.67	1.42	0.05	0.4650	2.05	2.20
1.2500	2.4404	0.6250	1580	1110	1.42	−0.12	▲LM67049A	LM67014	0.03	0.6600	1.46	1.42	0.05	0.4650	2.13	2.24
1.2500	2.4409	0.7150	1990	1190	1.67	−0.19	15123	15245	Spec.	0.7500	1.67	1.44	0.05	0.5625	2.17	2.28
1.2500	2.4409	0.7500	1990	1190	1.67	−0.23	15125	15245	0.14	0.8125	1.67	1.44	0.05	0.5625	2.17	2.28
1.2500	2.4409	0.7500	1990	1190	1.67	−0.23	15126	15245	0.03	0.8125	1.46	1.44	0.05	0.5625	2.17	2.28
1.3125	3.0000	1.1563					HM89443	HM89410		1.1250	1.83	1.75	0.13	0.9063	2.44	2.87
1.3125	3.0000	1.1563	3880	3630	1.07	−0.22	HM89444	HM89411	0.15	1.1250	2.09	1.75	0.03	0.9063	2.56	2.87
1.3125	3.4843	1.0000	3180	4250	0.75	0.09	44131	44348	0.08	0.9330	2.01	1.89	0.06	0.6875	2.95	3.31
1.3750	2.5625	0.7100	2140	1380	1.55	−0.15	▲LM48548	LM48510	Spec.	0.7200	1.81	1.57	0.05	0.5500	2.28	2.40
1.3750	2.5625	0.8300	2140	1380	1.55	−0.15	▲LM48548A	LM48511A	0.03	0.7200	1.59	1.66	0.06	0.6700	2.28	2.40
1.3750	2.6250	0.8125	2520	1520	1.66	−0.22	M38549	M38510	0.14	0.8125	1.83	1.57	0.09	0.6563	2.28	2.44
1.3750	2.6875	0.8125	2330	1410	1.66	−0.23	14585	14525	0.14	0.8125	1.81	1.57	0.09	0.6250	2.32	2.48
1.3750	2.7148	0.7813	2180	1420	1.53	−0.17	14137A	14274A	0.06	0.7710	1.65	1.57	0.13	0.6250	2.32	2.48
1.3750	2.7148	0.7813	2180	1420	1.53	−0.17	14138A	14274A	0.14	0.7710	1.81	1.57	0.13	0.6250	2.32	2.48
1.3750	2.8438	1.0000	3190	2980	1.07	−0.18	HM88649	HM88610	0.09	1.0000	1.91	1.69	0.09	0.7812	2.36	2.72
1.3750	2.8750	0.8750	2620	2030	1.29	−0.15	02877	02820	0.14	0.8750	1.91	1.65	0.13	0.6875	2.44	2.68
1.3750	2.8750	0.8750	2620	2030	1.29	−0.15	02878	02820	0.03	0.8750	1.67	1.65	0.13	0.6875	2.44	2.68

FIGURE 9-9
A portion of the TS bearing tables from the *Timken Engineering Journal,* Section 1.
The original page contains SI equivalents in red print below the English values.

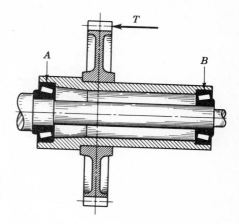

FIGURE 9-10
Tapered roller bearings applied to a gear-reduction unit. (*Courtesy of The Timken Company.*)

Since $F_{eB} < F_{rB}$, we use F_{rB}. Using Eq. (9-7) again, we find the L_{10} desired rating as

$$C_R = F\left[\left(\frac{L_D}{L_R}\right)\left(\frac{n_D}{n_R}\right)\right]^{1/a} = 625\left[\left(\frac{90}{3}\right)\left(\frac{150}{500}\right)\right]^{3/10} = 1210 \text{ lb}$$

This bearing is to have a bore of $1\frac{1}{4}$ in. Therefore, from Fig. 9-9 we select an 08125 cone and 08231 cup. The L_{10} rating is 1280 lb with $K = 1.23$. However, the actual load was used instead of the smaller equivalent load and so we need not recalculate.

////

9-6 LUBRICATION

The contacting surfaces in rolling bearings have a relative motion that is both rolling and sliding, and so it is difficult to understand exactly what happens. If the relative velocity of the sliding surfaces is high enough, then the lubricant action is hydrodynamic (see Chap. 10). *Elastohydrodynamic lubrication* (EHD) is the phenomenon that occurs when a lubricant is introduced between surfaces that are in pure rolling contact. The contact of gear teeth, rolling bearings, and cam-and-follower surfaces are typical examples. When a lubricant is trapped between two surfaces in rolling contact, a tremendous increase in the pressure within the lubricant film occurs. But viscosity is exponentially related to pressure and so a very large increase in viscosity occurs in the lubricant that is trapped between the surfaces. Leibensperger* observes that the change in viscosity in and out of contact pressure is equivalent to the difference between cold asphalt and light sewing machine oil.

* R. L. Leibensperger, When Selecting a Bearing, *Machine Design*, vol. 47, no. 8, pp. 142–147, April 3, 1975.

The purposes of an antifriction-bearing lubricant may be summarized as follows:

1. To provide a film of lubricant between the sliding and rolling surfaces
2. To help distribute and dissipate heat
3. To prevent corrosion of the bearing surfaces
4. To protect the parts from the entrance of foreign matter

Either oil or grease may be employed as a lubricant. The following rules may help in deciding between them:

Use grease when	Use oil when
1. The temperature is not over 200°F.	1. Speeds are high.
2. The speed is low.	2. Temperatures are high.
3. Unusual protection is required from the entrance of foreign matter.	3. Oiltight seals are readily employed.
4. Simple bearing enclosures are desired.	4. Bearing type is not suitable for grease lubrication.
5. Operation for long periods without attention is desired.	5. The bearing is lubricated from a central supply which is also used for other machine parts.

9-7 ENCLOSURE

To exclude dirt and foreign matter and to retain the lubricant, the bearing mountings must include a seal. The three principal methods of sealing are the felt seal, the commercial seal, and the labyrinth seal (Fig. 9-11).

Felt seals may be used with grease lubrication when the speeds are low. The rubbing surfaces should have a high polish. Felt seals should be protected from dirt by placing them in machined grooves or by using metal stampings as shields.

The commercial seal is an assembly consisting of the rubbing element and, generally, a spring backing, which are retained in a sheet-metal jacket. These seals are usually made by press-fitting them into a counterbored hole in the bearing cover. Since they obtain the sealing action by rubbing, they should not be used for high speeds.

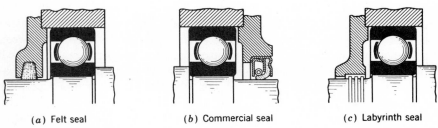

(a) Felt seal   (b) Commercial seal   (c) Labyrinth seal

FIGURE 9-11
Typical sealing methods. (*Courtesy of New-Departure-Hyatt Division, General Motors Corporation.*)

The labyrinth seal is especially effective for high-speed installations and may be used with either oil or grease. It is sometimes used with flingers. At least three grooves should be used, and they may be cut on either the bore or the outside diameter. The clearance may vary from 0.010 to 0.040 in, depending upon the speed and temperature.

9-8 SHAFT AND HOUSING DETAILS

There are so many methods of mounting antifriction bearings that each new design is a real challenge to the ingenuity of the designer. The housing bore and shaft outside diameter must be held to very close limits, which of course is expensive. There are usually one or more counterboring operations, several facing operations, and drilling, tapping, and threading operations, all of which must be performed on the shaft, housing, or cover plate. Each of these operations contributes to the cost of production, so that the designer, in ferreting out a trouble-free and low-cost mounting, is faced with a difficult and important problem. The various bearing manufacturers' handbooks give many mounting details in almost every design area. In a text of this nature, however, it is possible to give only the barest details.

The most frequently encountered mounting problem is that which requires one bearing at each end of a shaft. Such a design might use one ball bearing at each end, one tapered roller bearing at each end, or a ball bearing at one end and a straight roller bearing at the other. One of the bearings usually has the added function of positioning or axially locating the shaft. Figure 9-12 shows a very common solution to this problem. The inner rings are backed up against the shaft shoulders and are held in position by round nuts threaded onto the shaft. The outer ring of the left-hand bearing is backed up against a housing shoulder and is held in position by a device which is not shown. The outer ring of the right-hand bearing floats in the housing.

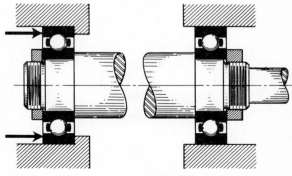

FIGURE 9-12
A common bearing mounting.

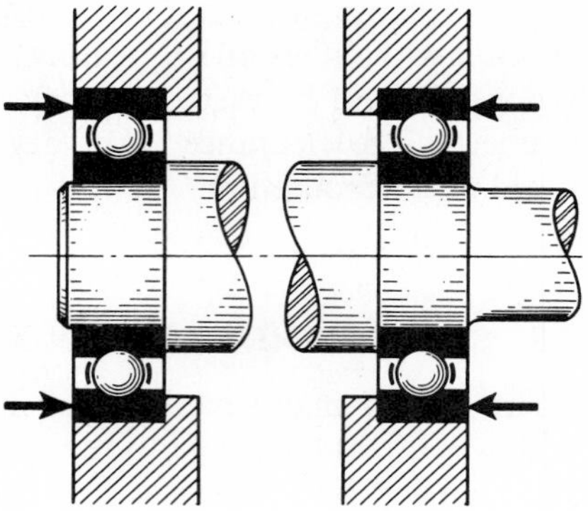

FIGURE 9-13
An alternative bearing mounting.

There are many variations possible to the method shown in Fig. 9-12. For example, the function of the shaft shoulder may be replaced by retaining rings, by the hub of a gear or pulley, or by spacing tubes or rings. The round nuts may be replaced by retaining rings or by washers locked in position by screws, cotters, or taper pins. The housing shoulder may be replaced by a retaining ring; the outer ring of the bearing may be grooved for a retaining ring, or a flanged outer ring may be used. The force against the outer ring of the left-hand bearing is usually applied by the cover plate, but if no thrust is present, the ring may be held in place by retaining rings.

Figure 9-13 shows an alternative method of mounting in which the inner races are backed up against the shaft shoulders as before but no retaining devices are required. With this method the outer races are completely retained. This eliminates the grooves or threads, which cause stress concentration on the overhanging end, but it requires accurate dimensions in an axial direction or the employment of adjusting means. This method has the disadvantage that, if the distance between the bearings is great, the temperature rise during operation may expand the shaft enough to wreck the bearings.

It is frequently necessary to use two or more bearings at one end of a shaft.

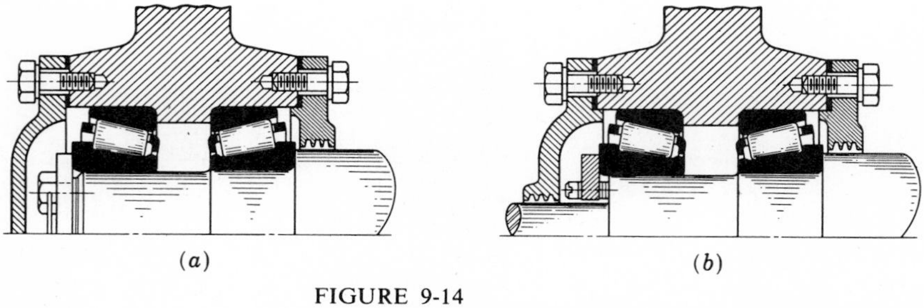

FIGURE 9-14
Two-bearing mountings. (*Courtesy of The Timken Company.*)

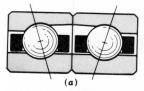 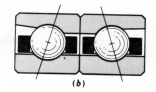 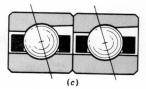

(a)   (b)   (c)

FIGURE 9-15

Duplex arrangements of angular-contact bearings. (a) DF mounting; (b) DB mounting; (c) DT mounting. (*Courtesy of Miniature Precision Bearings, Inc.*)

For example, two bearings could be used to obtain additional rigidity or increased load capacity or to cantilever a shaft. Several two-bearing mountings are shown in Fig. 9-14. These may be used with tapered roller bearings, as shown, or with ball bearings. In either case it should be noted that the effect of the mounting is to preload the bearings in an axial direction.

When maximum stiffness and resistance to shaft misalignment is desired, pairs of angular-contact ball bearings (Fig. 9-2) are often used in an arrangement called *duplexing*. Bearings manufactured for duplex mounting have their rings ground with an offset, so that when a pair of bearings is tightly clamped together, a preload is automatically established. As shown in Fig. 9-15, three mounting arrangements are used. The face-to-face mounting, called DF, will take heavy radial loads and thrust loads from either direction. The DB mounting (back-to-back) has the greatest aligning stiffness and is also good for heavy radial loads and thrust loads from either direction. The tandem arrangement, called the DT mounting, is used where the thrust is always in the same direction; since the two bearings have their thrust functions in the same direction, a preload, if required, must be obtained in some other manner.

Bearings are usually mounted with the rotating ring a press fit, whether it be the inner or outer ring. The stationary ring is then mounted with a push fit. This permits the stationary ring to creep in its mounting slightly, bringing new portions of the ring into the load-bearing zone to equalize wear.

PROBLEMS

9-1   The geared printing roll shown in the figure is driven at 300 rpm by a force $F = 200$ lb acting as shown. A uniform force distribution $w = 20$ lb/in acts against the bottom surface of roll 3 in the positive y direction. Radial contact 02-series ball bearings are to be selected for this application and mounted at O and A. Use an application factor of 1.2 and an L_{10} life of 30 kh and find the size of bearings to be used. Both bearings are to be the same size.

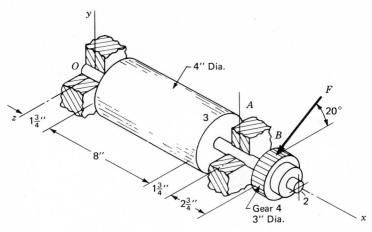

PROBLEM 9-1

9-2 The figure shows a geared countershaft with an overhanging pinion at C. You are to select a plain radial-contact ball bearing for mounting at O and a straight roller bearing for mounting at B. The force on gear A is $F_A = 600$ lb and the shaft is to run at a speed of 480 rpm. Use an application factor of 1.4 and an L_{10} life of 50 kh and determine the size of 02-series bearings that should be employed.

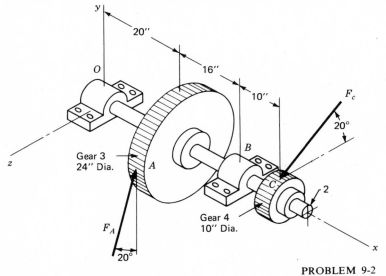

PROBLEM 9-2

9-3 The figure is a schematic drawing of a countershaft that supports two V-belt pulleys. Radial 02-series ball bearings are to be selected and located at O and E. The countershaft runs at 1100 rpm and the bearings are to have a life of 12 kh at 99 percent reliability using an application factor of unity. The belt tension on the loose side of pulley A is 15 percent of the tension on the tight side. What size bearings should be used if both are to be the same size?

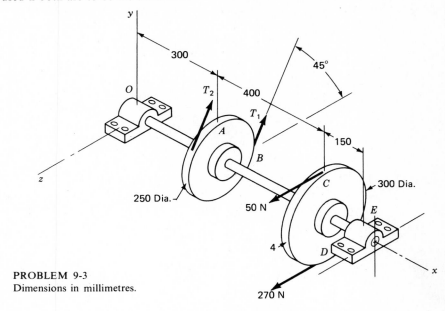

PROBLEM 9-3
Dimensions in millimetres.

9-4 Radial-contact 02-series ball bearings are to be selected and located at O and B to support the overhung countershaft shown in the figure. The belt tensions shown are parallel to each other. The tension on the loose side on pulley A is 20 percent of the tension on the right side. Shaft speed is 720 rpm. The bearings are to have a reliability of 90 percent corresponding to a life of 24 kh. Use an application factor of unity and the same size bearings at each location. Find an appropriate bearing size.

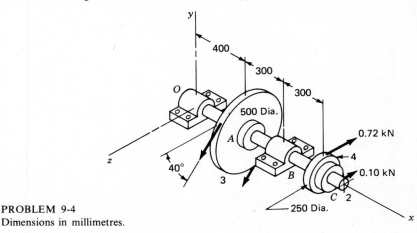

PROBLEM 9-4
Dimensions in millimetres.

9-5 The shaft in the figure has parallel belt pulls with the tension on the loose side of pulley 4 being 20 percent of the tension on the tight side. The shaft rotates at 840 rpm and the radial ball bearings, to be selected for locations at *O* and *B*, are to have a reliability of 99 percent corresponding to a life of 24 kh. Use an application factor of unity, same-size bearings, and specify the size of 02-series bearings needed.

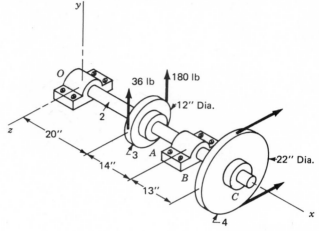

PROBLEM 9-5

9-6 Determine the required radial ratings for a pair of tapered roller bearings to be located in the two bearing housings supporting the geared shaft shown in the figure. The shaft rotates at 400 rpm. The bearings are to have an L_{10} life of 40 kh. Use unity for the application factor and 1.5 for *K*.

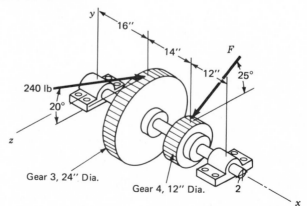

PROBLEM 9-6

9-7 Determine the required radial ratings in dekanewtons for a pair of radial tapered roller bearings for mounting in the housings at *D* and *C* of the gear-transmission shaft shown in the figure. The longitudinal dimensions locate the effective load centers of the bearings and gears. The bearings are to have a life of 24 kh corresponding to a reliability of 99.9 percent. The shaft speed is 360 rpm. Use 1.2 for the application factor and 1.5 for K.

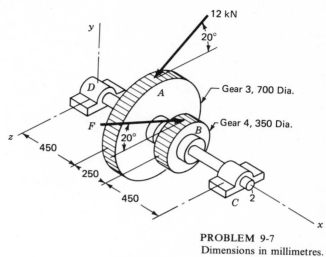

PROBLEM 9-7
Dimensions in millimetres.

9-8   The figure shows a shaft contained in a helical-gear reduction unit in which a force $\mathbf{F} = -1700\mathbf{i} + 6400\mathbf{j} - 2300\mathbf{k}$ lb is applied to gear B as shown. The forces $\mathbf{F}_A$ and $\mathbf{F}_C$, of equal magnitudes, resist the applied force. The directions of these two forces can be

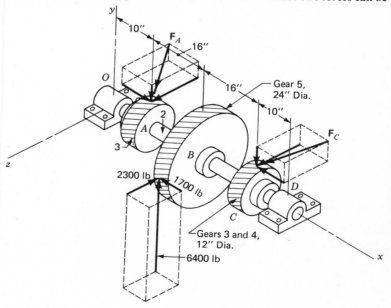

PROBLEM 9-8

indicated by the unit vectors $\hat{\mathbf{F}}_A = 0.470\mathbf{i} - 0.342\mathbf{j} + 0.814\mathbf{k}$ and $\hat{\mathbf{F}}_C = -0.470\mathbf{i} - 0.342\mathbf{j} + 0.814\mathbf{k}$. The notation $\hat{\mathbf{F}}$ means $\mathbf{F}/|\mathbf{F}|$. In this problem we wish to determine the required radial ratings of tapered roller bearings to be mounted in the housings at O and D. The shaft dimensions shown in the figure locate the effective load centers of the bearings and gears. The bearings are to have an L_{10} life of 60 kh. Use unity for the application factor and 1.5 for K. The shaft speed is 1200 rpm.

9-9 The figure shows a portion of a transmission containing an ordinary helical gear and an overhung bevel gear. Tapered roller bearings are to be mounted in housings at O and B with the bearing at O intended to take out the major thrust component. The dimensions are to the effective load centers of the gears and bearings. The vector bevel-gear force $\mathbf{F}_D$ can be expressed in the general form $\mathbf{F}_D = F_x\mathbf{i} + F_y\mathbf{j} + F_z\mathbf{k}$, or, for this particular bevel gear, in the form

$$\mathbf{F}_D = -0.242F_D\,\mathbf{i} - 0.242F_D\,\mathbf{j} + 0.940F_D\,\mathbf{k}$$

The bearings are to have an L_{10} life of 36 kh corresponding to a shaft speed of 900 rpm. Use 1.5 for K, unity for the application factor, and find the required radial rating of each bearing.

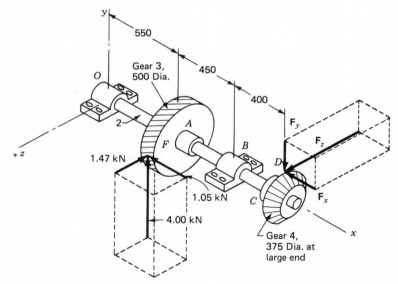

PROBLEMS 9-9 and 9-10
Dimensions in millimetres.

9-10 An angular-contact ball bearing with shallow angle is to be housed at O in the figure to take both radial and thrust loads. The bearing at B is to be a straight roller bearing. Determine the required radial ratings of each bearing based on an L_{10} life of 36 kh at a shaft speed of 900 rpm.

10

LUBRICATION AND JOURNAL BEARINGS

The object of lubrication is to reduce friction, wear, and heating of machine parts which move relative to each other. A lubricant is any substance which, when inserted between the moving surfaces, accomplishes these purposes. In a *sleeve bearing*, a shaft, or *journal*, rotates or oscillates within a sleeve, or *bearing*, and the relative motion is sliding. In an antifriction bearing, the main relative motion is rolling. A follower may either roll or slide on the cam. Gear teeth mate with each other by a combination of rolling and sliding. Pistons slide within their cylinders. All these applications require lubrication to reduce friction, wear, and heating.

The field of application for journal bearings is immense. The crankshaft and connecting-rod bearings of an automotive engine must operate for thousands of miles at high temperatures and under varying load conditions. The journal bearings used in the steam turbines of power-generating stations are said to have reliabilities approaching 100 percent. At the other extreme there are thousands of applications in which the loads are light and the service relatively unimportant. A simple, easily installed bearing is required, using little or no lubrication. In such cases an antifriction bearing might be a poor answer because of the cost, the elaborate enclosures, the close tolerances, the radial space required, the high speeds, or the increased inertial effects. Instead, a nylon bearing requiring no lubrication, a powder-metallurgy bearing with the lubrication "built in," or a bronze bearing with ring-oiled, wick-feed, solid-lubricant film, or grease lubrication might be a very satisfactory solution. Recent metallurgy developments in bearing materials, combined with increased knowledge of the lubrication process, now make it possible to design journal bearings with satisfactory lives and very good reliabilities.

Much of the material we have studied thus far in this book has been based on fundamental engineering studies, such as statics, dynamics, the mechanics of solids, metal processing, mathematics, and metallurgy. In the study of lubrication and journal bearings, additional fundamental studies, such as chemistry, fluid mechanics, thermodynamics, and heat transfer, must be utilized in developing the material. While we shall not utilize all of them in the material to be included here, you can now begin to appreciate better how the study of mechanical engineering design is really an integration of most of your previous studies and a directing of this total background toward the resolution of a single objective.

10-1 TYPES OF LUBRICATION

Five distinct forms of lubrication may be identified:

1 Hydrodynamic
2 Hydrostatic
3 Elastohydrodynamic
4 Boundary
5 Solid-film

Hydrodynamic lubrication means that the load-carrying surfaces of the bearing are separated by a relatively thick film of lubricant, so as to prevent metal-to-metal contact, and that the stability thus obtained can be explained by the laws of fluid mechanics. Hydrodynamic lubrication does not depend upon the introduction of the lubricant under pressure, though it may be; but it does require the existence of an adequate supply at all times. The film pressure is created by the moving surface itself pulling the lubricant into a wedge-shaped zone at a velocity sufficiently high to create the pressure necessary to separate the surfaces against the load on the bearing. Hydrodynamic lubrication is also called *full-film*, or *fluid*, lubrication.

Hydrostatic lubrication is obtained by introducing the lubricant, which is sometimes air or water, into the load-bearing area at a pressure high enough to separate the surfaces with a relatively thick film of lubricant. So, unlike hydrodynamic lubrication, motion of one surface relative to another is not required. We shall not deal with hydrostatic lubrication in this book,* but the subject should be considered in designing bearings where the velocities are small or zero and where the frictional resistance is to be an absolute minimum.

---

* See Oscar Pinkus and Beno Sternlicht, "Theory of Hydrodynamic Lubrication," chap. 6, McGraw-Hill Book Company, New York, 1961. See also Dudley D. Fuller, "Theory and Practice of Lubrication for Engineers," chaps. 3 and 4, John Wiley & Sons, Inc., New York, 1956.

Elastohydrodynamic lubrication is the phenomenon that occurs when a lubricant is introduced between surfaces which are in rolling contact, such as mating gears or rolling bearings. The mathematical explanation requires the Hertzian theory of contact stress and fluid mechanics.*

Insufficient surface area, a drop in the velocity of the moving surface, a lessening in the quantity of lubricant delivered to a bearing, an increase in the bearing load, or an increase in lubricant temperature resulting in a decrease in viscosity—any one of these—may prevent the build-up of a film thick enough for full-film lubrication. When this happens, the highest asperities may be separated by lubricant films only several molecular dimensions in thickness. This is called *boundary lubrication*. The change from hydrodynamic to boundary lubrication is not at all a sudden or abrupt one. It is probable that a mixed hydrodynamic- and boundary-type lubrication occurs first, and as the surfaces move closer together, the boundary-type lubrication becomes predominant. The viscosity of the lubricant is not of as much importance with boundary lubrication as is the chemical composition.

When bearings must be operated at extreme temperatures, a *solid-film lubricant* such as graphite or molybdenum disulfide must be used because the ordinary mineral oils are not satisfactory. Much research is currently being carried out in an effort, too, to find composite bearing materials with low wear rates as well as small frictional coefficients.

10-2 VISCOSITY†

In Fig. 10-1 let a plate A be moving with a velocity U on a film of lubricant of thickness h. We imagine the film as composed of a series of horizontal layers and the force F causing these layers to deform or slide on one another just like a deck of cards. The layers in contact with the moving plate are assumed to have a velocity U; those in contact with the stationary surface are assumed to have a zero velocity. Intermediate layers have velocities which depend upon their distances y from the stationary surface. Newton's law of viscous flow states that the shear stress in the fluid is proportional to the rate of change of velocity with respect to y. Thus

$$\tau = \frac{F}{A} = \mu \frac{du}{dy} \qquad (10\text{-}1)$$

where μ is the constant of proportionality and defines *absolute viscosity*. The derivative du/dy is the rate of change of velocity with distance and may be called the rate of shear, or the velocity gradient. The viscosity μ is thus a measure of the

---

* See A. Cameron, "Principles of Lubrication," chaps. 7–9, John Wiley & Sons, Inc., New York, 1966.
† For a complete discussion see any fluid mechanics text, for example, W. M. Swanson, "Fluid Mechanics," pp. 17–30 and 740, Holt, Rinehart and Winston, Inc., New York, 1970.

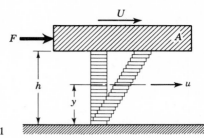

FIGURE 10-1

internal frictional resistance of the fluid. If the assumption is made that the rate of shear is a constant, then $du/dy = U/h$, and from Eq. (10-1),

$$\tau = \frac{F}{A} = \mu \frac{U}{h} \qquad (10\text{-}2)$$

The unit of viscosity in the IPS system is seen to be the pound-force-second per square inch; this is the same as stress or pressure multiplied by time. The IPS unit is called the *reyn* in honor of Sir Osborne Reynolds.

The absolute viscosity, also called *dynamic viscosity*, is measured by the pascal-second (Pa·s) in SI; this is the same as a newton-second per square metre. The conversion from IPS units to SI is the same as for stress. For example, multiply the absolute viscosity in reyns by 6890 to convert to units of Pa·s.

The American Society of Mechanical Engineers (ASME) has published a list of cgs units which are not to be used in ASME documents.* This list results from a recommendation by the International Committee of Weights and Measures (CIPM) that the use of cgs units with special names be discouraged. Included in this list is a unit of force called the *dyne* (dyn), a unit of dynamic viscosity called the *poise* (P), and a unit of kinematic viscosity called the *stoke* (St). All of these units have been, and still are, used extensively in lubrication studies.

The *poise* is the cgs unit of dynamic or absolute viscosity and its unit is the dyne-second per square centimetre (dyn·s/cm²). It has been customary to use the centipoise (cP) in analysis because its value is more convenient. When the viscosity is expressed in centipoises, it is designated by Z. The conversion from cgs units to SI and IPS units is as follows:

$$\mu(\text{Pa·s}) = (10)^{-3} Z(\text{cP})$$

$$\mu(\text{reyn}) = \frac{Z(\text{cP})}{6.89(10)^6}$$

The ASTM standard method for determining viscosity uses an instrument called the Saybolt Universal Viscosimeter. The method consists of measuring the

---

* *ASME Orientation and Guide for Use of Metric Units*, 2d ed., p. 13, American Society of Mechanical Engineers, 1972.

time in seconds for 60 ml of lubricant at a specified temperature to run through a tube 17.6 mm in diameter and 12.25 mm long. The result is called the kinematic viscosity and in the past the unit of square centimetre per second has been used. One square centimetre per second is defined as a *stoke*. By the use of the *Hagen-Poiseuille law** the kinematic viscosity based upon *seconds Saybolt*, also called *Saybolt Universal viscosity* (SUV) in seconds, is

$$Z_k = \left(0.22t - \frac{180}{t}\right) \qquad (10\text{-}3)$$

where Z_k is in centistokes (cSt) and t is the number of seconds Saybolt.

In SI the kinematic viscosity v has the unit of square metre per second (m²/s) and the conversion is

$$v(\text{m}^2/\text{s}) = 10^{-6}Z_k(\text{cSt})$$

Thus, Eq. (10-3) becomes

$$v = \left(0.22t - \frac{180}{t}\right)(10^{-6}) \qquad (10\text{-}4)$$

---

* See any text on fluid mechanics, for example, Chia-Shun Yih, "Fluid Mechanics," p. 314, McGraw-Hill Book Company, New York, 1969.

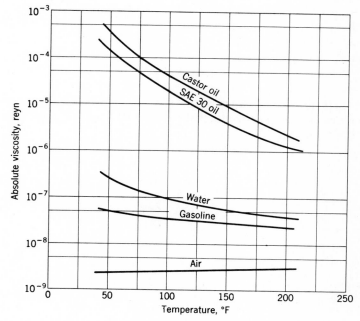

FIGURE 10-2
A comparison of the viscosities of various fluids.

To convert to dynamic viscosity we multiply v by the density in SI units. Designating the density as ρ with the unit of kilogram per cubic metre, we have

$$\mu = \rho\left(0.22t - \frac{180}{t}\right)(10^{-6}) \qquad (10\text{-}5)$$

where μ is in pascal-seconds.

Figure 10-2 shows the absolute viscosity in the IPS system of a number of fluids often used for lubrication purposes and their variation with temperature.

10-3   PETROFF'S LAW

The phenomenon of bearing friction was first explained by Petroff using the assumption that the shaft is concentric. Though we shall seldom make use of Petroff's method of analysis in the material to follow, it is important because it defines groups of dimensionless parameters and because the coefficient of friction predicted by this law turns out to be quite good even when the shaft is not concentric.

Let us now consider a vertical shaft rotating in a guide bearing. It is assumed that the bearing carries a very small load, that the clearance space c is completely filled with oil, and that leakage is negligible (Fig. 10-3). We denote the radius of the shaft by r, the radial clearance by c, and the length of the bearing by l, all dimensions being in inches. If the shaft rotates at N rps, then its surface velocity is $U = 2\pi r N$ in/s. Since the shearing stress in the lubricant is equal to the velocity gradient times the viscosity, from Eq. (10-2) we have

$$\tau = \mu\frac{U}{h} = \frac{2\pi r\mu N}{c} \qquad (a)$$

where the radial clearance c has been substituted for the distance h. The force required to shear the film is the stress times the area. The torque is the force times the lever arm. Thus

$$T = (\tau A)(r) = \left(\frac{2\pi r\mu N}{c}\right)(2\pi r l)(r) = \frac{4\pi^2 r^3 l\mu N}{c} \qquad (b)$$

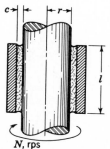

FIGURE 10-3 N, rps

If we now designate a small force on the bearing by W, in pounds-force, then the pressure P, in pounds-force per square inch of projected area, is $P = W/2rl$. The frictional force is fW, where f is the coefficient of friction, and so the frictional torque is

$$T = fWr = (f)(2rlP)(r) = 2r^2 flP \qquad (c)$$

Substituting the value of the torque from Eq. (c) into Eq. (b) and solving for the coefficient of friction, we find

$$f = 2\pi^2 \frac{\mu N}{P} \frac{r}{c} \qquad (10\text{-}6)$$

Equation (10-6) is called *Petroff's law* and was first published in 1883. The two quantities $\mu N/P$ and r/c are very important parameters in lubrication. Substitution of the appropriate dimensions in each parameter will show that they are dimensionless.

10-4 STABLE LUBRICATION

The difference between boundary and hydrodynamic lubrication can be explained by reference to Fig. 10-4. This plot of the change in the coefficient of friction versus the bearing characteristic $\mu N/P$ was obtained by the McKee brothers in an actual test of friction.* The plot is important because it defines stability of lubrication and helps us to understand hydrodynamic and boundary, or thin-film, lubrication.

Suppose we are operating to the right of ordinate BA and something happens, say, an increase in lubricant temperature. This results in a lower viscosity and hence a smaller value of $\mu N/P$. The coefficient of friction decreases, not as much heat is generated in shearing the lubricant, and consequently the lubricant

* S. A. McKee and T. R. McKee, Journal Bearing Friction in the Region of Thin Film Lubrication, *SAE Journal*, vol. 31, pp. (T) 371–377, 1932.

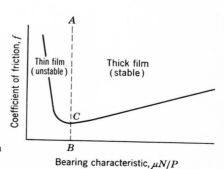

FIGURE 10-4
Variation of the coefficient of friction with $\mu N/P$.

temperature drops. Thus the region to the right of ordinate *BA* defines *stable lubrication* because variations are self-correcting.

To the left of ordinate *BA* a decrease in viscosity would increase the friction. A temperature rise would ensue, and the viscosity would be reduced still more. The result would be compounded. Thus the region to the left of ordinate *BA* represents *unstable lubrication*.

It is also helpful to see that a small viscosity, and hence a small $\mu N/P$, means that the lubricant film is very thin and that there will be a greater possibility of some metal-to-metal contact, and hence of more friction. Thus point *C* represents what is probably the beginning of metal-to-metal contact as $\mu N/P$ becomes smaller.

10-5 THICK-FILM LUBRICATION

Let us now examine the formation of a lubricant film in a journal bearing. Figure 10-5a shows a journal which is just beginning to rotate in a clockwise direction. Under starting conditions the bearing will be dry, or at least partly dry, and hence the journal will climb or roll up the right side of the bearing as shown in Fig. 10-5a. Under the conditions of a dry bearing, equilibrium will be obtained when the friction force is balanced by the tangential component of the bearing load.

Now suppose a lubricant is introduced into the top of the bearing as shown in Fig. 10-5b. The action of the rotating journal is to pump the lubricant around the bearing in a clockwise direction. The lubricant is pumped into a wedge-shaped space and forces the journal over to the other side. A *minimum film thickness* h_0 occurs, not at the bottom of the journal, but displaced clockwise from the bottom as in Fig. 10-5b. This is explained by the fact that a film pressure in the converging half of the film reaches a maximum somewhere to the left of the bearing center.

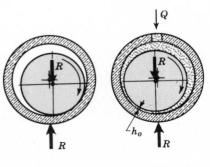

FIGURE 10-5
Formation of a film.   (*a*) Dry   (*b*) Lubricated

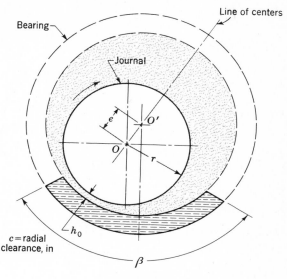

FIGURE 10-6
Nomenclature of a journal bearing.

Figure 10-5 shows how to decide whether the journal, under hydrodynamic lubrication, is eccentrically located on the right or on the left side of the bearing. Visualize the journal beginning to rotate. Find the side of the bearing upon which the journal tends to roll. Then, if the lubrication is hydrodynamic, place the journal on the opposite side.

The nomenclature of a journal bearing is shown in Fig. 10-6. The dimension c is the *radial clearance* and is the difference in the radii of the bearing and journal. In Fig. 10-6 the center of the journal is at O and the center of the bearing at O'. The distance between these centers is the *eccentricity* and is denoted by e. The *minimum film thickness* is designated by h_0, and it occurs at the line of centers. The film thickness at any other point is designated by h. We also define an *eccentricity ratio* ϵ as

$$\epsilon = \frac{e}{c}$$

The bearing shown in the figure is known as a *partial bearing*. If the radius of the bearing is the same as the radius of the journal, it is known as a *fitted bearing*. If the bearing encloses the journal, as indicated by the dashed lines, it becomes a *full bearing*. The angle β describes the angular length of a partial bearing. For example, a $120°$ partial bearing has the angle β equal to $120°$.

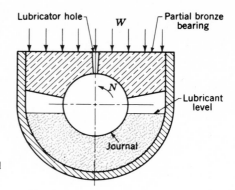

FIGURE 10-7
Schematic representation of the partial
bearing used by Tower.

10-6 HYDRODYNAMIC THEORY

The present theory of hydrodynamic lubrication originated in the laboratory of
Beauchamp Tower in the early 1880s in England. Tower had been employed to
study the friction in railroad journal bearings and learn the best methods of
lubricating them. It was an accident or error, during the course of this investigation, that prompted Tower to look at the problem in more detail and that resulted
in a discovery that eventually led to the development of the theory.

Figure 10-7 is a schematic drawing of the journal bearing which Tower
investigated. It is a partial bearing, 4 in in diameter by 6 in long, with a bearing
arc of 157°, and having bath-type lubrication, as shown. The coefficients of friction
obtained by Tower in his investigations on this bearing were quite low, which is
not now surprising. After testing this bearing, Tower later drilled a $\frac{1}{2}$-in-diameter
lubricator hole through the top. But when the apparatus was set in motion, oil
flowed out of this hole. In an effort to prevent this, a cork stopper was used, but
this popped out, and so it was necessary to drive a wooden plug into the hole.
When the wooden plug was pushed out too, Tower, at this point, undoubtedly
realized that he was on the verge of discovery. A pressure gauge connected to the
hole indicated a pressure of more than twice the unit bearing load. Finally, he
investigated the bearing film pressures in detail throughout the bearing width and
length and reported a distribution similar to that of Fig. 10-8.*

The results obtained by Tower had such regularity that Osborne Reynolds
concluded that there must be a definite law relating the friction, the pressure, and
the velocity. The present mathematical theory of lubrication is based upon Reynolds's work following the experiments by Tower.† The original differential equation, developed by Reynolds, was used by him to explain Tower's results. The

---

* Beauchamp Tower, First Report on Friction Experiments, *Proc. Inst. Mech. Eng.*, November
1883, pp. 632–666; Second Report, *ibid.*, 1885, pp. 58–70; Third Report, *ibid.*, 1888,
pp. 173–205; Fourth Report, *ibid.*, 1891, pp. 111–140.
† Osborne Reynolds, Theory of Lubrication, Part I, *Phil. Trans. Roy. Soc. London*, 1886.

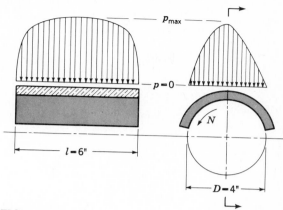

FIGURE 10-8
Approximate pressure-distribution curves obtained by Tower.

solution is a challenging problem which has interested many investigators ever since then, and it is still the starting point for lubrication studies.

Reynolds pictured the lubricant as adhering to both surfaces and being pulled by the moving surface into a narrowing, wedge-shaped space so as to create a fluid or film pressure of sufficient intensity to support the bearing load. One of the important simplifying assumptions resulted from Reynolds's realization that the fluid films were so thin in comparison with the bearing radius that the curvature could be neglected. This enabled him to replace the curved partial bearing with a flat bearing, called a *plane slider bearing*. Other assumptions made are:

1 The lubricant obeys Newton's law of viscous flow.
2 The forces due to the inertia of the lubricant are neglected.
3 The lubricant is assumed to be incompressible.
4 The viscosity is assumed to be constant throughout the film.
5 The pressure does not vary in the axial direction.

Figure 10-9*a* shows a journal rotating in the clockwise direction supported by a film of lubricant of variable thickness h on a partial bearing which is fixed. We specify that the journal has a constant surface velocity U. Using Reynolds's assumption that curvature can be neglected, we fix a right-handed xyz reference system to the stationary bearing. We now make the following additional assumptions:

6 The bearing and journal are assumed to extend infinitely in the z direction; this means there can be no lubricant flow in the z direction.
7 The film pressure is constant in the y direction. Thus the pressure depends only on the coordinate x.
8 The velocity of any particle of lubricant in the film depends only on the coordinates x and y.

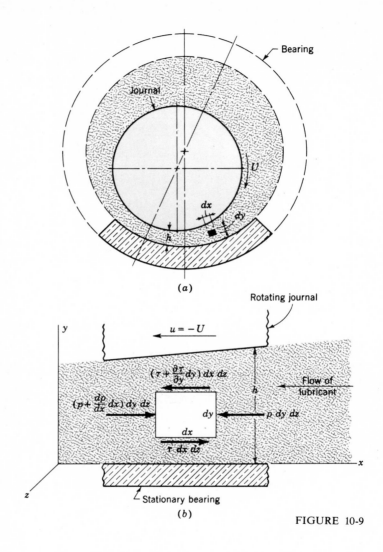

FIGURE 10-9

We now select an element of lubricant in the film (Fig. 10-9a) of dimensions dx, dy, and dz, and compute the forces which act on the sides of this element. As shown in Fig. 10-9b, normal forces, due to the pressure, act upon the right and left sides of the element, and shear forces, due to the viscosity and to the velocity, act upon the top and bottom sides. Summing the forces gives

$$\sum F = \left(p + \frac{dp}{dx}dx\right)dy\,dz + \tau\,dx\,dz - \left(\tau + \frac{\partial \tau}{\partial y}dy\right)dx\,dz - p\,dy\,dz = 0 \qquad (a)$$

This reduces to

$$\frac{dp}{dx} = \frac{\partial \tau}{\partial y} \qquad (b)$$

From Eq. (10-1) we have

$$\tau = \mu \frac{\partial u}{\partial y} \qquad (c)$$

where the partial derivative is used because the velocity u depends upon both x and y. Substituting Eq. (c) into (b), we obtain

$$\frac{dp}{dx} = \mu \frac{\partial^2 u}{\partial y^2} \qquad (d)$$

Holding x constant, we now integrate this expression twice with respect to y. This gives

$$\frac{\partial u}{\partial y} = \frac{1}{\mu} \frac{dp}{dx} y + C_1$$

$$u = \frac{1}{2\mu} \frac{dp}{dx} y^2 + C_1 y + C_2 \qquad (e)$$

Note that the act of holding x constant means that C_1 and C_2 can be functions of x. We now assume that there is no slip between the lubricant and the boundary surfaces. This gives two sets of boundary conditions for evaluating the constants C_1 and C_2:

$$\begin{array}{cc} y = 0 & y = h \\ u = 0 & u = -U \end{array} \qquad (f)$$

Notice, in the second condition, that h is a function of x. Substituting these conditions in Eq. (e) and solving for the constants gives

$$C_1 = -\frac{U}{h} - \frac{h}{2\mu} \frac{dp}{dx} \qquad C_2 = 0$$

or

$$u = \frac{1}{2\mu} \frac{dp}{dx} (y^2 - hy) - \frac{U}{h} y \qquad (10\text{-}7)$$

This equation gives the velocity distribution of the lubricant in the film as a function of the coordinate y and the pressure gradient dp/dx. The equation shows that the velocity distribution across the film (from $y = 0$ to $y = h$) is obtained by superposing a parabolic distribution (the first term) onto a linear distribution (the second term). Figure 10-10 shows the superposition of these two terms to obtain the velocity for particular values of x and dp/dx. In general, the parabolic term

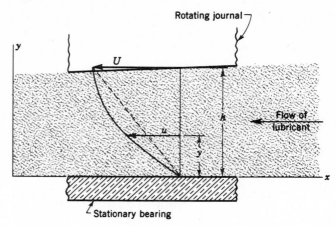

FIGURE 10-10
Velocity of the lubricant.

may be additive or subtractive to the linear term, depending upon the sign of the pressure gradient. When the pressure is maximum, $dp/dx = 0$ and the velocity is

$$u = -\frac{U}{h}y \qquad (g)$$

a linear relation.

We next define Q as the volume of lubricant flowing in the x direction per unit time. By using a width of unity in the z direction, the volume may be obtained by the expression

$$Q = \int_0^h u \, dy \qquad (h)$$

Substituting the value of u from Eq. (10-7) and integrating gives

$$Q = -\frac{Uh}{2} - \frac{h^3}{12\mu}\frac{dp}{dx} \qquad (i)$$

The next step uses the assumption of an incompressible lubricant and states that the flow is the same for any cross section. Thus

$$\frac{dQ}{dx} = 0$$

From Eq. (i),

$$\frac{dQ}{dx} = -\frac{U}{2}\frac{dh}{dx} - \frac{d}{dx}\left(\frac{h^3}{12\mu}\frac{dp}{dx}\right) = 0$$

or

$$\frac{d}{dx}\left(\frac{h^3}{\mu}\frac{dp}{dx}\right) = -6U\frac{dh}{dx} \qquad (10\text{-}8)$$

which is the classical Reynolds equation for one-dimensional flow. It neglects side leakage, that is, flow in the z direction. A similar development is used when side leakage is not neglected. The resulting equation is

$$\frac{\partial}{\partial x}\left(\frac{h^3}{\mu}\frac{\partial p}{\partial x}\right) - \frac{\partial}{\partial z}\left(\frac{h^3}{\mu}\frac{\partial p}{\partial z}\right) = -6U\frac{\partial h}{\partial x} \qquad (10\text{-}9)$$

There is no general solution to Eq. (10-9); approximate solutions have been obtained by using electrical analogies, mathematical summations, relaxation methods, and numerical and graphical methods. One of the important solutions is due to Sommerfeld and may be expressed in the form

$$\frac{r}{c}f = \phi\left[\left(\frac{r}{c}\right)^2\frac{\mu N}{P}\right] \qquad (10\text{-}10)$$

where ϕ indicates a functional relationship. Sommerfeld found the functions for half-bearings and full bearings by using the assumption of no side leakage.

10-7 DESIGN FACTORS

We may distinguish between two groups of variables in the design of sliding bearings. In the first group are those whose values are either given or are under the control of the designer. These are:

1 The viscosity μ
2 The load per unit of projected bearing area, P
3 The speed N
4 The bearing dimensions r, c, β, and l

Of these four variables, the designer usually has no control over the speed, because it is specified by the overall design of the machine. Sometimes the viscosity is specified in advance, as, for example, when the oil is stored in a sump and is used for lubricating and cooling a variety of bearings. The remaining variables, and sometimes the viscosity, may be controlled by the designer and are therefore the *decisions* he or she makes. In other words, when these four variables are defined, the design is complete.

In the second group are the dependent variables. The designer cannot control these except indirectly by changing one or more of the first group. These are:

1 The coefficient of friction f
2 The temperature rise ΔT
3 The flow of oil Q
4 The minimum film thickness h_0

We may regard this group as *design factors*, because it is necessary to set up limitations on their values. These limitations are specified by the characteristics of the bearing materials and the lubricant. The fundamental problem in bearing design, therefore, is to define satisfactory limits for the second group of variables and then to decide upon values for the first group such that these limitations are not exceeded.

10-8 THE RELATION OF THE VARIABLES

Before proceeding to the problem of design, it is necessary to establish the relationships between the variables. A. A. Raimondi and John Boyd, of Westinghouse Research Laboratories, used an iteration technique to solve Reynolds's equation on the digital computer.* This is the first time such extensive data have been available for use by designers, and consequently we shall employ them in this book.†

The Raimondi and Boyd papers were published in three parts and contain 45 detailed charts and 6 tables of numerical information. In all three parts, charts are used to define the variables for length-diameter (l/d) ratios of $1:4$, $1:2$, and 1 and for beta angles of 60 to 360°. Under certain conditions the solution to the Reynolds equation gives negative pressures in the diverging portion of the oil film. Since a lubricant cannot usually support a tensile stress, part III of the Raimondi–Boyd papers assumes that the oil film is ruptured when the film pressure becomes zero. Part III also contains data for the infinitely long bearing; since it has no ends, this means that there is no side leakage. The charts appearing in this book are from part III of the papers, and are for full journal bearings $(\beta = 360°)$ only. Space does not permit the inclusion of charts for partial bearings. This means that you must refer to the charts in the original papers when beta angles of less than 360° are desired. The notation is very nearly the same as in this book, and so no problems should arise.

Figures 10-11 and 10-12 relate the viscosity to the temperature for a number of SAE oils without the necessity for converting to a different set of units.

The *bearing-characteristic number*, or the *Sommerfeld number*, is defined by the equation

$$S = \left(\frac{r}{c}\right)^2 \frac{\mu N}{P} \qquad (10\text{-}11)$$

where S = bearing-characteristic number
r = journal radius, in
c = radial clearance, in
μ = absolute viscosity, reyn
N = relative speed of journal and bearing, rps
P = load per unit of projected bearing area, psi

* A. A. Raimondi and John Boyd, A Solution for the Finite Journal Bearing and Its Application to Analysis and Design, Parts I, II, and III, *Trans. ASLE*, vol. 1, no. 1, pp. 159–209, in "Lubrication Science and Technology," Pergamon Press, New York, 1958.

† A number of other sources of data are available; for a description of these see Fuller, *op. cit.*, pp. 150, 157, 175, 177, 195, and 201. See also the earlier companion paper, John Boyd and Albert A. Raimondi, Applying Bearing Theory to the Analysis and Design of Journal Bearings, Parts I and II, *J. Appl. Mechanics*, vol. 73, pp. 298–316, 1951.

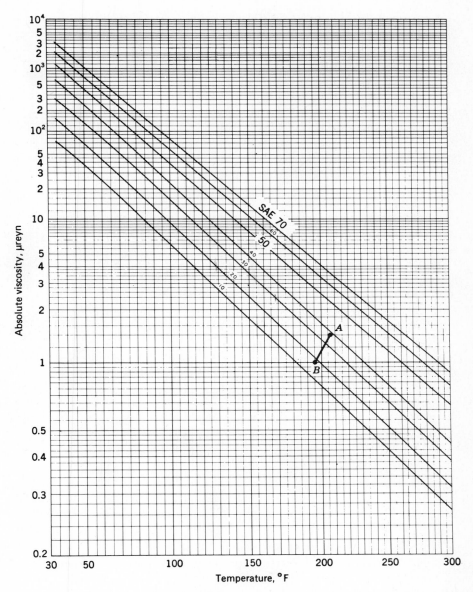

FIGURE 10-11
Viscosity-temperature chart in IPS units. (*Boyd and Raimondi.*)

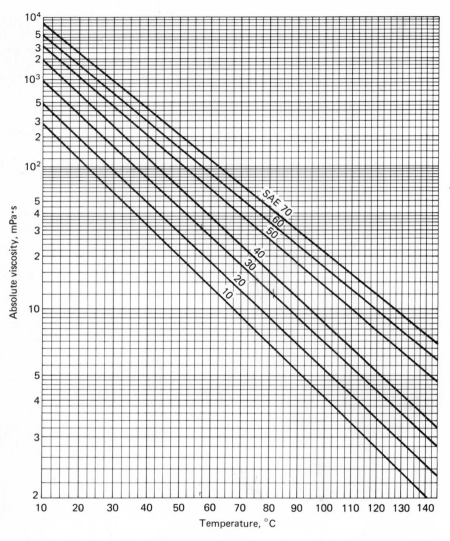

FIGURE 10-12
Viscosity-temperature chart in SI units. (*Adapted from Fig. 10-11.*)

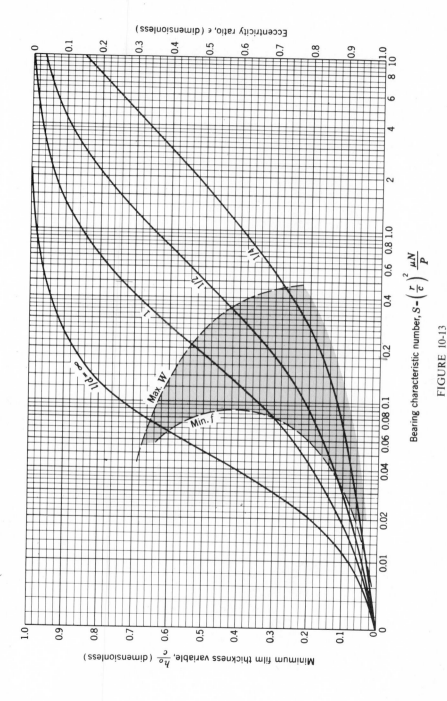

Eccentricity ratio, e (dimensionless)

Minimum film thickness variable, $\dfrac{h_0}{c}$ (dimensionless)

Bearing characteristic number, $S = \left(\dfrac{r}{c}\right)^2 \dfrac{\mu N}{P}$

FIGURE 10-13

Chart for minimum-film-thickness variable and eccentricity ratio. The left boundary of the shaded zone defines the optimum h_0 for minimum friction; the right boundary is the optimum h_0 for maximum load. (*Raimondi and Boyd.*)

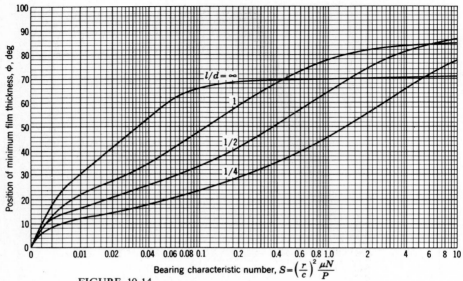

FIGURE 10-14
Chart for determining the position of the minimum film thickness h_0. For location of origin see Fig. 10-20. (*Raimondi and Boyd.*)

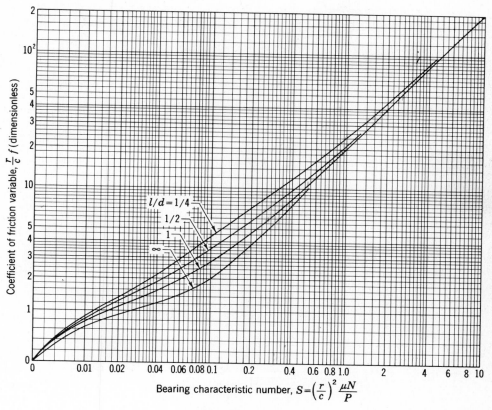

FIGURE 10-15
Chart for coefficient-of-friction variable. (*Raimondi and Boyd.*)

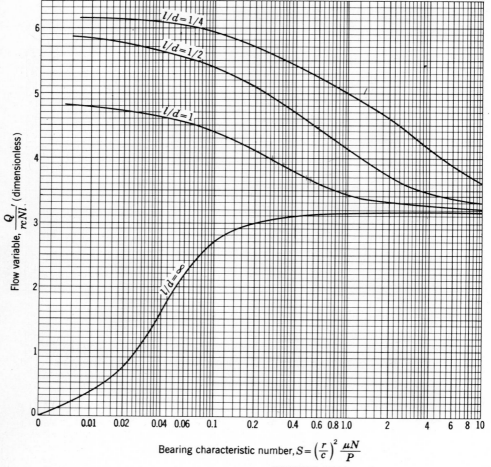

Bearing characteristic number, $S = \left(\dfrac{r}{c}\right)^2 \dfrac{\mu N}{P}$

FIGURE 10-16
Chart for flow variable. (*Raimondi and Boyd.*)

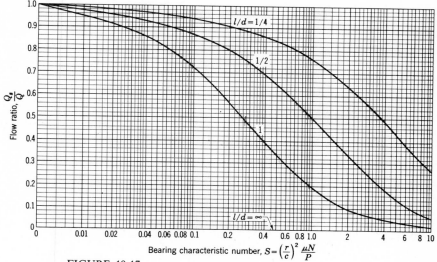

FIGURE 10-17
Chart for determining the ratio of side flow to total flow. (*Raimondi and Boyd.*)

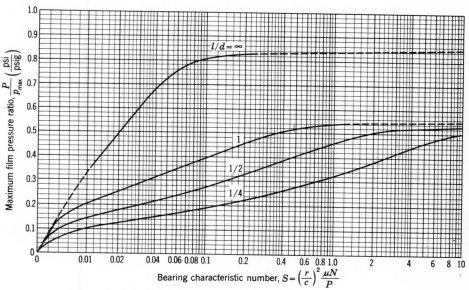

FIGURE 10-18
Chart for determining the maximum film pressure. (*Raimondi and Boyd.*)

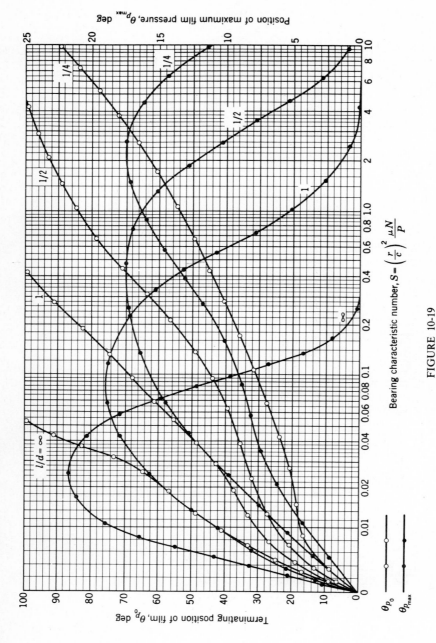

Bearing characteristic number, $S = \left(\dfrac{r}{c}\right)^2 \dfrac{\mu N}{P}$

FIGURE 10-19

Chart for finding the terminating position of the lubricant film and the position of maximum film pressure. (*Raimondi and Boyd.*)

The Sommerfeld number contains all the variables usually specified by the designer, and it is dimensionless, and so it has been used as the abscissa in the charts.

The *friction variable* $(r/c)f$ is plotted against S for various values of the length-diameter ratio l/d in Fig. 10-15. This chart is used as follows:

The following quantities are specified for a full journal bearing:

$\mu = 4$ μreyns
$N = 30$ rps (1800 rpm)
$W = 500$ lb (bearing load)
$r = 0.75$ in
$c = 0.0015$ in
$l = 1.50$ in

The unit load is

$$P = \frac{W}{2rl} = \frac{500}{(2)(0.75)(1.50)} = 222 \text{ psi}$$

From Eq. (10-11), the bearing-characteristic number is

$$S = \left(\frac{r}{c}\right)^2 \frac{\mu N}{P} = \left(\frac{0.75}{0.0015}\right)^2 \frac{(4)(10)^{-6}(30)}{222} = 0.135$$

For this bearing $l/d = 1.50/(2)(0.75) = 1$. Consequently, from Fig. 10-15, the friction variable is $(r/c)f = 3.50$. Therefore the coefficient of friction is

$$f = 3.50 \frac{c}{r} = 3.50\left(\frac{0.0015}{0.75}\right) = 0.007$$

With this known, other things can be learned about the performance. For example, the friction torque is

$$T = fWr = (0.007)(500)(0.75) = 2.62 \text{ lb·in}$$

The power lost in the bearing, in horsepower, is

$$\text{hp} = \frac{TN}{1050} = \frac{(2.62)(30)}{1050} = 0.0748$$

or, expressed in Btu, we have

$$H = \frac{2\pi TN}{(778)(12)} = \frac{2\pi(2.62)(30)}{(778)(12)} = 0.053 \text{ Btu/s}$$

One of the important variables in evaluating bearing performance is the minimum film thickness h_0. Should h_0 be less than a certain safe value, there is danger of metal-to-metal contact during overloads or of the film being so thin that any dirt contained in the oil cannot pass. In addition, the flow of oil depends upon the film thickness; with a small flow, the temperature rise may be excessive.

The *minimum-film-thickness variable* is h_0/c and is found from Fig. 10-13.

Another useful parameter is the eccentricity ratio $\epsilon = e/c$. As shown in Fig. 10-6, if $e = 0$, the bearing is centered and $h_0 = c$; this corresponds to a very light or zero load, and the eccentricity ratio is zero. As the load is increased, the journal is forced downward and the limiting position is reached when $h_0 = 0$ and $e = c$; that is, the journal is touching the bearing. For this condition the eccentricity ratio is unity. Since

$$h_0 = c - e \qquad (10\text{-}12)$$

we have, by dividing both sides by c,

$$\frac{h_0}{c} = 1 - \epsilon \qquad (10\text{-}13)$$

Thus the curve for the minimum-film-thickness variable may also be used for the eccentricity ratio, and this quantity too will be found on the chart of Fig. 10-13.

Design optimums frequently used are maximum load and minimum power loss. Dashed lines for both these conditions have been constructed in Fig. 10-13 so that optimum values of h_0 or ϵ can readily be found. The shaded zone between the boundaries defined by these two optimums may therefore be considered as a recommended operating zone.

From this discussion you should have concluded that lightly loaded bearings operate with a large Sommerfeld number, while heavily loaded bearings will operate at a small number.

Figure 10-14 gives the location of the minimum film thickness as defined by Fig. 10-20.

For the example and corresponding to $l/d = 1$ and $S = 0.135$, we find from Fig. 10-13 $h_0/c = 0.42$ and $\epsilon = 0.58$. So the minimum film thickness is

$$h_0 = 0.42c = (0.42)(0.0015) = 0.000\,63 \text{ in}$$

The *flow variable* $Q/rcNl$, found from the chart of Fig. 10-16, is used to find the amount of lubricant Q which is pumped into the converging space by the rotating journal. This chart is based on atmospheric pressure and no grooving. Thus the flow will be increased if the supply pressure is above atmospheric. The amount of oil supplied to the bearing must equal Q if the bearing is to perform according to the charts. Of the amount of oil Q required by the bearing, an amount Q_s flows out the ends, and hence is called the side leakage. The side leakage can be computed from the *flow ratio* Q_s/Q of Fig. 10-17.

Using the preceding example again, we enter Fig. 10-16 with $S = 0.135$ and $l/d = 1$. This gives $Q/rcNl = 4.28$. Therefore the total flow is

$$Q = 4.28rcNl = (4.28)(0.75)(0.0015)(30)(1.5) = 0.216 \text{ in}^3/\text{s}$$

From Fig. 10-17, $Q_s/Q = 0.655$, and so

$$Q_s = (0.655)(0.216) = 0.142 \text{ in}^3/\text{s}$$

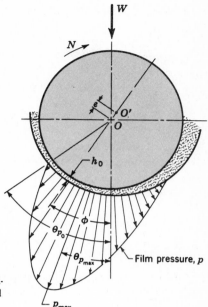

FIGURE 10-20
Polar diagram of film-pressure distribution, showing the notation used (*Raimondi and Boyd.*)

The maximum film pressure developed in the film can be computed from the *pressure ratio* P/p_{max} of Fig. 10-18. These data were obtained on the assumption that the ambient pressure was atmospheric. If the oil is supplied at a greater pressure, all points on the pressure-distribution diagram of Fig. 10-20 will be increased by the same amount.

The location of the maximum film pressure can be found by means of the chart of Fig. 10-19 and the notation of Fig. 10-20. This chart also contains data for locating the point of zero pressure, that is, the terminating point of the hydrodynamic film. Raimondi and Boyd state that neither of these position angles is as accurate as the other data because they were taken from graphs of the pressure distribution along the bearing centerline. Nevertheless, the information should be quite helpful in locating oil grooves.

From Fig. 10-18, the film-pressure ratio is $P/p_{max} = 0.42$. Consequently, $p_{max} = P/0.42 = 222/0.42 = 505$ psi. Then, from Fig. 10-19, we find $\theta_{p_{max}} = 18.5°$, and $\theta_{po} = 75°$.

Since the journal does work on the lubricant, heat is produced, as we have seen. This heat must be dissipated by conduction, convection, and radiation and carried away by the flow of oil. It is very difficult to calculate the rate of heat flow by each method with any accuracy. Later we shall examine this problem in more detail; but for the present we make the assumption that the oil flow carries away all the heat generated. Then, as far as oil temperature is concerned, we shall be on the conservative side.

The Raimondi-Boyd papers contain temperature-rise charts based on assumptions similar to these. Instead of presenting these charts in this book, we present an analytical approach based on information already obtained.

Let us use the following additional notation:

J = mechanical equivalent of heat, 9336 lbf-in per Btu
C_H = specific heat of lubricant, 0.42 Btu per lbf per °F being an average value for use
γ = weight per unit volume of the lubricant, at an average specific gravity of 0.86, $\gamma = (0.86)(62.4)/1728 = 0.0311$ lbf per in^3
ΔT_F = temperature rise, °F
$X = (r/c)f$ = friction variable
$Y = Q/rcNl$ = flow variable

The heat generated is

$$H = \frac{2\pi TN}{J} = \frac{2\pi f\, WrN}{J} \qquad (a)$$

Substituting $(c/r)X$ for f gives

$$H = \left(\frac{2\pi WNc}{J}\right)X \qquad (b)$$

We now assume that an oil flow Q is to carry away all the heat. Then the temperature rise of the oil will be

$$\Delta T_F = \frac{H}{\gamma C_H Q} \qquad (c)$$

If for Q we substitute $(rcNl)Y$, then

$$\Delta T_F = \frac{H}{(\gamma C_H rcNl)Y} \qquad (d)$$

We now multiply the numerator and denominator of Eq. (d) by the unit pressure P, noting that $P = W/2rl$, and substitute the value of H from Eq. (b). After canceling terms, this gives

$$\Delta T_F = \frac{4\pi P}{J\gamma C_H}\frac{X}{Y} \qquad (e)$$

If then we assume average lubrication conditions and substitute the values of J, γ, and C_H, we finally obtain

$$\Delta T_F = 0.103P\,\frac{(r/c)f}{Q/rcNl} \qquad (10\text{-}14)$$

where ΔT is in degrees Fahrenheit. This equation is valid when *all* the oil flow carries away all the heat generated. But some of the oil flows out the side of the bearing before the hydrodynamic film is terminated. If we assume that the temperature of the side flow is the mean of the inlet and the outlet temperatures, the temperature rise of the side flow is $\Delta T_F/2$. This means that the heat generated raises the temperature of the flow $Q - Q_s$ an amount ΔT_F, and the flow Q_s an amount $\Delta T_F/2$. Consequently,

$$\gamma C_H(Q - Q_s)\Delta T_F + \frac{\gamma C_H Q_s \Delta T_F}{2} = H \qquad (f)$$

and so

$$\Delta T_F = \frac{H}{\gamma C_H Q[(1 - \frac{1}{2}(Q_s/Q)]} \qquad (g)$$

Equation (10-14) therefore becomes

$$\Delta T_F = \frac{0.103P}{\left[1 - \frac{1}{2}(Q_s/Q)\right]} \frac{(r/c)f}{Q/rcNl} \qquad (10\text{-}15)$$

In this equation the pressure P is in IPS units and ΔT_F in degrees Fahrenheit. The corresponding equation using SI is

$$\Delta T_C = \frac{8.30P}{\left[1 - \frac{1}{2}(Q_s/Q)\right]} \frac{(r/c)f}{Q/rcNl} \qquad (10\text{-}16)$$

where P is in MPa and ΔT_C in degrees Celsius.

For the example problem, Eq. (10-15) gives a temperature rise of

$$\Delta T_F = \frac{(0.103)(222)}{1 - (0.5)(0.655)} \frac{3.50}{4.28} = 26.6°F$$

Interpolation

According to Raimondi and Boyd, interpolation of the chart data for other l/d ratios can be done by using the equation

$$y = \frac{1}{(l/d)^3}\left[-\frac{1}{8}\left(1 - \frac{l}{d}\right)\left(1 - 2\frac{l}{d}\right)\left(1 - 4\frac{l}{d}\right)y_\infty + \frac{1}{3}\left(1 - 2\frac{l}{d}\right)\left(1 - 4\frac{l}{d}\right)y_1\right.$$
$$\left. -\frac{1}{4}\left(1 - \frac{l}{d}\right)\left(1 - 4\frac{l}{d}\right)y_{1/2} + \frac{1}{24}\left(1 - \frac{l}{d}\right)\left(1 - 2\frac{l}{d}\right)y_{1/4}\right] \qquad (10\text{-}17)$$

where y is the desired variable within the interval $\infty > l/d > \frac{1}{4}$, and y_∞, y_1, $y_{1/2}$, and $y_{1/4}$ are the variables corresponding to l/d ratios of ∞, 1, $\frac{1}{2}$, and $\frac{1}{4}$, respectively.

Assumptions

Often you must use a given analytical approach to solve a problem, knowing in advance that the assumptions used in the analysis do not exactly fit the problem requirements. This is engineering—in fact, it is the "art" of engineering—the employment of judgment and experience in evaluating and altering the results of such an analysis so that it will predict the performance more exactly or will yield an optimum and reliable design. This is the reason why it is so necessary to be familiar with the assumptions used in any analysis.

In the Raimondi-Boyd analysis, some of the assumptions have already been stated and explained. These are:

1 It is assumed that the film ruptures after it has passed the point of minimum film thickness and is in the divergent zone.
2 The flow is based on lubricant supplied to the bearing at atmospheric pressure and on the absence of oil grooves or holes in the bearing.
3 The temperature rise of the lubricant is based on the assumption that all the heat generated raises the temperature of the lubricant.

We have seen that both the temperature and the pressure of the lubricant change as it passes through the bearing. Figure 10-11 shows that viscosity varies with temperature, and it is also affected by pressure. Consequently, another assumption used in this analysis is:

4 The viscosity of the lubricant is constant as it passes through the bearing.

This is a usual assumption; the viscosity used should probably be that which corresponds to the average of the inlet and outlet temperatures. In other words, the equation

$$T_{av} = T_1 + \frac{\Delta T}{2} \qquad (10\text{-}18)$$

where T_1 is the inlet temperature, gives the value of the temperature to be used in finding the viscosity when the lubricant flow is assumed to carry away all the generated heat.

Other assumptions implied by the analysis are that the lubricant is clean and is fluid (not a grease) and that the load is constant and fixed in direction.

10-9 TEMPERATURE AND VISCOSITY CONSIDERATIONS

In a *self-contained bearing* there is no method of circulating or cooling the lubricant; it passes through the bearing, heats up, and is stored in a sump. Heat is removed by convection, conduction, and radiation, and eventually the system reaches an equilibrium temperature.

In a *force-feed lubricating system* cool, clean lubricant is fed to the bearing from an external source.

For most problems it is possible to specify the inlet temperature, but, since the viscosity used in the analysis should correspond to the mean of the inlet and outlet temperatures, this does not yield a value of viscosity to use in the analysis. A solution to this problem, when the lubricant grade is specified, is to assume two trial values of viscosity. One of these should be somewhat lower than expected, the other higher. By using each of these viscosities, the temperature rise is computed and the average temperature determined from Eq. (10-18). When these pairs of results are plotted on Fig. 10-11, a straight line, such as AB, can be drawn between them, and the intersection of this line with the SAE grade of oil gives the correct viscosity to use in the analysis. It should be noted that a series of trial viscosities will yield a curved line instead of a straight line if their values differ considerably; therefore the viscosities chosen should not be too different from one another. The following example illustrates this procedure.

EXAMPLE 10-1   If an SAE 20 oil at an inlet temperature of 100°F is the lubricant for the example of the preceding section, what viscosity should be used in the analysis?

SOLUTION   We have already determined that a viscosity of 4 μreyn gives a temperature rise of 26.6°F. The mean temperature is

$$T_{av} = T_1 + \frac{\Delta T_F}{2} = 100 + \frac{26.6}{2} = 113.3°F$$

This yields one point on the viscosity-temperature chart, and when plotted, it is found to be below the SAE 20 line. Therefore we choose $\mu = 6$ μreyn for the second trial value. Computing the S number gives

$$S = \left(\frac{r}{c}\right)^2 \frac{\mu N}{P} = \left(\frac{0.75}{0.0015}\right)^2 \frac{(6)(10)^{-6}(30)}{222} = 0.202$$

Then, using Figs. 10-15, 10-16, and 10-17 gives $(r/c)f = 4.7$, $Q/rcNl = 4.1$, and $Q_s/Q = 0.56$. Equation (10-15) gives

$$\Delta T_F = \frac{0.103P}{1 - \frac{1}{2}(Q_s/Q)} \frac{(r/c)f}{Q/rcNl} = \frac{(0.103)(222)}{1 - (0.5)(0.56)} \frac{4.7}{4.1} = 36.4°F$$

Therefore the average temperature is

$$T_{av} = 100 + \frac{36.4}{2} = 118.2°F$$

When both pairs of points are plotted on Fig. 10-11 and joined, they cross the SAE 20 line at $\mu = 5.5$ μreyn and $T_{av} = 117°F$. Therefore this is the correct viscosity to use in completing the analysis. The temperature rise is twice 17, or 34°F.

////

10-10   OPTIMIZATION TECHNIQUES

In designing a journal bearing for thick-film lubrication the engineer must select the grade of oil to be used, together with suitable values for P, N, r, c, and l. A poor selection of these or inadequate control of them during manufacture or in use may

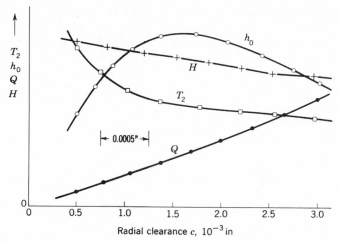

FIGURE 10-21

A plot of some performance characteristics of the bearing of Example 10-1 for radical clearances of 0.0005 to 0.003 in. The bearing outlet temperature is designated T_2. New bearings should be designed for the shaded zone because wear will move the operating point to the right.

result in a film that is too thin, so that the oil flow is insufficient, causing the bearing to overheat and, eventually, fail. Furthermore, the radial clearance c is difficult to hold accurate during manufacture, and it may increase because of wear. What is the effect of an entire range of radial clearances, expected in manufacture, and what will happen to the bearing performance if c increases because of wear? Most of these questions can be answered and the design optimized by plotting curves of the performance as functions of the quantities over which the designer has control.

Figure 10-21 shows the results obtained when the performance of a particular bearing is calculated for a whole range of radial clearances and is plotted with clearance as the independent variable. The bearing used for this graph is the one of Example 10-1, with SAE 20 oil at an inlet temperature of 100°F. The graph shows that if the clearance is too tight, the temperature will be too high and the minimum film thickness too low. High temperatures may cause the bearing to fail by fatigue. If the oil film is too thin, dirt particles may not pass without scoring or may embed themselves in the bearing. In either event, there will be excessive wear and friction, resulting in high temperatures and possible seizing.

It would seem that a large clearance will permit the dirt to pass through and also will permit a large flow of oil. This lowers the temperature and increases the life of the bearing. However, if the clearance becomes too large, the bearing becomes noisy and the minimum film thickness begins to decrease again.

When both the production tolerance and the future wear on the bearing are considered, it is seen, from Fig. 10-21, that the best compromise is a clearance range slightly to the left of the top of the minimum-film-thickness curve. In this way, future wear will move the operating point to the right and increase the film thickness and decrease the operating temperature.

Of course, if the tightest acceptable clearance or the maximum allowable temperature rise is specified, the statistical methods introduced in Chap. 4 may be used to determine the percentage of unacceptable bearings to be expected in a given lot or assembly.

10-11 PRESSURE-FED BEARINGS

When so much heat is generated by hydrodynamic action that the normal lubricant flow is insufficient to carry it away, an additional supply must be furnished under pressure. To force a maximum flow through the bearing and thus obtain the greatest cooling effect, a common practice is to use a circumferential groove at the center of the bearing, with an oil-supply hole located opposite the load-bearing zone. Such a bearing is shown in Fig. 10-22. The effect of the groove is to create two half-bearings, each having a smaller l/d ratio than the original. The groove divides the pressure-distribution curve into two lobes and reduces the minimum film thickness, but it has wide acceptance among lubrication engineers and carries more load without overheating.

To set up a method of solution for oil flow, we shall assume a groove ample enough so that the pressure drop in the groove itself is small. Initially we will neglect eccentricity and then apply a correction factor for this condition. The oil flow, then, is the amount which flows out of the two halves of the bearing in the direction of the concentric shaft. If we neglect the rotation of the shaft, we obtain the force situation shown in Fig. 10-23. Here we designate the supply pressure by p_s and the pressure at any point by p. Laminar flow is assumed, and we are interested in the static equilibrium of an element of width dx, thickness $2y$, and unit depth. Note particularly that the origin of the reference system has been

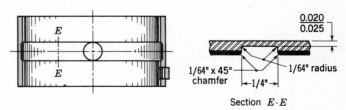

FIGURE 10-22
Centrally located, full annular groove. (*Courtesy of the Cleveland Graphite Bronze Company, Division of Clevite Corporation.*)

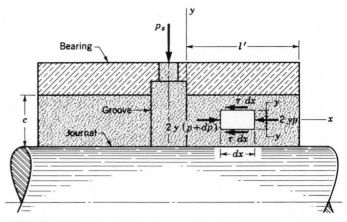

FIGURE 10-23
Flow of lubricant from a pressure-fed bearing having a central groove.

chosen at the midpoint of the clearance space.* The pressure is $p + dp$ on the left face and p on the right face, and the upper and lower surfaces are acted upon by the shear stresses τ. The equilibrium equation is

$$2y(p + dp) - 2yp - 2\tau \, dx = 0 \qquad (a)$$

Expanding and canceling terms, we find that

$$\tau = y\frac{dp}{dx} \qquad (b)$$

Newton's law for viscous flow [Eq. (10-1)] is

$$\tau = \mu\frac{du}{dy}$$

However, in this case we have taken τ in a negative direction. Also, du/dy is negative because u decreases as y increases. We therefore write Newton's law in the form

$$-\tau = \mu\left(-\frac{du}{dy}\right) \qquad (c)$$

Now eliminating τ from Eqs. (b) and (c) gives

$$\frac{du}{dy} = \frac{1}{\mu}\frac{dp}{dx}y \qquad (d)$$

---

* I am grateful to Professor Arthur W. Sear of California State College, Los Angeles, for suggestions concerning this analysis. J.E.S.

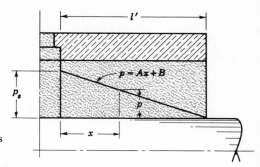

FIGURE 10-24
A linear oil-pressure distribution is assumed.

Treating dp/dx as a constant and integrating with respect to y gives

$$u = \frac{1}{2\mu}\frac{dp}{dx}y^2 + C_1 \qquad (e)$$

At the boundaries, where $y = \pm c/2$, the velocity u is zero. Using one of these conditions in Eq. (e) gives

$$0 = \frac{1}{2\mu}\frac{dp}{dx}\left(\frac{c}{2}\right)^2 + C_1$$

or

$$C_1 = -\frac{c^2}{8\mu}\frac{dp}{dx}$$

Substituting this constant into Eq. (e) yields

$$u = \frac{1}{8\mu}\frac{dp}{dx}(4y^2 - c^2) \qquad (f)$$

Next let us assume that the oil pressure varies linearly from the center to the end of the bearing as shown in Fig. 10-24. Since the equation of a straight line may be written

$$p = Ax + B$$

with $p = p_s$ at $x = 0$ and $p = 0$ at $x = l'$, then substituting these end conditions gives

$$A = -\frac{p_s}{l'} \qquad B = p_s$$

or

$$p = -\frac{p_s}{l'}x + p_s \qquad (g)$$

And therefore

$$\frac{dp}{dx} = -\frac{p_s}{l'} \qquad (h)$$

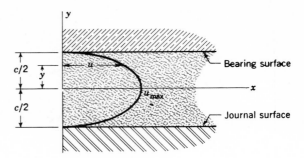

FIGURE 10-25
Parabolic distribution of the lubricant velocity.

We can now substitute Eq. (h) into Eq. (f) to get the relationship between the oil velocity and the coordinate y:

$$u = \frac{p_s}{8\mu l'}(c^2 - 4y^2) \qquad (10\text{-}19)$$

Figure 10-25 shows a graph of this relation fitted into the clearance space c so that you can see how the velocity of the lubricant varies from the journal surface to the bearing surface. The distribution is parabolic, as shown, with the maximum velocity occurring at the center, where $y = 0$. The magnitude is, from Eq. (10-19),

$$u_{max} = \frac{p_s c^2}{8\mu l'} \qquad (i)$$

The average ordinate of a parabola is two-thirds of the maximum, and so the average velocity is

$$u_{av} = \frac{2}{3}\frac{p_s c^2}{8\mu l'} = \frac{p_s c^2}{12\mu l'} \qquad (j)$$

We still have a long way to go in this analysis; so be patient. Now that we have an expression for the lubricant velocity, we can compute the amount of lubricant that flows out of the ends of the bearing. If the journal and bearing are concentric, as shown in Fig. 10-26a, then a close approximation to the area is

$$A = 2\pi rc$$

Since the lubricant flows out of both ends,

$$Q_s = 2u_{av} A = (2)\left(\frac{p_s c^2}{12\mu l'}\right)(2\pi rc) = \frac{\pi p_s c^3 r}{3\mu l'} \qquad (k)$$

in cubic inches per second because μ is in reyns. However, we must modify this relation to account for the fact that the journal is not concentric with the bearing.

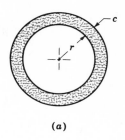

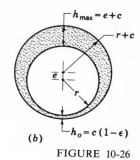

(a)   (b)

FIGURE 10-26

In Fig. 10-26b the minimum film thickness is

$$h_0 = c(1 - \epsilon) \qquad (l)$$

where $\epsilon = e/c$, the eccentricity ratio. The maximum film thickness is

$$h_{\max} = e + c = c(\epsilon + 1) \qquad (m)$$

Since the clearance in Eq. (k) is cubed, if we average the cubes of Eqs. (l) and (m), we get

$$\frac{h_0^3 + h_{\max}^3}{2} = c^3(1 + 3\epsilon^2) \qquad (n)$$

Thus the term $1 + 3\epsilon^2$ should be the correction factor necessary to modify Eq. (k) for nonconcentricity. However, Dennison* gives the eccentricity correction factor as $1 + 1.5\epsilon^2$. Using Dennison as the authority, Eq. (k) becomes

$$Q_s = (1 + 1.5\epsilon^2)\frac{\pi p_s c^3 r}{3\mu l'} \qquad (10\text{-}20)$$

In analyzing the performance of pressure-fed bearings, the bearing length should be taken as l' as defined in Fig. 10-23. Thus the unit load is

$$P = \frac{W/2}{2rl'} = \frac{W}{4rl'} \qquad (10\text{-}21)$$

because each half of the bearing carries half of the load.

The charts of Figs. 10-16 and 10-17 for flow variable and flow ratio, of course, do not apply to pressure-fed bearings. Also, to the maximum film pressure, given by Fig. 10-18, must be added the supply pressure p_s, to obtain the total film pressure.

* E. S. Dennison, Film Lubrication Theory and Engine-bearing Design, *Trans. ASME*, vol. 58, p. 25, 1936.

Since the oil flow has been increased by forced feed, Eq. (10-15) will give a temperature rise that is too high. From Eq. (c) of Sec. 10-8, we have

$$\Delta T_F = \frac{H}{\gamma C_H Q_s} \qquad (o)$$

and the heat loss is

$$H = \frac{2\pi f WrN}{J} \qquad (p)$$

Substituting Eqs. (10-20) and Eq. (p) into Eq. (o) and canceling terms gives

$$\Delta T_F = \frac{6\mu l' f WN}{(1 + 1.5\epsilon^2) J \gamma C_H p_s c^3} \qquad (10\text{-}22)$$

For average lubrication conditions, $\gamma = 0.0311$ lbf per in^3, $C_H = 0.42$ Btu per lbf per °F; also $J = 9336$ lbf·in per Btu, and so Eq. (10-22) can be written

$$\Delta T_F = \frac{0.0492 \mu l' f WN}{(1 + 1.5\epsilon^2) p_s c^3} \qquad (q)$$

Now multiply Eq. (q) by the Sommerfeld number S and divide by

$$S = \left(\frac{r}{c}\right)^2 \frac{\mu N}{P} = \left(\frac{r}{c}\right)^2 \frac{4r l' \mu N}{W} \qquad (r)$$

Upon rearranging the terms, we find

$$\Delta T_F = \frac{0.0123}{1 + 1.5\epsilon^2} \frac{[(r/c)f]SW^2}{p_s r^4} \qquad (10\text{-}23)$$

which is easier to solve than Eq. (q) because the number S must be computed anyway. Equation (10-23) is, of course, in the customary IPS units. The corresponding equation in SI is

$$\Delta T_C = \frac{978(10)^6}{1 + 1.5\epsilon^2} \frac{[(r/c)f]SW^2}{p_s r^4} \qquad (10\text{-}24)$$

where ΔT_C = temperature rise, °C
W = bearing load, kN
p_s = supply pressure, kPa
r = radius, mm

10-12 HEAT BALANCE

The case in which the lubricant carries away all the generated heat has already been discussed. We shall now investigate self-contained bearings, in which the lubricant is stored in the bearing housing itself. These bearings find many applications in industrial machinery; are often described as pedestal, or pillow-block, bearings; and are used on fans, blowers, pumps, motors, and the like. The problem is to balance the heat-dissipation capacity of the housing with the heat generated in the bearing itself.

The heat given up by the bearing housing may be approximated by the equation

$$H = CA(T_H - T_A) \qquad (10\text{-}25)$$

where H = heat dissipated, Btu per h

C = combined coefficient of radiation and convection, Btu per h per ft^2 per °F

A = surface area of housing, ft^2

T_H = surface temperature of housing, °F

T_A = temperature of surrounding air, °F

The coefficient C depends upon the material, color, geometry, and roughness of the housing, the temperature difference between the housing and the surrounding objects, and the temperature and velocity of the air. Equation (10-25) should be used only when ball-park answers are sufficient. Exact results can be obtained by experimentation under actual, not simulated, operating conditions and environment. With these limitations, assume C is a constant having the values

$$C = \begin{cases} 2 \text{ Btu/(h)(ft}^2)(°F) & \text{for still air} \\ 2.7 \text{ Btu/(h)(ft}^2)(°F) & \text{for average design practice} \\ 5.9 \text{ Btu/(h)(ft}^2)(°F) & \text{for air moving at 500 fpm} \end{cases}$$

An expression quite similar to Eq. (10-25) can be written for the temperature difference $T_L - T_H$ between the lubricant film and the housing. Because the type of lubricating system and the quality of the lubricant circulation affect this relationship, the resulting expression is even more approximate than that of Eq. (10-25). An *oil-bath lubrication system* in which a part of the journal is actually immersed in the lubricant would provide good circulation. A *ring-oiled bearing* in which oil rings ride on top of the journal, dip into the oil sump, and hence carry a moderate amount of lubricant into the load-bearing zone would provide satisfactory circulation for many purposes. On the other hand, if the lubricant is supplied by *wick-feeding* methods, the circulation is so inadequate that it is doubtful if any heat at all can be carried away by the lubricant. No matter what type of self-contained lubrication system is used, a great deal of engineering judgment is necessary in computing the heat balance. Based on these limitations, the equation

$$T_L - T_H = n(T_H - T_A) \qquad (a)$$

where T_L is the *average* film temperature and n is a constant which depends upon the lubrication system, may be used to get a rough estimate of the bearing temperature. Table 10-1 provides some guidance in deciding on a suitable value for n.

Since T_L and T_A are usually known, Eqs. (10-25) and (a) can be combined to give

$$H = \frac{CA}{n+1}(T_L - T_A)$$

Table 10-1

Lubrication system	Conditions	Range of n
Oil ring	Moving air	1–2
	Still air	$\frac{1}{2}$–1
Oil bath	Moving air	$\frac{1}{2}$–1
	Still air	$\frac{1}{5}$–$\frac{2}{5}$

In beginning a heat-balance computation, the film temperature, and hence the viscosity of the lubricant, in a self-contained bearing is unknown. Thus, finding the equilibrium temperatures is an iterative procedure which starts with an estimate of the film temperature and ends with verification or nonverification of this estimate. Since the computations are lengthy, a time-shared digital computer should be used to make them if at all possible.

10-13 BEARING DESIGN

A typical sleeve-bearing design situation is depicted in Fig. 10-27. Here a rotating shaft is to be supported by bearings at *A* and *B*. It is evident that some of the decisions have been fixed by other considerations, such as the shaft dimensions, heat treatment, speed, and general geometry. It is also evident that the problem is not completely revealed. What is the purpose of the shaft? What is it that causes the external forces? Is the shaft completely enclosed by a housing or is it in the open? Are the bearings to be self-contained or does the lubricant come from a sump, and is it also used for other purposes? Once the answers to these questions are known, the design process can begin.

The diameter and length of the bearing depend upon the magnitude of the unit load. While the experienced bearing designer will have a rather good idea of a satisfactory range, the beginner needs a starting point. Table 10-2 is presented to indicate the range of unit loads in current use. These values will have to be

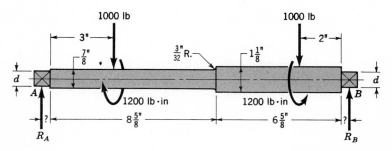

FIGURE 10-27

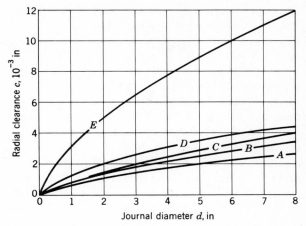

FIGURE 10-28

Recommended radial clearances for cast-bronze bearings. The curves are identified as follows:

A = precision spindles made of hardened ground steel, running on lapped cast-bronze bearings (8- to 16-μin rms finish), with a surface velocity less than 10 fps

B = precision spindles made of hardened ground steel, running on lapped cast-bronze bearings (8- to 16-μin rms finish), with a surface velocity more than 10 fps

C = electric motors, generators, and similar types of machinery using ground journals in broached or reamed cast-bronze bearings (16- to 32-μin rms finish)

D = general machinery which continuously rotates or reciprocates and uses turned or cold-rolled steel journals in bored and reamed cast-bronze bearings (32- to 63-μin rms finish)

E = rough-service machinery having turned or cold-rolled steel journals operating on cast-bronze bearings (65- to 125-μin rms finish)

modified upward or downward, depending upon the severity of the operating conditions, but they can be used to obtain a first-trial value of P. Having settled on a value of the unit load, suitable values for the bearing diameter d and the length l can be selected.

The next problem is the radial clearance. This depends somewhat on the material of the bearing as well as on the finish used and the relative velocity. For preliminary design, the following figures may be used:

Bearing material	Maximum clearance ratio r/c
Lead and tin-base	600–1000
Copper-lead	500–1000
Aluminum	400–500

As an additional guide, the Cast Bronze Bearing Institute (CBBI)* has published a list of recommended radial clearances for full bronze bearings with various degrees of finish. These recommendations, which permit up to 20 percent deviation, are summarized by the graph of Fig. 10-28.

The length-diameter ratio l/d of a bearing depends upon whether it is expected to run under thin-film-lubrication conditions. A long bearing (large l/d ratio) reduces the coefficient of friction and the side flow of oil and therefore is desirable where thin-film or boundary-value lubrication is present. On the other hand, where forced-feed or positive lubrication is present, the l/d ratio should be relatively small. The short bearing results in a greater flow of oil out of the ends, and thus keeps the bearing cooler. Current practice is to use an l/d ratio of about unity, in general, and then to increase this ratio if thin-film lubrication is likely to occur and to decrease it for thick-film lubrication or high temperatures. If shaft deflection is likely to be severe, a short bearing should be used to prevent metal-to-metal contact at the ends of the bearings.

One should always consider the use of a partial bearing if high temperatures are a problem, because relieving the nonload-bearing area of a bearing can very substantially reduce the heat generated.

With all these tentative decisions made, a lubricant can be selected and the hydrodynamic analysis made as already presented. The values of the various performance parameters, if plotted as in Fig. 10-20, for example, will then indicate whether a satisfactory design has been achieved or additional iterations are necessary.

* Harry C. Rippel, "Cast Bronze Bearing Design Manual," 2d ed., pp. 12 and 13, International Copper Research Association, Inc., 825 Third Ave., New York, NY 10022, 1965.

Table 10-2 RANGE OF UNIT LOADS IN CURRENT USE FOR SLEEVE BEARINGS

Application	Unit load	
	psi	MPa
Diesel engines:		
Main bearings	900–1700	6–12
Crankpin	1150–2300	8–15
Wristpin	2000–2300	14–15
Electric motors	120–250	0.8–1.5
Steam turbines	120–250	0.8–1.5
Gear reducers	120–250	0.8–1.5
Automotive engines:		
Main bearings	600–750	4–5
Crankpin	1700–2300	10–15
Air compressors:		
Main bearings	140–280	1–2
Crankpin	280–500	2–4
Centrifugal pumps	100–180	0.6–1.2

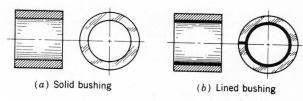

(a) Solid bushing   (b) Lined bushing

FIGURE 10-29
Sleeve bearings.

10-14 BEARING TYPES

A bearing may be as simple as a hole machined into a cast-iron machine member. It may still be simple yet require detailed design procedures, as, for example, the two-piece, grooved, pressure-fed, connecting-rod bearing in an automotive engine. Or it may be as elaborate as the large, water-cooled, ring-oiled bearings with built-in oil reservoirs used on heavy machinery.

Figure 10-29 shows two types of bearings which are often called bushings. The solid bushing is made by casting, by drawing and machining, or by using a powder-metallurgy process. The lined bushing is usually a split type. In one method of manufacture the molten lining material is cast continuously on thin strip steel. The babbitted strip is then processed through presses, shavers, and broaches, resulting in a lined bushing. Any type of grooving may be cut into the bushings. Bushings are assembled as a press fit and finished by boring, reaming, or burnishing.

Flanged and straight two-piece bearings are shown in Fig. 10-30. These are available in many sizes in both thick- and thin-wall types with or without lining material. A locking lug positions the bearing and effectively prevents axial or rotational movement of the bearing in the housing.

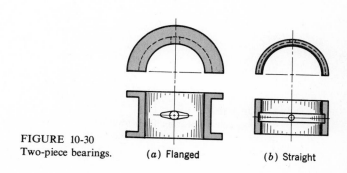

FIGURE 10-30
Two-piece bearings.   (a) Flanged   (b) Straight

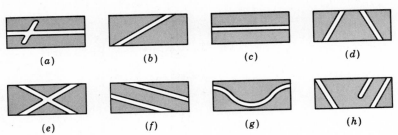

FIGURE 10-31
Developed views of typical groove patterns. (*Courtesy of the Cleveland Graphite Bronze Company, Division of Clevite Corporation.*)

Some typical groove patterns are shown in Fig. 10-31. In general, the lubricant may be brought in from the end of the bushing, through the shaft, or through the bushing. The flow may be intermittent or continuous. The preferred practice is to bring the oil in at the center of the bushing so that it will flow out both ends, thus increasing the flow and cooling action.

10-15 THRUST BEARINGS

This chapter is devoted to the study of the mechanics of lubrication and its application to the design and analysis of journal bearings. The design and analysis of thrust bearings is an important application of lubrication theory too. A detailed study of thrust bearings is not included here because it would not contribute anything significantly different and because of space limitations. Having studied

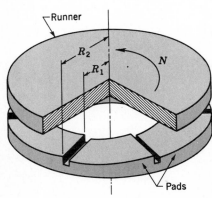

FIGURE 10-32
Fixed-pad thrust bearing. (*Courtesy of the Westinghouse Corporation, Research Laboratories.*)

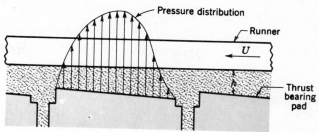

FIGURE 10-33
Pressure distribution of lubricant in a thrust bearing. (*Courtesy of International Copper Research Corporation.*)

this chapter, you should experience no difficulty in reading the literature on thrust bearings and applying that knowledge to actual design situations.*

Figure 10-32 shows a fixed-pad thrust bearing consisting essentially of a runner sliding over a fixed pad. The lubricant is brought into the radial grooves and pumped into the wedge-shaped space by the motion of the runner. Full-film, or hydrodynamic, lubrication is obtained if the speed of the runner is continuous and sufficiently high, if the lubricant has the correct viscosity, and if it is supplied in sufficient quantity. Figure 10-33 provides a picture of the pressure distribution under conditions of full-film lubrication.

We should note that bearings are frequently made with a flange, as shown in Fig. 10-34. The flange positions the bearing in the housing and also takes a thrust load. Even when it is grooved, however, and has adequate lubrication, such an arrangement is not a hydrodynamically lubricated thrust bearing. The reason for this is that the clearance space is not wedge-shaped but has a uniform thickness. Similar reasoning would apply to various designs of thrust washers.

* Harry C. Rippel, "Cast Bronze Thrust Bearing Design Manual," International Copper Research Association, Inc., 825 Third Ave., New York, NY 10022, 1967. CBBI, 14600 Detroit Ave., Cleveland, OH, 44107, 1967.

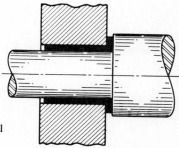

FIGURE 10-34
Flanged sleeve bearing takes both radial and thrust loads.

10-16   BOUNDARY LUBRICATION

When two surfaces slide relative to each other with only a partial lubricant film between them, *boundary lubrication* is said to exist. Boundary- or thin-film lubrication occurs in hydrodynamically lubricated bearings when they are starting or stopping, when the load increases, when the supply of lubricant decreases, or whenever other operating changes happen to occur. There are, of course, a very large number of cases in design in which boundary-lubricated bearings must be used because of the type of application or the competitive situation.

The coefficient of friction for boundary-lubricated surfaces may be greatly decreased by the use of animal or vegetable oils mixed with the mineral oil or grease. Fatty acids, such as stearic acid, palmitic acid, or oleic acid, or several of these, which occur in animal and vegetable fats, are called *oiliness agents*. These acids appear to reduce friction either because of their strong affinity for certain metallic surfaces or because they form a soap film which binds itself to the metallic surfaces by a chemical reaction. Thus the fatty-acid molecules bind themselves to the journal and bearing surfaces with such great strength that the metallic asperities of the rubbing metals do not weld or shear.

Fatty acids will break down at temperatures of 250°F or more, causing increased friction and wear in thin-film-lubricated bearings. In such cases the *extreme-pressure*, or EP, lubricants may be mixed with the fatty-acid lubricant. These are composed of chemicals such as chlorinated esters or tricresyl phosphate, which form an organic film between the rubbing surfaces. Though the EP lubricants make it possible to operate at higher temperatures, there is the added possibility of excessive chemical corrosion of the sliding surfaces.

10-17   BEARING MATERIALS

When we refer to a lubricated assembly, it is necessary to distinguish between the following elements:

1   The shaft, journal, or moving member
2   The lubricant, if it exists
3   The bearing, or fixed member
4   The housing in which the bearing is assembled

The two conflicting requirements of a good bearing material are that it must have a satisfactory compressive and fatigue strength to resist the externally applied loads and that it must be soft and have a low melting point and a low modulus of elasticity. The second set of requirements is necessary to permit the material to wear or break in, since the material can conform to slight irregularities and absorb and release foreign particles. The resistance to wear and the coefficient of friction are also important because all bearings must operate, at least for part of the time, with thin-film lubrication.

Additional considerations in the selection of a good bearing material are its

Table 10-3   COMPOSITION AND CHARACTERISTICS OF BEARING ALLOYS

Alloy name	Thickness, in	SAE number	% Cu	% Sn	% Pb	% Sb	Relative characteristics Load capacity	Relative characteristics Corrosion resistance
Tin-base babbitt	0.022	12	3.25	89		7.5	1.0	Excellent
Lead-base babbitt	0.022	15		1	83	15	1.2	Very good
Tin-base babbitt	0.004	12	3.25	89		7.5	1.5	Excellent
Lead-base babbitt	0.004	15		1	83	15	1.5	Very good
Leaded bronze	Solid	792	80	10	10		3.3	Very good
Copper-lead	0.022	480	65		35		1.9	Good
Aluminum alloy	Solid		1	6			3.0	Excellent
Silver plus overlay	0.013	17P					4.1	Excellent
Cadmium (1.5% Ni)	0.022	18					1.3	Good
Trimetal 88*							4.1	Excellent
Trimetal 77†							4.1	Very good

* This is a 0.008-in layer of copper-lead on a steel back plus 0.001 in of tin-base babbitt.
† This is a 0.013-in layer of copper-lead on a steel back plus 0.001 in of lead-base babbitt.

ability to resist corrosion and, of course, the cost of producing the bearing. Some of the commonly used materials are listed in Table 10-3, together with their composition and characteristics.

Bearing life can be increased very substantially by depositing a layer of babbitt, or other white metal, in thicknesses from 0.001 to 0.014 in over steel backup material. In fact, a copper-lead on steel to provide strength, with a babbitt overlay to provide surface and corrosion characteristics, makes an excellent bearing.

Small bushings and thrust collars are often expected to run with thin-film lubrication. When this is the case, improvements over a solid bearing material can be made to add significantly to the life. A powder-metallurgy bushing is porous and permits the oil to penetrate into the bushing material. Sometimes such a bushing may be enclosed by oil-soaked material to provide additional storage space. Bearings are frequently ball-indented to provide small basins for the storage of lubricant while the journal is at rest. This supplies some lubrication during starting. Another method of reducing friction is to indent the bearing wall and to fill these indentations with graphite.

Some of the materials often used when there is little or no lubrication are listed in Table 10-4, together with temperature, load, and PV limits.

10-18 DESIGN OF BOUNDARY-LUBRICATED BEARINGS

The performance of boundary-lubricated bearings may be analyzed using the formula

$$PV = \frac{k(T_B - T_A)}{f} \qquad (10\text{-}27)$$

Table 10-4 SOME MATERIALS FOR BOUNDARY-LUBRICATED BEARINGS
AND THEIR OPERATING LIMITS*

Material	Maximum load, psi	Maximum temperature, °F	Maximum speed, fpm	Maximum PV P = load, psi V = speed, fpm
Porous bronze	4 500	150	1 500	50 000
Porous iron	8 000	150	800	50 000
Phenolics	6 000	200	2 500	15 000
Nylon	1 000	200	1 000	3 000
Teflon	500	500	100	1 000
Reinforced Teflon	2 500	500	1 000	10 000
Teflon fabric	60 000	500	50	25 000
Delrin	1 000	180	1 000	3 000
Carbon-graphite	600	750	2 500	15 000
Rubber	50	150	4 000	
Wood	2 000	150	2 000	15 000

* E. R. Booser, The Bearings Book, chap. 4, p. 34, *Mach. Design*, June 13, 1963.

where V = surface velocity of journal, fpm
f = coefficient of friction
T_A = ambient air temperature, °F
T_B = bearing bore temperature, °F
k = proportionality constant
There is little or no flow of lubricant in a boundary-lubricated bearing, and so the heat generated must be dissipated by the bearing itself. Thus the constant k depends upon the ability of the bearing to dissipate heat. Table 10-4 lists a maximum PV value of 50 000 psi·fpm and a maximum temperature of 150°F. Using an ambient temperature of 75°F and a minimum coefficient of friction $f = 0.02$ enables us to solve Eq. (10-27), obtaining $k = 13.3$. Thus, for porous bronze bearings Eq. (10-27) becomes

$$PV = \frac{13.3(T_B - T_A)}{f} \qquad (10\text{-}28)$$

Temperature limitations and PV values for other materials are listed in Table 10-4.

The coefficient of friction to be used in Eqs. (10-27) and (10-28) depends upon the degree of lubrication and can be estimated from the following table:

Lubrication	Coefficient of friction f
Mixed-film	0.02–0.08
Thin-film	0.08–0.14
Dry	0.20–0.40

In this table, mixed-film means partial hydrodynamic and boundary lubrication such as might be obtained from wick- or drop-feed lubrication devices. Thin-film is boundary lubrication using grease or fatty-acid lubricants. And dry means the complete absence of a lubricant.

PROBLEMS

Section 10-8

10-1 A full journal bearing is 2 in long and 2 in in diameter. The bearing load is 700 lb, and the journal runs at 1200 rpm. Using a clearance of 0.001 in and an average viscosity of 20 μreyn, find the friction horsepower.

10-2 Purchase a quart of your favorite engine oil, say a multiviscosity grade, and determine the viscosity in your lubrication laboratory according to ASTM standards. Plot the absolute viscosity as a function of temperature on the chart of Fig. 10-11 for later use.

10-3 An 8-in-diameter bearing is 4 in long, has a load of 7500 lb, and runs at 900 rpm. Using a radial clearance of 0.004 in, find the friction horsepower for the following lubricants: SAE 10, 20, 30, and 40. Use an operating temperature of 160°F.

10-4 Repeat Prob. 10-3, but use an SAE 40 lubricant and the following radial clearances: 0.002, 0.003, 0.004, 0.005, and 0.006 in. Plot a curve showing the relation between the coefficient of friction and the clearance.

10-5 A 3-in-diameter bearing has a journal speed of 400 rpm, is 3 in long, and is subjected to a radial load of 600 lb. The bearing is lubricated with SAE 30 oil, which flows into the bearing at a temperature of 160°F. The radial clearance is 0.0014 in. Calculate the heat loss, the side flow, the total flow, the minimum film thickness, and the temperature rise.

10-6 A $1\frac{1}{4} \times 1\frac{1}{4}$-in sleeve bearing supports a load of 700 lb and has a journal speed of 3600 rpm. Using an SAE 10 oil at 160°F operating temperature, specify the radial clearance for an h_o/c value of 0.662.

10-7 A sleeve bearing has a diameter of 75 mm, a journal speed of 7 s^{-1}, and a length of 75 mm. The oil supply is SAE 30 at an inlet temperature of 70°C. The bearing carries a radial load of 2.7 kN and has a radial clearance of 35 μm. Calculate the heat loss, the side flow, the total flow, the minimum film thickness, and the temperature rise.

10-8 A sleeve bearing is 32 mm in diameter and 32 mm long and has a journal speed of 60 s^{-1} The bearing supports a radial load of 3 kN. SAE 10 oil at an average operating temperature of 60°C is used. Determine the radial clearance for an h_o/c value of 0.50.

10-9 A sleeve bearing is $\frac{3}{4}$ in long and $1\frac{1}{2}$ in in diameter. It has a clearance of 0.0015 in and uses SAE 20 lubricant at an operating temperature of 140°F. The bearing supports a load of 350 lb. Calculate the heat generated for speeds of 1000, 2000, 3000, and 4000 rpm, and construct a graph of the results.

Section 10-9

10-10 A bearing $1\frac{1}{2}$ in long and $1\frac{1}{2}$ in in diameter has a r/c ratio of 1000. The journal speed is 1200 rpm, the load is 600 lb, and the lubricant is SAE 40 at an inlet temperature of 100°F.
(a) Find the minimum film thickness and the oil outlet temperature.
(b) Determine the magnitude and location of the maximum film pressure.

10-11 An SAE 60 oil at 80°F inlet temperature is used to lubricate a sleeve bearing 6 in long and 2 in in diameter ($l/d = \infty$). The bearing load is 1800 lb and the journal speed is 160 rpm. Using a clearance ratio $r/c = 600$ find the temperature rise, the maximum film pressure, and the minimum film thickness.

10-12 A sleeve bearing is $\frac{3}{8}$ in in diameter and $\frac{3}{8}$ in long. SAE 10 oil at an inlet temperature of 120°F is used to lubricate the bearing. The radial clearance is 0.0003 in. If the journal speed is 3600 rpm and the radial load on the bearing is 15 lb, find the temperature rise of the lubricant and the minimum film thickness.

10-13 A sleeve bearing is $1\frac{1}{4}$ in in diameter and $1\frac{1}{4}$ in long. The shaft rotates at 1750 rpm and subjects the bearing to a radial load of 250 lb. The clearance is 0.000 75 in. Using SAE 30 oil at an initial temperature of 120°F, find the temperature rise and the minimum film thickness.

10-14 Repeat Prob. 10-13 for SAE 10, 20, and 40 oils, and compare the results. Which lubricant is the best to use?

10-15 A sleeve bearing is 38 mm in diameter and has an l/d ratio of unity. Other data include a clearance ratio of 1000, a radial load of 2.5 kN, and a journal speed of 20 s^{-1}. The bearing is supplied with SAE 40 lubricant at an inlet temperature of 35°C.
(a) Find the average oil temperature.
(b) What is the minimum film thickness?
(c) Find the maximum oil-film pressure.

10-16 A sleeve bearing 60 mm in diameter and 60 mm long is lubricated using SAE 30 oil at an inlet temperature of 40°C. The bearing supports a 4-kN radial load and has a journal speed of 1120 rpm. The radial clearance is 45 μm.
(a) Find both the temperature rise and the average temperature of the lubricant.
(b) Find the coefficient of friction.
(c) Find the magnitude and location of the minimum oil-film thickness.
(d) Find the side flow and the total flow.
(e) Determine the maximum oil-film pressure and its angular location.
(f) Find the terminating position of the oil film.

10-17 An SAE 20 oil is used to lubricate a sleeve bearing 3 in long and 3 in in diameter. The oil enters the bearing at a temperature of 100°F. The journal rotates at 1200 rpm, and the bearing supports a radial load of 1500 lb. The radial clearance is 0.0015 in.
(a) Find the magnitude and location of the minimum oil-film thickness.
(b) Find the coefficient of friction.
(c) Compute the side flow and the total oil flow.

(*d*) Find the maximum oil-film pressure and its location.

(*e*) Find the terminating position of the oil film.

(*f*) Find the average temperature of the oil flowing out the sides of the bearing and the temperature of the oil at the terminating position of the film.

10-18 A journal bearing has a diameter of $2\frac{1}{2}$ in and a length of $1\frac{1}{4}$ in. The journal is to operate at a speed of 1800 rpm and carry a load of 750 lb. If SAE 20 oil at an inlet temperature of 110°F is to be used, what should be the radial clearance for optimum load-carrying capacity?

Section 10-11

10-19 A $1\frac{3}{4}$-in-diameter bearing is 2 in long and has a central annular oil groove $\frac{1}{4}$ in wide which is fed by SAE 10 oil at 120°F and 30 psi supply pressure. The radial clearance is 0.0015 in. The journal rotates at 3000 rpm and the average load is 600 psi of projected area. Find the temperature rise, the minimum film thickness, and the maximum film pressure.

10-20 An eight-cylinder diesel engine has a front main bearing $3\frac{1}{2}$ in in diameter and 2 in long. The bearing has a central annular oil groove 0.250 in wide. It is pressure lubricated with SAE 30 oil at an inlet temperature of 180°F and at a supply pressure of 50 psi. Corresponding to a radial clearance of 0.0025 in, a speed of 2800 rpm, and a radial load of 4600 lb, find the temperature rise and the minimum oil-film thickness.

10-21 A 50-mm-diameter bearing is 55 mm long and has a central annular oil groove 5 mm wide which is fed by SAE 30 oil at 55°C and 200 kPa supply pressure. The radial clearance is 42 μm. The journal speed is 48 s^{-1} corresponding to a bearing load of 10 kN. Find the temperature rise of the lubricant, the total oil flow, and the minimum film thickness.

11

SPUR GEARS

We study gears because the transmission of rotary motion from one shaft to another occurs in nearly every machine one can imagine. Gears constitute one of the best of the various means available for transmitting this motion.

When you realize that the gears in, say, your automotive differential can be made to run 100 000 miles or more before they need to be replaced, and when you count the actual number of meshes or revolutions, you begin to appreciate the fact that the design and manufacture of these gears is really a remarkable accomplishment. People do not generally realize how highly developed the design, engineering, and manufacture of gearing has become because they are such ordinary machine elements. There are a great many lessons to be learned about engineering and design in general through the study of gears because both science and the art of engineering are employed. This is another reason for studying the design and analysis of gears. Maybe the lessons learned can be applied elsewhere.

You will find that this chapter consists essentially of four parts:

1 The kinematics of gear teeth and gear trains. In this part we shall learn something about the shape of the gear tooth itself, together with the problems caused by this shape and what to do about them. We shall also learn about the speed ratio of various kinds of gear trains. Students who have taken courses in mechanisms or kinematics of machines should use this part of the chapter as a review and a storage place for the nomenclature of gearing and pass on to the other parts.

2   The force analysis of gears and gear trains.

3   The design, that is, determination of the size, of gears based on the strength of the materials used.

4   The design of gears based on wear considerations.

Even though gears represent a high level of engineering achievement, design methods have been changing rapidly in recent years because, possibly, of the computer age. Committees made up of authorities in the field of gearing are constantly revising and changing the design codes. Our intent here is not to present a standardized approach which might become obsolete in a few years, but instead, to prepare you in the fundamentals, using the design background you have by now acquired. You will be able to read, understand, and utilize the existing design codes. It is much more important here to present the subject in such a manner as to enable you to assist in the development of the design codes of the future.

11-1   NOMENCLATURE

Spur gears are used to transmit rotary motion between parallel shafts; they are usually cylindrical in shape, and the teeth are straight and parallel to the axis of rotation.

The terminology of gear teeth is illustrated in Fig. 11-1, and most of the following definitions are shown:

The *pitch circle* is a theoretical circle upon which all calculations are usually based. The pitch circles of a pair of mating gears are tangent to each other.

A *pinion* is the smaller of two mating gears. The larger is often called the *gear*.

The *circular pitch p* is the distance, measured on the pitch circle, from a point on one tooth to a corresponding point on an adjacent tooth. Thus the circular pitch is equal to the sum of the *tooth thickness* and the *width of space*.

The *module m* is the ratio of the pitch diameter to the number of teeth. The customary unit of length used is the millimetre. The module is the index of tooth size in SI.

The *diametral pitch P* is the ratio of the number of teeth on the gear to the pitch diameter. Thus, it is the reciprocal of the module. Since diametral pitch is only used with English units, it is expressed as teeth per inch.

The *addendum a* is the radial distance between the *top land* and the pitch circle.

The *dedendum b* is the radial distance from the *bottom land* to the pitch circle.

The *whole depth* h_t is the sum of the addendum and dedendum.

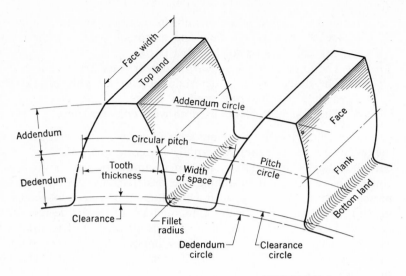

FIGURE 11-1
Nomenclature of gear teeth.

The *clearance circle* is a circle that is tangent to the addendum circle of the mating gear.

The *clearance c* is the amount by which the dedendum in a given gear exceeds the addendum of its mating gear.

The *backlash* is the amount by which the width of a tooth space exceeds the thickness of the engaging tooth measured on the pitch circles.

You should prove for yourself the validity of the following useful relations:

$$P = \frac{N}{d} \qquad (11\text{-}1)$$

where P = diametral pitch, teeth per inch
N = number of teeth
d = pitch diameter

$$m = \frac{d}{N} \qquad (11\text{-}2)$$

where m = module, mm
d = pitch diameter, mm

$$p = \frac{\pi d}{N} = \pi m \qquad (11\text{-}3)$$

where p = circular pitch

$$pP = \pi \qquad (11\text{-}4)$$

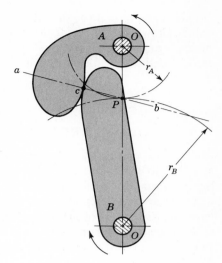

FIGURE 11-2

11-2 CONJUGATE ACTION

The following discussion assumes the teeth to be perfectly formed, perfectly smooth, and absolutely rigid. Such an assumption is, of course, unrealistic, because of the limitations of the machinery used to form the teeth and because the application of forces will cause deflections.

Mating gear teeth acting against each other to produce rotary motion are similar to cams. When the tooth profiles, or cams, are designed so as to produce a constant angular-velocity ratio during meshing, they are said to have *conjugate action*. In theory, at least, it is possible arbitrarily to select any profile for one tooth and then to find a profile for the meshing tooth which will give conjugate action. One of these solutions is the *involute* profile, which, with few exceptions, is in universal use for gear teeth and is the only one with which we shall be concerned.

When one curved surface pushes against another (Fig. 11-2), the point of contact occurs where the two surfaces are tangent to each other (point *c*), and the forces at any instant are directed along the common normal *ab* to the two curves. The line *ab*, representing the direction of action of the forces, is called the *line of action*. The line of action will intersect the line of centers *O-O* at some point *P*. The angular-velocity ratio between the two arms is inversely proportional to their radii to the point *P*. Circles drawn through point *P* from each center are called *pitch circles*, and the radius of each circle is called the *pitch radius*. Point *P* is called the *pitch point*.

To transmit motion at a constant angular-velocity ratio, the pitch point must remain fixed; that is, all the lines of action for every instantaneous point of contact must pass through the same point P. In the case of the involute profile it will be shown that all points of contact occur on the same straight line ab, that all normals to the tooth profiles at the point of contact coincide with the line ab, and thus, that these profiles transmit uniform rotary motion.

11-3 INVOLUTE PROPERTIES

An involute curve may be generated as shown in Fig. 11-3a. A partial flange B is attached to the cylinder A, around which is wrapped a cord def which is held tightly. Point b on the cord represents the tracing point, and as the cord is wrapped and unwrapped about the cylinder, point b will trace out the involute curve ac. The radius of curvature of the involute varies continuously, being zero at point a and a maximum at point c. At point b the radius is equal to the distance be, since point b is instantaneously rotating about point e. Thus the generating line de is normal to the involute at all points of intersection and, at the same time, is always tangent to the cylinder A. The circle on which the involute is generated is called the *base circle*.

Let us now examine the involute profile to see how it satisfies the requirement for the transmission of uniform motion. In Fig. 11-3b two gear blanks with fixed centers at O_1 and O_2 are shown having base circles whose respective radii are $O_1 a$ and $O_2 b$. We now imagine that a cord is wound clockwise around the base circle of gear 1, pulled tightly between points a and b, and wound counterclock-

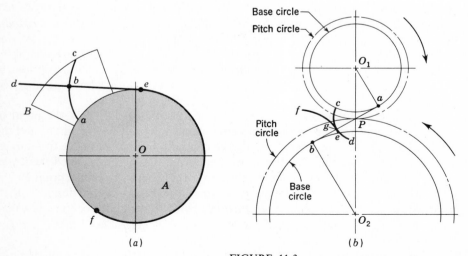

FIGURE 11-3
(a) Generation of an involute; (b) involute action.

wise around the base circle of gear 2. If, now, the base circles are rotated in different directions so as to keep the cord tight, a point g on the cord will trace out the involutes cd on gear 1 and ef on gear 2. The involutes are thus generated simultaneously by the tracing point. The tracing point, therefore, represents the point of contact, while the portion of the cord ab is the generating line. The point of contact moves along the generating line, the generating line does not change position because it is always tangent to the base circles, and since the generating line is always normal to the involutes at the point of contact, the requirement for uniform motion is satisfied.

11-4 FUNDAMENTALS

Among other things, it is necessary that you actually be able to draw the teeth on a pair of meshing gears. You should understand, however, that it is not necessary to draw the gear teeth for manufacturing or shop purposes. Rather, we make drawings of gear teeth to obtain an understanding of the problems involved in the meshing of the mating teeth.

First, it is necessary to learn how to construct an involute curve. As shown in Fig. 11-4, divide the base circle into a number of equal parts and construct radial lines OA_0, OA_1, OA_2, etc. Beginning at A_1, construct perpendiculars $A_1 B_1$, $A_2 B_2$, $A_3 B_3$, etc. Then along $A_1 B_1$ lay off the distance $A_1 A_0$, along $A_2 B_2$ lay off twice the distance $A_1 A_0$, etc., producing points through which the involute curve can be constructed.

To investigate the fundamentals of tooth action let us proceed step by step through the process of constructing the teeth on a pair of gears.

When two gears are in mesh, their pitch circles roll on one another without slipping. Designate the pitch radii as r_1 and r_2 and the angular velocities as ω_1 and ω_2, respectively. Then the pitch-line velocity is

$$V = r_1 \omega_1 = r_2 \omega_2$$

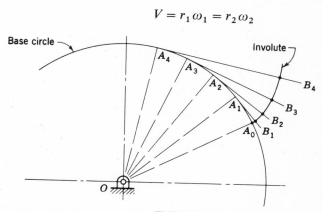

FIGURE 11-4
Construction of an involute curve.

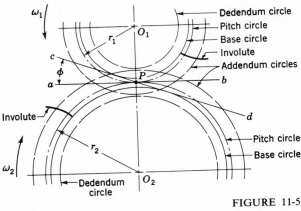

FIGURE 11-5
Gear layout.

Thus the relation between the radii and the angular velocities is

$$\frac{\omega_1}{\omega_2} = \frac{r_2}{r_1} \qquad (11\text{-}5)$$

Suppose now we wish to design a speed reducer such that the input speed is 1800 rpm and the output speed is 1200 rpm. This is a ratio of 3 : 2, and the pitch diameters would be in the same ratio, for example, a 4-in pinion driving a 6-in gear. The various dimensions found in gearing are always based on the pitch circles.

We next specify that an 18-tooth pinion is to mesh with a 30-tooth gear and that the diametral pitch of the gearset is to be 2 teeth per inch. Then, from Eq. (11-1) the pitch diameters of the pinion and gear are, respectively,

$$d_1 = \frac{N_1}{P} = \frac{18}{2} = 9 \text{ in} \qquad d_2 = \frac{N_2}{P} = \frac{30}{2} = 15 \text{ in}$$

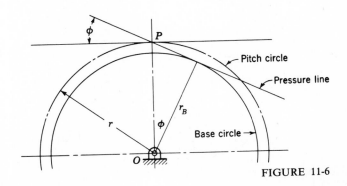

FIGURE 11-6

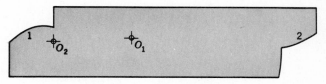

FIGURE 11-7
Template for drawing gear teeth.

The first step in drawing teeth on a pair of mating gears is shown in Fig. 11-5. The center distance is the sum of the pitch radii, in this case 12 in. So locate the pinion and gear centers O_1 and O_2, 12 in apart. Then construct the pitch circles of radii r_1 and r_2. These are tangent at P, the *pitch point*. Next draw line *ab*, the common tangent, through the pitch point. We now designate gear 1 as the driver, and since it is rotating counterclockwise, we draw a line *cd* through point P at an angle ϕ to the common tangent *ab*. The line *cd* has three names, all of which are in general use. It is called the *pressure line*, the *generating line*, and the *line of action*. It represents the direction in which the resultant force acts between the gears. The angle ϕ is called the *pressure angle*, and it usually has values of 20 or 25°, though $14\frac{1}{2}°$ was once used.

Next, on each gear draw a circle tangent to the pressure line. These circles are the *base circles*. Since they are tangent to the pressure line, the pressure angle determines their size. As shown in Fig. 11-6, the radius of the base circle is

$$r_b = r \cos \phi \qquad (11\text{-}6)$$

where r is the pitch radius.

Now generate an involute on each base circle as previously described and as shown in Fig. 11-5. This involute is to be used for one side of a gear tooth. It is not necessary to draw another curve in the reverse direction for the other side of the tooth because we are going to use a template which can be turned over to obtain the other side.

The addendum and dedendum distances for standard interchangeable teeth are, as we shall learn later, $1/P$ and $1.25/P$, respectively. Therefore, for the pair of gears we are constructing,

$$a = \frac{1}{P} = \frac{1}{2} = 0.500 \text{ in} \qquad b = \frac{1.25}{P} = \frac{1.25}{2} = 0.625 \text{ in}$$

Using these distances, draw the addendum and dedendum circles on the pinion and on the gear as shown in Fig. 11-5.

Next, using heavy drawing paper, or preferably, a sheet of 0.015- to 0.020-in clear plastic, cut a template for each involute, being careful to locate the gear centers properly with respect to each involute. Figure 11-7 is a reproduction of the template used to create some of the illustrations for this book. Note that only one

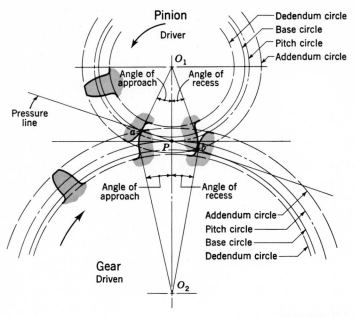

FIGURE 11-8
Tooth action.

side of the tooth profile is formed on the template. To get the other side, turn the template over. For some problems you might wish to construct a template for the entire tooth.

To draw a tooth we must know the tooth thickness. From Eq. (11-4) the circular pitch is

$$p = \frac{\pi}{P} = \frac{\pi}{2} = 1.57 \text{ in}$$

Therefore the tooth thickness is

$$t = \frac{p}{2} = \frac{1.57}{2} = 0.785 \text{ in}$$

measured on the pitch circle. Using this distance for the tooth thickness as well as the tooth space, draw as many teeth as are desired, using the template, after the points have been marked on the pitch circle. In Fig. 11-8 only one tooth has been drawn on each gear. You may run into trouble in drawing these teeth if one of the base circles happens to be larger than the dedendum circle. The reason for this is that the involute begins at the base circle and is undefined below this circle. So, in drawing gear teeth, we usually draw a radial line for the profile below the base

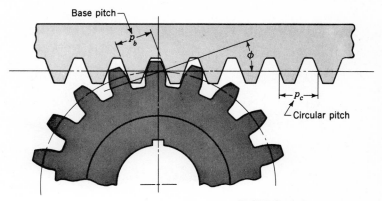

FIGURE 11-9
Involute pinion and rack.

circle. The actual shape, however, will depend upon the kind of machine tool used to form the teeth in manufacture, that is, how the profile is generated.

The portion of the tooth between the clearance circle and the dedendum circle is the fillet. In this instance the clearance is

$$c = b - a = 0.625 - 0.500 = 0.125 \text{ in}$$

The construction is finished when these fillets have been drawn.

Referring again to Fig. 11-8, the pinion with center at O_1 is the driver and turns counterclockwise. The pressure, or generating, line is the same as the cord used in Fig. 11-3a to generate the involute, and contact occurs along this line. The initial contact will take place when the flank of the driver comes into contact with the tip of the driven tooth. This occurs at point a in Fig. 11-8, where the addendum circle of the driven gear crosses the pressure line. If we now construct tooth profiles through point a and draw radial lines from the intersections of these profiles with the pitch circles to the gear centers, we obtain the *angles of approach.*

As the teeth go into mesh, the point of contact will slide up the side of the driving tooth so that the tip of the driver will be in contact just before contact ends. The final point of contact will therefore be where the addendum circle of the driver crosses the pressure line. This is point b in Fig. 11-8. By drawing another set of tooth profiles through b, we obtain the *angles of recess* for each gear in a manner similar to that of finding the angles of approach. The sum of the angle of approach and the angle of recess for either gear is called the *angle of action.* The line ab is called the *line of action.*

We may imagine a *rack* as a spur gear having an infinitely large pitch diameter. Therefore the rack has an infinite number of teeth and a base circle which is an infinite distance from the pitch point. The sides of involute teeth on a rack are straight lines making an angle to the line of centers equal to the pressure angle. Figure 11-9 shows an involute rack in mesh with a pinion.

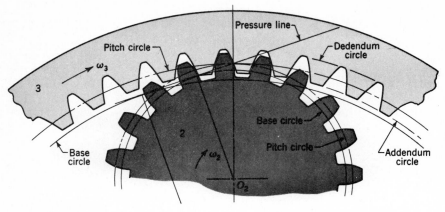

FIGURE 11-10
Internal gear and pinion.

Corresponding sides of involute teeth are parallel curves; the *base pitch* is the constant and fundamental distance between them along a common normal as shown in Fig. 11-9. The base pitch is related to the circular pitch by the equation

$$p_b = p_c \cos \phi \qquad (11\text{-}7)$$

where p_b is the base pitch.

Figure 11-10 shows a pinion in mesh with an *internal*, or *annular*, gear. Note that both of the gears now have their centers of rotation on the same side of the pitch point. Thus the positions of the addendum and dedendum circles with respect to the pitch circle are reversed; the addendum circle of the internal gear lies *inside* the pitch circle. Note, too, from Fig. 11-10, that the base circle of the internal gear lies inside the pitch circle near the addendum circle.

Another interesting observation concerns the fact that the operating diameters of the pitch circles of a pair of meshing gears need not be the same as the respective design pitch diameters of the gears, though this is the way they have been constructed in Fig. 11-8. If we increase the center distance we create two new operating pitch circles having larger diameters because they must be tangent to each other at the pitch point. Thus the pitch circles of gears really do not come into existence until a pair of gears is brought into mesh.

Changing the center distance has no effect on the base circles because these were used to generate the tooth profiles. Thus the base circle is basic to a gear. Increasing the center distance increases the pressure angle and decreases the length of the line of action but the teeth are still conjugate, the requirement for uniform motion transmission is still satisfied, and the angular-velocity ratio has not changed.

EXAMPLE 11-1 A gearset consists of a 16-tooth pinion driving a 40-tooth gear. The diametral pitch is 2, and the addendum and dedendum are $1/P$ and $1.25/P$, respectively. The gears are cut using a pressure angle of 20°.

(a) Compute the circular pitch, the center distance, and the radii of the base circles.

(b) In mounting these gears, the center distance was incorrectly made $\frac{1}{4}$ in larger. Compute the new values of the pressure angle and the pitch-circle diameters.

SOLUTION

a.
$$p = \frac{\pi}{P} = \frac{\pi}{2} = 1.57 \text{ in} \qquad Ans.$$

The pitch diameters of the pinion and gear are, respectively,

$$d_P = \tfrac{16}{2} = 8 \text{ in} \qquad d_G = \tfrac{40}{2} = 20 \text{ in}$$

Therefore the center distance is

$$\frac{d_P + d_G}{2} = \frac{8 + 20}{2} = 14 \text{ in} \qquad Ans.$$

Since the teeth were cut on the 20° pressure angle, the base-circle radii are found to be, using $r_b = r \cos \phi$,

$$r_b(\text{pinion}) = \tfrac{8}{2} \cos 20° = 3.76 \text{ in} \qquad Ans.$$

$$r_b(\text{gear}) = \tfrac{20}{2} \cos 20° = 9.40 \text{ in} \qquad Ans.$$

b. Designating d'_P and d'_G as the new pitch-circle diameters, the increase of $\frac{1}{4}$ in in the center distance requires that

$$\frac{d'_P + d'_G}{2} = 14.250 \qquad (1)$$

Also, the velocity ratio does not change, and hence

$$\frac{d'_P}{d'_G} = \frac{16}{40} \qquad (2)$$

Solving Eqs. (1) and (2) simultaneously yields

$$d'_P = 8.143 \text{ in} \qquad d'_G = 20.357 \text{ in} \qquad Ans.$$

Since $r_b = r \cos \phi$, the new pressure angle is

$$\phi' = \cos^{-1} \frac{r_b(\text{pinion})}{d'_P/2} = \cos^{-1} \frac{3.76}{8.143/2} = 22.56° \qquad Ans.$$

////

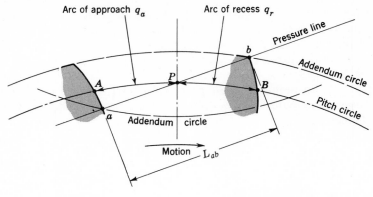

FIGURE 11-11
Definition of contact ratio.

11-5   CONTACT RATIO

The zone of action of meshing gear teeth is shown in Fig. 11-11. We recall that tooth contact begins and ends at the intersections of the two addendum circles with the pressure line. In Fig. 11-11 initial contact occurs at *a* and final contact at *b*. Tooth profiles drawn through these points intersect the pitch circle at *A* and *B*, respectively. As shown, the distance *AP* is called the *arc of approach* q_a, and the distance *PB*, the *arc of recess* q_r. The sum of these is the *arc of action* q_t.

Now, consider a situation in which the arc of action is exactly equal to the circular pitch, that is, $q_t = p$. This means that one tooth and its space will occupy the entire arc *AB*. In other words, when a tooth is just beginning contact at *a*, the previous tooth is simultaneously ending its contact at *b*. Therefore, during the tooth action from *a* to *b*, there will be exactly one pair of teeth in contact.

Next, consider a situation in which the arc of action is greater than the circular pitch, but not very much greater, say, $q_t \approx 1.2p$. This means that when one pair of teeth is just entering contact at *a*, another pair, already in contact, will not yet have reached *b*. Thus, for a short period of time, there will be two pairs of teeth in contact, one near the vicinity of *A* and another near *B*. As the meshing proceeds, the pair near *B* must cease contact, leaving only a single pair of contacting teeth, until the procedure repeats itself.

Because of the nature of this tooth action, either one or two pairs of teeth in contact, it is convenient to define the term *contact ratio* m_c as

$$m_c = \frac{q_t}{p} \qquad (11\text{-}8)$$

a number which indicates the average number of pairs of teeth in contact. Note that this ratio is also equal to the length of the path of contact divided by the base pitch. Gears should not generally be designed having contact ratios less than about 1.20 because inaccuracies in mounting might reduce the contact ratio even more, increasing the possibility of impact between the teeth as well as an increase in the noise level.

An easier way to obtain the contact ratio is to measure the line of action *ab* instead of the arc distance *AB*. Since *ab* in Fig. 11-11 is tangent to the base circle when extended, the base pitch p_b must be used to calculate m_c instead of the circular pitch as in Eq. (11-8). Designating the length of the line of action as L_{ab}, the contact ratio is

$$m_c = \frac{L_{ab}}{p \cos \phi} \qquad (11\text{-}9)$$

in which Eq. (11-7) was used for the base pitch.

11-6   INTERFERENCE

The contact of portions of tooth profiles which are not conjugate is called *interference*. Consider Fig. 11-12. Illustrated are two 16-tooth gears which have been cut using the now obsolete $14\frac{1}{2}°$ pressure angle. The driver, gear 2, turns clockwise. The initial and final points of contact are designated *A* and *B*, respectively, and are located on the pressure line. Now notice that the points of tangency of the pressure line with the base circles *C* and *D* are located *inside* of points *A* and *B*. Interference is present.

The interference is explained as follows. Contact begins when the tip of the driven tooth contacts the flank of the driving tooth. In this case the flank of the driving tooth first makes contact with the driven tooth at point *A*, and this occurs *before* the involute portion of the driving tooth comes within range. In other words, contact is occurring below the base circle of gear 2 on the *noninvolute* portion of the flank. The actual effect is that the involute tip or face of the driven gear tends to dig out the noninvolute flank of the driver.

In this example the same effect occurs again as the teeth leave contact. Contact should end at point *D* or before. Since it does not end until point *B*, the effect is for the tip of the driving tooth to dig out, or interfere with, the flank of the driven tooth.

When gear teeth are produced by a generation process, interference is automatically eliminated because the cutting tool removes the interfering portion of the flank. This effect is called *undercutting*; if undercutting is at all pronounced, the undercut tooth is considerably weakened. Thus the effect of eliminating interference by a generation process is merely to substitute another problem for the original one.

The importance of the problem of teeth which have been weakened by undercutting cannot be overemphasized. Of course, interference can be eliminated by using more teeth on the gears. However, if the gears are to transmit a given

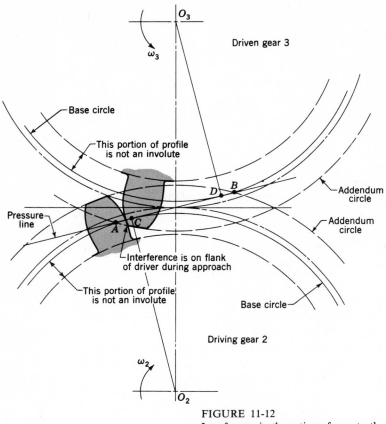

FIGURE 11-12
Interference in the action of gear teeth.

amount of power, more teeth can be used only by increasing the pitch diameter. This makes the gears larger, which is seldom desirable, and it also increases the pitch-line velocity. This increased pitch-line velocity makes the gears noisier and reduces the power transmission somewhat, although not in direct ratio. In general, however, the use of more teeth to eliminate interference or undercutting is seldom an acceptable solution.

Interference can also be reduced by using a larger pressure angle. This results in a smaller base circle, so that more of the tooth profile becomes involute. The demand for smaller pinions with fewer teeth thus favors the use of a 25° pressure angle even though the frictional forces and bearing loads are increased and the contact ratio decreased.

11-7 THE FORMING OF GEAR TEETH

There are a large number of ways of forming the teeth of gears, such as *sand casting, shell molding, investment casting, permanent-mold casting, die casting,* and *centrifugal casting.* They can be formed by using the *powder-metallurgy process*; or, by using *extrusion*, a single bar of aluminum may be formed and then sliced into gears. Gears which carry large loads in comparison with their size are usually made of steel and are cut with either *form cutters* or *generating cutters.* In form cutting, the tooth space takes the exact shape of the cutter. In generating, a tool having a shape different from the tooth profile is moved relative to the gear blank so as to obtain the proper tooth shape. One of the newest and most promising of the methods of forming teeth is called *cold forming*, or *cold rolling*, in which dies are rolled against steel blanks to form the teeth. The mechanical properties of the metal are greatly improved by the rolling process, and a high-quality generated profile is obtained at the same time.

Gear teeth may be shaped by milling, shaping, or hobbing. They may be finished by shaving, burnishing, grinding, or lapping.

Milling

Gear teeth may be cut with a form milling cutter shaped to conform to the tooth space. With this method it is theoretically necessary to use a different cutter for each gear, because a gear having 25 teeth, for example, will have a different-shaped tooth space than one having, say, 24 teeth. Actually, the change in space is not too great, and it has been found that eight cutters may be used to cut with reasonable accuracy any gear in the range of 12 teeth to a rack. A separate set of cutters is, of course, required for each pitch.

Shaping

Teeth may be generated with either a pinion cutter or a rack cutter. The pinion cutter (Fig. 11-13) reciprocates along the vertical axis and is slowly fed into the gear blank to the required depth. When the pitch circles are tangent, both the cutter and blank rotate slightly after each cutting stroke. Since each tooth of the cutter is a cutting tool, the teeth are all cut after the blank has completed one revolution.

The sides of an involute rack tooth are straight. For this reason, a rack-generating tool provides an accurate method of cutting gear teeth. This is also a shaping operation and is illustrated by the drawing of Fig. 11-14. In operation, the cutter reciprocates and is first fed into the gear blank until the pitch circles are tangent. Then, after each cutting stroke, the gear blank and cutter roll slightly on their pitch circles. When the blank and cutter have rolled a distance equal to the circular pitch, the cutter is returned to the starting point, and the process is continued until all the teeth have been cut.

FIGURE 11-13
Generating a spur gear with a pinion cutter. (*Courtesy of Boston Gear Works, Inc.*)

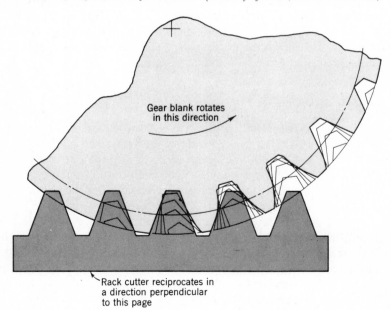

Gear blank rotates
in this direction

Rack cutter reciprocates in
a direction perpendicular
to this page

FIGURE 11-14
Shaping teeth with a rack cutter.

FIGURE 11-15
Hobbing a worm gear. (*Courtesy of Boston Gear Works, Inc.*).

Hobbing

The hobbing process is illustrated in Fig. 11-15. The hob is simply a cutting tool which is shaped like a worm. The teeth have straight sides, as in a rack, but the hob axis must be turned through the lead angle in order to cut spur-gear teeth. For this reason, the teeth generated by a hob have a slightly different shape than those generated by a rack cutter. Both the hob and the blank must be rotated at the proper angular-velocity ratio. The hob is then fed slowly across the face of the blank until all the teeth have been cut.

Finishing

Gears which run at high speeds and transmit large forces may be subjected to additional dynamic forces due to errors in tooth profiles. These errors may be diminished somewhat by finishing the tooth profiles. The teeth may be finished, after cutting, either by shaving or burnishing. Several shaving machines are available which cut off a minute amount of metal, bringing the accuracy of the tooth profile within the limits of 250 μin.

Burnishing, like shaving, is used with gears which have been cut but not heat-treated. In burnishing, hardened gears with slightly oversize teeth are run in mesh with the gear until the surfaces become smooth.

Grinding and lapping are used for hardened gear teeth after heat-treatment. The grinding operation employs the generating principle and produces very accurate teeth. In lapping, the teeth of the gear and lap slide axially so that the whole surface of the teeth is abraded equally.

11-8 TOOTH SYSTEMS

A *tooth system* is a *standard** which specifies the relationships involving addendum, dedendum, working depth, tooth thickness, and pressure angle, to attain interchangeability of gears of all tooth numbers but of the same pressure angle and pitch. You should be aware of the advantages and disadvantages of the various systems so that you can choose the optimum tooth for a given design and have a basis of comparison when departing from a standard tooth profile.

Table 11-1 lists the tooth proportions for completely interchangeable English-system gears and for operation on standard center distances. No standard has been established in this country for tooth systems based wholly upon the use of SI units. In fact it is probable that a number of years will elapse before agreement can be reached; the problems to be solved are complex as well as expensive. Even in England, where they have been ahead of this country in the metric changeover, the inch system is still predominantly used for gearing. Merritt† states that among the reasons is that new standards had been approved and adopted shortly before metrication began.

The addenda listed in Table 11-1 are for gears having tooth numbers equal to or greater than the minimum numbers listed, and for these numbers there will be no undercutting. For fewer numbers of teeth a modification called the *long and short addendum system* should be used.‡ In this system the addendum of the gear is decreased just sufficiently so that contact does not begin before the interference point (point *a* in Fig. 11-11). The addendum of the pinion is then increased a

---

* Standardized by the American Gear Manufacturers Association (AGMA) and the American National Standards Institute (ANSI). The AGMA Standards may be quoted or extracted in their entirety provided an appropriate credit line is included, for example, "Extracted from AGMA Information Sheet—Strength of Spur, Helical, Herringbone, and Bevel Gear Teeth (AGMA 225.01), with permission of the publisher, the American Gear Manufacturers Association, 1330 Massachusetts Avenue, N.W., Washington, DC 20005." These standards have been used extensively in this chapter and in the chapter which follows. In each case the information sheet number is given. Table 11-1 is from AGMA publication 201.02 and 201.02A, but see also 207.04. Write the AGMA for a complete list of standards because changes and additions are made from time to time.

† H. E. Merritt, "Gear Engineering," John Wiley & Sons, New York, 1971.

‡ For an explanation of these modifications and others, see Joseph E. Shigley, "Kinematic Analysis of Mechanisms," 2nd ed., pp. 268–278, McGraw-Hill Book Company, New York, 1969.

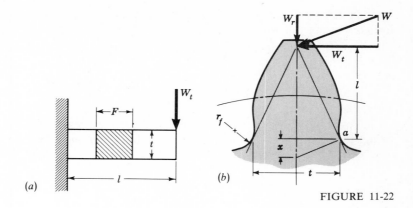

FIGURE 11-22

distributed across the distance F. The section modulus is $I/c = Ft^2/6$, and therefore the bending stress is

$$\sigma = \frac{M}{I/c} = \frac{6W_t l}{Ft^2} \qquad (a)$$

Referring now to Fig. 11-22b, we assume that the maximum stress in a gear tooth occurs at point a. By similar triangles, one can write

$$\frac{t/2}{x} = \frac{l}{t/2} \qquad \text{or} \qquad x = \frac{t^2}{4l} \qquad (b)$$

By rearranging Eq. (a),

$$\sigma = \frac{6W_t l}{Ft^2} = \frac{W_t}{F} \frac{1}{t^2/6l} = \frac{W_t}{F} \frac{1}{t^2/4l} \frac{1}{\frac{4}{6}} \qquad (c)$$

If we now substitute the value of x from Eq. (b) into Eq. (c) and multiply the numerator and denominator by the circular pitch p, we find

$$\sigma = \frac{W_t p}{F(\frac{2}{3})xp} \qquad (d)$$

Letting $y = 2x/3p$, we have

$$\sigma = \frac{W_t}{Fpy} \qquad (11\text{-}17)$$

This completes the development of the original Lewis equation. The factor y is called the *Lewis form factor*, and it may be obtained by a graphical layout of the gear tooth or by digital computation.

In using this equation, most engineers prefer to employ the diametral pitch in determining the stresses. This is done by substituting $P = \pi/p$ and $Y = \pi y$ in Eq. (11-17). This gives

$$\sigma = \frac{W_t P}{F Y} \qquad (11\text{-}18)$$

Values of the form factor Y are given in Table 11-3.

Equation (11-18) can be used to get a quick estimate of gear size by substituting the strength of the material divided by a suitable factor of safety for the

Table 11-3 VALUES OF THE FORM FACTOR Y FOR VARIOUS TOOTH SYSTEMS

Number of teeth	$14\frac{1}{2}°$ composite and involute (obsolete)	$20°$ full depth	Small pinions, $20°$ full depth	$20°$ stub	Internal drives, $20°$ full depth	
					Pinion	Gear
5	—	—	0.320	—	0.322	
6	—	—	0.301	—	0.322	
7	—	—	0.282	—	0.322	
8	—	—	0.264	—	0.324	
9	—	—	0.264	—	0.324	
10	—	—	0.264	—	0.324	
11	—	—	0.264	—	0.326	
12	0.211	0.245	0.264	0.312	0.326	
13	0.223	0.261	0.270	0.324	0.326	
14	0.236	0.277	0.277	0.340	0.330	
15	0.245	0.290	—	0.350	0.330	
16	0.254	0.296	—	0.362	0.333	
17	0.264	0.303	—	0.368	0.342	
18	0.270	0.309	—	0.378	0.348	
19	0.277	0.314	—	0.388	0.358	
20	0.283	0.322	—	0.394	0.364	
21	0.289	0.328	—	0.400	0.370	
22	0.292	0.331	—	0.406	0.374	
24	0.299	0.337	—	0.416	0.383	
26	0.308	0.346	—	0.425	0.393	
28	0.314	0.353	—	0.432	0.399	0.691
30	0.318	0.359	—	0.438	0.405	0.678
34	0.327	0.371	—	0.447	0.414	0.659
38	0.333	0.384	—	0.457	0.424	0.643
43	0.340	0.397	—	0.463	0.430	0.628
50	0.346	0.409	—	0.476	0.436	0.612
60	0.355	0.422	—	0.485	0.446	0.596
75	0.361	0.435	—	0.497	0.452	0.581
100	0.367	0.447	—	0.507	0.461	0.565
150	0.374	0.460	—	0.520	0.468	0.549
300	0.383	0.472	—	0.535	0.477	0.533
Rack	0.390	0.485	—	0.552		

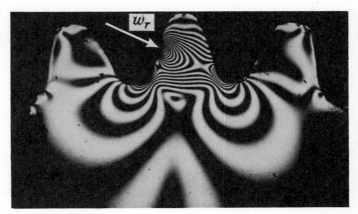

FIGURE 11-23
Stress distribution in a gear tooth determined by photoelastic procedures.

bending stress σ. It should not be used for final design purposes, however, for, as will be shown in later sections, considerable refinement is necessary in order to make the equation yield reliable, high-performance gears.

Assumptions

1   The Lewis equation is solved by using the tangential component of the load. If the radial component is considered, this would produce a uniform compressive stress to which must be added the bending stress. The effect of the radial component, therefore, is to increase the compression and decrease the tension. This is shown clearly by the photograph of Fig. 11-23, in which the stress on the compression side is seen to be larger.

2   The greatest stress is assumed to occur when the load is at the tip of the tooth. If gears are cut with sufficient accuracy, the tip-load condition is not the worst, because another pair of teeth will be in contact when this condition occurs. Examination of run-in teeth will show that the heaviest loads occur near the middle of the tooth. Therefore the maximum stress probably occurs while a single pair of teeth is carrying the full load, at a point where another pair of teeth is nearly ready to come into contact. If size and weight are important, this condition should be assumed.

3   The tangential load W_t is assumed to be uniformly distributed across the full face of the gear. However, the gears and their supporting shafts are made of elastic materials which deflect under the application of loads. Therefore we can expect to find deflection of the gear teeth, torsional deflection of the gear blank, and bending deflection of the supporting shaft. The effect of these deformations is to cause a nonuniform distribution of the load. When the ratio of the face width to the circular pitch (F/p) is large, say, over 6, these deformations should probably be considered.

4 The effects of stress concentration are neglected. Stress-concentration factors were not used in Lewis's day, but recent investigations indicate the advisability of doing so.

11-12 ESTIMATING GEAR SIZE

In order to analyze a gearset to determine the reliability corresponding to a specified life or to determine the factor of safety guarding against a failure, it is necessary that we know the size of the gears and the materials of which they are made. In this section we are concerned mostly with getting a preliminary estimate of the size of gears required to carry a given load. The results will then be useful as a starting point for a more sophisticated analysis. Alternately, the method can be used to get a quick or rough estimate of the strength of a gear in resisting bending.

We first modify the Lewis equation [Eq. (11-18)] by including a velocity factor K_v in the denominator. This gives the bending stress in the tooth as

$$\sigma = \frac{W_t P}{K_v F Y} \qquad (11\text{-}19)$$

The purpose of the velocity factor is to account for the fact that the instantaneous force acting between meshing teeth is somewhat greater than the transmitted load W_t due to inaccuracies in the tooth profiles, and the dynamic effects due to the elasticities of the teeth and shafting during operation. Many designers use the Barth equation for K_v when estimating gear sizes. This equation is

$$K_v = \frac{600}{600 + V} \qquad (11\text{-}20)$$

where V is the pitch-line velocity in feet per minute.

In solving Eq. (11-19) the form factor Y is selected for the pinion if both gears are of the same material. If the gear is of a weaker material than the pinion, then the equation must be solved twice, once for the pinion and once for the gear.

For estimating gear sizes use a factor of safety between 3 and 5 based on the yield strength for ordinary gear problems. Larger factors of safety should be used if shock or vibration is present.

Generally the face width F should be from 3 to 5 times the circular pitch.

Unless kinematic requirements dictate otherwise, always use the minimum number of teeth listed in Table 11-1. This will give the smallest gearset and avoids interference or undercutting of the teeth.

EXAMPLE 11-4   A pair of 4 : 1 reduction gears is desired for a 100-hp 1120-rpm motor. The gears are to be 20° full depth and made of UNS G10400 steel heat-treated and drawn to 1000°F. Make a preliminary estimate of the size of gears required assuming that the starting torque is no more than the full-load torque at rated speed.

SOLUTION   From Table 11-1 we find the minimum number of pinion teeth to avoid undercutting is 18. Thus we shall choose a 72-tooth gear to go with the 18-tooth pinion to give a 4 : 1 reduction.

Equation (11-19) cannot be solved directly for the gear size because both W_t and K_v depend upon the pitch P. Thus it is necessary to try various values of P until a satisfactory result is obtained. For the first series of determinations let us use $P = 4$ teeth per inch. Then the pinion diameter is $d_2 = N_2/P = 18/4 = 4.5$ in. The pitch-line velocity is

$$V = \frac{\pi dn}{12} = \frac{\pi(4.5)(1120)}{12} = 1320 \text{ fpm}$$

Then, from Eq. (11-15), we have the transmitted load as

$$W_t = \frac{33(10)^3 \text{ hp}}{V} = \frac{33(10)^3(100)}{1320} = 2500 \text{ lb}$$

Next, find $Y = 0.309$ from Table 11-3. Then, using Eq. (11-20), we find the velocity factor to be

$$K_v = \frac{600}{600 + V} = \frac{600}{600 + 1320} = 0.312$$

From Table A-17 we find $S_y = 84$ kpsi. Choosing a factor of safety of 4, the design stress to be used is

$$\sigma = \frac{S_y}{n} = \frac{84}{4} = 21 \text{ kpsi}$$

where n is the factor of safety. We now use Eq. (11-19) with $\sigma = 21(10)^3$ psi, $W_t = 2500$ lb, $P = 4$ teeth/in, $K_v = 0.312$, and $Y = 0.309$, to get

$$21(10)^3 = \frac{2500(4)}{0.312(0.309)F}$$

Solving for the face width F gives

$$F = \frac{2500(4)}{21(10)^3(0.312)(0.309)} = 4.94 \text{ in}$$

To see whether this is a good design we first solve for the circular pitch p. This is $p = \pi/P = \pi/4 = 0.785$ in.   Then $3p = 2.36$ in   and $5p = 3.92$ in.   Since the computed face width is not between $3p$ and $5p$, we deem it not very satisfactory.

For a second series of determinations suppose we try $P = 3$ teeth/in. Using the same procedure we find the following:

$d_2 = 6$ in
$V = 1760$ fpm
$W_t = 1875$ lb
$K_v = 0.254$

We again substitute these values into Eq. (11-19). The result is $F = 2.28$ in. But in this case $3p = 3.15$ in, and so a 3-pitch gear is not satisfactory either.

A solution to this problem can be obtained in one of two ways. A 3-pitch gear can be used with a weaker material, or a 4-pitch gear can be used with a stronger material. Since the 4-pitch gear is smaller, this solution is preferred. Though the higher-strength material may cost more, a small gear is cheaper to machine. Furthermore the smaller gears will allow for a smaller housing which is a further saving.

By going to a UNS G10500 steel heat-treated and drawn to 900°F, we find $S_y = 130$ kpsi in Table A-17. Based upon this strength, the design stress is

$$\sigma = \frac{S_y}{n} = \frac{130}{4} = 32.5 \text{ kpsi}$$

When this design stress is used with the other determinations for the 4-pitch gear, the Lewis equation yields $F = 3.19$ in for the face width and this is within the imposed limits of $3p$ and $5p$.

Note that UNS G10400 steel could also be heat-treated to about this same strength. ////

11-13 TOOTH FATIGUE STRESS

A second modification can be made to the Lewis equation by exchanging the form factor Y for a geometry factor J. The equation is then written as

$$\sigma = \frac{W_t P}{K_v F J} \quad (11\text{-}21)$$

We now wish to investigate the factors K_v and J in more detail so as to cause Eq. (11-21) to yield much more accurate results than are possible with the relations of the preceding section.

Dynamic Loading

The velocity factor K_v, also called the *dynamic factor* by AGMA* is intended to account for

1 The effect of tooth spacing and profile errors
2 The effect of pitch-line speed and rpm
3 The inertia and stiffness of all rotating elements
4 The transmitted load per inch of face width
5 The tooth stiffness

---

* AGMA Information Sheet for Strength of Spur, Helical, Herringbone, and Bevel Gear Teeth, AGMA 225.01, American Gear Manufacturers Association, Washington, D.C.

For spur gears whose teeth are finished by hobbing or shaping, AGMA recommends the formula

$$K_v = \frac{50}{50 + \sqrt{V}} \quad (11\text{-}22)$$

But use Eq. (11-20) for inaccurate teeth, such as gears having milled teeth.

If the gears have high-precision shaved or ground teeth and if an appreciable dynamic load is developed, then the AGMA dynamic factor is

$$K_v = \sqrt{\frac{78}{78 + \sqrt{V}}} \quad (11\text{-}23)$$

In both of these equations V is the pitch-line velocity in feet per minute.

If the gears have high-precision shaved or ground teeth and there is no appreciable dynamic load, then the AGMA recommends the dynamic factor $K_v = 1$. Thus, if the design involves high-accuracy gears, then you must decide whether there is an appreciable dynamic load or not. To do this, examine the driving and driven machinery. If the gears are between a motor and a fan, it is doubtful if much of a dynamic load will be developed. On the other hand, one would expect a considerable dynamic load if the gears were between, say, a one-cylinder engine and the blade of a chain saw.

Stress Concentration

A photoelastic investigation by Dolan and Broghamer conducted over 30 years ago still constitutes the primary source of information on stress concentration.[*] For 20° involute helical- and spur-gear teeth, they established the formula for geometric stress-concentration factor as

$$K_t = 0.18 + \left(\frac{t}{r_f}\right)^{0.15} \left(\frac{t}{l}\right)^{0.45} \quad (11\text{-}24)$$

and for 25° involute teeth,

$$K_t = 0.14 + \left(\frac{t}{r_f}\right)^{0.11} \left(\frac{t}{l}\right)^{0.50} \quad (11\text{-}25)$$

where the meaning of the quantities t, r_f, and l is shown in Fig. 11-22b. These quantities may be obtained by a graphical layout of the tooth outline or by digital computation.

In most cases of gear design, $K_f \approx K_t$ because of the high notch sensitivity q of the gear materials. And as observed in Chap. 5, if there is any doubt whether to use K_t or K_f in analysis, one can always err on the safe side by using K_t.

---

[*] T. J. Dolan and E. I. Broghamer, A Photoelastic Study of the Stresses in Gear Tooth Fillets, *Univ. Ill. Eng. Expt. Sta. Bull.* 335, March 1942.

We shall discover in our investigation of the contact-ratio effect that values of K_f have already been worked out for $20°$ full-depth teeth. No matter how unpleasant the task may be, it is necessary to solve Eq. (11-24) or (11-25) when other tooth forms must be investigated. However, there is a less accurate method of getting a quick estimate of K_f. Begin by assuming that the tooth thickness at the base is half the circular pitch, that is,

$$t \approx \frac{p}{2} \qquad (a)$$

Then, from Table 11-1, note that the standard fillet radius is

$$r_f = \frac{0.300}{P} = \frac{0.300}{\pi/p} = 0.0956p \qquad (b)$$

so that, from Eqs. (a) and (b),

$$\frac{r_f}{t} = \frac{0.0956p}{0.5p} = 0.191$$

We now refer to Table A-25-6 and find $K_t \approx 1.53$ corresponding to $r/d = 0.191$ and $D/d = 3$. Also, using $q = 0.95$ from Fig. 5-19, we find

$$K_f = 1 + q(K_t - 1) = 1 + 0.95(1.53 - 1) = 1.50$$

which represents a safe value of K_f to use when tabulated values are not available and when you cannot afford to use Eq. (11-24) or (11-25).

Geometry factor

The shape of the tooth, the point of application of the critical load, stress concentration, and the method by which the load is shared by meshing pairs of teeth are all accounted for by the use of a geometry factor.

Figure 11-24 shows a pair of pitch circles in contact at P. The pinion or driver rotates clockwise, driving the gear counterclockwise about O_3. Contact will begin at point A where the flank of the pinion contacts the tip of the gear tooth. If the contact ratio is greater than unity, then another pair of teeth will already be in contact at some point such as D. As rotation proceeds, the pair in contact at D will move to B where they leave contact while the pair at A move to C. Thus, from C to D only a single pair of teeth is in contact.

Point C is called the *lowest point of single-tooth contact* on the pinion. Point D is called the *highest point of single-tooth contact*.

If the teeth are formed with the highest precision and with a high-quality surface finish, then it is quite likely that the two pairs of meshing teeth will share the load during the period that both pairs are in contact. This means that the critical load will be exerted on a pinion tooth at D, the highest point of single-tooth contact.

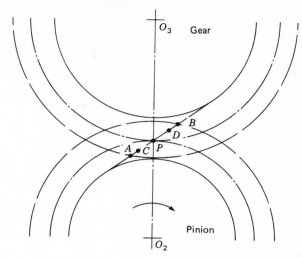

FIGURE 11-24

If the teeth are not of the highest precision, then it is not likely that the teeth will share the load at all. This means that the greatest load will be exerted on the tip of the pinion tooth at *B*.

The AGMA has defined a method of determining whether teeth share the load or not. This method is based on the accuracy with which the teeth are cut as well as the probable tooth deflection during rotation.* However, the method requires a detailed knowledge of all machining or finishing errors as well as considerable judgment. Thus, in this book we shall usually assume tip loading unless it is definitely known that the gears are made with the highest precision.

Since stress concentration depends upon the tooth geometry, the AGMA *geometry factor J* is defined as

$$J = \frac{Y}{K_f} \qquad (11\text{-}26)$$

where *Y* is similar to the Lewis form factor but also includes the bending and compressive components of the load. The AGMA sheet includes a procedure for determining this factor.† They have used this procedure to determine the data for the charts of Fig. 11-25 which will be satisfactory for most tooth forms.

---

* AGMA Sheet 225.01
† A programmable desk calculator or computer can also be used to obtain *J*. See K. R. Gitchel, Computed Strength and Durability Geometry Factors for External Spur and Helical Gears with Tooling Check, *ASME Paper* No. 72-PTG-18, 1972.

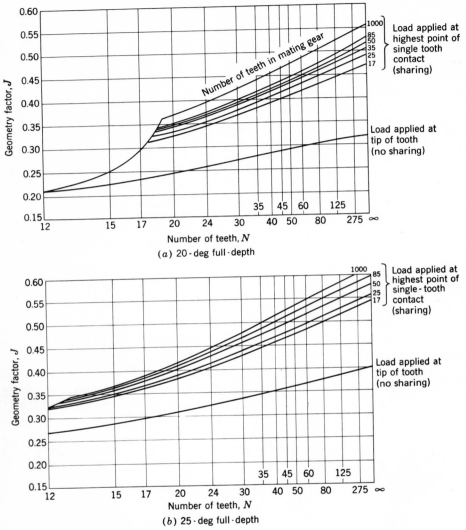

FIGURE 11-25
Spur-gear geometry factors J based on $r_f = 0.35/P$. The abscissa is the number of teeth on the particular gear whose geometry factor is desired. (*AGMA Information Sheet 225.01*.)

For unusual tooth forms, the estimate $K_f = 1.5$ together with a Y factor from Table 11-3 can be used when time is not available to compute accurate values. The approximate geometry factor is then

$$J = \frac{Y}{K_f} = \frac{Y}{1.5} = 0.667Y \qquad (11\text{-}27)$$

EXAMPLE 11-5   In Example 11-4 a set of reduction gears were designed. Based on a 100-hp motor running at 1120 rpm, a 4 : 1 reduction was obtained by selecting an 18-T pinion to mate with a 72-T gear. A diametral pitch of 4 teeth/in was selected as the probable optimum and, based on this size, the pitch-line velocity was found to be 1320 fpm and the transmitted load was $W_t = 2500$ lb. If a face width $F = 3\frac{1}{4}$ in is selected, what is the fatigue stress using the methods of Sec. 11-13?

SOLUTION   Equation (11-22) gives the velocity factor as

$$K_v = \frac{50}{50 + \sqrt{V}} = \frac{50}{50 + \sqrt{1320}} = 0.579$$

Next, using the curve for tip loading in Fig. 11-25a we find $J = 0.23$. Substituting these and the given values into Eq. (11-21) gives the fatigue stress as

$$\sigma = \frac{W_t P}{K_v F J} = \frac{2500(4)}{(0.579)(3.25)(0.23)} = 23.1(10)^3 \text{ psi} \qquad Ans.$$

////

11-14 BENDING STRENGTH

Having obtained the stress using the modified Lewis equation, we will wish to compare it with the strength of the tooth to determine if an adequate factor of safety exists.

If the static strength of the tooth is desired, use S_y if the material has a yield strength; otherwise use S_{ut}.

Certain simplifications are available in computing the endurance limits of steels for gears, and so we repeat Eq. (5-30) here as a convenience.

$$S_e = k_a k_b k_c k_d k_e k_f S_e' \qquad (11-28)$$

where S_e = endurance limit of the gear tooth, psi
   S_e' = endurance limit of the rotating-beam specimen, psi
   k_a = surface factor
   k_b = size factor
   k_c = reliability factor
   k_d = temperature factor
   k_e = modifying factor for stress concentration
   k_f = miscellaneous-effects factor

Surface Finish

The surface factor k_a should always correspond to a machined finish even when the flank of the tooth is ground or shaved. The reason for this is that the bottom land is usually not ground, probably because of the weakening effects. For convenience, a chart of the corresponding surface factors from Fig. 5-17 is included here as Fig. 11-26.

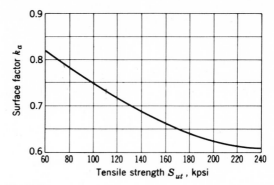

FIGURE 11-26
Surface-finish factors k_a for cut, shaved, and ground gear teeth.

Size

You will recall that the size factor k_b depends upon whether the element under consideration is larger or smaller than the standard rotating-beam specimen ($d = 0.30$ in). For gear teeth this transition occurs at a diametral pitch of 5 teeth/in. Therefore use

$$k_b = \begin{cases} 1.00 & P > 5 \\ 0.85 & P \leq 5 \end{cases} \quad (11\text{-}29)$$

Reliability

The reliability factors are applied exactly as in Chap. 5, and so a part of Table 5-2 is reproduced for convenience here as Table 11-4.

Temperature

The lubricant or gear temperature is often a factor in gear design. Until more information becomes available, the empirical relationship

$$k_d = \begin{cases} 1.00 & T \leq 160°\text{F} \\ \dfrac{620}{460 + T} & T > 160°\text{F} \end{cases} \quad (11\text{-}30)$$

Table 11-4   RELIABILITY FACTORS

Reliability R	0.50	0.90	0.99	0.999	0.999 9	0.999 99
Factor k_c	1.000	0.897	0.814	0.753	0.702	0.659

is suggested, where T is the maximum lubricant temperature in degrees Fahrenheit. This relation is recommended for both spur and helical gears.

Stress Concentration

In most of our previous studies the stress-concentration factor has been used as a strength-reducing factor by adopting a value for k_e that is less than unity. Since we have used K_f as a stress-increasing factor in the modified Lewis equation, $k_e = 1.00$ for gears.

Miscellaneous Effects

Gears that always rotate in the same direction and are not idlers are subjected to a tooth force that always acts on the same side of the tooth. Thus the fatigue load is repeated but not reversed, and so the tooth is said to be subjected to one-way bending. By constructing a set of modified Goodman diagrams for steels having Brinell hardnesses from 140 to 450, it is found that the effective or one-way endurance limit varies only between 40 and 50 percent more than the endurance limit S_e. And so it is convenient in the analysis of gears to utilize the miscellaneous effects factor k_f to modify the endurance limit for cases of one-way bending of gear teeth. Thus

$$k_f = \begin{cases} 1.00 & \text{reversed or two-way bending} \\ 1.40 & \text{repeated or one-way bending} \end{cases} \quad (11\text{-}31)$$

Completely reversed or two-way bending occurs when the gears are idlers and on any gear when rotation is possible in either direction.

Cast Iron

For cast-iron gears use the endurance limits given in Table A-20. These values are fully corrected for surface finish but not for size, temperature, or miscellaneous effects. The low grades of cast iron should probably not be used if high reliabilities are desired. In any event the reliability factor k_c for cast irons should be evaluated in a laboratory testing program because the variance of the mechanical properties may be quite large.

11-15 FACTOR OF SAFETY

The formula

$$n_G = K_o K_m n \quad (11\text{-}32)$$

may be used to compute the factor of safety n_G for gears. In this formula K_o is the overload factor. Values recommended by AGMA are listed in Table 11-5. Factor

K_m is an **AGMA** load-distribution factor which accounts for the possibility that the tooth force may not be uniformly distributed across the full face width. Use Table 11-6 for K_m. The factor n in Eq. (11-32) is the usual factor of safety as defined in Chap. 1. The AGMA practice is to use $n \geq 2$ for fatigue loads.

EXAMPLE 11-6 In Examples 11-4 and 11-5 an 18-T 20° full-depth pinion having a diametral pitch of 4 and a face width of $3\frac{1}{4}$ in was tentatively selected for a motor-driven speed reducer. The fatigue stress was found from Eq. (11-21) to be 23.1 kpsi. The material selected was a UNS G10500 steel, heat-treated and drawn to 900°F. Based on a reliability of 90 percent, compute the factors of safety n_G and n.

SOLUTION Using Table A-17, we find the tensile strength to be $S_u = 155$ kpsi. Therefore $S'_e = 0.50S_u = 0.50(155) = 77.5$ kpsi. From Fig. 11-26 $k_a = 0.66$. Equation (11-29) gives $k_b = 0.85$. And, from Table 11-4, $k_c = 0.897$. Then $k_d = k_e = 1$, and $k_f = 1.40$. Therefore, the endurance limit of the teeth is

$$S_e = k_a k_b k_c k_d k_e k_f S'_e = 0.66(0.85)(0.897)(1)(1)(1.40)(77.5) = 54.6 \text{ kpsi}$$

and so the factor of safety is

$$n_G = \frac{S_e}{\sigma} = \frac{54.6}{23.1} = 2.36 \qquad Ans.$$

Next, we find $K_o = 1.00$ from Table 11-5, by assuming no shock loads exist. By assuming average conditions of mounting, a load-distribution factor $K_m = 1.7$ is selected from Table 11-6. Rearranging Eq. (11-32) we then find the factor of safety n to be

$$n = \frac{n_G}{K_o K_m} = \frac{2.36}{1.00(1.7)} = 1.39 \qquad Ans.$$

which is a trifle small, as noted in Sec. 11-15. ////

Table 11-5 OVERLOAD CORRECTION FACTOR K_o^*

	Driven machinery		
Source of power	Uniform	Moderate shock	Heavy shock
Uniform	1.00	1.25	1.75
Light shock	1.25	1.50	2.00
Medium shock	1.50	1.75	2.25

* Darle W. Dudley (ed.), "Gear Handbook," p. 13–20, McGraw-Hill Book Company, New York, 1962.

Table 11-6   LOAD-DISTRIBUTION FACTOR K_m FOR SPUR GEARS*

Characteristics of support	Face width, in			
	0 to 2	6	9	16 up
Accurate mountings, small bearing clearances, minimum deflection, precision gears	1.3	1.4	1.5	1.8
Less rigid mountings, less accurate gears, contact across full face	1.6	1.7	1.8	2.2
Accuracy and mounting such that less than full-face contact exists		Over 2.2		

* Darle W. Dudley (ed.), "Gear Handbook," p. 13–21, McGraw-Hill Book Company, New York, 1962.

11-16   SURFACE DURABILITY

The preceding sections have been concerned with the stress and strength of a gear tooth subjected to bending action and how to guard against the possibility of tooth breakage by static overloads or by fatigue action. In this section we are interested in the failure of the *surfaces* of gear teeth, generally called *wear*. *Pitting*, as explained in Sec. 5-24, is a surface fatigue failure due to many repetitions of high contact stresses. Other surface failures are *scoring*, which is a lubrication failure, or *abrasion*, which is wear due to the presence of foreign material.

To assure a satisfactory life, the gears must be designed so that the dynamic surface stresses are within the surface-endurance limit of the material. In many cases the first visible evidence of wear is seen near the pitch line; this seems reasonable because the maximum dynamic load occurs near this area.

To obtain an expression for the surface-contact stress, we shall employ the Hertz theory. In Eq. (2-89) it was shown that the contact stress between two cylinders may be computed from the equation

$$p_{max} = -\frac{2F}{\pi b l} \qquad (11-33)$$

where p_{max} = surface compressive stress, psi
  F = force pressing the two cylinders together, lb
  l = length of cylinders, in
and b is obtained from the equation [Eq. (2-88)]

$$b = \sqrt{\frac{2F}{\pi l}\frac{[(1-\mu_1^2)/E_1] + [(1-\mu_2^2)/E_2]}{(1/d_1) + (1/d_2)}} \qquad (11-34)$$

where μ_1, μ_2, E_1, and E_2 are the elastic constants and d_1 and d_2 are the diameters, respectively, of the two cylinders.

To adapt these relations to the notation used in gearing, we replace F by $W_t/\cos\phi$, d by $2r$, and l by the face width F. With these changes we can substitute the value of b as given by Eq. (11-34) into Eq. (11-33). Replacing p_{max} by σ_H, the *surface compressive stress (Hertzian stress)* is found to be

$$\sigma_H^2 = \frac{W_t}{\pi F \cos \phi} \frac{(1/r_1) + (1/r_2)}{[(1 - \mu_1^2)/E_1] + [(1 - \mu_2^2)/E_2]} \tag{11-35}$$

where r_1 and r_2 are the instantaneous values of the radii of curvature on the pinion- and gear-tooth profiles, respectively, at the point of contact. By accounting for load sharing in the value of W_t used, Eq. (11-35) can be solved for the Hertzian stress for any or all points from the beginning to the end of tooth contact. Of course, pure rolling exists only at the pitch point. Elsewhere, the motion is a mixture of rolling and sliding. Equation (11-35) does not account for any sliding action in the evaluation of stress.

As an example of the use of this equation let us find the contact stress when a pair of teeth is in contact at the pitch point. The radii of curvature r_1 and r_2 of the tooth profiles, when they are in contact at the pitch point, are

$$r_1 = \frac{d_P \sin \phi}{2} \qquad r_2 = \frac{d_G \sin \phi}{2} \tag{a}$$

where ϕ is the pressure angle. Then

$$\frac{1}{r_1} + \frac{1}{r_2} = \frac{2}{\sin \phi}\left(\frac{1}{d_P} + \frac{1}{d_G}\right) \tag{b}$$

Defining the *speed ratio* m_G as

$$m_G = \frac{N_G}{N_P} = \frac{d_G}{d_P} \tag{11-36}$$

we can write Eq. (b) as

$$\frac{1}{r_1} + \frac{1}{r_2} = \frac{2}{\sin \phi} \frac{m_G + 1}{m_G d_P} \tag{c}$$

After some rearranging and the use of Eq. (c), Eq. (11-35) becomes

$$\sigma_h = \sqrt{\frac{W_t}{F d_P} \frac{1}{\pi\left(\dfrac{1 - \mu_P^2}{E_P} + \dfrac{1 - \mu_G^2}{E_G}\right)} \frac{1}{\dfrac{\cos \phi \sin \phi}{2}} \frac{m_G}{m_G + 1}} \tag{11-37}$$

where the subscripts P and G applied to μ and E refer to the pinion and gear, respectively.

The second term under the radical is called the *elastic coefficient* C_p. Thus the formula for C_p is

$$C_p = \sqrt{\frac{1}{\pi\left(\dfrac{1 - \mu_P^2}{E_P} + \dfrac{1 - \mu_G^2}{E_G}\right)}} \tag{11-38}$$

Values of C_p have been worked out for various combinations of materials, and these are listed in Table 11-7.

The *geometry factor* I for spur gears is the denominator of the third term under the radical of Eq. (11-37). Thus

$$I = \frac{\cos \phi \sin \phi}{2} \frac{m_G}{m_G + 1} \qquad (11\text{-}39)$$

which is valid for external spur gears. For internal gears, the factor is

$$I = \frac{\cos \phi \sin \phi}{2} \frac{m_G}{m_G - 1} \qquad (11\text{-}40)$$

Now recall that a velocity factor K_v was used in the bending-stress equation to account for the fact that the force between the teeth is actually more than the transmitted load because of the dynamic effect. This factor must also be used in the equation for surface-compressive stress for exactly the same reasons. When used here, the velocity factor is designated as C_v, but it has the same values and so $C_v = K_v$; the same formulas are used.

With Eqs. (11-38) to (11-40) and the addition of the velocity factor, Eq. (11-37) can be written in the more useful form

$$\sigma_H = C_p \sqrt{\frac{W_t}{C_v F d_p I}} \qquad (11\text{-}41)$$

Table 11-7   VALUES OF THE ELASTIC COEFFICIENT C_p FOR SPUR AND HELICAL GEARS WITH NONLOCALIZED CONTACT AND FOR $\mu = 0.30$. IN EACH CASE THE MODULUS OF ELASTICITY IS IN MPSI.*

Pinion	Gear			
	Steel	Cast iron	Aluminum bronze	Tin bronze
Steel, $E = 30$	2300	2000	1950	1900
Cast iron, $E = 19$	2000	1800	1800	1750
Aluminum bronze, $E = 17.5$	1950	1800	1750	1700
Tin bronze, $E = 16$	1900	1750	1700	1650

* Darle W. Dudley (ed.), "Gear Handbook," p. 13–22, McGraw-Hill Book Company, New York, 1962.

11-17 SURFACE FATIGUE STRENGTH

In our investigations of the strength of contacting surfaces in Sec. 5-24, we found that the surface-endurance limit in kpsi for steels is found from the equation

$$S_{fe} = 0.4H_B - 10 \text{ kpsi} \qquad (11\text{-}42)$$

where H_B is the Brinell hardness number of the softer of the two contacting surfaces.

The AGMA recommends that the surface endurance limit be modified in a manner quite similar to that used for the bending endurance limit. The equation is

$$S_H = \frac{C_L C_H}{C_T C_R} S_{fe} \qquad (11\text{-}43)$$

where S_H = corrected surface endurance limit, or Hertzian strength
C_L = life factor
C_H = hardness-ratio factor; use 1.0 for spur gears
C_T = temperature factor; use 1.0 for temperatures less than 250°F
C_R = reliability factor

The *life modification factor* C_L is used to increase the strength when the gear is to be used for short periods of time; use Table 11-8. The *reliability modification factor* C_R, as presented by AGMA, is rather vague. It is believed that the AGMA meant the values of C_R to be about as listed in Table 11-8.

The *hardness-ratio factor* C_H was probably included by the AGMA to account for differences in strength due to the fact that one of the mating gears might be softer than the other. However, for spur gears $C_H = 1$.

The AGMA makes no recommendations on values to use for the *temperature factor* C_T when the temperature exceeds 250°F except to imply that a value $C_T > 1.0$ should probably be used.

Factors of safety to guard against surface failures should be selected using the guidelines outlined in Sec. 11-15 and Eq. (11-32). The AGMA uses C_o and C_m to designate the overload and load-distribution factors, but their values are the same as those for K_o and K_m. These factors should be used in the numerator of Eq. (11-41) as load-multiplying factors.

As we have observed many times in this book, there is no satisfactory substitute for a thorough laboratory testing program to verify analytical results. This is particularly true in the design of gearing for a long life. The analytical

Table 11-8 LIFE AND RELIABILITY MODIFICATION FACTORS

Cycles of life	Life factor C_L	Reliability R	Reliability factor C_R
10^4	1.5	Up to 0.99	0.80
10^5	1.3	0.99 to 0.999	1.00
10^6	1.1	0.999 up	1.25 up
10^8 up	1.0		

methods presented here are useful in getting a ball-park answer and in predicting possible remedies when trouble is encountered. They are not intended to produce exact results.

11-18 HEAT DISSIPATION

The power loss at each tooth mesh for spur gears is usually less than 1 percent of the power transmitted. The magnitude of this loss depends upon the gear materials, the tooth system, the lubrication, the character of the tooth surface, and the pitch-line velocity. In addition, there is a power loss at the bearings which may reach 1 or 2 percent. When the gearset is mounted in a housing, it is suggested that the power loss of the gears be added to that for the bearings and that Eq. (10-25) be used.

It is sometimes necessary to direct a stream of cooling oil against the teeth in order to carry away the generated heat. A rule of thumb which is occasionally employed is to use 1 gpm of cooling oil for each 400 hp transmitted.

11-19 GEAR MATERIALS

Gears are commonly made of steel, cast iron, bronze, or phenolic resins. Recently nylon, Teflon, titanium, and sintered iron have been used successfully. The great variety of materials available provides the designer with the opportunity of obtaining the optimum material for any particular requirement, whether it be high strength, a long wear life, quietness of operation, or high reliability.

In many applications, steel is the only satisfactory material because it combines both high strength and low cost. Gears are made of both plain carbon and alloy steels, and there is no one best material. In many cases the choice will depend upon the relative success of the heat-treating department with the various steels. When the gear is to be quenched and tempered, a steel with 40 to 60 points of carbon is used. If it is to be case-hardened, one with 20 points or less of carbon is used. The core as well as the surface properties must always be considered.

Cast iron is a very useful gear material because it has such good wear resistance. It is easy to cast and machine and transmits less noise than steel. The tensile strengths of AGMA grades of cast irons are the same as the ASTM grades listed in the Appendix.

Bronzes may be used for gears when corrosion is a problem, and they are quite useful for reducing friction and wear when the sliding velocity is high, as in worm-gear applications. The AGMA lists five tin bronzes containing small percentages of nickel, lead, or zinc which are suitable gear materials. Their hardness varies from 70 to 85 Bhn.

Nonmetallic gears are mated with steel or cast-iron gears to obtain the greatest load-carrying capacity. To secure good wear resistance, the metal gear should have a hardness of at least 300 Bhn. A nonmetallic gear will carry almost as much load as a good cast-iron or mild-steel gear, even though the strength is much lower, because of the low modulus of elasticity. This low modulus permits the nonmetallic gear to absorb the effects of tooth errors so that a dynamic load is not created. A nonmetallic gear also has the important advantage of operating well on marginal lubrication.

Thermosetting laminates are widely used for gears. They are made of sheet materials composed of a fibrous or woven material, together with a resin binder, or are cast. Both nylon and Teflon, as gear materials, have given excellent results in service.

11-20   GEAR-BLANK DESIGN

Gear blanks are produced by casting, forging, machining from a solid blank, and fabricating. Some typical fabrication methods are shown in Fig. 11-27. When the

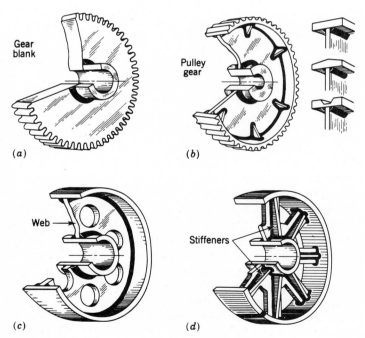

FIGURE 11-27
Methods of fabricating gears as weldments. (*a*) Solid gear blank to which hub is welded. (*b*) The gear has a solid web with stiffeners, giving additional support to the rim. (*c*) A satisfactory construction for small-diameter gears with short face widths. (*d*) A fabricated gear blank with spokes. (*Courtesy of Lincoln Electric Company.*)

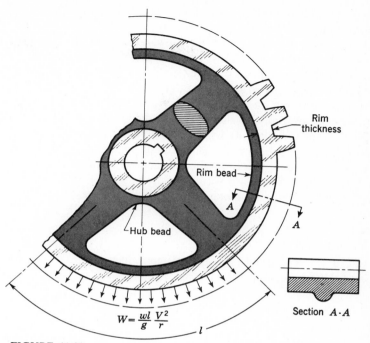

$$W = \frac{wl}{g}\frac{V^2}{r}$$

FIGURE 11-28
A cast-iron gear showing how bending is produced by the centrifugal force.

pinion is small, it is frequently made integral with the shaft, thus eliminating the key as well as an axial-locating device.

In designing a gear blank, rigidity is almost always a prime consideration. The hub must be thick enough to maintain a proper fit with the shaft and to provide sufficient metal for the key slot. This thickness must also be large enough so that the torque may be transmitted through the hub to the web or spokes without serious stress concentrations. The hub must have length so that the gear will rotate in a single plane without wobble. The arms or web and the rim must also have rigidity, but not too much, because of its effect upon the dynamic load.

There are no general rules for the design of hubs. If they are designed with sufficient rigidity, the stresses are usually quite low, especially when compared with the tooth stresses. The length of the hub should be at least equal to the face width, or greater if this does not give sufficient key length. Sometimes two keys are used. If the clearance between the bore and the shaft is large, the hub should have a length which is at least twice the bore diameter, because a slight inaccuracy here is magnified at the rim. Many designers prefer to make a scale drawing of the gear; the hub dimensions can then be adjusted by eye to obtain the necessary rigidity.

Figure 11-28 is a drawing of a portion of a cast-iron gear. The hub bead is used to brace the arms and reduce the stress concentration caused by the torque transmitted from the hub to the arms. The arms are shown with an elliptical section, but they may also be designed with an H or I section, or any other shape, depending upon the stiffness and strength desired. The rim bead gives additional rigidity and strength to the rim.

If a gear rotates at a high pitch-line velocity, the weight of the rim and teeth may be sufficient to cause large bending stresses in the portion of the rim contained between any two arms. When the gear is made of steel, these stresses are usually not serious, but when cast iron is used, this stress should be checked. Although the problem is complicated, an approximation may be obtained by assuming that the rim is a uniformly loaded beam fixed at the ends by the spokes. The length of the beam would be the length of arc measured at the mean rim diameter between the spoke centerlines. Under these assumptions, the total bending load W is

$$W = \frac{wl}{g} \frac{V^2}{r} \qquad (11\text{-}44)$$

where w = unit weight of rim and teeth, lb per in
  V = pitch-line velocity, fps
  g = acceleration due to gravity, fps^2
The maximum bending moment occurs at the arms and is

$$M_{max} = \frac{Wl}{12} \qquad (11\text{-}45)$$

The stress may then be obtained by substituting the maximum moment and the section modulus into the equation for bending stress, $\sigma = Mc/I$. This solution neglects the curvature of the rim; the tensile, compressive, or bending forces in the rim due to the transfer of torque between the arms and the rim; and the stress-concentration effect where the arm joins the rim. In addition, we cannot be sure of the accuracy of the assumption that the ends are fixed.

The rim must also have rigidity in the direction parallel to the axis of the gear (Fig. 11-29a) to maintain a uniform load across the face. This means that the arm or web must be thick enough for adequate support.

The loading on the arms of a gear is complicated. The transmitted torque produces bending, the centrifugal force on the rim produces a combination of bending and tension, and the dynamic load acting between the teeth produces a vibrating bending force. An approximation can be obtained by neglecting all these except the bending produced by the transmitted torque. Then the bending force is (Fig. 11-29b)

$$F = \frac{T}{rn} \qquad (11\text{-}46)$$

where T = transmitted torque, lb·in
  r = length of spoke, in
  n = number of spokes

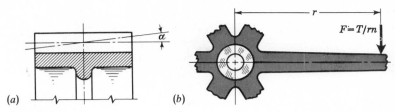

FIGURE 11-29
(*a*) Section of rim; the rim must have sufficient rigidity to avoid deflection through the angle α. (*b*) Bending force on a spoke.

The stress may then be determined by finding the maximum moment for a cantilever beam and substituting this, together with the section modulus, in the bending-stress equation, $\sigma = Mc/I$. A high factor of safety should be used because this method is only a rough approximation and stress concentration is present.

The analytical methods investigated above are not used every time a gear is to be designed. In many cases the loads and velocities are not high, and the gear can be designed on the drawing board by using pleasing proportions. On the other hand, cases sometimes occur where the loads are extremely high, or where the weight of the gear is a very important consideration; in these situations it may be desirable to make a much more thorough investigation than is indicated here.

11-21 INVOLUTE SPLINES

Splines are used to couple two shafts together or to replace the action of a key in transferring torque to a gear, pulley, flywheel, or the like. The use of splines is the best solution whenever large amounts of torque are to be transmitted. Although splines resemble gear teeth and are cut on the same machines, their action is somewhat different, for there is no rolling action, the teeth all fit together, and the absence of relative motion means that wear is no problem in their design.

The USASI lists five standards for involute splines,* with tooth forms numbered from 1 to 5, depending upon the pressure angle employed:

* See Darle W. Dudley, Involute Splines, *Prod. Eng.*, vol. 28, p. 75, October 1957.

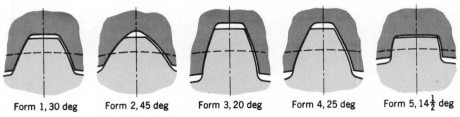

| Form 1, 30 deg | Form 2, 45 deg | Form 3, 20 deg | Form 4, 25 deg | Form 5, 14$\frac{1}{2}$ deg |

FIGURE 11-30
Tooth forms for standard involute splines. (*Courtesy of Product Engineering.*)

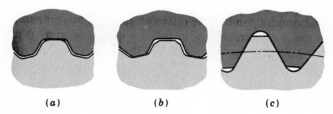

(a) (b) (c)

FIGURE 11-31
Three types of spline fits: (a) Major-diameter fit; (b) minor-diameter fit; (c) side-bearing fit. (*Courtesy of Product Engineering.*)

The form 1 spline (Fig. 11-30) has a 30° pressure angle and a tooth depth of 50 percent of that for full-depth gear teeth and can be manufactured with as few as six teeth.

The form 2 spline, with a pressure angle of 45°, is a stub tooth about 50 percent as deep as a standard involute gear tooth. The high pressure angle produces a very narrow top land, but it is normally used with large numbers of teeth where indexing may be necessary or to obtain a stronger shaft.

The form 3 spline has a 20° pressure angle and a 75 percent tooth depth, though 50 percent is used in automotive industries. It is easy to machine and has more bearing area than forms 1 and 2.

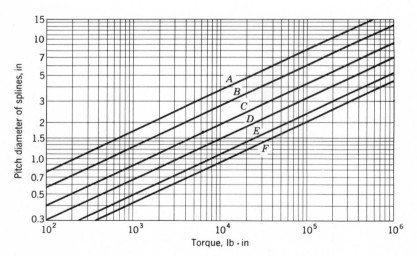

FIGURE 11-32
Chart for determining the approximate pitch diameter of a spline to carry a given torque. A is a commercial spline, B a high-capacity spline, and C an aircraft-quality spline; all three are for flexibly coupling two shafts together. D, E, and F represent three grades of fixed splines, with F the limit of spline design at a shaft torsional stress of 65 kpsi. (*Courtesy of Product Engineering.*)

The form 4 tooth has a 25° pressure angle and a 70 percent tooth depth.

The form 5 tooth with only a $14\frac{1}{2}°$ pressure angle and a 30 percent tooth depth has the largest top land of them all, as can be seen from Fig. 11-30. The shorter tooth provides space for a larger shaft, which is often an important consideration.

Three types of fits are illustrated in Fig. 11-31; the major-diameter fit is easiest to obtain and control and gives better centering characteristics than the side-bearing fit. Mating parts may be shrink-fitted together, or they may have a light push or even a loose fit, depending upon the requirements of the design. Even a loose-fitted assembly will center itself when torque is applied.

The chart of Fig. 11-32 was prepared by Dudley* and can be used to determine an approximate value for the pitch diameter. The face width to be used depends on whether the spline is fixed or flexible. For flexible splines, used to couple two shafts together, the face width can be anywhere between one-sixth and one-half the pitch diameter, the former value being used for severe misalignment. For fixed splines, there seems to be no advantage in making the face width any greater than the pitch diameter.

Neither wear nor bending stress is important in the design of spline teeth. The most frequent type of failure is a torsional failure of the shaft, and this can be guarded against by applying the usual methods of stress and strength analysis. If strength is an important consideration, the shear stress corresponding to the pitch diameter should be checked; but note that all teeth must fail on both members before shearing of the teeth will interrupt the drive. Other stresses which must sometimes be checked are the compressive stress on the sides of the teeth and the bursting, or circumferential, tensile stress in the outer ring of a flexible coupling.

PROBLEMS

Sections 11-1 to 11-8

11-1   A 22-tooth pinion has a diametral pitch of 4 teeth/in, runs at 1200 rpm, and drives a gear at 660 rpm. Find the number of teeth on the gear and the theoretical center distance.

11-2   A pair of gears has an angular-velocity ratio of 3.20. There are 20 teeth on the driver, and the circular pitch is 3 in. Find the number of teeth on the driven gear, the diametral pitch, and the theoretical center distance.

11-3   A 24-tooth pinion has a module of 2 mm and runs at a speed of 1800 rpm. The driven gear is to operate at 450 rpm. Find the circular pitch, the number of teeth on the gear, and the theoretical center distance.

11-4   A 24-tooth pinion mates with a 36-tooth gear and has a diametral pitch of 4 teeth/in and a pressure angle of 20°. Make a drawing of the gears showing one tooth on each gear. Find and tabulate the following results: the addendum, dedendum, clearance,

Ibid.

circular pitch, tooth thickness, and base-circle diameters; the arcs of approach, recess, and action; and the base pitch and contact ratio.

11-5 A 17-tooth pinion paired with a 50-tooth gear has a diametral pitch of $2\frac{1}{2}$ teeth/in and a 20° pressure angle. Make a drawing of the gears showing one tooth on each gear. Find the arcs of approach, recess, and action and the contact ratio.

11-6 Draw a 26-tooth pinion in mesh with a rack having a diametral pitch of 2 teeth/in and a pressure angle of 20°.

  (a) Find the arcs of approach, recess, and action and the contact ratio.

  (b) Draw a second rack in mesh with the same gear but offset $\frac{1}{8}$ in away from the pinion center. Determine the new contact ratio. Has the pressure angle changed?

11-7 A 15-tooth pinion having a 25° pressure angle and a diametral pitch of 3 teeth/in is to drive an 18-tooth gear. Without drawing the teeth, make a full-scale drawing and show the pitch circles, base circles, addendum circles, dedendum circles, and the pressure line. Locate both interference points and show the amount of interference, if it exists. Locate the initial and final points of contact and label them. Compute the base pitch and find the contact ratio.

11-8 We wish to establish a new gear-tooth system having teeth in which the addendum is still $1/P$ but such that a 12-tooth pinion will not be undercut when it is generated. We also wish to use the smallest possible pressure angle. Calculate the value of this angle if the mating gear is a rack.

Section 11-9

11-9 In part a of the figure, shaft a is attached to the planet carrier, shaft b is keyed to sun gear 5, and sun gear 2 is fixed to the housing. Planet gears 3 and 4 are attached to each other. Shaft a is driven at 800 rpm ccw; find the speed and direction of rotation of shaft b.

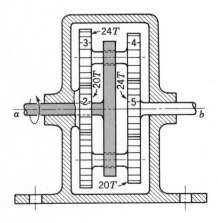
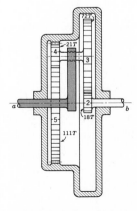

PROBLEMS 11-9 and 11-10.

11-10 The epicyclic train shown in part *b* of the figure has the arm connected to shaft *a*, and sun gear 2 connected to shaft *b*. Gear 5, having 111 teeth, is a part of the frame, and planet gears 3 and 4 are both keyed to the same shaft. Note that gear 5 is an internal gear and that the teeth shown in the cross-sectional view are on the inside of the housing.

(*a*) Find the speed and direction of rotation of shaft *a* if the frame is held stationary and shaft *b* is driven at 120 rpm cw.

(*b*) Shaft *b* is held stationary, and the arm rotates at 8 rpm cw. Find the speed and direction of rotation of the housing.

11-11 Part *a* of the figure shows a reverted planetary train. Gear 2 is fastened to its shaft and is driven at 250 rpm in a clockwise direction. Gears 4 and 5 are planet gears which are joined to each other but are free to turn on the shaft carried by the arm. Gear 6 is stationary. Find the rpm and direction of rotation of the arm.

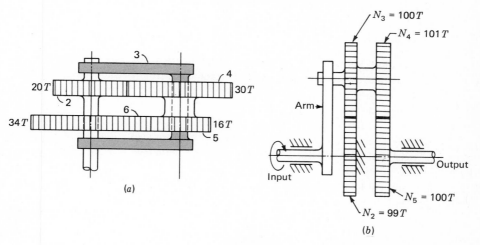

PROBLEMS 11-11 and 11-12

11-12 By using nonstandard gears, it is possible to mate a 99-*T* gear with a 100-*T* gear at the same distance between centers as would be required to mate a 100-*T* gear with a 101-*T* gear. The planetary train shown in part *b* of the figure is based on this idea.

(*a*) Find the ratio of the speed of the output shaft to the speed of the input shaft.

(*b*) The housing for this planetary train is cylindrical, with the axis of the cylinder coincident with the axis of the input and output shafts. If the pitch of gears 4 and 5 is 10 teeth per inch of pitch diameter, and if these gears have standard addendums, what should be the inside diameter of this housing? Allow $\frac{1}{2}$ in of radial clearance.

(*c*) Suppose you designed an ordinary two-gear nonplanetary train having the same speed ratio. How many teeth would the gear have if the pinion had 20 teeth and was 10 diametral pitch? What size cylinder would be needed to enclose such a gear train? Use the same clearance as in part *b*.

Section 11-10

11-13 The gears shown in part *a* of the figure are 3 diametral pitch and 20° pressure angle, and are in the same plane or can be assumed to be. The pinion rotates counterclockwise at 600 rpm and transmits 25 hp through the idler to the 28-*T* gear on shaft *c*. Calculate the resulting shaft reaction on the 36-*T* idler.

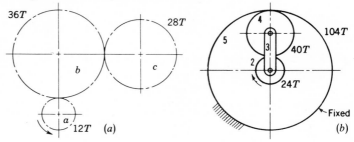

PROBLEMS 11-13 and 11-14.

11-14 The 24-tooth 8-pitch 20° pinion shown in the figure for part *b* rotates clockwise at 900 rpm and transmits 3 hp to the planetary gear train. What torque can arm 3 deliver to its output shaft? Draw a free-body diagram of the arm and of each gear, and show all forces which act upon them.

11-15 Part *a* of the figure illustrates a double-reduction gear train. Shaft *a* is driven by a 1.25-kW source at 1720 rpm. The reduction between shafts *a* and *b* is 3.5 : 1 and between shafts *b* and *c* it is 4 : 1. The pinion on shaft *a* has 24 teeth, and the gear on shaft *c* has 160 teeth.
(*a*) Find the tooth numbers for the gears on shaft *b*.
(*b*) Find the speed of shafts *b* and *c*.
(*c*) If the power loss is 4 percent at each mesh, find the torque at each shaft.

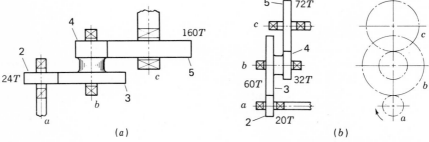

PROBLEMS 11-15 and 11-16.

11-16 In part *b* of the figure, the gears connecting shafts *a* and *b* have a module of 2.5 mm and a 20° pressure angle. The mating gears on shafts *b* and *c* have a module of 3 mm with a 20° pressure angle. The pinion transmits 1.5 kW at a speed of 20 s⁻¹. Find the resulting shaft reactions by assuming that the forces acting on gears 3 and 4 are in the same plane.

11-17 In part *a* of the figure, gear 2 on shaft *a* is the driver and it transmits 10 hp at 600 rpm to gear 3 on shaft *b*. Gear 2 is 12-pitch and has 20 teeth and a 25° pressure angle. Gear 3 has 60 teeth. Gear 4 on shaft *b* has 24 teeth, is 6-pitch, and has a pressure angle of 20°. Gear 5 has 56 teeth.

(*a*) Find the shaft center distances.

(*b*) Find the force with which each gear pushes on its shaft.

(*c*) Assume gears 3 and 4 are in the same plane and find the resultant force on shaft *b*.

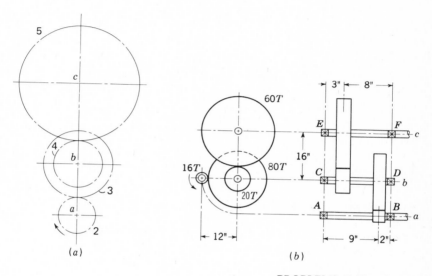

(*a*)   (*b*)

PROBLEMS 11-17 and 11-18.

11-18 The 16-tooth pinion drives the double-reduction gear train shown in part *b* of the figure. All gears have 25° pressure angles. The pinion rotates counterclockwise at 1200 rpm and transmits 50 hp to the gear train. Calculate the magnitude and direction of the radial force exerted by each bearing on its shaft.

11-19 A 10-hp motor drives gear 2 in the figure at 1800 rpm. A flat belt on pulley 4 drives pulley 5. The shaft to which pulley 5 is attached drives a blower.

(*a*) Find the speed and the torque input to the blower, assuming no frictional losses.

(*b*) The belt tension on the loose side is 20 percent of the tension on the tight side. Assuming the belt tensile forces are vertical, find the tension in both sides.

(*c*) Find the torque in countershaft *b*.

(*d*) Compute the force exerted by pulley 4 on the countershaft.

(*e*) Compute the *y* and *z* components of the force with which the bearings at *A* and *D* push against the shaft.

(*f*) Locate and find the maximum bending moment on the countershaft.

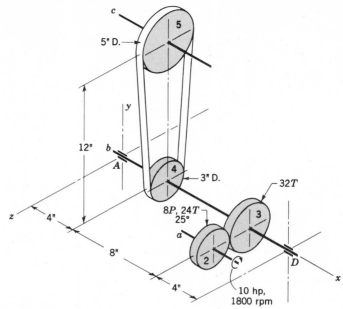

PROBLEM 11-19.

Section 11-12

11-20 A gearset consists of a 16-tooth pinion that mates with a 32-tooth gear. The gears have a face width of $1\frac{1}{4}$ in and a diametral pitch of 10; they are cut by hobbing using the 20° full-depth system. The material is UNS G10350 cold-drawn steel. Estimate the horsepower capacity of this gearset based on the yield strength of the material, a factor of safety of 5, and a pinion speed of 1720 rpm.

11-21 A steel pinion has a pitch of 6 teeth/in, a 20° pressure angle, full-depth teeth, and 22 teeth. This pinion runs at 900 rpm and transmits 12.5 horsepower to a 60-tooth gear. Compute the bending stress on the pinion teeth using Eq. (11-19) based on a face width of $1\frac{1}{2}$ in.

11-22 A 20° full-depth steel pinion is to be designed to transmit a steady load corresponding to 0.05 hp at 900 rpm. The pinion hub is to be $\frac{1}{2}$-in OD and bored to fit a $\frac{3}{16}$-in shaft. Estimate the pinion size by determining the pitch, pitch diameter, and face width, based on stamping the pinion from UNS G10180 sheet steel and keying it to the hub. Use an even-numbered U.S. Standard sheet-metal gauge size from Table A-24. Also use a factor of safety of 4 based on the yield strength. The steel is cold-drawn.

11-23 Estimate the required face width and pitch needed for an 8 : 1 reduction gearset to be connected between a 30-hp gas turbine running at 3600 rpm and a centrifugal pump. The gears are to be 20° full-depth, and made of UNS G61500 steel heat-treated and drawn to 1000°F. Use a factor of safety of 5.

Sections 11-13 to 11-15

11-24 A 12-tooth, 16-pitch pinion is paired with a 64-tooth gear. The gears have a face width of 1 in and are hobbed using the 25° full-depth system. The pinion is cut from hot-rolled UNS G10350 steel and the gear is made of ASTM no. 30 cast iron. The application is such that light shock loads can be expected. Use a factor of safety $n = 2.5$, 50 percent reliability, average mounting conditions, and a pitch-line velocity of 750 fpm and determine the safe horsepower capacity of this gearset.

11-25 A 14-tooth, 25° full-depth, precision-made pinion with shaved teeth is to drive a 21-tooth gear. The pinion is to rotate at 1150 rpm and transmit 25 hp under steady load conditions from an electric motor as the power source. Both gears are to be forged to UNS G10400 steel and heat-treated to a hardness of 235 Bhn. Based upon accurate bearing mounting, 99 percent reliability, and a factor of safety $n = 2$, determine suitable values for the diametral pitch and face width.

11-26 A 15-tooth, 3-pitch, 25° full-depth pinion has hobbed teeth and is forged of UNS G10350 steel and heat-treated to 220 Bhn. The pinion is to drive a 75-tooth gear made of ASTM no. 35 cast iron. These gears have a face width of 4 in. The pinion is driven at 575 rpm by a motor. The gear is connected to machinery that subjects the gear to moderate shock loads. Based on a factor of safety $n = 3$ for fatigue loading and a reliability of 90 percent, specify the safe horsepower capacity of this gearset. Use $K_m = 1.60$.

Sections 11-16 and 11-17

11-27 A pair of mating 8-pitch, 25° full-depth gears have a face width of 2 in. Both gears are made of UNS G10400 steel heat-treated to 235 Bhn, with shaved teeth. The 14-tooth pinion rotates at 1150 rpm and drives a 21-tooth gear. The bearing mountings are accurate, input and output loading is steady, and 99 percent reliability is desired. Based upon a factor of safety $n = 2$, what horsepower can this gearset safely transmit?

11-28 A hobbed steel pinion has a pitch of 6 teeth/in, a 20° pressure angle, full-depth teeth, and 22 teeth. This pinion rotates at 575 rpm and transmits 10 hp to a 60-tooth steel gear. Determine the critical surface-compressive stress on the teeth based on a face width of $1\frac{1}{2}$ in.

11-29 A 12-tooth 10-pitch steel pinion, hobbed on the 25° full-depth system is mated with a 72-tooth cast-iron gear. Compute the critical surface-contact stress that would be obtained if 2 hp were transferred from the pinion to the gear at a pinion speed of 1800 rpm. The face widths are each $1\frac{1}{4}$ in.

12

HELICAL, WORM, AND BEVEL GEARS

In the force analysis of spur gears, the forces are assumed to act in a single plane. In this chapter we shall study gears in which the forces have three dimensions. The reason for this, in the case of helical gears, is that the teeth are not parallel to the axis of rotation. And in the case of bevel gears, the rotational axes are not parallel to each other. There are also other reasons, as we shall learn.

In this chapter we shall rely heavily upon the fundamentals introduced in Chap. 11, especially the tables, charts, and graphs. And for each type of gearing the same general plan of presentation will be employed—kinematics, force analysis, bending strength, and surface strength, in that order.

12-1 PARALLEL HELICAL GEARS—KINEMATICS

Helical gears, used to transmit motion between parallel shafts, are shown in Fig. 12-1. The helix angle is the same on each gear, but one gear must have a right-hand helix and the other a left-hand helix. The shape of the tooth is an involute helicoid and is illustrated in Fig. 12-2. If a piece of paper cut in the shape of a parallelogram is wrapped around a cylinder, the angular edge of the paper

FIGURE 12-1
A pair of helical gears. (*Courtesy of Fellows Gear Shaper Company.*)

becomes a helix. If we unwind this paper, each point on the angular edge generates an involute curve. The surface obtained when every point on the edge generates an involute is called an *involute helicoid*.

The initial contact of spur-gear teeth is a line extending all the way across the face of the tooth. The initial contact of helical-gear teeth is a point which changes into a line as the teeth come into more engagement. In spur gears the line of contact is parallel to the axis of rotation; in helical gears the line is diagonal across the face of the tooth. It is this gradual engagement of the teeth and the smooth transfer of load from one tooth to another which give helical gears the ability to transmit heavy loads at high speeds. Because of the nature of contact

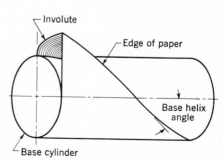

FIGURE 12-2
An involute helicoid.

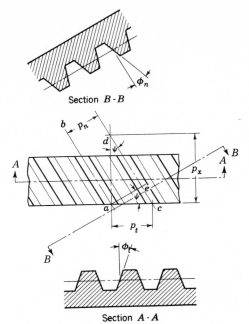

FIGURE 12-3
Nomenclature of helical gears.

between helical gears, the contact ratio is of only minor importance, and it is the contact area, which is proportional to the face width of the gear, that becomes significant.

Helical gears subject the shaft bearings to both radial and thrust loads. When the thrust loads become high or are objectionable for other reasons, it may be desirable to use double helical gears. A double helical gear (herringbone) is equivalent to two helical gears of opposite hand, mounted side by side on the same shaft. They develop opposite thrust reactions and thus cancel out the thrust load.

When two or more single helical gears are mounted on the same shaft, the hand of the gears should be selected so as to produce the minimum thrust load.

Figure 12-3 represents a portion of the top view of a helical rack. Lines *ab* and *cd* are the centerlines of two adjacent helical teeth taken on the pitch plane. The angle ψ is the *helix angle*. The distance *ac* is the *transverse circular pitch* p_t in the plane of rotation (usually called the *circular pitch*). The distance *ae* is the *normal circular pitch* p_n and is related to the transverse circular pitch as follows:

$$p_n = p_t \cos \psi \qquad (12\text{-}1)$$

The distance *ad* is called the *axial pitch* p_x and is related by the expression

$$p_x = \frac{p_t}{\tan \psi} \qquad (12\text{-}2)$$

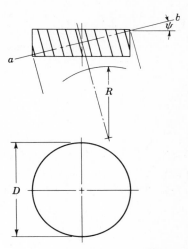

FIGURE 12-4
A cylinder cut by an oblique plane.

Since $p_n P_n = \pi$, the *normal diametral pitch* is

$$P_n = \frac{P_t}{\cos \psi} \qquad (12\text{-}3)$$

The pressure angle ϕ_n in the normal direction is different from the pressure angle ϕ_t in the direction of rotation, because of the angularity of the teeth. These angles are related by the equation

$$\cos \psi = \frac{\tan \phi_n}{\tan \phi_t} \qquad (12\text{-}4)$$

Figure 12-4 illustrates a cylinder cut by an oblique plane ab at an angle ψ to a right section. The oblique plane cuts out an arc having a radius of curvature of R. For the condition that $\psi = 0$, the radius of curvature is $R = D/2$. If we imagine the angle ψ to be slowly increased from zero to $90°$, we see that R begins at a value of $D/2$ and increases until, when $\psi = 90°$, $R = \infty$. The radius R is the apparent pitch radius of a helical-gear tooth when viewed in the direction of the tooth elements. A gear of the same pitch and with the radius R will have a greater number of teeth, because of the increased radius. In helical-gear design this is called the *virtual number of teeth*. It can be shown by analytical geometry that the virtual number of teeth is related to the actual number by the equation

$$N' = \frac{N}{\cos^3 \psi} \qquad (12\text{-}5)$$

where N' is the virtual number of teeth and N is the actual number of teeth. It is necessary to know the virtual number of teeth in applying the Lewis equation and also, sometimes, in cutting helical teeth. This apparently larger radius of curvature means that fewer teeth may be used on helical gears, because there will be less undercutting.

12-2 HELICAL GEARS—TOOTH PROPORTIONS

Except for fine-pitch gears (20 diametral pitch and finer), there is no standard for the proportions of helical-gear teeth. One reason for this is that it is cheaper to change the design slightly than it is to purchase special tooling. Since helical gears are rarely used interchangeably anyway, and since many different designs will work well together, there is really little advantage in having them interchangeable.

As a general guide, tooth proportions should be based on a normal pressure angle of 20°. Most of the proportions tabulated in Table 11-1 can then be used. The tooth dimensions should be calculated by using the normal diametral pitch. These proportions are suitable for helix angles from 0 to 30°, and all helix angles may be cut with the same hob. Of course the normal diametral pitch of the hob and the gear must be the same.

An optional set of proportions can be based on a transverse pressure angle of 20° and the use of the transverse diametral pitch. For these the helix angles are generally restricted to 15, 23, 30, or 45°. Angles greater than 45° are not recommended. The normal diametral pitch must still be used to compute the tooth dimensions. The proportions shown in Table 11-1 will usually be satisfactory.

Many authorities recommend that the face width of helical gears be at least two times the axial pitch $(F = 2p_x)$ to obtain helical-gear action. Exceptions to this rule are automotive gears which have a face width considerably less, and marine reduction gears which often have a face width much greater.

12-3 HELICAL GEARS—FORCE ANALYSIS

Figure 12-5 is a three-dimensional view of the forces acting against a helical-gear tooth. The point of application of the forces is in the pitch plane and in the center of the gear face. From the geometry of the figure, the three components of the total (normal) tooth force W are

$$W_r = W \sin \phi_n$$
$$W_t = W \cos \phi_n \cos \psi \qquad (12\text{-}6)$$
$$W_a = W \cos \phi_n \sin \psi$$

where W = total force
W_r = radial component
W_t = tangential component; also called transmitted load
W_a = axial component; also called thrust load

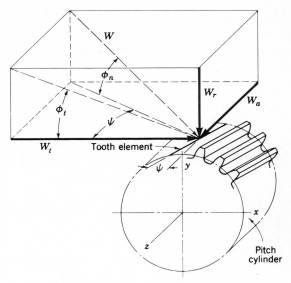

FIGURE 12-5
Tooth forces acting on a right-hand helical gear.

Usually W_t is given and the other forces are desired. In this case, it is not difficult to discover that

$$W_r = W_t \tan \phi_t$$

$$W_a = W_t \tan \psi \qquad (12\text{-}7)$$

$$W = \frac{W_t}{\cos \phi_n \cos \psi}$$

EXAMPLE 12-1   In Fig. 12-6 a 1-hp electric motor runs at 1800 rpm in the clockwise direction, as viewed from the positive x axis. Keyed to the motor shaft is an 18-tooth helical pinion having a normal pressure angle of 20°, a helix angle of 30°, and a normal diametral pitch of 12 teeth/in. The hand of the helix is shown in the figure. Make a three-dimensional sketch of the motor shaft and pinion and show the forces acting on the pinion and the bearing reactions at A and B. The thrust should be taken out at A.

SOLUTION   From Eq. (12-4) we find

$$\phi_t = \tan^{-1} \frac{\tan \phi_n}{\cos \psi} = \tan^{-1} \frac{\tan 20°}{\cos \underset{30°}{\cancel{15°}}} = 22.8°$$

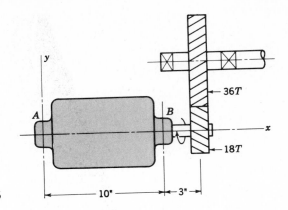

FIGURE 12-6

Also $P_t = P_n \cos \psi = 12 \cos 30° = 10.4$ teeth/in. Therefore the pitch diameter of the pinion is $d_P = 18/10.4 = 1.73$ in. The pitch-line velocity is

$$V = \frac{\pi d n}{12} = \frac{\pi (1.73)(1800)}{12} = 815 \text{ fpm}$$

The transmitted load is

$$W_t = \frac{33\,000\,\text{P}}{V} = \frac{(33\,000)(1)}{815} = 40.5 \text{ lb}$$

From Eq. (12-7) we find

$$W_r = W_t \tan \phi_t = (40.5)(0.422) = 17.1 \text{ lb}$$

$$W_a = W_t \tan \psi = (40.5)(0.577) = 23.4 \text{ lb}$$

$$W = \frac{W_t}{\cos \phi_n \cos \psi} = \frac{40.5}{(0.940)(0.866)} = 49.8 \text{ lb}$$

These three forces, W_r in the $-y$ direction, W_a in the $-x$ direction, and W_t in the $+z$ direction, are shown acting at point C in Fig. 12-7. We assume bearing reactions at A and B as shown. Then $F_A^x = W_a = 23.4$ lb. Taking moments about the z axis,

$$-(17.1)(13) + (23.4)\left(\frac{1.73}{2}\right) + 10F_B^y = 0$$

or $F_B^y = 20$ lb. Summing forces in the y direction then gives $F_A^y = 2.9$ lb. Taking moments about the y axis, next,

$$10F_B^z - (40.5)(13) = 0$$

or $F_B^z = 52.6$ lb. Summing forces in the z direction and solving gives $F_A^z = 12.1$ lb. Also, the torque is $T = W_t d_P/2 = (40.5)(1.73/2) = 35$ lb·in.   ////

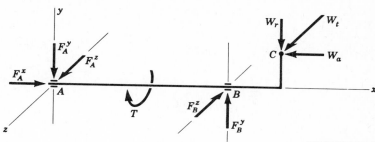

FIGURE 12-7

EXAMPLE 12-2   Solve Example 12-1 using vectors.

SOLUTION   The force at C is

$$W = -23.4i - 17.1j + 40.5k$$

Position vectors to B and C from origin A are

$$R_B = 10i \qquad R_C = 13i + 0.865j$$

Taking moments about A, we have

$$R_B \times F_B + T + R_C \times W = 0$$

Using the directions assumed in Fig. 12-7 and substituting values gives

$$10i \times (F_B^y j - F_B^z k) - Ti + (13i + 0.865j) \times (-23.4i - 17.1j + 40.5k) = 0$$

When the cross products are formed, we get

$$(10F_B^y k + 10F_B^z j) - Ti + (35i - 526j - 200k) = 0$$

whence $T = 35$ lb·in, $F_B^y = 20$ lb, and $F_B^z = 52.6$ lb.
    Next, $F_A = -F_B - W$, and so $F_A = 23.4i - 2.9j + 12.1k$ lb.   ////

12-4   HELICAL GEARS—STRENGTH ANALYSIS

We repeat here the equation for bending and surface stresses in spur gears because they also apply to helical gears.

$$\sigma = \frac{W_t P_t}{K_v F J} \tag{12-8}$$

$$\sigma_H = C_p \sqrt{\frac{W_t}{C_v F d_p I}} \tag{12-9}$$

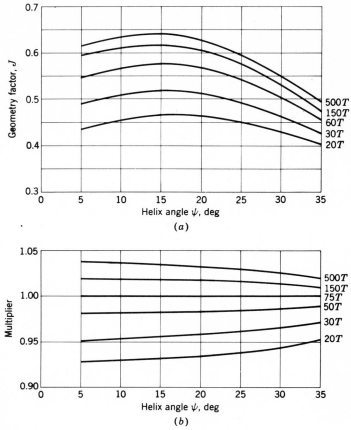

FIGURE 12-8
Geometry factors for helical and herringbone gears having a normal pressure angle
of 20°. (*a*) Geometry factors for gears mating with a 75-tooth gear. (*b*) *J*-factor
multipliers when tooth numbers other than 75 are used in the mating gear. (*AGMA
Information Sheet 225.01.*)

where σ = bending stress, psi
  σ_H = surface compressive stress, psi
  W_t = transmitted load, lb
  P_t = transverse diametral pitch, teeth/in
  $K_v = C_v$ = dynamic, or velocity, factor
  d_P = pitch diameter of pinion, in
  J = geometry factor (bending)
  I = geometry factor (surface durability)

For helical gears the velocity factor is usually taken as

$$K_v = C_v = \sqrt{\frac{78}{78 + \sqrt{V}}} \qquad (12\text{-}10)$$

where V is the pitch-line velocity in feet per minute (fpm).

Geometry factors for helical gears must account for the fact that contact takes place along a diagonal line across the tooth face and that we are usually dealing with the transverse pitch instead of the normal pitch. The worst loading occurs when the line of contact intersects the tip of the tooth, but the unloaded end strengthens the tooth.

The J factors for $\phi_n = 20°$ can be found in Fig. 12-8. The AGMA also publishes J factors for $\phi_n = 15°$ and $\phi_n = 22°$.

Geometry factors I for helical and herringbone gears are calculated from the equation*

$$I = \frac{\sin \phi_t \cos \phi_t}{2m_N} \frac{m_G}{m_G + 1} \qquad (12\text{-}11)$$

for external gears. (Use a minus sign in the denominator of the second term for internal gears.) In this equation ϕ_t is the transverse pressure angle and m_N is the load-sharing ratio and is found from the equation

$$m_N = \frac{p_N}{0.95Z} \qquad (12\text{-}12)$$

Here p_N is the normal base pitch; it is related to the *normal circular pitch* p_n by the relation

$$p_N = p_n \cos \phi_n \qquad (12\text{-}13)$$

The quantity Z is the length of the line of action in the transverse plane. It is best obtained from a layout of the two gears, but may also be found from the equation†

$$Z = \sqrt{(r_P + a)^2 - r_{bP}^2} + \sqrt{(r_G + a)^2 - r_{bG}^2} - (r_P + r_G) \sin \phi_t \qquad (12\text{-}14)$$

where r_P and r_G are the pitch radii and r_{bP} and r_{bG} the base-circle radii, respectively, of the pinion and gear. Certain precautions must be taken in using Eq. (12-14). The tooth profiles are not conjugate below the base circle, and consequently, if either $\sqrt{(r_P + a)^2 - r_{bP}^2}$ or $\sqrt{(r_G + a)^2 - r_{bG}^2}$ is larger than

* See AGMA Information Sheet 211.02, February 1969.
† For a development see Joseph E. Shigley, "Kinematic Analysis of Mechanisms," 2d ed., p. 266, McGraw-Hill Book Company, New York, 1969.

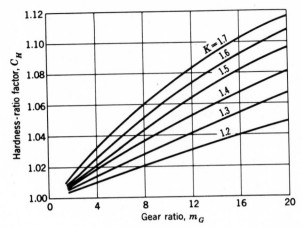

FIGURE 12-9
Hardness-ratio factor C_H for helical gears. The factor K is the Brinell hardness of the pinion divided by the Brinell hardness of the gear. Use $C_H = 1.00$ when $K < 1.2$. (*AGMA Information Sheet 215.01.*)

$(r_P + r_G) \sin \phi_t$, that term should be replaced by $(r_P + r_G) \sin \phi_t$. In addition, the effective outside radius is sometimes less than $r + a$ owing to removal of burrs or rounding of the tips of the teeth. When this is the case, always use the effective outside radius instead of $r + a$.

The modification and correction factors for helical gears are the same as for spur gears, except for the load-distribution factors K_m and C_m (Table 12-1) and the hardness-ratio factor C_H (Fig. 12-9). With these changes Eq. (11-28) gives the endurance limit in bending, Eq. (11-32) the factor of safety, and Eqs. (11-42) and (11-43) the surface endurance limit.

Table 12-1 LOAD-DISTRIBUTION FACTORS C_m AND K_m FOR HELICAL GEARS*

Characteristics of support	Face width, in			
	0–2	6	9	16 up
Accurate mountings, small bearing clearances, minimum deflection, precision gears	1.2	1.3	1.4	1.7
Less rigid mountings, less accurate gears, contact across full face	1.5	1.6	1.7	2.0
Accuracy and mounting such that less than full-face contact exists	Over 2.0			

* Darle W. Dudley (ed.), "Gear Handbook," p. 13–23, McGraw-Hill Book Company, New York, 1962.

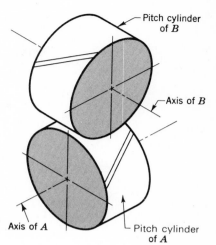

FIGURE 12-10
View of the pitch cylinders of a pair of crossed-helical gears.

12-5 CROSSED-HELICAL GEARS

Crossed-helical, or spiral, gears are those in which the shaft centerlines are neither parallel nor intersecting. They are essentially nonenveloping worm gears, because the gear blanks have a cylindrical form. This class of gears is illustrated in Fig. 12-10.

The teeth of crossed-helical gears have "point contact" with each other, which changes to "line contact" as the gears wear in. For this reason they will carry only very small loads. Crossed-helical gears are for instrumental applications, and they are definitely not recommended for use in the transmission of power.

There is no difference between a crossed-helical gear and a helical gear until they are mounted in mesh with each other. They are manufactured in the same way. A pair of meshed crossed-helical gears usually have the same hand; that is, a right-hand driver goes with a right-hand driven. The relation between thrust, hand, and rotation for crossed-helical gears is shown in Fig. 12-11.

When specifying tooth sizes, the normal pitch should always be used. The reason for this is that, when different helix angles are used for the driver and driven, the transverse pitches are not the same. The relation between the shaft and helix angles is as follows:

$$\Sigma = \psi_1 \pm \psi_2 \qquad (12\text{-}15)$$

where Σ is the shaft angle. The plus sign is used when both helix angles are of the same hand, and the minus sign when they are of opposite hand. Opposite-hand crossed-helical gears are used when the shaft angle is small.

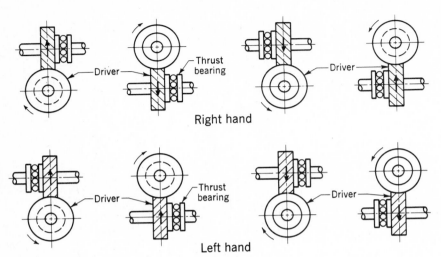

FIGURE 12.11
Thrust, rotation, and hand relations for crossed-helical gears. (*Courtesy of Boston Gear Works, Inc.*)

The pitch diameter is obtained from the equation

$$d = \frac{N}{P_n \cos \psi} \qquad (12\text{-}16)$$

where N = number of teeth
$\quad\quad P_n$ = normal diametral pitch
$\quad\quad \psi$ = helix angle

Since the pitch diameters are not directly related to the tooth numbers, they cannot be used to obtain the angular-velocity ratio. This ratio must be obtained from the ratio of the tooth numbers.

In the design of crossed-helical gears, the minimum sliding velocity is obtained when the helix angles are equal. However, when the helix angles are not equal, the gear with the larger helix angle should be used as the driver if both gears have the same hand.

There is no standard for crossed-helical gear-tooth proportions. Many different proportions give good tooth action. Since the teeth are in point contact, an effort should be made to obtain a contact ratio of 2 or more. For this reason, crossed-helical teeth are usually cut with a low pressure angle and a deep tooth.

12-6 WORM GEARING—KINEMATICS

Figure 12-12 shows a worm and a worm gear. Note that the shafts do not intersect and that the shaft angle is 90°; this is the usual shaft angle, though other angles can be used. The worm is the screwlike member in the figure, and you can see that

FIGURE 12-12
A single-enveloping worm and worm gear. (*Courtesy of Horsburgh and Scott Company, Cleveland.*)

it has, perhaps, five or six teeth (threads). A one-toothed worm would resemble an Acme screw thread very closely.

Worm gearsets are either single- or double-enveloping. A single-enveloping gearset is one in which the gear wraps around or partially encloses the worm, as in Fig. 12-12. A gearset in which each element partially encloses the other is, of course, a double-enveloping worm gearset. The important difference between the two is that *area contact* exists between the teeth of double-enveloping gears and only *line contact* between those of single-enveloping gears.

The nomenclature of a worm and worm gear is shown in Fig. 12-13. The worm and worm gear of a set have the same hand of helix as for crossed-helical gears, but the helix angles are usually quite different. The helix angle on the worm is generally quite large, and that on the gear very small. Because of this, it is usual to specify the lead angle λ on the worm and the helix angle ψ_G on the gear; the two angles are equal for a 90° shaft angle. The worm lead angle is the complement of the worm helix angle, as shown in Fig. 12-13.

In specifying the pitch of worm gearsets, it is customary to state the *axial pitch* p_x of the worm and the *transverse circular pitch* p_t, often simply called the circular pitch, of the mating gear. These are equal if the shaft angle is 90°. The pitch diameter of the gear is the diameter measured on a plane containing the worm axis, as shown in Fig. 12-13; it is the same as for spur gears and is

$$d_G = \frac{N_G p_t}{\pi} \qquad (12\text{-}17)$$

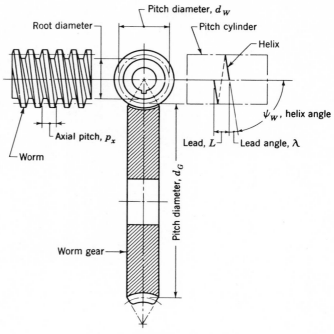

FIGURE 12-13
Nomenclature of a single-enveloping worm gearset.

Since it is not related to the number of teeth, the worm may have any pitch diameter; this diameter should, however, be the same as the pitch diameter of the hob used to cut the worm-gear teeth. Generally, the pitch diameter of the worm should be selected so as to fall into the range

$$\frac{C^{0.875}}{3.0} \leq d_W \leq \frac{C^{0.875}}{1.7} \qquad (12\text{-}18)$$

where C is the center distance. These proportions appear to result in optimum horsepower capacity of the gearset.

Table 12-2 RECOMMENDED PRESSURE ANGLES AND TOOTH DEPTHS FOR WORM GEARING

Lead angle λ, degrees	Pressure angle ϕ_n, degrees	Addendum a	Dedendum b_G
0–15	$14\frac{1}{2}$	$0.3683p_x$	$0.3683p_x$
15–30	20	$0.3683p_x$	$0.3683p_x$
30–35	25	$0.2865p_x$	$0.3314p_x$
35–40	25	$0.2546p_x$	$0.2947p_x$
40–45	30	$0.2228p_x$	$0.2578p_x$

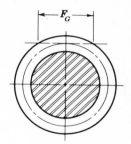

FIGURE 12-14

The *lead L* and the *lead angle* λ of the worm have the following relations:

$$L = p_x N_W \qquad (12\text{-}19)$$

$$\tan \lambda = \frac{L}{\pi d_W} \qquad (12\text{-}20)$$

Tooth forms for worm gearing have not been highly standardized, perhaps because there has been less need for it. The pressure angles used depend upon the lead angles and must be large enough to avoid undercutting of the worm-gear tooth on the side at which contact ends. A satisfactory tooth depth, which remains in about the right proportion to the lead angle, may be obtained by making the depth a proportion of the axial circular pitch. Table 12-2 summarizes what may be regarded as good practice for pressure angle and tooth depth.

The *face width F_G* of the worm gear should be made equal to the length of a tangent to the worm pitch circle between its points of intersection with the addendum circle, as shown in Fig. 12-14.

12-7   WORM GEARING—FORCE ANALYSIS

If friction is neglected, then the only force exerted by the gear will be the force W, shown in Fig. 12-15, having the three orthogonal components W^x, W^y, and W^z. From the geometry of the figure we see

$$W^x = W \cos \phi_n \sin \lambda$$
$$W^y = W \sin \phi_n \qquad (12\text{-}21)$$
$$W^z = W \cos \phi_n \cos \lambda$$

We now use the subscripts W and G to indicate forces acting against the worm and gear, respectively. We note that W^y is the separating, or radial, force for both the worm and gear. The tangential force on the worm is W^x and is W^z on the gear, assuming a 90° shaft angle. The axial force on the worm is W^z, and on the gear,

W^x. Since the gear forces are opposite to the worm forces, we can summarize these relations by writing

$$W_{Wt} = -W_{Ga} = W^x$$
$$W_{Wr} = -W_{Gr} = W^y \qquad (12\text{-}22)$$
$$W_{Wa} = -W_{Gt} = W^z$$

It is helpful in using Eq. (12-21) and also Eq. (12-22) to observe that *the gear axis is parallel to the x direction* and *the worm axis is parallel to the z direction* and that we are employing a right-handed coordinate system.

In our study of spur-gear teeth we have learned that the motion of one tooth relative to the mating tooth is primarily a rolling motion; in fact, when contact occurs at the pitch point, the motion is pure rolling. In contrast, the relative motion between worm and worm-gear teeth is pure sliding, and so we must expect that friction plays an important role in the performance of worm gearing. By introducing a coefficient of friction μ, we can develop another set of relations similar to those of Eq. (12-21). In Fig. 12-15 we see that the force W acting normal to the worm-tooth profile produces a frictional force $W_f = \mu W$, having a component $\mu W \cos \lambda$ in the negative x direction and another component $\mu W \sin \lambda$ in the positive z direction. Equation (12-21) therefore becomes

$$W^x = W(\cos \phi_n \sin \lambda + \mu \cos \lambda)$$
$$W^y = W \sin \phi_n \qquad (12\text{-}23)$$
$$W^z = W(\cos \phi_n \cos \lambda - \mu \sin \lambda)$$

Equation (12-22), of course, still applies.

If we substitute W^z into the third part of Eq. (12-22) and multiply both sides by μ, we find the frictional force to be

$$W_f = \mu W = \frac{\mu W_{Gt}}{\mu \sin \lambda - \cos \phi_n \cos \lambda} \qquad (12\text{-}24)$$

Another useful relation can be obtained by solving the first and third parts of Eq. (12-22) simultaneously to get a relation between the two tangential forces. The result is

$$W_{Wt} = W_{Gt} \frac{\cos \phi_n \sin \lambda + \mu \cos \lambda}{\mu \sin \lambda - \cos \phi_n \cos \lambda} \qquad (12\text{-}25)$$

Efficiency η can be defined by using the equation

$$\eta = \frac{W_{Wt}(\text{without friction})}{W_{Wt}(\text{with friction})} \qquad (a)$$

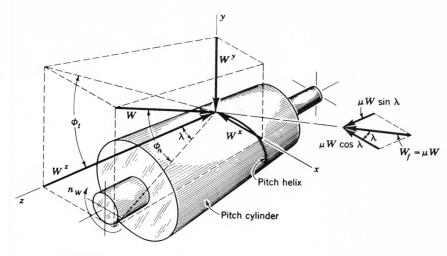

FIGURE 12-15
Drawing of the pitch cylinder of a worm, showing the forces exerted upon it by the worm gear.

Substitute Eq. (12-25) with $\mu = 0$ in the numerator of Eq. (*a*) and the same equation in the denominator. After some rearranging you will find the efficiency to be

$$\eta = \frac{\cos \phi_n - \mu \tan \lambda}{\cos \phi_n + \mu \cot \lambda} \quad (12\text{-}26)$$

Selecting a typical value of the coefficient of friction, say $\mu = 0.05$, and the pressure angles shown in Table 12-2, we can use Eq. (12-26) to get some useful design information. Solving this equation for helix angles from 1 to $30°$ gives the interesting results shown in Table 12-3.

Table 12-3 EFFICIENCY OF WORM GEAR-SETS FOR $\mu = 0.05$

Helix angle ψ, degrees	Efficiency η, percent
1.0	25.2
2.5	46.8
5.0	62.6
7.5	71.2
10.0	76.8
15.0	82.7
20.0	86.0
25.0	88.0
30.0	89.2

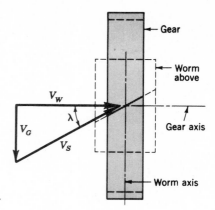

FIGURE 12-16
Velocity components in worm gearing.

Many experiments have shown that the coefficient of friction is dependent on the relative or sliding velocity. In Fig. 12-16, V_G is the pitch-line velocity of the gear and V_W the pitch-line velocity of the worm. Vectorially, $\mathbf{V}_W = \mathbf{V}_G + \mathbf{V}_S$; consequently,

$$V_S = \frac{V_W}{\cos \lambda} \quad (12\text{-}27)$$

Published values of the coefficient of friction vary as much as 20 percent, undoubtedly because of the differences in surface finish, materials, and lubrication. The values on the chart of Fig. 12-17 are representative and indicate the general trend.

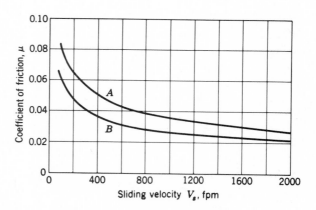

FIGURE 12-17
Representative values of the coefficient of friction for worm gearing. These values are based on good lubrication. Use curve B for high-quality materials, such as a case-hardened worm mating with a phosphor-bronze gear. Use curve A when more friction is expected, as for a cast-iron worm and worm gear.

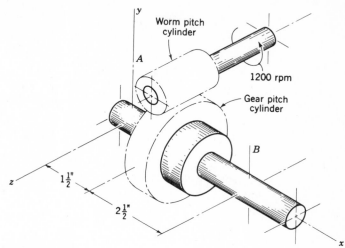

FIGURE 12-18

EXAMPLE 12-3 A 2-tooth right-hand worm transmits 1 hp at 1200 rpm to a 30-tooth worm gear. The gear has a transverse diametral pitch of 6 teeth/in and a face width of 1 in. The worm has a pitch diameter of 2 in and a face width of $2\frac{1}{2}$ in. The normal pressure angle is $14\frac{1}{2}°$. The materials and workmanship are such that curve B of Fig. 12-17 should be used to obtain the coefficient of friction.

(a) Find the axial pitch, the center distance, the lead, and the lead angle.

(b) Figure 12-18 is a drawing of the worm gear oriented with respect to the coordinate system described in Sec. 12-7; the gear is supported by bearings A and B. Find the forces exerted by the bearings against the worm-gear shaft, and the output torque.

SOLUTION (a) The axial pitch is the same as the transverse circular pitch of the gear, which is

$$p_t = \frac{\pi}{P} = \frac{\pi}{6} = 0.5236 \text{ in} \qquad Ans.$$

The pitch diameter of the gear is $d_G = N_G/P = 30/6 = 5$ in. Therefore the center distance is

$$C = \frac{d_W + d_G}{2} = \frac{2 + 5}{2} = 3.5 \text{ in} \qquad Ans.$$

From Eq. (12-19) the lead is

$$L = p_x N_W = (0.5236)(2) = 1.0472 \text{ in} \qquad Ans.$$

Also, using Eq. (12-20), find

$$\lambda = \tan^{-1}\frac{L}{\pi d_w} = \tan^{-1}\frac{1.0472}{\pi(2)} = 9.47° \qquad Ans.$$

(b) Using the right-hand rule for the rotation of the worm, you will see that your thumb points in the positive z direction. Now use the bolt-and-nut analogy (the worm is right-handed, as is the screw thread of a bolt), and turn the bolt clockwise with the right hand while preventing nut rotation with the left. The nut will move axially along the bolt toward your right hand. Therefore the surface of the gear (Fig. 12-18) in contact with the worm will move in the negative z direction. Thus the gear rotates clockwise about x, with your right thumb pointing in the negative x direction.

The pitch-line velocity of the worm is

$$V_W = \frac{\pi d_w n_W}{12} = \frac{\pi(2)(1200)}{12} = 628 \text{ fpm}$$

The speed of the gear is $n_G = (\frac{2}{30})(1200) = 80$ rpm. So the pitch-line velocity is

$$V_G = \frac{\pi d_G n_G}{12} = \frac{\pi(5)(80)}{12} = 105 \text{ fpm}$$

Then, using Eq. (12-27), the sliding velocity V_S is found to be

$$V_S = \frac{V_W}{\cos \lambda} = \frac{628}{\cos 9.47°} = 638 \text{ fpm}$$

Using Fig. 12-17, we find $\mu = 0.03$. We shall also require the normal pressure angle ϕ_n. Since the helix angle of the gear is the same as the lead angle of the worm, we can use Eq. (12-4). Thus

$$\phi_n = \tan^{-1}(\tan \phi_i \cos \psi) = \tan^{-1}(\tan 14.5° \cos 9.47°) = 14.3°$$

Getting to the forces now, we begin with the horsepower formula

$$W_{W_t} = \frac{33\,000 \text{ P}}{V_W} = \frac{(33\,000)(1)}{628} = 52.5 \text{ lb}$$

This force acts in the negative x direction, the same as in Fig. 12-15. Using the first part of Eq. (12-23), we next find

$$W = \frac{W^x}{\cos \phi_n \sin \lambda + \mu \cos \lambda}$$

$$= \frac{52.5}{\cos 14.3° \sin 9.47° + 0.03 \cos 9.47°} = 278 \text{ lb}$$

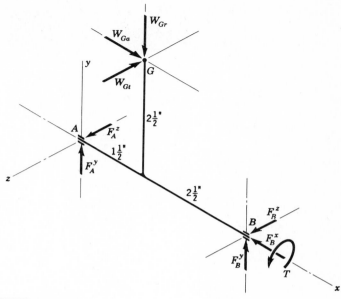

FIGURE 12-19

Also, from Eq. (12-23),

$$W^y = W \sin \phi_n = 278 \sin 14.3° = 68.6 \text{ lb}$$

$$W^z = W(\cos \phi_n \cos \lambda - \mu \sin \lambda)$$

$$= 278(\cos 14.3° \cos 9.47° - 0.03 \sin 9.47°) = 264 \text{ lb}$$

We now identify the components acting upon the gear as

$$W_{Ga} = -W^x = 52.5 \text{ lb}$$

$$W_{Gr} = -W^y = 68.6 \text{ lb} \qquad W_{Gt} = -W^z = -264 \text{ lb}$$

At this point a three-dimensional line drawing should be made in order to simplify the work to follow. An isometric sketch, such as the one in Fig. 12-19, is easy to make and will help you to avoid errors. Note that the y axis is vertical, while the x and z axes make angles of 30° with the horizontal. The illusion of depth is enhanced by sketching lines parallel to each of the coordinate axes through every point of interest.

We shall make B a thrust bearing in order to place the gear shaft in compression. Thus, summing forces in the x direction gives

$$F_B^x = -52.5 \text{ lb} \qquad Ans.$$

Taking moments about the z axis,

$$-(52.5)(2.5) - (68.6)(1.5) + 4F_B^y = 0 \qquad F_B^y = 58.6 \text{ lb} \qquad Ans.$$

Taking moments about the y axis,

$$(264)(1.5) - 4F_B^z = 0 \qquad F_B^z = 99 \text{ lb} \qquad Ans.$$

These three components are now inserted on the sketch as shown at B in Fig. 12-19. Summing forces in the y direction,

$$-68.6 + 58.6 + F_A^y = 0 \qquad F_A^y = 10 \text{ lb} \qquad Ans.$$

Similarly, summing forces in the z direction,

$$-264 + 99 + F_A^z = 0 \qquad F_A^z = 165 \text{ lb} \qquad Ans.$$

These two components can now be placed at A on the sketch. We still have one more equation to write. Summing moments about x,

$$-(264)(2.5) + T = 0 \qquad T = 660 \text{ lb·in} \qquad Ans.$$

It is because of the frictional loss that this output torque is less than the product of the gear ratio and the input torque. $////$

EXAMPLE 12-4   Solve (b) of Example 12-3 using vectors.

SOLUTION   Using Fig. 12-19, write

$$\mathbf{W}_G = 52.5\mathbf{i} - 68.6\mathbf{j} - 264\mathbf{k}$$

Then define the position vectors

$$\mathbf{R}_G = 1.5\mathbf{i} + 2.5\mathbf{j} \qquad \mathbf{R}_B = 4\mathbf{i}$$

Writing the moment equation about A gives

$$\mathbf{R}_G \times \mathbf{W}_G + \mathbf{T} + \mathbf{R}_B \times \mathbf{F}_B = 0 \qquad (1)$$

Substituting the known values,

$$(1.5\mathbf{i} + 2.5\mathbf{j}) \times (52.5\mathbf{i} - 68.6\mathbf{j} - 264\mathbf{k}) + T\mathbf{i} + (4\mathbf{i}) \times (F_B^x\mathbf{i} + F_B^y\mathbf{j} + F_B^z\mathbf{k}) = 0$$

When the cross products are formed, we have

$$(-660\mathbf{i} + 396\mathbf{j} - 234\mathbf{k}) + T\mathbf{i} + (-4F_B^z\mathbf{j} + 4F_B^y\mathbf{k}) = 0 \qquad (2)$$

Thus

$$T = 660\mathbf{i} \text{ lb·in} \qquad Ans.$$

$$F_B^y = 58.6 \text{ lb} \qquad F_B^z = 99 \text{ lb}$$

Taking a summation of forces, next, yields

$$\mathbf{F}_A + \mathbf{F}_B + \mathbf{W}_G = 0 \qquad (3)$$

or substituting known values,

$$(F_A^y \mathbf{j} + F_A^z \mathbf{k}) + (F_B^x \mathbf{i} + 58.6\mathbf{j} + 99\mathbf{k}) + (52.5\mathbf{i} - 68.6\mathbf{j} - 264\mathbf{k}) = 0 \qquad (4)$$

whence

$$F_B^x = -52.5 \text{ lb}$$

and so

$$\mathbf{F}_B = -52.5\mathbf{i} + 58.6\mathbf{j} + 99\mathbf{k} \text{ lb} \qquad \textit{Ans.}$$

Also, from Eq. (4),

$$\mathbf{F}_A = 10\mathbf{j} + 165\mathbf{k} \text{ lb} \qquad \textit{Ans.}$$

$$////$$

12-8   POWER RATING OF WORM GEARING

When worm gearsets are used intermittently or at slow gear speeds, the bending strength of the gear tooth may become a principal design factor. Since the worm teeth are inherently stronger than the gear teeth, they are usually not considered, though the methods of Chap. 6 can be used to compute worm-tooth stresses. The teeth of worm gears are thick and short at the two edges of the face and thin in the central plane, and this makes it difficult to determine the bending stress. Buckingham* adapts the Lewis equation as follows:

$$\sigma = \frac{W_{Gt}}{p_n F_G y} \qquad (12\text{-}28)$$

$$p_n = p_x \cos \lambda \qquad (12\text{-}29)$$

where σ = bending stress, psi
$\quad W_{Gt}$ = transmitted load, lb
$\quad p_n$ = normal circular pitch, in
$\quad p_x$ = axial circular pitch, in
$\quad F_G$ = face width of gear, in
$\quad y$ = Lewis form factor referred to the circular pitch
$\quad \lambda$ = lead angle

Since the equation is only a rough approximation, stress concentration is not considered. Also, for this reason, the form factors are not referred to the number of teeth, but only to the normal pressure angle. The values of y are listed in Table 12-4.

The AGMA equation for *input-horsepower rating* of worm gearing is

$$P = \frac{W_{Gt} d_G n_W}{126\,000 m_G} + \frac{V_s W_f}{33\,000} \qquad (12\text{-}30)$$

---

* Earle Buckingham, "Analytical Mechanics of Gears," p. 495, McGraw-Hill Book Company, New York, 1949.

Table 12-4   VALUES OF y FOR WORM GEARS

Normal pressure angle ϕ_n, degrees	Form factor y
$14\frac{1}{2}$	0.100
20	0.125
25	0.150
30	0.175

The first term on the right is the *output horsepower*, and the second is the *power loss*. The *permissible* transmitted load W_{Gt} is computed from the equation

$$W_{Gt} = K_s d_G^{0.8} F_e K_m K_v \qquad (12\text{-}31)$$

The notation of Eqs. (12-30) and (12-31) is as follows:

W_{Gt} = transmitted load, lb
d_G = pitch diameter of gear, in
n_W = speed of worm, rpm
m_G = gear ratio = N_G/N_W
V_S = sliding velocity at mean worm diameter, fpm
W_f = frictional force, lb
K_s = materials and size-correction factor
F_e = effective face width of gear; the effective face width is the face width
of the gear or two-thirds of the worm pitch diameter, whichever is least
K_m = ratio-correction factor
K_v = velocity factor

Table 12-5   MATERIALS FACTOR K_s FOR CYLINDRICAL WORM GEARING*

Face width of gear F_G, in	Sand-cast bronze	Static chill-cast bronze	Centrifugal-cast bronze
Up to 3	700	800	1000
4	665	780	975
5	640	760	940
6	600	720	900
7	570	680	850
8	530	640	800
9	500	600	750

* Darle W. Dudley (ed.), "Gear Handbook," p. 13–38, McGraw-Hill Book Company, New York, 1962.
     For copper-tin and copper-tin-nickel bronze gears operating with steel worms case-hardened to Rockwell 58C minimum.

Table 12-6 RATIO-CORRECTION FACTOR K_m*

Ratio m_G	K_m	Ratio m_G	K_m	Ratio m_G	K_m
3.0	0.500	8.0	0.724	30.0	0.825
3.5	0.554	9.0	0.744	40.0	0.815
4.0	0.593	10.0	0.760	50.0	0.785
4.5	0.620	12.0	0.783	60.0	0.745
5.0	0.645	14.0	0.799	70.0	0.687
6.0	0.679	16.0	0.809	80.0	0.622
7.0	0.706	20.0	0.820	100.0	0.490

* Darle W. Dudley (ed.), "Gear Handbook," p. 13–38, McGraw-Hill Book Company, New York, 1962.

Table 12-7 VELOCITY FACTOR K_v*

Velocity V_S, fpm	K_v	Velocity V_S, fpm	K_v	Velocity V_S, fpm	K_v
1	0.649	300	0.472	1400	0.216
1.5	0.647	350	0.446	1600	0.200
10	0.644	400	0.421	1800	0.187
20	0.638	450	0.398	2000	0.175
30	0.631	500	0.378	2200	0.165
40	0.625	550	0.358	2400	0.156
60	0.613	600	0.340	2600	0.148
80	0.600	700	0.310	2800	0.140
100	0.588	800	0.289	3000	0.134
150	0.558	900	0.269	4000	0.106
200	0.528	1000	0.258	5000	0.089
250	0.500	1200	0.235	6000	0.079

* Darle W. Dudley (ed.), "Gear Handbook," p. 13–39, McGraw-Hill Book Company, New York, 1962.

Values of the materials factor for hardened steel worms mating with bronze gears are listed in Table 12-5. Note the effect of the size-correction factor as the face width increases.

Values of the ratio-correction factor K_m and the velocity factor K_v are tabulated in Tables 12-6 and 12-7, respectively.

12-9 STRAIGHT BEVEL GEARS—KINEMATICS

When gears are to be used to transmit motion between intersecting shafts, some form of bevel gear is required. A bevel gearset is shown in Fig. 12-20. Although bevel gears are usually made for a shaft angle of 90°, they may be produced for almost any angle. The teeth may be cast, milled, or generated. Only the generated teeth may be classed as accurate.

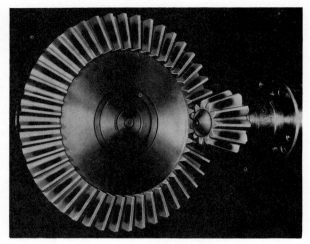

FIGURE 12-20
A straight bevel gear and pinion. (*Courtesy of Gleason Works, Rochester, N.Y.*)

The terminology of bevel gears is illustrated in Fig. 12-21. The pitch of bevel gears is measured at the large end of the tooth, and both the circular pitch and the pitch diameter are calculated in the same manner as for spur gears. It should be noted that the clearance is uniform. The pitch angles are defined by the pitch cones meeting at the apex, as shown in the figure. They are related to the tooth numbers as follows:

$$\tan \gamma = \frac{N_P}{N_G} \qquad \tan \Gamma = \frac{N_G}{N_P} \qquad (12\text{-}32)$$

Table 12-8   TOOTH PROPORTIONS FOR 20° STRAIGHT BEVEL-GEAR TEETH

Item	Formula
Working depth	$h_k = 2.0/P$
Clearance	$c = (0.188/P) + 0.002$ in
Addendum of gear	$a_G = \dfrac{0.54}{P} + \dfrac{0.460}{P(m_{90})^2}$
Gear ratio	$m_G = N_G/N_P$
Equivalent 90° ratio	$m_{90} = m_G$ when $\sum = 90°$
	$m_{90} = \sqrt{m_G \dfrac{\cos \gamma}{\cos \Gamma}}$ when $\sum \neq 90°$
Face width	$F = \dfrac{A_o}{3}$ or $F = \dfrac{10}{P}$, whichever is smaller

Minimum number of teeth:	Pinion	16	15	14	13
	Gear	16	17	20	30

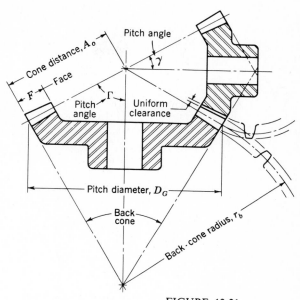

Pitch angle

Cone distance, A_o

Face

F

Pitch angle

Γ

Uniform clearance

γ

Pitch diameter, D_G

Back cone

Back-cone radius, r_b

FIGURE 12-21
Terminology of bevel gears.

where the subscripts P and G refer to the pinion and gear, respectively, and where γ and Γ are, respectively, the pitch angles of the pinion and gear.

Figure 12-21 shows that the shape of the teeth, when projected on the back cone, is the same as in a spur gear having a radius equal to the back-cone distance r_b. This is called Tredgold's approximation. The number of teeth in this imaginary gear is

$$N' = \frac{2\pi r_b}{p} \qquad (12\text{-}33)$$

where N' is the *virtual number of teeth* and p is the circular pitch measured at the large end of the teeth.

Standard straight-tooth bevel gears are cut by using a 20° pressure angle, unequal addenda and dedenda, and full-depth teeth. This increases the contact ratio, avoids undercut, and increases the strength of the pinion. Table 12-8 lists the standard tooth proportions at the large end of the teeth.

12-10   BEVEL GEARS—FORCE ANALYSIS

In determining shaft and bearing loads for bevel-gear applications, the usual practice is to use the tangential or transmitted load which would occur if all the force were concentrated at the midpoint of the tooth. While the actual resultant

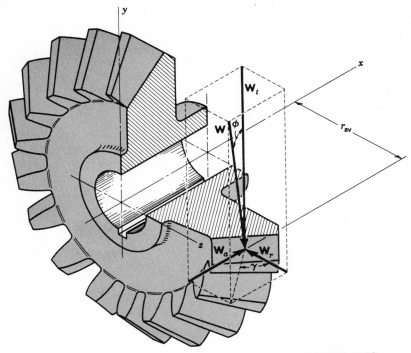

FIGURE 12-22
Bevel-gear tooth forces.

occurs somewhere between the midpoint and the large end of the tooth, there is only a small error in making this assumption. For the transmitted load, this gives

$$W_t = \frac{T}{r_{av}} \quad (12\text{-}34)$$

where T is the torque and r_{av} is the pitch radius of the gear under consideration at the midpoint of the tooth.

The forces acting at the center of the tooth are shown in Fig. 12-22. The resultant force W has three components, a tangential force W_t, a radial force W_r, and an axial force W_a. From the trigonometry of the figure,

$$W_r = W_t \tan \phi \cos \gamma \quad (12\text{-}35)$$

$$W_a = W_t \tan \phi \sin \gamma \quad (12\text{-}36)$$

The three forces W_t, W_r, and W_a are at right angles to each other and can be used to determine the bearing loads by using the methods of statics.

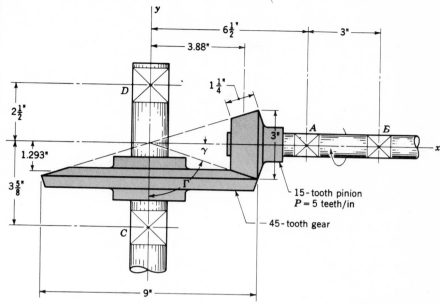

y

$6\frac{1}{2}"$

3"

3.88"

D

$1\frac{1}{4}"$

$2\frac{1}{2}"$

3"

A

E

γ

1.293"

x

$3\frac{5}{8}"$

15-tooth pinion
P = 5 teeth/in

Γ

45-tooth gear

C

9"

FIGURE 12-23

EXAMPLE 12-5 The bevel pinion in Fig. 12-23 rotates at 600 rpm in the direction shown and transmits 5 hp to the gear. The mounting distances, the location of all bearings, and the average pitch radii of the pinion and gear are shown in the figure. For simplicity, the teeth have been replaced by the pitch cones. Bearings A and C should take the thrust loads. Find the bearing forces on the gearshaft.

SOLUTION The pitch angles are

$$\gamma = \tan^{-1}\left(\tfrac{3}{9}\right) = 18.4° \qquad \Gamma = \tan^{-1}\left(\tfrac{9}{3}\right) = 71.6°$$

The pitch-line velocity corresponding to the average pitch radius is

$$V = \frac{2\pi r_P n}{12} = \frac{2\pi(1.293)(600)}{12} = 406 \text{ fpm}$$

Therefore the transmitted load is

$$W_t = \frac{33\,000\,P}{V} = \frac{(33\,000)(5)}{406} = 406 \text{ lb}$$

which acts in the positive z direction, as shown in Fig. 12-24. We next have

$$W_r = W_t \tan \phi \cos \Gamma = 406 \tan 20° \cos 71.6° = 46.6 \text{ lb}$$

$$W_a = W_t \tan \phi \sin \Gamma = 406 \tan 20° \sin 71.6° = 140 \text{ lb}$$

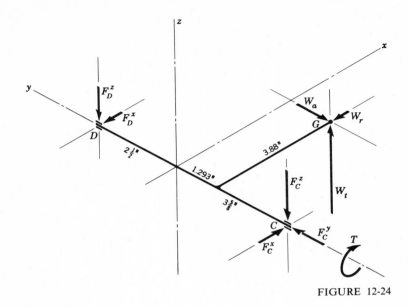

FIGURE 12-24

where W_r is in the $-x$ direction and W_a in the $-y$ direction, as illustrated in the isometric sketch of Fig. 12-24.

In preparing to take a sum of the moments about bearing D, define the position vector from D to G as

$$\mathbf{R}_G = 3.88\mathbf{i} - (2.5 + 1.293)\mathbf{j} = 3.88\mathbf{i} - 3.793\mathbf{j}$$

We shall also require a vector from D to C.

$$\mathbf{R}_C = -(2.5 + 3.625)\mathbf{j} = -6.125\mathbf{j}$$

Then, summing moments about D gives

$$\mathbf{R}_G \times \mathbf{W} + \mathbf{R}_C \times \mathbf{F}_C + \mathbf{T} = 0 \qquad (1)$$

When we place the details in Eq. (1), we get

$$(3.88\mathbf{i} - 3.793\mathbf{j}) \times (-46.6\mathbf{i} - 140\mathbf{j} + 406\mathbf{k})$$
$$+ (6.125\mathbf{j}) \times (F_C^x\mathbf{i} + F_C^y\mathbf{j} + F_C^z\mathbf{k}) + T\mathbf{j} = 0 \qquad (2)$$

After forming the two cross products, the equation becomes

$$(-1504\mathbf{i} - 1580\mathbf{j} - 721\mathbf{k}) + (-6.125F_C^z\mathbf{i} + 6.125F_C^x\mathbf{k}) + T\mathbf{j} = 0$$

from which

$$\mathbf{T} = 1580\mathbf{j} \text{ lb·in} \qquad F_C^x = 118 \text{ lb} \qquad F_C^z = -246 \text{ lb}$$

Now sum the forces to zero. Thus

$$\mathbf{F}_D + \mathbf{F}_C + \mathbf{W} = 0 \qquad (4)$$

When the details are inserted, Eq. (4) becomes

$$(F_D^x \mathbf{i} + F_D^z \mathbf{k}) + (118\mathbf{i} + F_C^y \mathbf{j} - 246\mathbf{k}) + (-46.6\mathbf{i} - 140\mathbf{j} + 406\mathbf{k}) = 0 \qquad (5)$$

First we see that $F_C^y = 140$ lb, and so

$$\mathbf{F}_C = 118\mathbf{i} + 140\mathbf{j} - 246\mathbf{k} \text{ lb} \qquad Ans.$$

Then, from Eq. (5),

$$\mathbf{F}_D = -71\mathbf{i} - 160\mathbf{k} \text{ lb} \qquad Ans.$$

These are all shown in Fig. 12-24 in the proper directions. The analysis for the pinion shaft is quite similar. ////

12-11   BEVEL GEARING— BENDING STRESS AND STRENGTH

In a typical bevel-gear mounting, Fig. 12-23 for example, one of the gears is often mounted outboard of the bearings. This means that shaft deflections can be more pronounced and have a greater effect on the contact of the teeth. Another difficulty which occurs in predicting the stress in bevel-gear teeth is the fact that the teeth are tapered. Thus, to achieve perfect line contact passing through the cone center, the teeth ought to bend more at the large end than at the small end. To obtain this condition requires that the load be proportionately greater at the large end. Because of this varying load across the face of the tooth, it is desirable to have a fairly short face width.

The equation for bending stress in spur gears is used for bevel gears, too, and is repeated here for convenience.

$$\sigma = \frac{W_t P}{K_v F J} \qquad (12\text{-}37)$$

where all relations are based on the large end of the teeth.

Caution: The transmitted load W_t must be computed using the pitch radius at the large end of the teeth in Eq. (12-37). Note that this is not the same transmitted load used in force analysis (Sec. 12-10), though the symbol is the same.

Table 12-9   APPROXIMATE BEVEL-GEAR LOAD-DISTRIBUTION FACTORS K_m AND C_m*

Application	Both gears inboard	One gear outboard	Both gears outboard
General industrial	1.00–1.10	1.10–1.25	1.25–1.40
Automotive	1.00–1.10	1.10–1.25	
Aircraft	1.00–1.25	1.10–1.40	1.25–1.50

* AGMA Information Sheet 225.01, 1967, table 4.

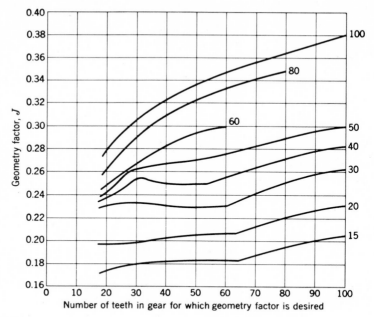

FIGURE 12-25

Geometry factors J for straight bevel gears; these are for a 90° shaft angle, 20° pressure angle, and a clearance of $c = 0.240/P$ in. (*AGMA Information Sheet 225.01.*)

The geometry factor J is different for bevel gears because the long-and-short addendum system is used and because the teeth are tapered. Use Fig. 12-25.

The modification and correction factors for bevel gears are the same as for spur gears except for the load-distribution factor K_m (Table 12-9).*

12-12 BEVEL GEARING—SURFACE DURABILITY

The Hertzian contact stress for bevel gears is given by the equation

$$\sigma_H = C_p \sqrt{\frac{W_t}{C_v F d_p I}} \qquad (12\text{-}38)$$

where, again, all values correspond to the large end of the teeth.

Since the contact of bevel-gear teeth tends to be localized, the elastic coefficient C_p must be based on a Hertzian analysis of contacting spheres rather than cylinders. This yields slightly different values. Thus, use Table 12-10.

* The AGMA uses a different size factor for bevel gears than for others. However, they compensate for this by recommending a different set of allowable stresses. See AGMA Information Sheet 225.01, 1967.

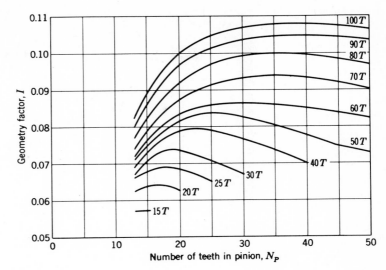

FIGURE 12-26
Geometry factors I for straight bevel gears of 20° pressure angle mounted at a 90° shaft angle. (*AGMA Information Sheet 212.02.*)

Figure 12-26 is a chart of the geometry factor I for bevel gears. All other factors may be obtained using the methods of Chap. 11.

12-13 SPIRAL BEVEL GEARS

Straight bevel gears are easy to design and simple to manufacture and give very good results in service if they are mounted accurately and positively. As in the case of spur gears, however, they become noisy at the higher values of the pitch-line velocity. In these cases it is often good design practice to go to the spiral bevel

Table 12-10   VALUES OF THE ELASTIC COEFFICIENT C_p FOR BEVEL GEARS AND OTHERS WITH LOCALIZED CONTACT. IN EACH CASE THE MODULUS OF ELASTICITY IS IN MPSI.*

| | Gear | | | |
Pinion	Steel	Cast iron	Aluminum bronze	Tin bronze
Steel, $E = 30$	2800	2450	2400	2350
Cast iron, $E = 19$	2450	2250	2200	2150
Aluminum bronze,				
$E = 17.5$	2400	2200	2150	2100
Tin bronze, $E = 16$	2350	2150	2100	2050

* AGMA Sheet 212.02

FIGURE 12-27
Spiral bevel gears. (*Courtesy Gleason Works, Rochester, N.Y.*)

gear, which is the bevel counterpart of the helical gear. Figure 12-27 shows a mating pair of spiral bevel gears, and it can be seen that the pitch surfaces and the nature of contact are the same as for straight bevel gears, except for the differences brought about by the spiral-shaped teeth.

The *hand* of the spiral is found using the right-hand rule, with the thumb pointing along the axis of rotation. In Fig. 12-27, a left-hand pinion is in mesh with a right-hand gear.

Table 12-11   GEAR ADDENDUM FOR 1-DIAMETRAL-PITCH SPIRAL BEVEL GEARS*

Ratios			Ratios			Ratios		
From	To	Addendum	From	To	Addendum	From	To	Addendum
1.00	1.00	0.850	1.23	1.26	0.710	1.82	1.90	0.570
1.00	1.02	0.840	1.26	1.28	0.700	1.90	1.99	0.560
1.02	1.03	0.830	1.28	1.31	0.690	1.99	2.10	0.550
1.03	1.05	0.820	1.31	1.34	0.680	2.10	2.23	0.540
1.05	1.06	0.810	1.34	1.37	0.670	2.23	2.38	0.530
1.06	1.08	0.800	1.37	1.41	0.660	2.38	2.58	0.520
1.08	1.09	0.790	1.41	1.44	0.650	2.58	2.82	0.510
1.09	1.11	0.780	1.44	1.48	0.640	2.82	3.17	0.500
1.11	1.13	0.770	1.48	1.52	0.630	3.17	3.67	0.490
1.13	1.15	0.760	1.52	1.57	0.620	3.67	4.56	0.480
1.15	1.17	0.750	1.57	1.63	0.610	4.56	7.00	0.470
1.17	1.19	0.740	1.63	1.68	0.600	7.00	∞	0.460
1.19	1.21	0.730	1.68	1.75	0.590			
1.21	1.23	0.720	1.75	1.82	0.580			

* Gleason Works, Rochester, N.Y.

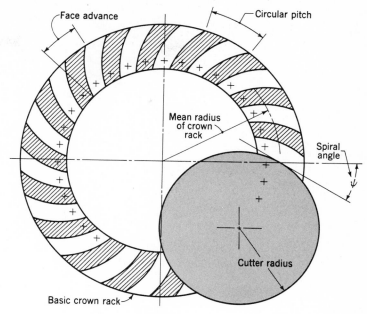

FIGURE 12-28
Cutting spiral-gear teeth on the basic crown rack.

Spiral-bevel-gear teeth are conjugate to a basic crown rack ($2\Gamma = 180°$), which is generated, as shown in Fig. 12-28, by using a circular cutter. The spiral angle ψ is measured at the mean radius of the gear. As in the case of helical gears, spiral bevel gears give a much smoother tooth action than straight bevel gears, and hence are useful where high speeds are encountered. Antifriction bearings should be used to take the thrust loads, however, because these loads are larger than for straight bevel gears. The *face-contact ratio*, which is the face advance divided by the circular pitch (Fig. 12-28), should be at least 1.25 to obtain true spiral-tooth action.

Pressure angles used with spiral bevel gears are generally $14\frac{1}{2}$ to $20°$, while the spiral angle is usually $35°$. The hand of the spiral should be selected so as to cause the gears to separate from each other and not to force them together, which might cause jamming. For example, the left-hand pinion of Fig. 12-27 will be forced into the teeth of the gear if it is rotated in the direction of the fingers of your right hand when your thumb is pointing from left to right. In any event, the supporting bearings should always be designed so that there is no looseness or play in the axial direction.

Tooth proportions for spiral bevel gears having a $20°$ pressure angle, a $35°$ spiral angle, and a stub height are given in Table 12-11. For these the working depth is $1.700/P$ and the clearance $0.188/P$.

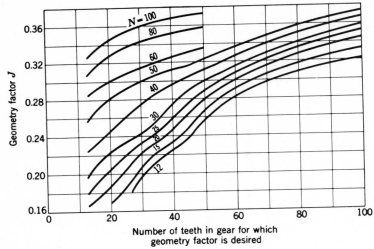

FIGURE 12-29
Geometry factors J for spiral bevel gears of 20° pressure angle and 35° spiral angle.
(*By permission, from "Gear Handbook," p. 13-36, McGraw-Hill Book Company, New York, 1962.*)

The total force W acting normal to the pinion tooth and assumed concentrated at the average radius of the pitch cone may be divided into three perpendicular components. These are the transmitted, or tangential, load W_t; the axial, or thrust, component W_a; and the separating, or radial, component W_r. The force W_t may, of course, be computed from the equation

$$W_t = \frac{T}{r_{av}} \qquad (12\text{-}39)$$

where T is the input torque and r_{av} is the average radius of the pinion pitch cone. The forces W_a and W_r depend upon the hand of the spiral and the direction of rotation. Thus there are four possible cases to consider. For a *right-hand pinion spiral with clockwise pinion rotation* and for a *left-hand spiral with counterclockwise rotation*, the equations are

$$W_a = \frac{W_t}{\cos \psi} (\tan \phi_n \sin \gamma - \sin \psi \cos \gamma)$$
$$W_r = \frac{W_t}{\cos \psi} (\tan \phi_n \cos \gamma + \sin \psi \sin \gamma) \qquad (12\text{-}40)$$

The other two cases are a *left-hand spiral with clockwise rotation* and a *right-hand spiral with counterclockwise rotation*. For these two cases the equations are

$$W_a = \frac{W_t}{\cos \psi} (\tan \phi_n \sin \gamma + \sin \psi \cos \gamma)$$
$$W_r = \frac{W_t}{\cos \psi} (\tan \phi_n \cos \gamma - \sin \psi \sin \gamma) \qquad (12\text{-}41)$$

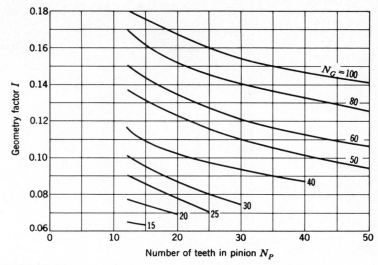

FIGURE 12-30
Geometry factors I for spiral bevel gears of 20° pressure angle and 35° spiral angle. (*By permission, from "Gear Handbook," p. 13-27, McGraw-Hill Book Company, New York, 1962.*)

where ψ = spiral angle
     γ = pinion pitch angle
     ϕ_n = normal pressure angle

and the rotation is observed from the input end of the pinion shaft. Equations (12-40) and (12-41) give the forces exerted by the gear on the pinion. A positive sign for either W_a or W_r indicates that it is directed away from the cone center.

The forces exerted by the pinion on the gear are equal and opposite. Of course, the opposite of an axial pinion load is a radial gear load, and the opposite of a radial pinion load is an axial gear load.

Except for the geometry factors I and J, the same stress and strength equations apply for both bending and wear as for straight bevel gears. Figures 12-29 and 12-30 are used to obtain the factors J and I, respectively.

The *Zerol bevel gear* is a patented gear having curved teeth but with a zero spiral angle. It thus can be generated using the same tooling as for a regular spiral bevel gear. The curved teeth give somewhat better tooth action than can be obtained with straight-tooth bevel gears. In design it is probably best to proceed as for straight-tooth bevel gears and then simply substitute a Zerol bevel gear.

It is frequently desirable, as in the case of automotive differential applications, to have gearing similar to bevel gears but with the shafts offset. Such gears are called *hypoid gears* because their pitch surfaces are hyperboloids of revolution. The tooth action between such gears is a combination of rolling and sliding along

FIGURE 12-31
Hypoid gears. (*Gleason Works, Rochester, N. Y.*)

a straight line and has much in common with that of worm gears. Figure 12-31 shows a pair of hypoid gears in mesh.

Figure 12-32 is included to assist in the classification of spiral-type bevel gearing. It is seen that the hypoid gear has a relatively small shaft offset. For larger offsets the pinion begins to resemble a tapered worm, and the set is then called *spiroid gearing*.

EXAMPLE 12-6 Let the pinion of Example 12-5 (Fig. 12-23) be cut with a left-hand 35° spiral. Using the remaining data of the example, find the forces exerted on the gear shaft by the bearings at C and D.

SOLUTION Using $W_t = 406$ lb, we find, from Eq. (12-41),

$$W_a = \frac{406}{\cos 35°} (\tan 20° \sin 18.4° + \sin 35° \cos 18.4°) = 326 \text{ lb}$$

$$W_r = \frac{406}{\cos 35°} (\tan 20° \cos 18.4° - \sin 35° \sin 18.4°) = 81 \text{ lb}$$

These are the forces exerted by the gear on the pinion. Referring to Fig. 12-23, W_t is in the $-z$ direction, W_a in the $+x$ direction, and W_r in the $+y$ direction. The corresponding component forces on the gear are the same as in Fig. 12-24. Therefore, for the gear, we write

$$\mathbf{W} = -326\mathbf{i} - 81\mathbf{j} + 406\mathbf{k} \text{ lb}$$

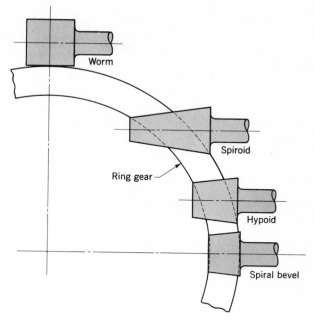

FIGURE 12-32
Comparison of intersecting- and offset-shaft bevel-type gearing. (*By permission, from "Gear Handbook," p. 2-14, McGraw-Hill Book Company, New York, 1962.*)

The procedure is now the same as in Example 12-5. The results are

$$\mathbf{F}_C = 252\mathbf{i} + 81\mathbf{j} - 251\mathbf{k} \text{ lb} \qquad Ans.$$

$$\mathbf{F}_D = 74\mathbf{i} - 155\mathbf{k} \text{ lb} \qquad Ans.$$

////

PROBLEMS

Sections 12-1, 12-2, and 12-3

12-1   A parallel helical gearset consists of an 18-tooth pinion driving a 32-tooth gear. The pinion has a left-hand helix angle of 25°, a normal pressure angle of 20°, and a normal diametral pitch of 8 teeth/in.
   (*a*) Find the normal, transverse, axial, and the normal base-circular pitches.
   (*b*) Find the transverse diametral pitch and the transverse pressure angle.
   (*c*) Find the addendum and dedendum and the pitch diameters of both gears.

12-2 The same as Prob. 12-1 except the normal diametral pitch is 6, the pinion has 16 teeth with a right-hand helix angle of 30°, and the gear has 48 teeth.

12-3 A pair of mating parallel helical gears has an angular-velocity ratio of 3.20. The driver has 20 teeth, a transverse diametral pitch of 4, a 15° left-hand helix, and a transverse pressure angle of 20°.

(*a*) Compute all circular and diametral pitches and the normal pressure angle.

(*b*) Find the addendum and dedendum and the pitch diameters of both gears.

12-4 A 14-tooth helical pinion runs at a speed of 2400 rpm and drives a gear on a parallel shaft at a speed of 600 rpm. The pinion has a normal diametral pitch of 12, and a 23° left-hand helix angle. The normal pressure angle is 25°.

(*a*) Compute all the circular and diametral pitches and the transverse pressure angle.

(*b*) Find the whole depth of the teeth based on shaved profiles using the proportions in Table 11-1.

(*c*) What are the pitch diameters of both gears?

12-5 The gears shown in the figure have a normal diametral pitch of 6, a normal pressure angle of 20°, and a helix angle of 30°. For both gear trains the transmitted load is 400 lb. Find the force exerted by each gear on its shaft. The direction of pinion rotation is shown.

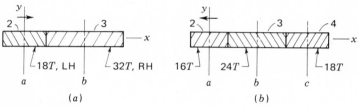

PROBLEM 12-5

12-6 A gear train is composed of four helical gears with the shaft axes in a single plane. The gears have a 20° transverse pressure angle and a 30° helix angle. In the figure, gear 2 drives gear 3 with a transmitted load of 500 lb. Find the radial and thrust forces exerted on each shaft. The gears on shaft *b* have 56 and 14 teeth and are each 7-diametral-pitch in the normal plane.

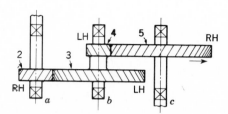

PROBLEM 12-6

12-7 Gear 2, in the figure, has 16 teeth, a 20° transverse pressure angle, a 15° helix angle, and a normal diametral pitch of 8 teeth/in. Gear 2 drives the idler on shaft *b*, which has 36 teeth. The driven gear on shaft *c* has 28 teeth. If the driver rotates at 1720 rpm and transmits $7\frac{1}{2}$ hp, find the radial and thrust loads on each shaft.

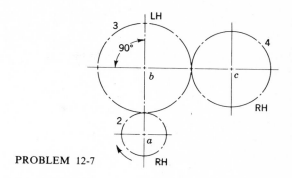

PROBLEM 12-7

12-8 The figure shows a double-reduction helical gearset. Pinion 2 is the driver, and it receives a torque of 1200 lb·in from its shaft in the direction shown. Pinion 2 has a normal diametral pitch of 8 teeth/in, 14 teeth, and a normal pressure angle of 20° and is cut right-handed with a helix angle of 30°. The mating gear 3 on shaft *b* has 36 teeth. Gear 4, which is the driver for the second pair of gears in the train, has a normal diametral pitch of 5 teeth/in, 15 teeth, and a normal pressure angle of 20° and is cut right-handed with a helix angle of 15°. Mating gear 5 has 45 teeth. Find the magnitude and direction of the force exerted by the bearings at *C* and *D* on shaft *b* if bearing *C* can take only radial load while bearing *D* is mounted to take both radial and thrust load.

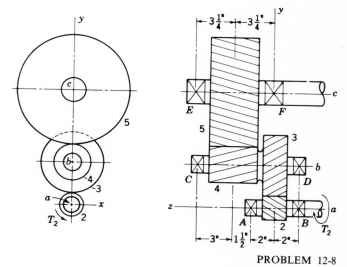

PROBLEM 12-8

12-9 The double-reduction helical gearset shown in the figure is driven through shaft *a* and receives 7½ hp at a speed of 900 rpm. Gears 2 and 3 have a transverse diametral pitch of 10 teeth/in, a 30° helix angle, and a transverse pressure angle of 20°. Pinion 2 is cut with 14 teeth, left-handed; gear 3 has 54 teeth. Each of the second pair of gears in the train, 4 and 5, has a transverse diametral pitch of 6 teeth/in, a 23° helix angle, and a transverse pressure angle of 20°. Gear 4 is left-handed and has 16 teeth; gear 5 has 36 teeth. The gears are supported by bearings located as shown in the figure. Good design dictates that the thrust bearing should be located so as to load the shaft in compression. The thrust bearing then takes both radial and thrust loads, while the second bearing on the shaft is subject to pure radial load. In accordance with this convention, determine the magnitude and direction of the radial and thrust loads which bearings C and D exert on shaft b.

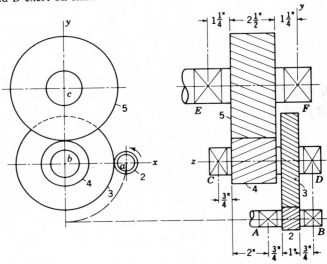

PROBLEM 12-9

Section 12-4

12-10 A parallel-shaft helical gearset is composed of a driving pinion with 20 teeth and a driven gear with 96 teeth. The gears have a transverse diametral pitch of 20 teeth/in, a 25° helix angle, and a pressure angle $\phi_n = 20°$ and are cut full depth. The face width is $\frac{3}{4}$ in, and the gears run at a pitch-line velocity of 1600 fpm. The driver is cut from cold-drawn UNS G10180 steel without heat-treatment, and the driven of grade 30 cast iron. Basing computations on bending strength, a factor of safety $n_G = 4$, and unity for the life, temperature, and reliability factors, find the maximum safe capacity in horsepower for general industrial use.

12-11 A speed reducer is composed of a helical driver of 25 teeth and a mating gear of 64 teeth. The gears have a normal diametral pitch of 16, a 30° helix angle, and a 20° normal pressure angle. The face width is 1 in, and the maximum pinion speed is 2300 rpm. Both gears are made of nickel-alloy steel, case-hardened to 560 Bhn. Determine the maximum tangential load that these gears can safely transmit based on bending strength and a reliability of 99.9 percent or better. Use unity for overload and temperature factors and a factor of safety $n_G = 3.00$.

12-12 Find the horsepower capacity of the gears of Prob. 12-10 based upon surface durability.

12-13 Find the maximum tangential load for the gears of Prob. 12-11 based on surface durability.

12-14 The first pair of gears in a three-stage herringbone reducer consists of a 26-tooth driver and a 104-tooth driven. The gears have a normal diametral pitch of 14 teeth/in, a normal pressure angle of 20°, and a helix angle of 25°. The face width of each half is $1\frac{1}{4}$ in. The gears are forged of UNS G46500 steel and heat-treated to 410 Bhn. For a pinion speed of 3600 rpm and infinite life with 99 percent reliability, find the capacity in horsepower of the first pair of gears on the basis of bending strength. Use an overall factor of safety $n_G = 3.20$.

Section 12-5

12-15 Find the rpm and direction of rotation of gear 7 in the figure.

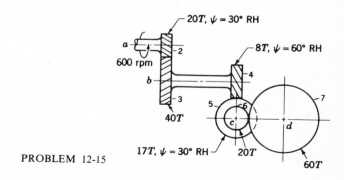

PROBLEM 12-15

Section 12-7

12-16 A 4-tooth right-hand worm transmits $\frac{1}{3}$ hp at 750 rpm to a 40-tooth gear having a transverse diametral pitch of 10 teeth/in. The worm has a pitch diameter of $1\frac{1}{4}$ in and a face width of $1\frac{3}{8}$ in. The normal pressure angle is $14\frac{1}{2}°$. The worm is made of low-carbon steel, without heat-treatment, and the gear of cast iron. Find the input torque, the output torque, and the efficiency. Use $\mu = 0.07$.

12-17 A right-hand single-tooth hardened-steel worm (hardness not specified) has a cata-log rating of 2.70 hp at 600 rpm when meshed with a 48-tooth cast-iron gear. The transverse diametral pitch is 3 teeth/in, the transverse pressure angle is $14\frac{1}{2}°$, the pitch diameter of the worm is 4 in, and the face widths of the worm and gear are, respectively, 4 and 2 in. The figure shows bearings A and B on the worm shaft symmetrically located with respect to the worm and 8 in apart. Determine which should be the thrust bearing, and find the magnitudes and directions of the forces exerted by both bearings.

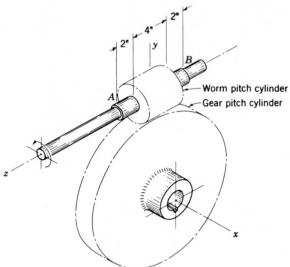

PROBLEM 12-17

12-18 The hub diameter and projection for the gear of Prob. 12-17 are 4 and $1\frac{1}{2}$ in, respectively. The face width of the gear is 2 in. Locate bearings C and D on opposite sides, spacing them 1 in from the gear, and then find the magnitudes and directions of the forces exerted by the bearings on the gearshaft. Also, find the output torque.

Section 12-8

12-19 A 2-tooth case-hardened worm of forged steel is to drive a 40-tooth sand-cast bronze worm gear. The diametral pitch is 10 teeth/in. The worm has a $14\frac{1}{2}°$ normal pressure angle, a lead angle of 9.083° a pitch diameter of 1.25 in, and a face width of 2 in, and its teeth are ground and polished. The gear has a $\frac{5}{8}$-in face and a pitch diameter of 4.00 in. For a worm speed of 1720 rpm find the maximum safe horsepower output and the efficiency of this gearset.

12-20 A 5-diametral-pitch 3-tooth steel worm is case-hardened and has ground and polished teeth. The worm has a 20° normal pressure angle, a 8.62° lead angle, a 2.30-in pitch diameter, and a 3-in face width. The worm rotates at 1740 rpm and drives a 50-tooth $1\frac{3}{8}$-in-face sand-cast bronze worm gear. Find the maximum horsepower that can be transmitted by this gearset.

12-21 A 16-diametral-pitch 4-tooth case-hardened steel worm has a lead angle of 21.8° and a $\frac{5}{8}$-in pitch diameter. It drives a centrifugal-cast-bronze worm gear having a $\frac{5}{16}$-in face, 40 teeth, and a 20° normal pressure angle. The worm is right-handed and rotates at 600 rpm. Calculate the maximum rated horsepower and the efficiency of this gearset.

Section 12-9

12-22 The figure shows a gear train consisting of bevel gears, spur gears, and a worm and worm gear. The bevel pinion is mounted on a shaft driven by a V belt and pulleys. If pulley 2 rotates at 1200 rpm in the direction shown, find the speed and direction of rotation of gear 9.

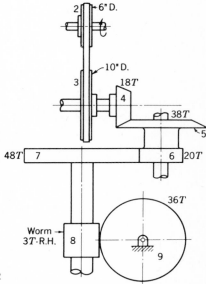

PROBLEM 12-22

12-23 The tooth numbers for the automotive differential shown in the figure are $N_2 = 17$, $N_3 = 54$, $N_4 = 11$, $N_5 = N_6 = 16$. The drive shaft turns at 1200 rpm. What is the speed of the right wheel if it is jacked up and the left wheel is resting on the road surface?

12-24 A vehicle using the differential shown in the figure turns to the right at a speed of 30 mph on a curve of 80-ft radius. Use the same tooth numbers as in Prob. 12-23. The tires are 15 in in diameter. Use 60 in as the center-to-center distance between treads.

(*a*) Calculate the speed of each rear wheel.

(*b*) What is the speed of the ring gear?

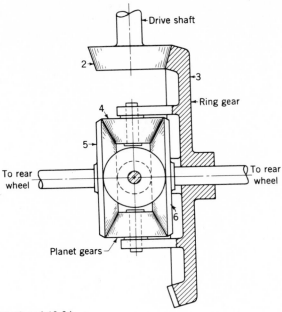

PROBLEMS 12-23 and 12-24

Section 12-10

12-25 The figure shows a gear train composed of a pair of helical gears and a pair of straight bevel gears. Shaft *c* is the output of the train, and it delivers 6 hp to the load at a speed of 370 rpm. The bevel gears have a pressure angle of 20°. If bearing *E* is to take both thrust and radial load while bearing *F* is to take only radial, determine the magnitude and direction of the forces which these bearings exert against shaft *c*.

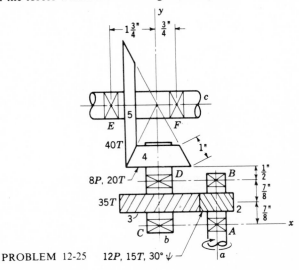

PROBLEM 12-25   12P, 15T, 30° ψ

12-26 Using the data of Prob. 12-25, find the forces exerted by bearings C and D on shaft b. Which of these bearings should take the thrust load if the shaft is to be loaded in compression? The helical gears have a 20° transverse pressure angle.

12-27 The gear train shown in the figure is composed of a right-hand helical pinion on shaft a which is cut with a $17\frac{1}{2}$° normal pressure angle, the mating gear on shaft b, a straight-tooth miter gear on shaft b, and its mating miter gear on shaft c. The miter gears operate at a 20° pressure angle. The torque input to shaft a is 875 lb·in in the direction shown. Find the forces exerted by bearings C and D on shaft b. Specify which bearing is to take the thrust load.

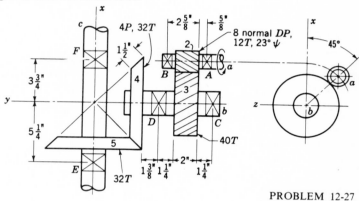

PROBLEM 12-27

12-28 Using the data of Prob. 12-27, find the forces exerted by bearings at E and F on shaft c. Which of these two bearings should take the thrust component?

Section 12-11

12-29 A straight-tooth bevel pinion has 15 teeth and a diametral pitch of 6 teeth/in, is made of cold-drawn UNS G10180 steel without heat treatment, and is formed by hobbing the teeth. The pinion drives a 60-tooth gear made of grade 30 cast iron. The shaft angle is 90°, the face widths $1\frac{1}{4}$ in, and the pressure angle 20°. The pinion rotates at 900 rpm and is mounted outboard of its bearings, while the gear is straddle-mounted. Based on bending strength, 50 percent reliability, and an overall factor of safety $n_G = 4$, specify the maximum safe capacity of the drive in horsepower for general industrial use.

12-30 Based on surface durability and an overall factor of safety $n_G = 1.5$, find the horsepower rating of the gears of Prob. 12-29.

12-31 A 16-tooth bevel pinion has a diametral pitch of 4 teeth/in, teeth cut by hobbing and then shaving using the 20° system, and a face width of $1\frac{1}{2}$ in. The pinion drives a 32-tooth gear with a shaft angle of 90°. The pinion and gear are made of UNS G43400 forged steel, heat-treated to 350 Bhn. Based on bending strength, a reliability of 99 percent, and a pitch-line velocity of 1500 fpm, determine the horsepower capacity of this pair of gears. Use $n_G = 3.50$.

13

SHAFTS

The design of shafts is a basic design problem. It utilizes most, if not all, of the fundamentals discussed in the first five chapters of this book. You have by now gained some degree of maturity in applying these fundamentals to the analysis and design of various mechanical elements, and this, coupled with your knowledge of bearings and gears, makes it possible to reexamine shafting with somewhat more sophistication.

13-1   INTRODUCTION

A *shaft* is a rotating or stationary member, usually of circular cross section, having mounted upon it such elements as gears, pulleys, flywheels, cranks, sprockets, and other power-transmission elements. Shafts may be subjected to bending, tension, compression, or torsional loads, acting singly or in combination with one another. When they are combined, one may expect to find both static and fatigue strength to be important design considerations, since a single shaft may be subjected to static stresses, completely reversed stresses, and repeated stresses, all acting at the same time.

The word "shaft" covers numerous variations, such as axles and spindles. An *axle* is a shaft, either stationary or rotating, not subjected to a torsion load. A short rotating shaft is often called a *spindle*.

When either the lateral or the torsional deflection of a shaft must be held to close limits, the shaft should be sized on the basis of deflection before analyzing the stresses. The reason for this is that, if the shaft is made stiff enough so that the deflection is not too large, it is probable that the resulting stresses will be safe. But by no means should the designer *assume* that they are safe; it is almost always necessary to calculate them so that he *knows* that they are within acceptable limits.

Whenever possible, the power-transmission elements, such as gears or pulleys, should be located close to the supporting bearings. This reduces the bending moment, and hence the deflection and bending stress.

In the following sections the design and analysis of shafting are discussed for the case when stress is the governing design consideration. When deflection is important, you should use the methods of Chap. 3 first, to size the members tentatively.

The design methods which follow differ in various respects from each other. Some are quite conservative while others are useful because they provide quick results. But you must not expect the methods to produce identical results.

13-2 DESIGN FOR STATIC LOADS

The stresses at the surface of a solid round shaft subjected to combined loading of bending and torsion are

$$\sigma_x = \frac{32M}{\pi d^3} \qquad \tau_{xy} = \frac{16T}{\pi d^3} \qquad (13\text{-}1)$$

where σ_x = bending stress
τ_{xy} = torsional stress
d = shaft diameter
M = bending moment at critical section
T = torsional moment at critical section

By the use of a Mohr's circle it is found that the maximum shear stress is

$$\tau_{max} = \sqrt{\left(\frac{\sigma_x}{2}\right)^2 + \tau_{xy}^2} \qquad (a)$$

Eliminating σ_x and τ_{xy} from Eq. (*b*) gives

$$\tau_{max} = \frac{16}{\pi d^3} \sqrt{M^2 + T^2} \qquad (b)$$

The maximum shear-stress theory of static failure states that $S_{sy} = S_y/2$. By employing a factor of safety n, we can now write Eq. (*b*) as

$$\frac{S_y}{2n} = \frac{16}{\pi d^3} \sqrt{M^2 + T^2} \qquad (c)$$

or

$$d = \left[\left(\frac{32n}{\pi S_y} \right) (M^2 + T^2)^{1/2} \right]^{1/3} \qquad (13\text{-}2)$$

A similar approach using the distortion-energy theory yields

$$d = \left[\frac{32n}{\pi S_y} \left(M^2 + \frac{3T^2}{4} \right)^{1/2} \right]^{1/3} \qquad (d)$$

13-3 REVERSED BENDING AND STEADY TORSION

Any rotating shaft loaded by stationary bending and torsional moments will be stressed by completely reversed bending stress, because of shaft rotation, but the torsional stress will remain steady. This is a very common situation and probably occurs more often than any other loading. By using the subscript a to mean alternating stress and the subscript m for mean stress, Eqs. (13-1) can be expressed as

$$\sigma_a = \frac{32M}{\pi d^3} \qquad \tau_m = \frac{16T}{\pi d^3} \qquad (a)$$

Similar expressions can be written for shafts made of tubing.

Sines[*] states that experimental evidence shows that bending-fatigue strength is not affected by the existence of torsional mean stress until the torsional yield strength is exceeded by about 50 percent. This disclosure furnishes a very simple means of designing for the special case of combined reversed bending and steady torsion.

Designating S_e as the completely corrected endurance limit (see Sec. 5-13) and n as the factor of safety, then the design equation is simply

$$\frac{S_e}{n} = \sigma_a \qquad (b)$$

By substituting σ_a from Eq. (a) and solving the result for the diameter, we obtain

$$d = \left(\frac{32Mn}{\pi S_e} \right)^{1/3} \qquad (13\text{-}3)$$

because the presence of τ_m does not affect the bending-endurance limit according to Sines.

A word of caution with respect to the use of Eq. (13-3) is necessary. To be certain that a shaft whose diameter is obtained from Eq. (13-3) will not yield, compute the static factor of safety n using the methods of Sec. 13-2.

---

[*] George Sines in George Sines (ed.), "Metal Fatigue," p. 158, McGraw-Hill Book Company, New York, 1959.

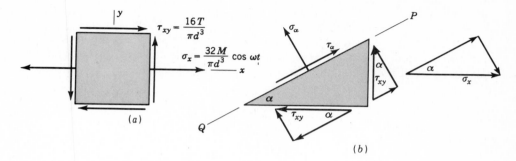

FIGURE 13-1
(a) Shaft stress element of unit depth has a steady shear stress τ_{xy} and an alternating stress σ_x due to rotation; (b) element cut at angle α.

13-4   THE SODERBERG APPROACH*

In the simplest applications of a Soderberg diagram (Fig. 5-24), the diagram is used to determine the required dimensions of a machine part which must carry a steady stress and an alternating stress of the same kind. In the following discussion, an example will be given to show how a Soderberg diagram can be used to determine the required dimensions for a shaft subjected to a combination of steady torque and alternating bending, a common type of loading for shafts. The procedure will suggest how more complicated cases of loading might be treated. Both the example given below and the more general case where both the torque and the bending moment have steady and alternating components are treated in a paper by C. R. Soderberg.† In fact, the derivation given below follows exactly that published by Soderberg, except that some of Soderberg's equations are omitted and the purpose for which he uses these equations is accomplished graphically. Also, the symbols used here differ from those used in Soderberg's paper.

Figure 13-1a shows a stress element on the surface of a solid round shaft whose rotational speed is ω rad/s. Now suppose that a plane PQ is passed through the lower right-hand corner of the element. Then, below plane PQ there will be a wedge-shaped element, as shown in Fig. 13-1b. The angle α shown in the figure is the angle between plane PQ and a horizontal plane. We shall consider all possible values of α to see whether or not we can decide what it will be for those planes on which failure occurs.

Since both bending and torsion are involved in this problem, it is necessary to decide upon a theory of failure. Though failures due to variable stresses do not

* Except for slight changes in notation, this section is a faithful reproduction of the notes used at the University of California, Berkeley, and is reproduced with the permission of the engineering-design staff.
† C. R. Soderberg, Working Stresses, *J. Appl. Mechanics*, vol. 57, p. A-106, 1935.

occur on planes of maximum shear, we shall use the maximum-shear-stress theory because both ratios S_{se}/S_e and S_{sy}/S_y are only slightly greater than 0.50.

Since we have decided to use the maximum-shear theory of failure, we are interested in the value of the shear stress on the inclined face of the element. By taking an equilibrium equation for all forces in the direction of τ_α, we get

$$\tau_\alpha + \sigma_x \sin \alpha \cos \alpha + \tau_{xy} \sin^2 \alpha - \tau_{xy} \cos^2 \alpha = 0$$

or
$$\tau_\alpha = \tau_{xy}(\cos^2 \alpha - \sin^2 \alpha) - \sigma_x \sin \alpha \cos \alpha \qquad (a)$$

Substituting the values of τ_{xy} and σ_x in Eq. (a) and using several trigonometric identities yields

$$\tau_\alpha = \frac{16T}{\pi d^3} \cos 2\alpha - \frac{16M}{\pi d^3} \sin 2\alpha \cos \omega t \qquad (b)$$

In other words, for any plane making an angle α with the horizontal plane, the shear stress has a mean value of

$$\tau_{\alpha m} = \frac{16T}{\pi d^3} \cos 2\alpha \qquad (c)$$

and an alternating component with amplitude

$$\tau_{\alpha a} = \frac{16M}{\pi d^3} \sin 2\alpha \qquad (d)$$

Shown in Fig. 13-2 is the Soderberg diagram for shear strength. Alternating shear stresses are plotted on the vertical axis, while static or mean shear stresses are plotted on the horizontal axis. As shown, the Soderberg line is a straight line between the completely corrected shear endurance limit S_{se} and the yield strength in shear S_{sy}. Note, particularly, that the shear endurance limit is the endurance limit of the machine element after size, surface finish, reliability, life, stress concentration, and the like, have been accounted for, by using Eq. (5-41).*

To determine whether or not failure will occur on certain planes making an angle α with the horizontal, we plot a point in Fig. 13-2 for each value of α. The coordinates of the point are $(\tau_{\alpha m}, \tau_{\alpha a})$, as determined above. For example, for horizontal planes ($\alpha = 0$), the coordinates of the point are

$$\left(\frac{16T}{\pi d^3}, 0 \right)$$

For vertical planes, the coordinates are

$$\left(-\frac{16T}{\pi d^3}, 0 \right)$$

---

* In the original paper, modification for many of these effects was made in the final equation.

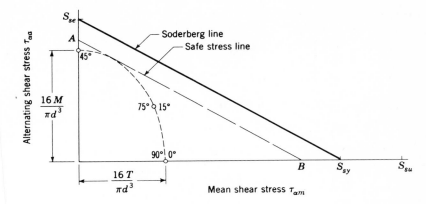

FIGURE 13-2
Soderberg diagram showing how the line of safe stress AB is drawn parallel to the
Soderberg line and tangent to the ellipse.

but this is really the same as the point for $\alpha = 0$. For $\alpha = 45°$, the point is

$$\left(0, \frac{16M}{\pi d^3}\right)$$

It is recommended that additional points be plotted so as to verify the fact that all
points fall on a quarter ellipse, as indicated in Fig. 13-2.

Consideration of Fig. 13-2 leads to the conclusion that the factor of safety
should be taken to correspond to that point of the ellipse which is closest to the
failure line. The problem is to draw a line parallel to the failure line and tangent to
the ellipse. With such a line the factor of safety n can be determined graphically.
Such a complete graphical solution would be in every way acceptable. However,
by means of analytic geometry, it can be shown that the value of n thus obtained
will be

$$n = \frac{\pi d^3}{16\sqrt{(T/S_{sy})^2 + (M/S_{se})^2}} \qquad (13\text{-}4)$$

For use as a design formula, this equation can be rewritten

$$d = \left\{\frac{16n}{\pi}\left[\left(\frac{T}{S_{sy}}\right)^2 + \left(\frac{M}{S_{se}}\right)^2\right]^{1/2}\right\}^{1/3} \qquad (13\text{-}5)$$

or more conveniently,

$$d = \left\{\frac{32n}{\pi}\left[\left(\frac{T}{S_y}\right)^2 + \left(\frac{M}{S_e}\right)^2\right]^{1/2}\right\}^{1/3} \qquad (13\text{-}6)$$

because the maximum-shear-stress theory was assumed. In view of the fact that
the maximum-shear theory is only approximately true, Eq. (13-6) is safer to use.

For the most general case, where the bending stress and the torsion stress each contain both a steady component and a variable component, the equation corresponding to Eq. (13-6) for the simple case above is

$$d = \left\{ \frac{32n}{\pi} \left[\left(\frac{T_a}{S_e} + \frac{T_m}{S_y} \right)^2 + \left(\frac{M_a}{S_e} + \frac{M_m}{S_y} \right)^2 \right]^{1/2} \right\}^{1/3} \tag{13-7}$$

Equation (13-7) is sometimes referred to (perhaps in a less complete form) as the Westinghouse code formula.

By noting that $I/c = (\frac{1}{2})(J/c) = \pi d^3/32$, Eq. (13-7) can be solved for the factor of safety in terms of the stresses and strengths. The result is

$$n = \frac{1}{\sqrt{\left(\dfrac{\tau_a}{0.5S_e} + \dfrac{\tau_m}{0.5S_y} \right)^2 + \left(\dfrac{\sigma_a}{S_e} + \dfrac{\sigma_m}{S_y} \right)^2}} \tag{13-8}$$

which may be more convenient for purposes of analysis.

EXAMPLE 13-1   The integral pinion shaft shown in Fig. 13-3a is to be mounted in bearings at the locations shown, and is to have a gear (not shown) mounted on the right-hand or overhanging end. The loading diagram (Fig. 13-3b) shows that the pinion force at A and the gear force at C are in the same xy plane. Equal and opposite torques T_A and T_C are assumed to be concentrated at A and C as were the forces.

The bending-moment diagram of Fig. 13-3c shows a maximum moment at A, but the diameter of the pinion is quite large and so the stress due to this moment can be neglected. On the other hand, another moment M_B, almost as large, occurs at the center of the right-hand bearing. Find the diameter d of the shaft at the right-hand bearing based upon a material having a yield strength $S_y = 66$ kpsi, a fully corrected endurance limit $S_e = 20$ kpsi, and a factor of safety of 1.80.

SOLUTION   Based on static loads only, Eq. (13-2) yields

$$d = \left[\frac{32n}{\pi S_y} (M^2 + T^2)^{1/2} \right]^{1/3} = \left\{ \frac{(32)(1.80)}{\pi(66)(10)^3} [(1920)^2 + (3300)^2]^{1/2} \right\}^{1/3}$$

$$= 1.02 \text{ in} \qquad\qquad Ans.$$

But, if we consider fatigue, and use Eq. (13-3), we get

$$d = \left(\frac{32Mn}{\pi S_e} \right)^{1/3} = \left[\frac{(32)(1920)(1.80)}{\pi(20)(10)^3} \right]^{1/3} = 1.21 \text{ in} \qquad Ans.$$

This result is greater than the previous one which means that a 1.21-in-diameter shaft is safe for fatigue loading and static loading.

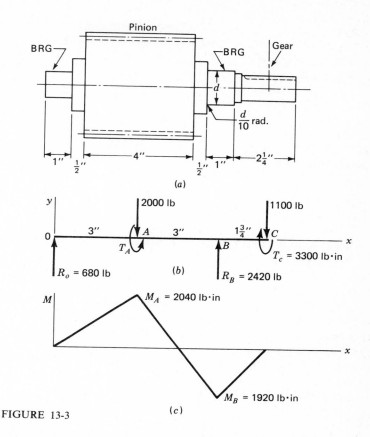

FIGURE 13-3

The Soderberg approach is more conservative. Using Eq. (13-6), we get

$$d = \left\{ \frac{32n}{\pi} \left[\left(\frac{T}{S_y} \right)^2 + \left(\frac{M}{S_e} \right)^2 \right]^{1/2} \right\}^{1/3}$$

$$= \left\{ \frac{(32)(1.80)}{\pi} \left[\left(\frac{3300}{66\,000} \right)^2 + \left(\frac{1920}{20\,000} \right)^2 \right]^{1/2} \right\}^{1/3} = 1.26 \text{ in} \qquad Ans.$$

////

13-5 THE GENERAL BIAXIAL STRESS PROBLEM

The previous sections have dealt with the kinds of shaft-design problems that are most often encountered. The methods discussed give reliable and conservative results when appropriate factors of safety are used. They can be used for the great majority of shaft-design problems.

A more general type of problem is the one in which two normal stresses, σ_x and σ_y say, as well as the shear stress τ_{xy} have both alternating and mean components. When this is the case, it is possible to construct two Mohr's circles, one for the alternating components and another for the mean components, and obtain the corresponding principal stresses from each circle. The von Mises stresses can then be obtained from the equations

$$\sigma'_a = \sqrt{\sigma_{1a}^2 - \sigma_{1a}\sigma_{2a} + \sigma_{2a}^2}$$

$$\sigma'_m = \sqrt{\sigma_{1m}^2 - \sigma_{1m}\sigma_{2m} + \sigma_{2m}^2}$$

(13-9)

as explained in Sec. 5-23. However, instead of going to the trouble of constructing and analyzing the results by the use of two Mohr's circles, it is possible to obtain the same results by the use of the equations

$$\sigma'_a = \sqrt{\sigma_{xa}^2 - \sigma_{xa}\sigma_{ya} + \sigma_{ya}^2 + 3\tau_{xya}^2}$$

$$\sigma'_m = \sqrt{\sigma_{xm}^2 - \sigma_{xm}\sigma_{ym} + \sigma_{ym}^2 + 3\tau_{xym}^2}$$

(13-10)

These equations can be derived from a general Mohr's circle by the use of the distortion-energy criterion [Eq. (5-15)]. Of course, the subscripts a and m in Eqs. (13-9) and (13-10) refer to the alternating and mean stress components, respectively.

The distortion-energy method combined with the modified Goodman diagram was suggested in Chap. 5 as a conservative solution to the general problem posed here. This method is illustrated using a number of different stress situations in Example 5-6.

Dr. L. D. Mitchell and D. T. Vaughn of the Virginia Polytechnic Institute have developed a generalized approach to this problem using equations which combine the modified Goodman diagram with any theory of failure.* Equations are provided that give specific solutions using any desired theory of failure, but they are too lengthy to be included here.

Because the distortion-energy method combined with the modified Goodman diagram is quite conservative, researchers have been working to obtain another approach that would yield results closer to experimental observations. George Sines, of the University of California at Los Angeles and Dimitri Kececioglu of the University of Arizona have each proposed solutions which have been verified by many experiments.†, ‡ Both of these methods will be shown in the sections which follow. As will be seen, the Sines approach cannot be illustrated, except for special cases, on the usual Goodman-type fatigue diagram. However, the modified Goodman line and the Kececioglu curve are shown in Fig. 13-4. The Kececioglu curve is the mean of test results from 527 specimens tested. Comparison of these with the Sines approach will be shown by other methods.

---

* L. D. Mitchell and D. T. Vaughn, A General Method for the Fatigue-Resistant Design of Mechanical Components, Part I, Graphical; Part II, Analytical, *ASME* Papers, 74-WA/DE-4 and 74-WA/DE-5, 1974.

† George Sines, "Elasticity and Strength," pp. 72–78, Allyn and Bacon, Inc., Boston, 1969.

‡ Dimitri Kececioglu, Louie B. Chester, and Thomas M. Dodge, Combined Bending-Torsion Fatigue Reliability of AISI 4340 Steel Shafting with $K_t = 2.34$, *ASME* paper no. 74-WA/DE-12, 1974.

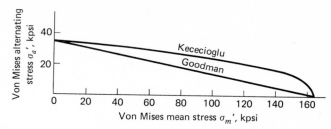

FIGURE 13-4

Fatigue diagram showing a comparison between the Kececioglu curve and the modified Goodman line. $S_{ut} = 165.1$ kpsi, $S_e = 33.7$ kpsi.

13-6 THE SINES APPROACH

To explain Sines' method, it is first necessary to introduce the concept of octahedral stresses. Visualize a principal stress element having the stresses σ_1, σ_2, and σ_3. Now cut the stress element by a plane that forms equal angles with each of the three principal stresses. The plane ABC in Fig. 13-5 satisfies this requirement. It is called an *octahedral plane*. Note that the solid cut from the stress element retains one of the original corners. Since there are eight of these corners in all, there is a total of eight of these planes.

Figure 13-5 may be regarded as a free-body diagram when each of the stress components shown is multiplied by the area over which it acts to obtain the corresponding force. It is possible, then, to sum these forces to zero in each of the three coordinate directions. When this is done, it is found that a force exists on plane ABC, called the *octahedral force*. When this force is divided by the area of face ABC, it is found that the resulting stress has two components σ_{oct} and τ_{oct}, called, respectively, the *octahedral normal* and the *octahedral shear stresses*. The magnitudes of these stresses are*

$$\sigma_{oct} = \frac{\sigma_1 + \sigma_2 + \sigma_3}{3} \tag{13-11}$$

$$\tau_{oct} = \tfrac{1}{3}\sqrt{(\sigma_1 - \sigma_2)^2 + (\sigma_2 - \sigma_3)^2 + (\sigma_3 - \sigma_1)^2} \tag{13-12}$$

The Sines concept of fatigue failure is based on his observation that a linear relation must exist between the magnitude of the alternating octahedral shear stress components and the sum of the mean components of the principal stresses. This relation is expressed as

$$\tfrac{1}{3}\sqrt{(\sigma_{1a} - \sigma_{2a})^2 + (\sigma_{2a} - \sigma_{3a})^2 + (\sigma_{3a} - \sigma_{1a})^2} = A - \alpha(\sigma_{1m} + \sigma_{2m} + \sigma_{3m})$$

$$\tag{13-13}$$

---

* For a complete derivation see J. O. Smith and O. M. Sidebottom, "Elementary Mechanics of Deformable Bodies," p. 131, The Macmillan Company, 1969.

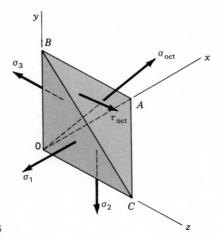

FIGURE 13-5

The constants A and α are to be determined from two S-N diagrams, or two fatigue-testing programs, in which the mean stresses are significantly different. For example, the rotating-beam fatigue test has $S_e = \sigma_{1a}$ and σ_{2a}, σ_{3a}, σ_{1m}, σ_{2m}, and σ_{3m} are all zero. Thus Eq. (13-13) gives

$$A = \frac{\sqrt{2}}{3} S_e \qquad (13\text{-}14)$$

Another test suggested by Sines is the zero-tension fluctuating-stress cycle. In this cycle $\sigma_{1a} = \sigma_{1m}$ and σ_{2a}, σ_{3a}, σ_{2m}, and σ_{3m} are zero. Let us designate the strength as $S_A = \sigma_{1a} = \sigma_{1m}$. Substituting these in Eq. (13-13) gives

$$\frac{\sqrt{2}}{3} S_A = A - \alpha S_A \qquad (a)$$

If we now substitute the value of A and solve for the constant α, we get

$$\alpha = \frac{\sqrt{2}}{3}\left(\frac{S_e}{S_A} - 1\right) \qquad (13\text{-}15)$$

For biaxial stresses σ_{3a} and σ_{3m} are zero and Eq. (13-13) simplifies to

$$\frac{\sqrt{2}}{3} \sigma_a' = A - \alpha(\sigma_{1m} + \sigma_{2m}) \qquad (b)$$

where σ_a' is the von Mises alternating-stress component. Equation (b) gives the limiting value of the alternating-stress component σ_a' due to the mean stresses σ_{1m} and σ_{2m}. Thus, it is convenient to replace σ_a' with S_a, using the notation proposed in Example 5-5. If we also substitute the constants found in Eqs. (13-14) and (13-15) and simplify the result, we obtain

$$S_a = S_e - \left(\frac{S_e}{S_A} - 1\right)(\sigma_{1m} + \sigma_{2m}) \qquad (13\text{-}16)$$

13-7 THE KECECIOGLU APPROACH

The equation

$$\left(\frac{S_a}{S_e}\right)^a + \left(\frac{S_m}{S_{ut}}\right)^2 = 1 \qquad (13\text{-}17)$$

is used by Kececioglu as the criteria of fatigue failure when combined alternating and mean stresses are present.

Based on 50 percent reliability, that is, the mean values of the strengths, Kececioglu found the exponent to be $a = 2.606$ for AISI 4340 steel, heat-treated to a Brinell hardness range of 323–370 with $K_t = 2.34$. But, using a higher reliability three standard deviations removed, he found the exponent to be $a = 2.750$.

EXAMPLE 13-2   In Example 13-1 a shaft diameter of $1\frac{1}{4}$ in seems to be a conservative choice for size. This was based on an alternating bending stress due to a moment of 1920 lb·in, a steady torsion of 3300 lb·in, and a factor of safety of 1.80. The material has a yield strength of 66 kpsi and a fully corrected endurance limit of 20 kpsi.

(a) Find the factor of safety using the distortion-energy theory and the modified Goodman diagram. $S_{ut} = 100$ kpsi.

(b) Find the factor of safety using the Sines method. Use $S_A = 17$ kpsi (this is the value that would be obtained from a modified Goodman diagram using $S_e = 20$ kpsi, $S_{ut} = 100$ kpsi, and $\sigma_a/\sigma_m = 1$).

(c) Find the factor of safety using the Kececioglu equation with the exponent $a = 2.61$.

SOLUTION   (a) The alternating-stress component is

$$\sigma_a' = \sigma_a = \frac{32M}{\pi d^3} = \frac{32(1920)}{\pi(1.25)^3} = 10(10)^3 \text{ psi}$$

The steady shear stress is

$$\tau_m = \frac{16T}{\pi d^3} = \frac{16(3300)}{\pi(1.25)^3} = 8.6(10)^3 \text{ psi}$$

Consequently, the von Mises mean stress is

$$\sigma_m' = \sqrt{3\tau_m^2} = \sqrt{3(8.6)^2}\,(10)^3 = 14.9(10)^3 \text{ psi}$$

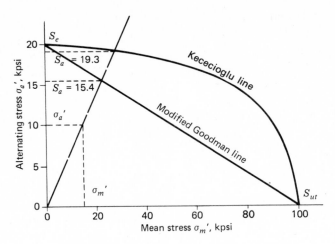

FIGURE 13-6
Note that the two axes are plotted using different scales.

We now construct the modified Goodman diagram as shown in Fig. 13-6. The alternating component of the strength is found to be $S_a = 15.4$ kpsi. Consequently the factor of safety is

$$n = \frac{S_a}{\sigma'_a} = \frac{15.4}{10} = 1.54 \qquad Ans.$$

(b) Since $\tau_m = 8.6$ kpsi is a pure shear stress, $\sigma_{1m} = -\sigma_{2m} = 8.6$ kpsi, and so $\sigma_{1m} + \sigma_{2m} = 0$. Therefore Sines' equation reduces to

$$S_a = S_e$$

Thus the factor of safety is

$$n = \frac{S_a}{\sigma'_a} = \frac{20}{10} = 2.00 \qquad Ans.$$

(c) We first rearrange the Kececioglu equation as follows:

$$S_a = S_e \left[1 - \left(\frac{S_m}{S_{ut}} \right)^2 \right]^{1/a}$$

Then, substitute $S_e = 20$ kpsi, $S_{ut} = 100$ kpsi, and $a = 2.61$ to get

$$S_a = 20 \left[1 - \left(\frac{S_m}{100} \right)^2 \right]^{1/2.61}$$

Now solve this equation for S_a using various values of S_m. Some of the results are

S_m, kpsi	0	20	40	60	80	100
S_a, kpsi	20	19.7	18.7	16.9	13.5	0

When these data are plotted in Fig. 13-6, the Kececioglu line is obtained. The intersection of a line drawn through the coordinates $\sigma_a' \sigma_m'$ is seen to be located at $S_a = 19.3$ kpsi. Therefore the factor of safety is

$$n = \frac{S_a}{\sigma_a'} = \frac{19.3}{10} = 1.93 \qquad Ans.$$

////

13-8 FORMULAS FOR STRESS-CONCENTRATION FACTORS

The use of graphs and tabulated material is both inconvenient and time-consuming when one is using computer-aided design techniques and also when one is designing by use of the programmable calculator. Empirical curve-fitting techniques can often be employed to reduce such material to a programmable routine for use in these design procedures. For example, most of the data from the stress-concentration chart of Table A-25-9 for a round shaft with a shoulder fillet in bending can be obtained from the equation

$$K_t = 1.68 + 1.24 \cos^{-1}\frac{d}{D} - \frac{r}{d}\left(2.82 + 11.6 \cos^{-1}\frac{d}{D}\right)$$

$$+ \left(\frac{r}{d}\right)^2\left(5.71 + 29.6 \cos^{-1}\frac{d}{D}\right) \qquad (13\text{-}18)$$

This equation is valid for the ranges $0.04 \le r/d \le 0.20$ and $1.1 \le D/d \le 3$. The calculator radian-degree switch should be on radians. The results obtained by use of this equation will agree with those obtained from the chart within about 5 percent accuracy.

PROBLEMS

Sections 13-2 to 13-4

13-1 The small geared industrial roll shown in the figure is driven at 300 rpm by a force $F = 200$ lb acting on the 3-in gear as shown. The material transported by the roll exerts a resultant uniform force distribution of $w = 20$ lb/in on the roll in the positive z direction. The roll shaft is to be a solid round UNS G10180 cold-drawn steel bar of uniform diameter. Determine the size of shaft required to the nearest $\frac{1}{16}$ in using a factor of safety of 3.50. The solution should be based on the shaft stresses at bearing A. Use a reliability factor corresponding to $R = 0.50$.

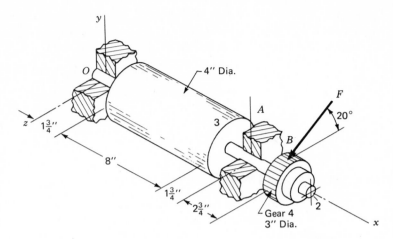

13-2 Suppose a UNS G10180 cold-drawn round steel tubing 1-in OD by ¼-in wall thickness were selected for the shaft of Prob. 13-1. Determine the factors of safety guarding against both a static failure and a fatigue failure using the methods of Secs. 13-2 and 13-3.

13-3 The resultant gear force $F_A = 600$ lb acts at an angle of 20° from the y axis of the overhanging countershaft shown in the figure. The shaft is to be made of a solid round UNS G10400 cold-drawn steel shafting cut to length. The factor of safety is to be 2.60 corresponding to 99 percent reliability. Determine the diameter of shaft required to the nearest safe $\frac{1}{16}$ in. Use the methods of Secs. 13-2 and 13-3.

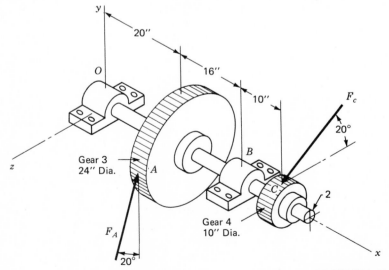

13-4 Using the method of Sec. 13-3, select from Table A-10 a cold-drawn UNS G10400 steel tube for use as the shaft of Prob. 13-3. If two or more sizes are found to be safe, choose the size having the smallest OD. Also find the factor of safety guarding against yielding.

13-5 The figure is a schematic drawing of a countershaft that supports two *V*-belt pulleys. All belt pulls are horizontal and parallel. For pulley *A* the loose belt tension is 15 percent of the tension on the tight side. A cold-drawn UNS G10180 steel shaft of uniform diameter is to be selected for this application. Use a reliability of 90 percent and a factor of safety of 1.90 and determine the diameter to the next higher nearest millimetre. Use the methods of Secs. 13-2 and 13-3.

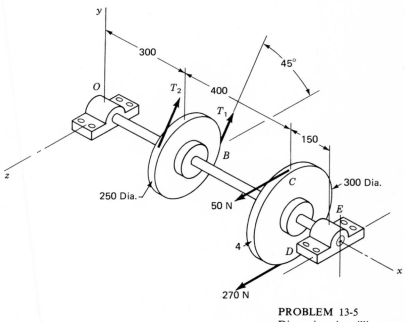

PROBLEM 13-5
Dimensions in millimetres.

13-6 The figure shows a countershaft on which are mounted two spur gears having teeth cut on the 20° system. A uniform-diameter steel shaft of UNS G10350 steel heat-treated and tempered to 800°F with a machined finish is to be used. The factor of safety is to be 1.60 corresponding to a reliability of 99 percent. Find the diameter required to safely carry the load to the next higher even millimetre using the methods of Sec. 13-3. What factor of safety guards against yielding?

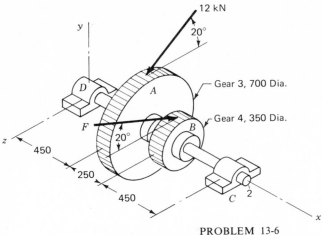

PROBLEM 13-6
Dimensions in millimetres.

Sections 13-5 to 13-7

13-7   The speed-reducer shaft shown in the figure is designed to support a pulley and a worm. The shaft is made of plain carbon steel, heat-treated to $H_B = 226$ resulting in the strengths $S_{ut} = 113$ kpsi and $S_y = 86$ kpsi. All surfaces of the shaft are finished by grinding. The pulley subjects the shaft to a radial bending load of 618 lb and a torque of 3640 lb·in. The worm receives the torque and, in addition, subjects the shaft to a 3200-lb axial load, which is to be taken by the right-hand bearing, and a radial bending load of 335 lb. Note that this is a special case; in rotating machinery two radial bending loads are seldom in the same plane. This is evident from the previous problems. The shaft rotates at 60 rpm and is to have a life of 10 h corresponding to 99 percent reliability. Compute the factor of safety using the von Mises-Hencky-Goodman approach.

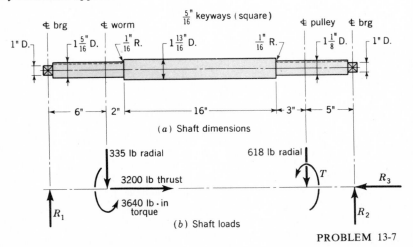

PROBLEM 13-7

13-8 Part *a* of the figure is a schematic drawing of a shaft, gear, and bearing subassembly that is a part of a helical-gear reduction unit. A force $\mathbf{F} = -1700\mathbf{i} + 6400\mathbf{j} - 2300\mathbf{k}$ lb is applied to gear *B* as shown. The forces $\mathbf{F}_A$ and $\mathbf{F}_C$, of equal magnitudes, have directions given by the unit vectors $\hat{\mathbf{F}}_A = 0.470\mathbf{i} - 0.342\mathbf{j} + 0.814\mathbf{k}$ and $\hat{\mathbf{F}}_C = -0.470\mathbf{i} - 0.342\mathbf{j} + 0.814\mathbf{k}$. Assume no torque is lost due to friction or windage. Part *b* of the figure shows the dimensions tentatively selected for the shaft. The material is a UNS G43400 steel, heat-treated and tempered to 1000°F with all important surfaces

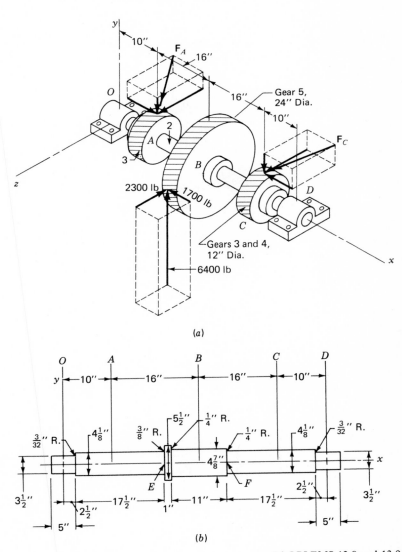

(a)

(b)

PROBLEMS 13-8 and 13-9.

ground. Sled-runner keyways 8 in long are milled in the shaft at *A*, *B*, and *C*; they extend 4 in each way from these points. Analyze this shaft for safety, using a reliability of 99.999 percent and basing your analysis on the possibility of a fatigue failure initiated by the keyway at *A* or by the fillet on the shaft shoulder at *E*. As the failure mechanism use the distortion-energy theory combined with the modified Goodman diagram.

13-9 The same as Prob. 13-8, except determine the factors of safety for possible failure in the keyway at *C* and at the shoulder fillet designated *F*.

13-10 The figure illustrates a countershaft subjected to alternating bending together with steady torsion and axial loads because of the action of the two gears. Part *b* of the figure shows all shaft dimensions. Sled-runner keyways are used at *A* and *C*, but

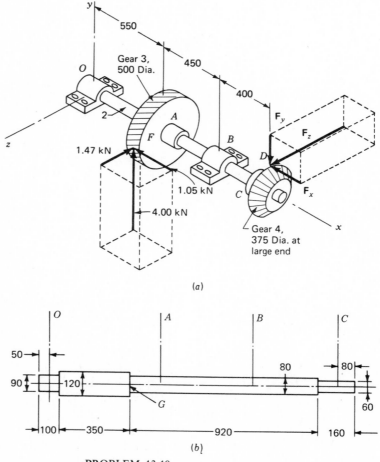

(a)

(b)

PROBLEM 13-10
Dimensions in millimetres; all shoulder fillets are 5 mm radius.

these do not extend completely to the shaft shoulders. Note that the shaft design is such that only the bearing at O can take the axial load. The bevel-gear force is $\mathbf{F}_D = -0.242\mathbf{F}_D\mathbf{i} - 0.242\mathbf{F}_D\mathbf{j} + 0.940\mathbf{F}_D\mathbf{k}$. The shaft is made of UNS G10500 steel, heat-treated and drawn to $900°F$, and all important surfaces are finished by grinding. The shaft is to be analyzed to determine how safe it is using a reliability of 99 percent and the von Mises-Hencky-Goodman approach.

(a) Determine the factor of safety at shoulder G.

(b) Determine the factor of safety at A where the keyway is located.

(c) Though the bending moment may be quite large at bearing B, there is no stress concentration here. Find the factor of safety at this point.

14

CLUTCHES, BRAKES, AND COUPLINGS

Because of the similarity of their functions, clutches, brakes, and couplings are treated together in this book. A simplified dynamic representation of a friction clutch, or brake, is shown in Fig. 14-1. Two inertias I_1 and I_2 traveling at the respective angular velocities ω_1 and ω_2, one of which may be zero in the case of brakes, are to be brought to the same speed by engaging the clutch or brake. Slippage occurs because the two elements are running at different speeds and energy is dissipated during actuation, resulting in a temperature rise. In analyzing the performance of these devices we shall be interested in:

1. The actuating force
2. The torque transmitted
3. The energy loss
4. The temperature rise

The torque transmitted is related to the actuating force, the coefficient of friction, and to the geometry of the clutch or brake. This is a problem in statics which will have to be studied separately for each geometric configuration. However, temperature rise is related to energy loss and can be studied without regard to the type of brake or clutch because the geometry of interest is the heat-dissipating surfaces.

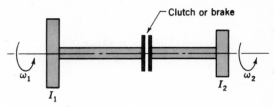

FIGURE 14-1
Dynamic representation of a clutch or brake.

The various types of devices to be studied may be classified as follows:

1 Rim types with internally expanding shoes
2 Rim types with externally contracting shoes
3 Band types
4 Disk or axial types
5 Cone types
6 Miscellaneous types

14-1 STATICS

The analysis of all types of friction clutches and brakes uses the same general procedure. The following steps are necessary:

1 Assume or determine the distribution of pressure on the frictional surfaces.
2 Find a relation between the maximum pressure and the pressure at any point.
3 Apply the conditions of static equilibrium to find (*a*) the actuating force, (*b*) the torque, and (*c*) the support reactions.

Let us now apply these steps to the theoretical problem shown in Fig. 14-2. The figure shows a short shoe hinged at *A*, having an actuating force *F*, a normal force *N* pushing the surfaces together, and a frictional force *fN*, *f* being the coefficient of friction. The body is moving to the right, and the shoe is stationary. We designate the pressure at any point by *p* and the maximum pressure by p_a. The area of the shoe is designated by *A*.

Step 1 Since the shoe is short, we assume the pressure is uniformly distributed over the frictional area.

Step 2 From step 1 it follows that

$$p = p_a \qquad (a)$$

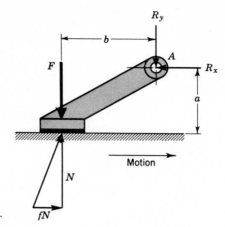

FIGURE 14-2
Forces acting upon a hinged friction shoe.

Step 3 Since the pressure is uniformly distributed, we may replace the unit normal forces by an equivalent normal force. Thus

$$N = p_a A \qquad (b)$$

We now apply the conditions of static equilibrium by taking a summation of moments about the hinge pin. This gives

$$\sum M_A = Fb - Nb + fNa = 0 \qquad (c)$$

Substituting $p_a A$ for N and solving Eq. (c) for the actuating force,

$$F = \frac{p_a A(b - fa)}{b} \qquad (d)$$

Taking a summation of forces in the horizontal and vertical directions gives the hinge-pin reactions:

$$\sum F_x = 0 \qquad R_x = fp_a A \qquad (e)$$

$$\sum F_y = 0 \qquad R_y = p_a A - F \qquad (f)$$

This completes the analysis of the problem.

The preceding analysis is very useful when the dimensions of the clutch or brake are known and the characteristics of the friction material are specified. In design, however, we are interested more in synthesis than in analysis; that is, our purpose is to select a set of dimensions to obtain the best brake or clutch within the limitations of the frictional material we have specified.

In the preceding problem (Fig. 14-2) we make good use of the frictional material because the pressure is a maximum at all points of contact. Let us examine the dimensions. In Eq. (d), if we make $b = fa$, the numerator becomes zero, and no actuating force is required. This is the condition for *self-locking*. We are usually not interested in designing a brake to be self-locking, but on the other hand, we should take full advantage of the *self-energizing* effect. This can be done by selecting for the frictional material a value of f which will never be exceeded, even under the most adverse conditions. One way to do this is to increase the manufacturer's specification for the coefficient of friction by, say, 25 to 50 percent. Thus, if we let $f' = 1.25f$ to $1.50f$, the equation

$$b = f'a$$

may be used to obtain the dimensions of a and b to give the degree of self-energization desired.

14-2 INTERNAL-EXPANDING RIM CLUTCHES AND BRAKES

The internal-shoe rim clutch shown in Fig. 14-3 consists essentially of three elements: the mating frictional surfaces, the means of transmitting the torque to and from the surfaces, and the actuating mechanism. To analyze an internal-shoe device, refer to Fig. 14-4, which shows a shoe pivoted at point A, with the actuating force acting at the other end of the shoe. Since the shoe is long, we cannot make the assumption that the distribution of normal forces is uniform. The mechanical arrangement permits no pressure to be applied at the heel, and we will therefore assume the pressure at this point to be zero.

FIGURE 14-3
An internal-expanding centrifugal-acting rim clutch. (*Courtesy of the Hilliard Corporation.*)

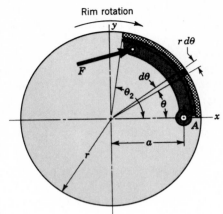

FIGURE 14-4
Internal friction shoe.

It is the usual practice to omit the friction material for a short distance away from the heel. This eliminates interference, and it would have contributed little to the performance anyway. In some designs the hinge pin is made movable to provide additional heel pressure. This gives the effect of a floating shoe. (Floating shoes will not be treated in this book, although their design follows the same general principles.)

Let us consider the unit pressure p acting upon an element of area of the frictional material located at an angle θ from the hinge pin (Fig. 14-4). We designate the maximum pressure by p_a located at the angle θ_a from the hinge pin. We now make the assumption (step 1) that the pressure at any point is proportional to the vertical distance from the hinge pin. This vertical distance is proportional to $\sin \theta$, and (step 2) the relation between the pressures is

$$\frac{p}{\sin \theta} = \frac{p_a}{\sin \theta_a} \qquad (a)$$

Rearranging,

$$p = p_a \frac{\sin \theta}{\sin \theta_a} \qquad (14\text{-}1)$$

From Eq. (14-1) we see that p will be a maximum when $\theta = 90°$, or if the toe angle θ_2 is less than $90°$, then p will be a maximum at the toe.

When $\theta = 0$, Eq. (14-1) shows that the pressure is zero. The frictional material located at the heel therefore contributes very little to the braking action and might as well be omitted. A good design would concentrate as much frictional material as possible in the neighborhood of the point of maximum pressure. Such a design is shown in Fig. 14-5. In this figure the frictional material begins at an angle θ_1, measured from the hinge pin A, and ends at an angle θ_2. Any arrangement such as this will give a good distribution of the frictional material.

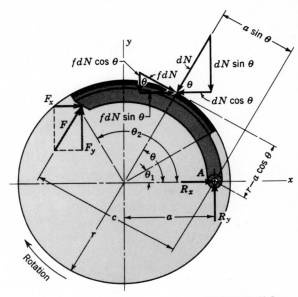

FIGURE 14-5
Forces on the shoe.

Proceeding now to step 3 (Fig. 14-5), the hinge-pin reactions are R_x and R_y. The actuating force F has components F_x and F_y and operates at distance c from the hinge pin. At any angle θ from the hinge pin there acts a differential normal force dN whose magnitude is

$$dN = pbr \; d\theta \qquad (b)$$

where b is the face width (perpendicular to the paper) of the friction material. Substituting the value of the pressure from Eq. (14-1), the normal force is

$$dN = \frac{p_a br \sin \theta \; d\theta}{\sin \theta_a} \qquad (c)$$

The normal force dN has horizontal and vertical components $dN \cos \theta$ and $dN \sin \theta$, as shown in the figure. The frictional force $f \, dN$ has horizontal and vertical components whose magnitudes are $f \, dN \sin \theta$ and $f \, dN \cos \theta$, respectively. By applying the conditions of statical equilibrium, we may find the actuating force F, the torque T, and the pin reactions R_x and R_y.

We shall find the actuating force F, using the condition that the summation of the moments about the hinge pin is zero. The frictional forces have a moment arm about the pin of $r - a \cos \theta$. The moment M_f of these frictional forces is

$$M_f = \int f \, dN(r - a \cos \theta) = \frac{fp_a br}{\sin \theta_a} \int_{\theta_1}^{\theta_2} \sin \theta(r - a \cos \theta) \; d\theta \qquad (14\text{-}2)$$

which is obtained by substituting the value of dN from Eq. (c). It is convenient to integrate Eq. (14-2) for each problem, and we shall therefore retain it in this form. The moment arm of the normal force dN about the pin is $a \sin \theta$. Designating the moment of the normal forces by M_N and summing these about the hinge pin give

$$M_N = \int dN(a \sin \theta) = \frac{p_a bra}{\sin \theta_a} \int_{\theta_1}^{\theta_2} \sin^2 \theta \, d\theta \qquad (14\text{-}3)$$

The actuating force F must balance these moments. Thus

$$F = \frac{M_N - M_f}{c} \qquad (14\text{-}4)$$

We see here that a condition for zero actuating force exists. In other words, if we make $M_N = M_f$, self-locking is obtained, and no actuating force is required. This furnishes us with a method for obtaining the dimensions for some self-energizing action. Thus, by using f' instead of f in Eq. (14-2), we may solve for a from the relation

$$M_N = M_{f'} \qquad (14\text{-}5)$$

where, as before, f' is made about 1.25 to 1.50f.

The torque T, applied to the drum by the brake shoe, is the sum of the frictional forces $f \, dN$ times the radius of the drum:

$$T = \int f \, dNr = \frac{fp_a br^2}{\sin \theta_a} \int_{\theta_1}^{\theta_2} \sin \theta \, d\theta$$

$$= \frac{fp_a br^2(\cos \theta_1 - \cos \theta_2)}{\sin \theta_a} \qquad (14\text{-}6)$$

The hinge-pin reactions are found by taking a summation of the horizontal and vertical forces. Thus, for R_x, we have

$$R_x = \int dN \cos \theta - \int f \, dN \sin \theta - F_x$$

$$= \frac{p_a br}{\sin \theta_a} \left(\int_{\theta_1}^{\theta_2} \sin \theta \cos \theta \, d\theta - f \int_{\theta_1}^{\theta_2} \sin^2 \theta \, d\theta \right) - F_x \qquad (14\text{-}7)$$

The vertical reaction is found in the same way:

$$R_y = \int dN \sin \theta + \int f \, dN \cos \theta - F_y$$

$$= \frac{p_a br}{\sin \theta_a} \left(\int_{\theta_1}^{\theta_2} \sin^2 \theta \, d\theta + f \int_{\theta_1}^{\theta_2} \sin \theta \cos \theta \, d\theta \right) - F_y \qquad (14\text{-}8)$$

The direction of the frictional forces is reversed if the rotation is reversed. Thus, for counterclockwise rotation the actuating force is

$$F = \frac{M_N + M_f}{c} \qquad (14\text{-}9)$$

and since both moments have the same sense, the self-energizing effect is lost. Also, for counterclockwise rotation the signs of the frictional terms in the equations for the pin reactions change, and Eqs. (14-7) and (14-8) become

$$R_x = \frac{p_a br}{\sin \theta_a} \left(\int_{\theta_1}^{\theta_2} \sin \theta \cos \theta \, d\theta + f \int_{\theta_1}^{\theta_2} \sin^2 \theta \, d\theta \right) - F_x \qquad (14\text{-}10)$$

$$R_y = \frac{p_a br}{\sin \theta_a} \left(\int_{\theta_1}^{\theta_2} \sin^2 \theta \, d\theta - f \int_{\theta_1}^{\theta_2} \sin \theta \cos \theta \, d\theta \right) - F_y \qquad (14\text{-}11)$$

In using these equations the reference system always has its origin at the center of the drum. The positive x axis is taken through the hinge pin. The positive y axis is always in the direction of the shoe, even if this should result in a left-handed system.

The following assumptions are implied by the preceding analysis:

1   The pressure at any point on the shoe was assumed to be proportional to the distance from the hinge pin, being zero at the heel. This should be considered from the standpoint that pressures specified by manufacturers are averages rather than maximums.

2   The effect of centrifugal force has been neglected. In the case of brakes, the shoes are not rotating, and no centrifugal force would exist. In clutch design, the effect of this force must be considered in writing the equations of statical equilibrium.

3   The shoe was assumed to be rigid. Since this cannot be true, some deflection will occur, depending upon the load, pressure, and stiffness of the shoe. The resulting pressure distribution may be different from that which has been assumed.

4   The entire analysis has been based upon a coefficient of friction which does not vary with pressure. Actually, the coefficient may vary with a number of conditions, including temperature, wear, and environment.

EXAMPLE 14-1   The brake shown in Fig. 14-6 is 300 mm in diameter and is actuated by a mechanism that exerts the same force F on each shoe. The shoes are identical and have a face width of 32 mm. The lining is a molded asbestos having a coefficient of friction of 0.32 and a pressure limitation of 1000 kPa.

(a) Determine the actuating force F.
(b) Find the braking capacity.
(c) Calculate the hinge-pin reactions.

SOLUTION   (a) The right-hand shoe is self-energizing and so the force F is found on the basis that the maximum pressure will occur on this shoe. Here $\theta_1 = 0$, $\theta_2 = 126°$, $\theta_a = 90°$, and $\sin \theta_a = 1$. Also

$$a = \sqrt{(112)^2 + (50)^2} = 123 \text{ mm}$$

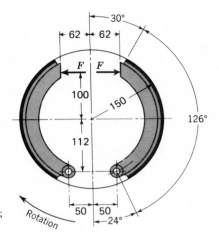

FIGURE 14-6
Brake with internal-expanding shoes; dimensions in millimetres.

Integrating Eq. (14-2) from zero to θ_2 yields

$$M_f = \frac{fp_a br}{\sin \theta_a} \left\{ \left[-r \cos \theta \right]_0^{\theta_2} - a \left[\frac{1}{2} \sin^2 \theta \right]_0^{\theta_2} \right\}$$

$$= \frac{fp_a br}{\sin \theta_a} \left(r - r \cos \theta_2 - \frac{a}{2} \sin^2 \theta_2 \right)$$

Changing all lengths to metres, we have

$$M_f = (0.32)[1000(10)^3](0.032)(0.150)[0.150 - 0.150 \cos 126°$$
$$- 0.123 \sin^2 126°]$$

$$= 243 \text{ N·m}$$

The moment of the normal forces is obtained from Eq. (14-3). Integrating from 0 to θ_2 gives

$$M_N = \frac{p_a bra}{\sin \theta_a} \left[\frac{\theta}{2} - \frac{1}{4} \sin 2\theta \right]_0^{\theta_2}$$

$$= \frac{p_a bra}{\sin \theta_a} \left(\frac{\theta_2}{2} - \frac{1}{4} \sin 2\theta_2 \right)$$

$$= [1000(10)^3](0.032)(0.150)(0.123) \left[\frac{\pi}{2} \frac{126}{180} - \frac{1}{4} \sin (2)(126°) \right]$$

$$= 790 \text{ N·m}$$

From Eq. (14-4), the actuating force is

$$F = \frac{M_N - M_f}{c} = \frac{790 - 243}{100 + 112} = 2.58 \text{ kN} \qquad Ans.$$

(b) From Eq. (14-6), the torque applied by the right-hand shoe is

$$T_R = \frac{fp_a br^2(\cos\theta_1 - \cos\theta_2)}{\sin\theta_a}$$

$$= \frac{0.32[1000(10)^3](0.032)(0.150)^2(\cos 0° - \cos 126°)}{1} = 366 \text{ N·m}$$

The torque contributed by the left-hand shoe cannot be obtained until we learn its maximum operating pressure. Equations (14-2) and (14-3) indicate that the frictional and normal moments are proportional to this pressure. Thus, for the left-hand shoe

$$M_N = \frac{790p_a}{1000} \qquad M_f = \frac{243p_a}{1000}$$

Then, from Eq. (14-9),

$$F = \frac{M_N + M_f}{c}$$

or

$$2.58 = \frac{(790/1000)p_a + (243/1000)p_a}{100 + 112}$$

Solving gives $p_a = 529$ kPa. Then, from Eq. (14-6), the torque on the left-hand shoe is

$$T_L = \frac{fp_a br^2(\cos\theta_1 - \cos\theta_2)}{\sin\theta_a}$$

Since $\sin\theta_a = 1$, we have

$$T_L = 0.32[529(10)^3](0.032)(0.150)^2(\cos 0° - \cos 126°) = 194 \text{ N·m}$$

The braking capacity is the total torque:

$$T = T_R + T_L = 366 + 194 = 560 \text{ N·m} \qquad Ans.$$

(c) The hinge-pin reactions for the right-hand shoe are found from Eqs. (14-7) and (14-8). Substituting $\sin\theta_a = 1$ and $\theta_1 = 0$ and integrating, we have

$$R_x = p_a br\left\{ \left[\frac{1}{2}\sin^2\theta\right]_0^{\theta_2} - f\left[\frac{\theta}{2} - \frac{1}{4}\sin 2\theta\right]_0^{\theta_2} \right\} - F_x$$

$$= p_a br\left(\frac{1}{2}\sin^2\theta_2 - \frac{f\theta_2}{2} + \frac{f}{4}\sin 2\theta_2 \right) - F_x$$

$$= (1000)(0.032)(0.150)\left[\frac{1}{2}\sin^2 126° - \frac{0.32\pi(126)}{2(180)} + \frac{0.32}{4}\sin (2)(126°) \right]$$

$$- 2.58\sin 24°$$

$$= -1.53 \text{ kN}$$

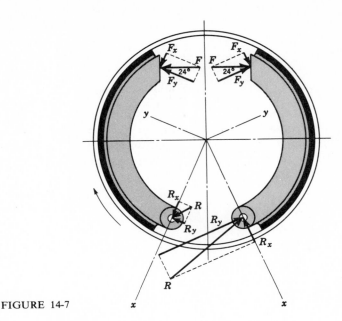

FIGURE 14-7

In the same way we find

$$R_y = 4.56 \text{ kN}$$

The resultant on this hinge pin is

$$R = \sqrt{(1.53)^2 + (4.56)^2} = 4.81 \text{ kN} \qquad \text{Ans.}$$

The reactions at the hinge pin of the left-hand shoe are found using Eqs. (14-10) and (14-11) for a pressure of 529 kPa. They are found to be $R_x = 0.868$ kN and $R_y = 0.773$ kN. The resultant is

$$R = \sqrt{(0.868)^2 + (0.773)^2} = 1.16 \text{ kN} \qquad \text{Ans.}$$

The reactions for both hinge pins, together with their directions, are shown in Fig. 14-7.

This example dramatically shows the benefit to be gained by arranging the shoes to be self-energizing. If the left-hand shoe were turned over so as to place the hinge pin at the top, it could apply the same torque as the right-hand shoe. This would make the capacity of the brake (2)(366) = 732 N·m instead of the present 560 N·m, a 30 percent improvement. In addition, some of the friction material at the heel could be eliminated without seriously affecting the capacity, because of the low pressure in this area. This change might actually improve the overall design because the additional rim exposure would improve the heat-dissipation capacity. *////*

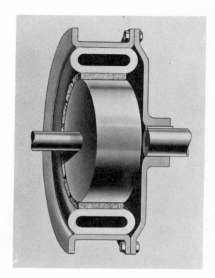

FIGURE 14-8
An external-contracting clutch-brake that
is engaged by expanding the flexible tube
with compressed air. (*Courtesy of Twin
Disc Clutch Company.*)

14-3 EXTERNAL-CONTRACTING RIM CLUTCHES AND BRAKES

The patented clutch-brake of Fig. 14-8 has external-contracting friction elements,
but the actuating mechanism is pneumatic. Here we shall study only pivoted
external shoe brakes and clutches, though the methods presented can easily be
adapted to the clutch-brake of Fig. 14-8.

The notation for external-contracting shoes is shown in Fig. 14-9. The
moments of the frictional and normal forces about the hinge pin are the same as
for the internal-expanding shoes. Equations (14-2) and (14-3) apply and are
repeated here for convenience:

$$M_f = \frac{f p_a b r}{\sin \theta_a} \int_{\theta_1}^{\theta_2} \sin \theta (r - a \cos \theta) \, d\theta \quad (14\text{-}2)$$

$$M_N = \frac{p_a b r a}{\sin \theta_a} \int_{\theta_1}^{\theta_2} \sin^2 \theta \, d\theta \quad (14\text{-}3)$$

Both these equations give positive values for clockwise moments (Fig. 14-9) when
used for external-contracting shoes. The actuating force must be large enough to
balance both moments:

$$F = \frac{M_N + M_f}{c} \quad (14\text{-}12)$$

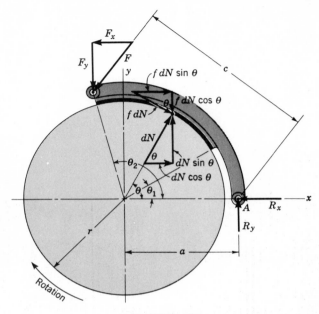

FIGURE 14-9
Notation for external-contracting shoe.

The horizontal and vertical reactions at the hinge pin are found in the same manner as for internal-expanding shoes. They are

$$R_x = \int dN \cos \theta + \int f \, dN \sin \theta - F_x$$

$$= \frac{p_a br}{\sin \theta_a} \left(\int_{\theta_1}^{\theta_2} \sin \theta \cos \theta \, d\theta + f \int_{\theta_1}^{\theta_2} \sin^2 \theta \, d\theta \right) - F_x \qquad (14\text{-}13)$$

$$R_y = \int f \, dN \cos \theta - \int dN \sin \theta + F_y$$

$$= \frac{p_a br}{\sin \theta_a} \left(f \int_{\theta_1}^{\theta_2} \sin \theta \cos \theta \, d\theta - \int_{\theta_1}^{\theta_2} \sin^2 \theta \, d\theta \right) + F_y \qquad (14\text{-}14)$$

If the rotation is counterclockwise, the sign of the frictional term in each equation is reversed. Thus Eq. (14-12) for the actuating force becomes

$$F = \frac{M_N - M_f}{c} \qquad (14\text{-}15)$$

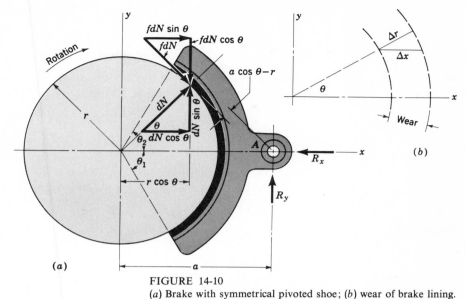

FIGURE 14-10

(*a*) Brake with symmetrical pivoted shoe; (*b*) wear of brake lining.

and self-energization exists for counterclockwise rotation. The horizontal and vertical reactions are

$$R_x = \frac{p_a br}{\sin \theta_a} \left(\int_{\theta_1}^{\theta_2} \sin \theta \cos \theta \, d\theta - f \int_{\theta_1}^{\theta_2} \sin^2 \theta \, d\theta \right) - F_x \qquad (14\text{-}16)$$

$$R_y = \frac{p_a br}{\sin \theta_a} \left(-f \int_{\theta_1}^{\theta_2} \sin \theta \cos \theta \, d\theta - \int_{\theta_1}^{\theta_2} \sin^2 \theta \, d\theta \right) + F_y \qquad (14\text{-}17)$$

It should be noted that, when external-contracting designs are used as clutches, the effect of centrifugal force is to decrease the normal force. Thus, as the speed increases, a larger value of the actuating force F is required.

A special case arises when the pivot is symmetrically located and also placed so that the moment of the friction forces about the pivot is zero. The geometry of such a brake will be similar to that of Fig. 14-10*a*. To get a pressure-distribution relation we assume that the lining wears so as always to retain its cylindrical shape. This means that the wear Δx in Fig. 14-10*b* is constant regardless of the angle θ. Thus the radial wear of the shoe is $\Delta r = \Delta x \cos \theta$. If, on any elementary area of the shoe, we assume that the energy or frictional loss is proportional to the radial pressure, and if we also assume that the wear is directly related to the frictional loss, then by direct analogy,

$$p = p_a \cos \theta \qquad (a)$$

and p is maximum at $\theta = 0$.

Proceeding to the force analysis, we observe that (Fig. 14-10a)

$$dN = pbr \, d\theta \qquad (b)$$

or

$$dN = p_a br \cos \theta \, d\theta \qquad (c)$$

The distance a to the pivot is to be chosen such that the moment of the frictional forces M_f is zero. Symmetry means that $\theta_1 = \theta_2$, and so

$$M_f = 2 \int_0^{\theta_2} (f \, dN)(a \cos \theta - r) = 0$$

Substituting Eq. (c) gives

$$2 f p_a br \int_0^{\theta_2} (a \cos^2 \theta - r \cos \theta) \, d\theta = 0$$

from which

$$a = \frac{4r \sin \theta_2}{2\theta_2 + \sin 2\theta_2} \qquad (14\text{-}18)$$

With the pivot located according to this equation, the moment about the pin is zero, and the horizontal and vertical reactions are

$$R_x = 2 \int_0^{\theta_2} dN \cos \theta = \frac{p_a br}{2} (2\theta_2 + \sin 2\theta_2) \qquad (14\text{-}19)$$

where, because of symmetry,

$$\int f \, dN \sin \theta = 0$$

Also

$$R_y = 2 \int_0^{\theta_2} f \, dN \cos \theta = \frac{p_a brf}{2} (2\theta_2 + \sin 2\theta_2) \qquad (14\text{-}20)$$

where

$$\int dN \sin \theta = 0$$

also because of symmetry. Note, too, that $R_x = -N$ and $R_y = -fN$, as might be expected for the particular choice of the dimension a. Therefore the torque is

$$T = afN \qquad (14\text{-}21)$$

14-4 BAND-TYPE CLUTCHES AND BRAKES

Flexible clutch and brake bands are used in power excavators and in hoisting and other machinery. The analysis follows the notation of Fig. 14-11.

Because of friction and the rotation of the drum, the actuating force P_2 is less than the pin reaction P_1. Any element of the band, of angular length $d\theta$, will be in

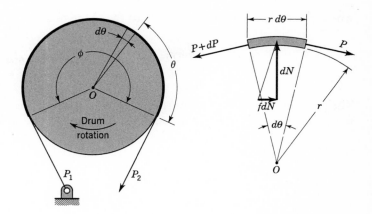

FIGURE 14-11
Forces on a brake band.

equilibrium under the action of the forces shown in the figure. Summing these forces in the vertical direction, we have

$$(P + dP) \sin \frac{d\theta}{2} + P \sin \frac{d\theta}{2} - dN = 0 \qquad (a)$$

$$dN = P \, d\theta \qquad (b)$$

since for small angles $\sin d\theta/2 = d\theta/2$. Summing the forces in the horizontal direction gives

$$(P + dP) \cos \frac{d\theta}{2} - P \cos \frac{d\theta}{2} - f \, dN = 0 \qquad (c)$$

$$dP - f \, dN = 0 \qquad (d)$$

Substituting the value of dN from Eq. (b) into (d) and integrating,

$$\int_{P_2}^{P_1} \frac{dP}{P} = f \int_0^\phi d\theta \qquad \ln \frac{P_1}{P_2} = f\phi$$

and

$$\frac{P_1}{P_2} = e^{f\phi} \qquad (14\text{-}22)$$

The torque may be obtained from the equation

$$T = (P_1 - P_2)\frac{D}{2} \qquad (14\text{-}23)$$

The normal force dN acting on an element of area of width b and length $r \, d\theta$ is

$$dN = pbr \, d\theta \qquad (e)$$

where p is the pressure. Substitution of the value of dN from Eq. (b) gives

$$P \, d\theta = pbr \, d\theta$$

Therefore
$$p = \frac{P}{br} = \frac{2P}{bD} \qquad (14\text{-}24)$$

The pressure is therefore proportional to the tension in the band. The maximum pressure p_a will occur at the toe and has the value

$$p_a = \frac{2P_1}{bD} \qquad (14\text{-}25)$$

14-5 FRICTIONAL-CONTACT AXIAL CLUTCHES

An axial clutch is one in which the mating frictional members are moved in a direction parallel to the shaft. One of the earliest of these is the cone clutch, which is simple in construction and quite powerful. However, except for relatively simple installations, it has been largely displaced by the disk clutch employing one or more disks as the operating members. Advantages of the disk clutch include the freedom from centrifugal effects, the large frictional area which can be installed in a small space, the more effective heat-dissipation surfaces, and the favorable pressure distribution. One very successful disk-clutch design is shown in Fig. 14-12. Let us now determine the capacity of such a clutch or brake in terms of the material and dimensions.

Figure 14-13 shows a friction disk having an outside diameter D and an inside diameter d. We are interested in obtaining the axial force F necessary to produce a certain torque T and pressure p. Two methods of solving the problem, depending upon the construction of the clutch, are in general use. If the disks are rigid, then, at first, the greatest amount of wear will occur in the outer areas, since the work of friction is greater in those areas. After a certain amount of wear has taken place, the pressure distribution will change so as to permit the wear to be uniform. This is the basis of the first method of solution.

Another method of construction employs springs to obtain a uniform pressure over the area. It is this assumption of uniform pressure that is used in the second method of solution.

Uniform Wear

After initial wear has taken place and the disks have worn down to the point where uniform wear becomes possible, the greatest pressure must occur at $r = d/2$ in order for the wear to be uniform. Denoting the maximum pressure by p_a, we can then write

$$pr = p_a \frac{d}{2} \qquad \text{or} \qquad p = p_a \frac{d}{2r} \qquad (a)$$

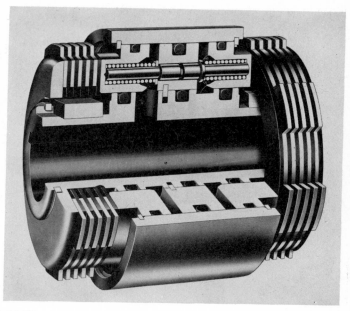

FIGURE 14-12
An oil-actuated multiple-disk clutch-brake or clutch-clutch for enclosed operation in an oil bath or spray. It is especially useful for rapid cycling. (*Courtesy of Twin Disc Clutch Company.*)

which is the condition for having the same amount of work done at radius r as is done at radius $d/2$. Referring to Fig. 14-13, we have an element of area of radius r and thickness dr. The area of this element is $2\pi r\,dr$, so that the normal force acting upon this element is $dF = 2\pi pr\,dr$. We can find the total normal force by letting r vary from $d/2$ to $D/2$ and integrating. Thus

$$F = \int_{d/2}^{D/2} 2\pi pr\,dr = \pi p_a d \int_{d/2}^{D/2} dr = \frac{\pi p_a d}{2}(D - d) \qquad (14\text{-}26)$$

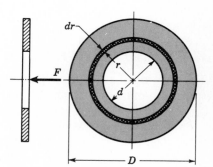

FIGURE 14-13
Disk friction member.

The torque is found by integrating the product of the frictional force and the radius:

$$T = \int_{d/2}^{D/2} 2\pi f p r^2 \, dr = \pi f p_a d \int_{d/2}^{D/2} r \, dr = \frac{\pi f p_a d}{8}(D^2 - d^2) \quad (14\text{-}27)$$

By substituting the value of F from Eq. (14-26) we may obtain a more convenient expression for the torque. Thus

$$T = \frac{Ff}{4}(D + d) \quad (14\text{-}28)$$

In use, Eq. (14-26) gives the actuating force per friction surface pair for the selected maximum pressure p_a. Equation (14-28) is then used to obtain the torque capacity per friction surface.

Uniform Pressure

When uniform pressure can be assumed over the area of the disk, the actuating force F is simply the product of the pressure and the area. This gives

$$F = \frac{\pi p_a}{4}(D^2 - d^2) \quad (14\text{-}29)$$

As before, the torque is found by integrating the product of the frictional force and the radius:

$$T = 2\pi f p \int_{d/2}^{D/2} r^2 \, dr = \frac{2\pi f p}{24}(D^3 - d^3) \quad (14\text{-}30)$$

Since $p = p_a$, we can rewrite Eq. (14-30) as

$$T = \frac{Ff}{3}\frac{D^3 - d^3}{D^2 - d^2} \quad (14\text{-}31)$$

It should be noted for both equations that the torque is for a single pair of mating surfaces. This value must therefore be multiplied by the number of pairs of surfaces in contact.

14-6 CONE CLUTCHES AND BRAKES

The drawing of a *cone clutch* in Fig. 14-14 shows that it consists of a *cup* keyed or splined to one of the shafts, a *cone* which must slide axially on splines or keys on the mating shaft, and a helical spring to hold the clutch in engagement. The clutch is disengaged by means of a fork which fits into the shifting groove on the friction cone. The *cone angle* α and the diameter and face width of the cone are the important geometric design parameters. If the cone angle is too small, say, less

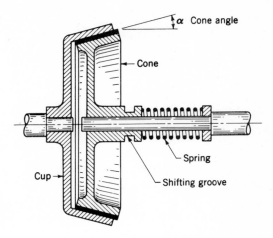

FIGURE 14-14
A cone clutch.

than about 8°, then the force required to disengage the clutch may be quite large. And the wedging effect lessens rapidly when larger cone angles are used. Depending upon the characteristics of the friction materials, a good compromise can usually be found using cone angles between 10 and 15°.

To find a relation between the operating force F and the torque transmitted, designate the dimensions of the friction cone as shown in Fig. 14-15. As in the case of the axial clutch, we can obtain one set of relations for a uniform-wear and another set for a uniform-pressure assumption.

Uniform Wear

The pressure relation is the same as for the axial clutch:

$$p = p_a \frac{d}{2r} \qquad (a)$$

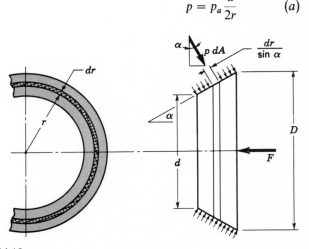

FIGURE 14-15

Next, referring to Fig. 14-15, we see that we have an element of area dA of radius r and width $dr/\sin \alpha$. Thus $dA = 2\pi r\, dr/\sin \alpha$. As shown in Fig. 14-15, the operating force will be the integral of the axial component of the differential force $p\, dA$. Thus

$$F = \int p\, dA \sin \alpha = \int_{d/2}^{D/2} \left(p_a \frac{d}{2r}\right)\left(\frac{2\pi r\, dr}{\sin \alpha}\right)(\sin \alpha)$$

$$= \pi p_a d \int_{d/2}^{D/2} dr = \frac{\pi p_a d}{2}(D - d) \tag{14-32}$$

the same as Eq. (14-26).

The differential friction force is $fp\, dA$, and the torque is the integral of the product of this force with the radius. Thus

$$T = \int rfp\, dA = \int_{d/2}^{D/2} (rf)\left(p_a \frac{d}{2r}\right)\left(\frac{2\pi r\, dr}{\sin \alpha}\right)$$

$$= \frac{\pi f p_a d}{\sin \alpha} \int_{d/2}^{D/2} r\, dr = \frac{\pi f p_a d}{8 \sin \alpha}(D^2 - d^2) \tag{14-33}$$

Note that Eq. (14-27) is a special case of (14-33) with $\alpha = 90°$. Using Eq. (14-32), we find the torque can also be written

$$T = \frac{Ff}{4 \sin \alpha}(D + d) \tag{14-34}$$

Uniform Pressure

Using $p = p_a$, the actuating force is found to be

$$F = \int p_a\, dA \sin \alpha = \int_{d/2}^{D/2} (p_a)\left(\frac{2\pi r\, dr}{\sin \alpha}\right)(\sin \alpha) = \frac{\pi p_a}{4}(D^2 - d^2) \tag{14-35}$$

The torque is

$$T = \int rfp_a\, dA = \int_{d/2}^{D/2} (rfp_a)\left(\frac{2\pi r\, dr}{\sin \alpha}\right) = \frac{\pi f p_a}{12 \sin \alpha}(D^3 - d^3) \tag{14-36}$$

Or using Eq. (14-35) in (14-36),

$$T = \frac{Ff}{3 \sin \alpha}\frac{D^3 - d^3}{D^2 - d^2} \tag{14-37}$$

14-7   MISCELLANEOUS CLUTCHES AND COUPLINGS

The square-jaw clutch shown in Fig. 14-16a is one form of positive-contact clutch. These clutches have the following characteristics:

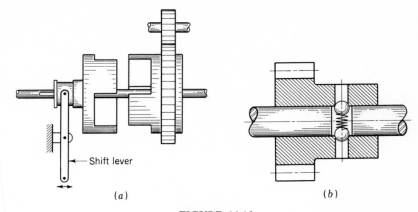

FIGURE 14-16
(a) Square-jaw clutch; (b) overload-release clutch.

1 They do not slip.
2 No heat is generated.
3 They cannot be engaged at high speeds.
4 Sometimes they cannot be engaged when both shafts are at rest.
5 Engagement at any speed is accompanied by shock.

The greatest differences between the various types of positive clutches are concerned with the design of the jaws. To provide a longer period of time for shift action during engagement, the jaws may be ratchet-shaped, spiral-shaped, or gear-tooth-shaped. Sometimes a great many teeth or jaws are used, and they may be cut either circumferentially, so that they engage by cylindrical mating, or on the faces of the mating elements.

Although positive clutches are not used to the extent of the frictional-contact types, they do have important applications where synchronous operation is required, as, for example, in power presses or rolling-mill screw-downs.

Devices such as linear drives or motor-operated screw drivers must run to a definite limit and then come to a stop. An overload-release type of clutch is required for these applications. Figure 14-16b is a schematic drawing illustrating the principle of operation of such a clutch. These clutches are usually spring-loaded so as to release at a predetermined torque. The clicking sound which is heard when the overload point is reached is considered to be a desirable signal.

Both fatigue and shock loads must be considered in obtaining the stresses and deflections of the various portions of positive clutches. In addition, wear must generally be considered. The application of the fundamentals discussed in Part 1 is usually sufficient for the complete design of these devices.

An overrunning clutch or coupling permits the driven member of a machine to "freewheel" or "overrun" because the driver is stopped or because another source of power increases the speed of the driven mechanism. The construction uses rollers or balls mounted between an outer sleeve and an inner member having cam flats machined around the periphery. Driving action is obtained by wedging the rollers between the sleeve and the cam flats. This clutch is therefore equivalent to a pawl and ratchet with an infinite number of teeth.

There are many varieties of overrunning clutches available, and they are built in capacities up to hundreds of horsepower. Since no slippage is involved, the only power loss is that due to bearing friction and windage.

14-8 FRICTION MATERIALS

A brake or clutch friction material should have the following characteristics to a degree which is dependent upon the severity of the service:

1 A high and uniform coefficient of friction
2 Properties which are not affected by environmental conditions, such as moisture
3 The ability to withstand high temperatures, together with good heat conductivity
4 Good resiliency
5 High resistance to wear, scoring, and galling

Table 14-1 lists some of the commonly employed friction materials, together with their characteristics. In selecting a coefficient of friction for design purposes, only one-half to three-quarters of the value listed should be used. This will provide some margin of safety against wear, dirt, and other unfavorable conditions.

Some of the materials may be run wet by allowing them to dip in oil or to be sprayed by oil. This reduces the coefficient of friction somewhat but carries away more heat and permits higher pressures to be used.

14-9 ENERGY CONSIDERATIONS

When the rotating members of a machine are caused to stop by means of a brake, the kinetic energy of rotation must be absorbed by the brake. This energy appears in the brake in the form of heat. In the same way, when the members of a machine which are initially at rest are brought up to speed, slipping must occur in the clutch until the driven members have the same speed as the driver. Kinetic energy is absorbed during slippage of either a clutch or brake, and this energy appears as heat.

We have seen how the torque capacity of a clutch or brake depends upon the coefficient of friction of the material and upon a safe normal pressure. However, the character of the load may be such that, if this torque value is permitted, the clutch or brake may be destroyed by its own generated heat. The capacity of a

Table 14-1 FRICTION MATERIALS FOR CLUTCHES*

Contact surfaces		Friction coefficient‡		Maximum temperature °F	Maximum pressure psi	Relative cost	Comment
Wearing	Opposing†	Wet	Dry				
Cast bronze	Cast iron or steel	0.05	—	300	80–120	Low	Subject to seizing
Cast iron	Cast iron	0.05	0.15–0.2	600	150–250	Very low	Good at low speeds
	Steel	0.06	—	500	120–200	Very low	Fair at low speeds
Hard steel	Hard steel	0.05	—	500	100	Moderate	Subject to galling
	Hard steel, chromium-plated	0.03	—	500	200	High	Durable combination
Hard-drawn phosphor bronze	Hard steel, chromium-plated	0.03	—	500	150	High	Good wearing qualities
Powder metal¶	Cast iron or steel	0.05–0.1	0.1–0.4	1000	150	High	Good wearing qualities
	Hard steel, chromium-plated	0.05–0.1	0.1–0.3	1000	300	Very high	High energy absorption
Wood	Cast iron or steel	0.16	0.2–0.35	300	60–90	Lowest	Unsuitable at high speed
Leather	Cast iron or steel	0.12–0.15	0.3–0.5	200	10–40	Very low	Subject to glazing
Cork	Cast iron or steel	0.15–0.25	0.3–0.5	200	8–14	Very low	Cork-insert preferred
Felt	Cast iron or steel	0.18	0.22	280	5–10	Low	Resilient engagement
Vulcanized fiber or paper	Cast iron or steel	—	0.3–0.5	200	10–40	Very low	Low speeds, light duty
Woven asbestos¶	Cast iron or steel	0.1–0.2	0.3–0.6	350–500	50–100	Low	Prolonged slip service ratings given
	Cast iron or steel	0.1–0.2	—	500	100–200	Low	This rating for short infrequent engagements
Molded asbestos¶	Hard steel, chromium-plated	0.1	—	—	1200	Moderate	Used in Napier Sabre engine
	Cast iron or steel	0.08–0.12	0.2–0.5	500	50–150	Very low	Wide field of applications
Impregnated asbestos	Cast iron or steel	0.12	0.32	500–750	150	Moderate	For demanding applications
Carbon graphite	Steel	0.05–0.1	0.25	700–1000	300	High	For critical requirements
Molded phenolic plastic, macerated cloth base	Cast iron or steel	0.1–0.15	0.25	300	100	Low	For light special service

* A. F. Gagne, Jr., Clutches, *Machine Design*, vol. 24, no. 8, p. 136, August 1952. Reproduced with the permission of *Machine Design*.
† Steel, where specified, should have a carbon content of approximately 0.70 percent. Surfaces should be ground true and smooth.
‡ Conservative values should be used to allow for possible glazing of clutch surfaces in service and for adverse operating conditions.
¶ For a specific material within this group, the coefficient usually is maintained within plus or minus 5 percent.

clutch is therefore limited by two factors, the characteristics of the material and the ability of the clutch to dissipate heat. In this section we shall consider the amount of heat generated by a clutching or braking operation. If the heat is generated faster than it is dissipated, we have a temperature-rise problem; that is the subject of the next section.

If the velocity is constant, the kinetic energy of a translating body is

$$E_k = \tfrac{1}{2}mv^2 \qquad (14\text{-}38)$$

where E_k = kinetic energy
m = mass
v = velocity
and the kinetic energy of a rotating body is

$$E_k = \tfrac{1}{2}I\omega^2 \qquad (14\text{-}39)$$

where I = mass moment of inertia
ω = angular velocity
If the initial speed is zero in the case of a clutch or if the final speed is zero in the case of a brake, then Eq. (14-38) or (14-39), whichever is applicable, gives the kinetic energy that must be absorbed. If the clutching or braking operation merely changes the speed, the kinetic energy absorbed is the difference in the energies computed separately at each speed.

14-10   HEAT DISSIPATION

The temperature rise of the clutch plates or brake drum may be approximated by the classic expression

$$\Delta T = \frac{E_k}{Cm} \qquad (14\text{-}40)$$

where ΔT = temperature rise, °C
E_k = energy absorbed, J
C = specific heat capacity; use 500 J/(kg·°C) for steel or cast iron
m = mass of clutch plates or brake drum, kg
It may be that the operating frequency is small enough so that the elements cool off after each cycle is completed. If this is not the case, the temperature will climb in sawtooth fashion until, finally, an equilibrium condition is established. If the numerical details of each cycle are known, the computer can be used to predict the final temperature, using Eq. (14-40) repeatedly. It may be, however, that the operating conditions vary so much that the only satisfactory approach is to build a prototype and test it in the laboratory.

Another approach to the problem, especially useful in preliminary design, is to specify limiting values of the product of pressure and velocity. These are called

pV values and are roughly proportional to the energy absorbed per unit time. Recommended values for preliminary or prototype design lie in the range

$$1000 \leq pV \leq 3000 \qquad (14\text{-}41)$$

where p is in megapascals (MPa) and V is in metres per second (m/s). Values higher than $pV = 3000$ can be used if the load is not applied continuously or if the heat-dissipation capability is considered good.

PROBLEMS

Section 14-2

14-1 An internal-expanding rim-type brake is shown in the figure. The brake drum has an inside diameter of 12 in and the radius to the hinge pins is $R = 5$ in. The shoes have a face width of $1\frac{1}{2}$ in, both of which are actuated by the same force F. The design coefficient of friction is 0.28 with a maximum pressure of 120 psi.
   (a) Find the actuating force F.
   (b) Calculate the torque capacity.
   (c) Find the hinge-pin reactions.

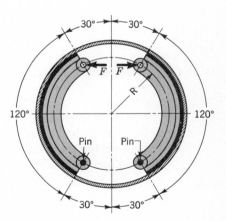

PROBLEM 14-1

14-2 If the actuating forces may differ from each other in Prob. 14-1, what values should they have in order that each shoe will be loaded by the maximum pressure of 120 psi?

14-3 In the figure for Prob. 14-1 $R = 100$ mm, the drum diameter is 250 mm, and the face width of the shoes is 28 mm. Calculate the actuating force F, the torque capacity, and the hinge-pin reactions if $p_a = 600$ kPa and $f = 0.32$.

14-4 The figure shows a 400-mm-diameter brake drum with four internally expanding shoes. Each of the hinge pins A and B supports a pair of shoes. The actuating mechanism is to be arranged to produce the same force F on each shoe. The face width of the shoes is 75 mm. The material used permits a coefficient of friction of 0.24 and a maximum pressure of 1.0 MPa.

(a) Determine the actuating force.

(b) Calculate the braking capacity.

(c) Noting that rotation may be in either direction, compute the hinge-pin reactions.

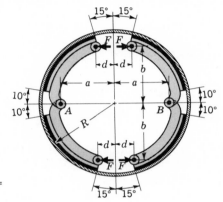

PROBLEM 14-4
The dimensions in millimetres are: $a = 150$, $b = 165$, $R = 200$, $d = 50$.

Section 14-3

14-5   The block-type hand brake shown in the figure has a face width of 45 mm. The frictional material permits a maximum pressure of 550 kPa with a coefficient of friction of 0.24.

(a) Determine the force F.

(b) What is the torque capacity?

(c) If the speed is 100 rpm and the brake is applied for 5 s at full capacity to bring the shaft to stop, how much heat is generated?

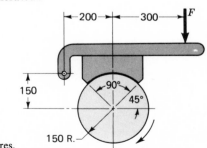

PROBLEM 14-5
Dimensions in millimetres.

14-6   The brake whose dimensions are shown in the figure has a coefficient of friction of 0.30 and is to have a maximum pressure of 150 psi against the friction material.

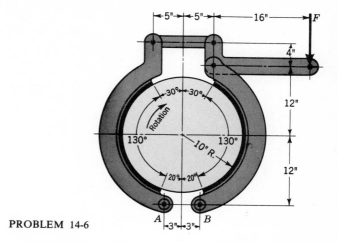

PROBLEM 14-6

(*a*) Using an actuating force of 400 lb, determine the width of face of the shoes. (Both shoes are to have the same width.)

(*b*) What torque will the brake absorb?

14-7 Solve Prob. 14-6 for counterclockwise drum rotation.

Section 14-4

14-8 The band brake shown in the figure is to have a maximum lining pressure of 600 kPa. The drum is 350 mm in diameter, and the band width is 100 mm. The coefficient of friction is 0.25, and the angle of wrap 270°.

(*a*) Find the band tensions.

(*b*) Calculate the torque capacity.

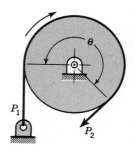

PROBLEM 14-8

14-9 The brake shown in the figure has a coefficient of friction of 0.30 and is to operate at a maximum band pressure of 1000 kPa. The width of the band is 50 mm.

(*a*) What is the limiting value of the force F?

(*b*) Determine the torque capacity of the brake.

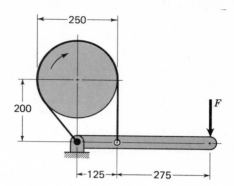

PROBLEM 14-9
Dimensions in millimetres.

14-10 Solve Prob. 14-9, for counterclockwise drum rotation.

14-11 The figure shows a 16-in differential band brake. The maximum pressure is to be 60 psi, with a coefficient of friction of 0.26 and a band width of 4 in. Determine the band tensions and the actuating force for clockwise rotation.

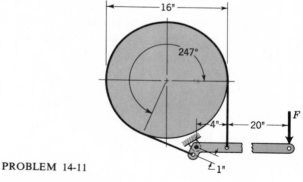

PROBLEM 14-11

14-12 Solve Prob. 14-11 for counterclockwise rotation.

Section 14-5

14-13 A plate clutch has a single pair of mating friction surfaces 300 mm OD by 225 mm ID. The coefficient of friction is 0.25 and the maximum pressure is 825 kPa. Find the torque capacity using the uniform-wear assumption.

14-14 Use the data of Prob. 14-13 to find the torque capacity of a similar clutch employing the uniform-pressure assumption.

14-15 A plate clutch has a single pair of mating friction surfaces, is 200 mm OD by 100 mm ID, and has a coefficient of friction of 0.30. What is the maximum pressure corresponding to an actuating force of 15 kN? Use the uniform-wear method.

14-16 Solve Prob. 14-15 using the uniform-pressure method.

14-17 A disk clutch has four pairs of mating friction surfaces, is 5 in OD × 3 in ID, and has a coefficient of friction of 0.10.

(*a*) What actuating force is required for a pressure of 120 psi?

(*b*) What is the torque rating?

(Use the uniform-pressure method.)

14-18 A plate clutch has two pairs of mating friction surfaces 14 in OD × 6 in ID. The coefficient of friction is 0.24, and the pressure is not to exceed 80 psi.

(*a*) Determine the actuating force.

(*b*) Calculate the torque capacity.

(Use the uniform-wear method.)

Section 14-6

14-19 A leather-faced cone clutch is to transmit 1200 lb·in of torque. The cone angle is 10°, the mean diameter of the friction surface is 12 in, and the face width is 2 in. Based on a coefficient of friction of 0.25, find the operating force and pressure.

(*a*) Use the uniform-pressure assumption.

(*b*) Use the uniform-wear assumption.

14-20 A cone clutch has a cone angle of 11.5°, a mean frictional diameter of 320 mm, and a face width of 60 mm. The clutch is to transmit a torque of 200 N·m. The coefficient of friction is 0.26. Find the actuating force and pressure using the assumption of uniform pressure.

14-21 A one-cylinder two-cycle engine develops its maximum torque at a speed of 3400 rpm when delivering 12 hp. The tentative design of a cone clutch to couple this engine to its load has $\alpha = 13°$, $D = 4$ in, and $p_a = 50$ psi, with $f = 0.20$, corresponding to a woven asbestos friction facing. Determine the required face width and operating force. Use the uniform-wear assumption.

Sections 14-7 to 14-10

14-22 A two-jaw clutch has the dimensions shown in the figure and is made of ductile steel. The clutch has been designed to transmit 2 kW at 500 rpm. Find the bearing and shear stress in the key and in the jaws.

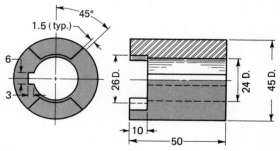

PROBLEM 14-22
Dimensions in millimetres.

14-23 A 0.500-m externally contracting brake drum has heat-dissipating surfaces whose mass is 20 kg. The brake has a rated torque capacity of 300 N·m. The brake is used to bring a rotating inertia to a complete stop. Determine the temperature rise if the inertia has an initial speed of 1800 rpm and is brought to rest in 8 s using the full torque capacity of the brake.

14-24 A flywheel is made from a steel disk 250 mm in diameter and 20 mm thick. It rotates at 500 rpm and is to be brought to rest in 0.40 s by a brake. Compute the total amount of energy to be absorbed and the required torque capacity of the brake.

14-25 A brake permits a mass of 250 kg to be lowered at a velocity of 3 m/s from a drum, as shown in the figure. The drum is 400 mm in diameter, weighs 1.40 kN, and has a radius of gyration of 180 mm.

    (*a*) Compute the energy of the system.

    (*b*) How much additional braking torque must be applied if the load is to be stopped in 0.50 s?

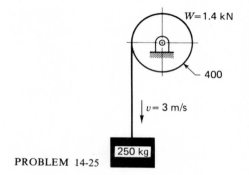

PROBLEM 14-25

15

FLEXIBLE MECHANICAL ELEMENTS

Flexible machine elements, such as belts, ropes, or chains, are used for the transmission of power over comparatively long distances. When these elements are employed, they usually replace a group of gears, shafts, and bearings or similar power-transmission devices. They thus greatly simplify a machine and consequently are a major cost-reducing element.

In addition, since these elements are elastic and usually long, they play an important part in absorbing shock loads and in damping out the effects of vibrating forces. Although this advantage is important as concerns the life of the driving machine, the cost-reduction element is generally the major factor in the selection of this means of power transmission.

15-1 BELTS

Ordinarily, belts are used to transmit power between two parallel shafts. The shafts must be separated a certain minimum distance, which is dependent upon the type of belt used, in order to work most efficiently. Belts have the following characteristics:

FIGURE 15-1
Link V belts successfully applied to a press. The use of endless belts would necessitate dismantling the press to replace belts. (*Courtesy of Manheim Manufacturing and Belting Company.*)

1 They may be used for long center distances.

2 Because of the slip and creep of belts, the angular-velocity ratio between the two shafts is neither constant nor exactly equal to the ratio of the pulley diameters.

3 When using flat belts, clutch action may be obtained by shifting the belt from a loose to a tight pulley.

4 When V belts are used, some variation in the angular-velocity ratio may be obtained by employing a small pulley with spring-loaded sides. The diameter of the pulley is then a function of the belt tension and may be varied by changing the center distance.

5 Some adjustment of the center distance is usually necessary when belts are used.

6 By employing step pulleys, an economical means of changing the velocity ratio may be obtained.

Flat belts are usually made of either oak-tanned leather or a fabric, such as cotton or rayon, which has been impregnated with rubber. They find their greatest use where the center distances are fairly long. Because of the clutching action which can be obtained and their adaptability to fairly long distances, flat belts are

FIGURE 15-2
A timing-belt drive for textile machinery.
(*Courtesy of U.S. Rubber Company.*)

very useful in group-drive installations. Because of the convenience and appearance of unit drives, most machines manufactured today have a built-in drive; hence the use of flat belts has greatly decreased in recent years. Nevertheless flat belts are very efficient for high speeds, they can transmit large amounts of power, they are very flexible, they don't require large pulleys, and they can transmit power around corners.

A *V belt* is made of fabric and cord, usually cotton or rayon, and impregnated with rubber. In contrast to flat belts, V belts may be operated with smaller pulleys or sheaves and at shorter center distances. In addition, a number of them may be used on a single sheave, thus making a multiple drive. They are endless, thus eliminating the joint which must be made in flat belts.

A *link V belt* is composed of a large number of rubberized-fabric links joined by suitable metal fasteners. This type of belt may be disassembled at any point and adjusted to any length by removing some of the links. This eliminates the necessity for adjustable centers and simplifies the installation. It makes it possible to change the tension for maximum efficiency and also reduces the inventory of belt sizes which would usually be stocked. A typical application of the link V belt is shown in Fig. 15-1.

A *timing belt* is a patented belt, made of rubberized fabric and steel wire, having teeth which fit into grooves cut on the periphery of the pulleys (Fig. 15-2).

The timing belt does not stretch or slip and consequently transmits power at a constant angular-velocity ratio. The fact that the belt is toothed provides several advantages over ordinary belting. One of these is that no initial tension is necessary, so that fixed center drives may be used. Another is the elimination of the restriction on speeds; the teeth make it possible to run the belts at nearly any speed, slow or fast. Disadvantages are the first cost of the belt and the necessity of grooving the pulleys.

15-2   FLAT-BELT DRIVES

Materials used for flat belts are rubberized fabric, rubberized cord, rubberized cord and fabric, reinforced plastic or rubber, and leather. Some of these materials may be spliced to obtain the desired loop size while others are made only as a continuous loop. Leather belts will carry large amounts of power at moderate speeds for a long life but they may either stretch or shrink and they are expensive. The reinforced plastic and rubber belts can carry power loads as high as 3 kW per mm of belt width at belt speeds up to 200 m/s. Other factors which influence the selection of belt materials are the desired reliability and life, pulley sizes, and cost.

Figure 15-3 illustrates open and crossed belts and gives equations for the angle of contact θ and the total belt length L for each case. When a horizontal open-belt arrangement is used, the driver should rotate so that the slack side is on top. This makes for a larger angle of contact on both pulleys. When the drive is vertical or the center distance is short, a larger angle of contact may be obtained by using an idler tension pulley.

Let us now investigate the relation between the belt tensions and the horsepower which is transmitted. When a belt is running and transmitting power, there exist a tension P_1 on the tight side and a lesser tension P_2 on the slack side. If the centrifugal force on the belt is neglected, the relation between these tensions is the same as for a band brake. From Eq. (14-22), we have

$$\frac{P_1}{P_2} = e^{f\theta} \qquad (15\text{-}1)$$

where f is the coefficient of friction between the belt and the pulley and θ is the angle of contact. The horsepower transmitted is

$$\text{hp} = \frac{(P_1 - P_2)V}{33\,000} \qquad (15\text{-}2)$$

where V is the belt velocity in feet per minute.

The effect of centrifugal force is not shown in Eq. (15-1). By including the centrifugal force in the development of the equation, it can be shown that

$$\frac{P_1 - F_c}{P_2 - F_c} = e^{f\theta} \qquad (15\text{-}3)$$

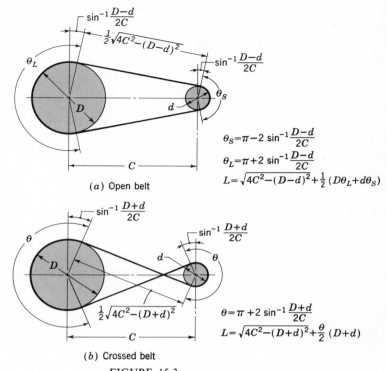

FIGURE 15-3
Belt lengths and contact angles for open and crossed belts.

where

$$F_c = \frac{wv^2}{g} \qquad (15\text{-}4)$$

and where w = weight of belt, lb/ft

    v = belt velocity, ft/s

    g = acceleration due to gravity, ft/s^2

Leather weighs 0.035 lb/in^3.

In SI units the power transmitted in watts·(W) is the difference in belt tensions measured in newtons (N) times the belt velocity in metres per second (m/s).

These relations can be used for belt-drive design by limiting the maximum tension P_1 according to the permissible tensile stress in the belt material. For leather, a conservative figure is 1750 kPa (about 250 psi). The coefficient of friction used depends upon both the belt and pulley material. Average design values are: leather on cast iron, 0.30; on wood pulleys, 0.45; on molded paper or plastic, 0.40 to 0.55.

Table 15-1 STANDARD V-BELT SECTIONS

Belt section	Width a in	Thickness b in	Minimum sheave diameter, in	Hp range one or more belts
A	$\frac{1}{2}$	$\frac{11}{32}$	3.0	$\frac{1}{4}$–10
B	$\frac{21}{32}$	$\frac{7}{16}$	5.4	1–25
C	$\frac{7}{8}$	$\frac{17}{32}$	9.0	15–100
D	$1\frac{1}{4}$	$\frac{3}{4}$	13.0	50–250
E	$1\frac{1}{2}$	1	21.6	100 and up

15-3 V BELTS

The cross-sectional dimensions of V belts have been standardized by manufacturers, with each section designated by a letter of the alphabet. The dimensions, minimum sheave diameters, and the horsepower range for each section are listed in Table 15-1.

To specify a V belt, give the belt-section letter, followed by the inside circumference in inches. For example, B75 is a B-section belt having an inside circumference of 75 in.

Calculations involving the belt length usually are based on the pitch length. For any given belt section the pitch length is obtained by adding a quantity to the inside circumference (Table 15-2). For example, a B75 belt has a pitch length of 76.8 in. Similarly, calculations of the velocity ratios are made using the pitch diameters of the sheaves, and for this reason the stated diameters are usually understood to be the pitch diameters even though they are not always so specified.

The groove angle of a sheave is made somewhat less than the belt-section angle. This causes the belt to wedge itself into the groove, thus increasing the friction. The exact value of this angle depends upon the belt section, the sheave diameter, and the angle of contact. If it is made too much smaller than the belt, the force required to pull the belt out of the groove as the belt leaves the pulley will be excessive. Optimum values are given in the commercial literature.

The design of a V-belt drive is similar to that of a flat-belt drive, both of them requiring the same initial information.

The minimum sheave diameters have been listed in Table 15-1. For best results a V belt should run quite fast; 4000 fpm is a good speed. Trouble may be encountered if the belt runs much faster than 5000 fpm or much slower than 1000 fpm. Therefore, when possible, the pulleys should be sized for a belt speed in the neighborhood of 4000 fpm.

Table 15-2 CONVERSION QUANTITIES—INSIDE CIRCUMFERENCE TO PITCH LENGTH

Belt section	A	B	C	D	E
Quantity to be added to inside circumference, in	1.3	1.8	2.9	3.3	4.5

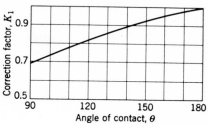

FIGURE 15-4
Correction factor K_1 for angle of contact.
Multiply the rated horsepower per belt by
this factor to obtain the corrected
horsepower.

The length of a V belt is obtained in the same way as for a flat belt. In the case of flat belts there is virtually no limit to the center distance. Long center distances, however, are not recommended for V belts because the excessive vibration of the slack side will shorten the belt life materially. In general, the center distance should not be greater than three times the sum of the sheave diameters nor less than the diameter of the larger sheave. Link-type V belts have less vibration, because of better balance, and hence may be used with longer center distances.

The selection of V belts is based on obtaining a long and trouble-free life. Table 15-3 gives the horsepower capacity of standard single V belts for various sheave diameters and belt speeds corresponding to a satisfactory life. These ratings are based on a 180° contact angle. For smaller angles this rating must be reduced. Figure 15-4 gives the values of the correction factor K_1 which is used to reduce the rated horsepower when the angle of contact is less than 180°.

For a given sheave speed, the hours of life of a short belt are less than those of a long belt, because the short belt is subjected to the action of the load a greater number of times. For this reason it is necessary to apply a second factor, K_2, which is called a belt-length correction factor. These factors are itemized in Table 15-4 for various belt sections and lengths. The rated belt horsepower must be multiplied by this factor to obtain the corrected horsepower.

The characteristics of both the driving and driven machinery must be considered in selecting the belt. If, for example, the load is started frequently by a source of power which develops 200 percent of full-load torque in starting (such as a squirrel-cage motor), then the full-load horsepower must be multiplied by an overload service factor. The characteristics of the driven machinery are considered in a similar manner. Manufacturers of V belts list these overload service factors in considerable detail. Space does not permit their inclusion here. Table 15-5 may be used to obtain this factor if the percentage overload is known.

EXAMPLE 15-1   A 10-hp split-phase motor running at 1750 rpm is to be used to drive a rotary pump which operates 24 h per day. The pump should run at approximately 1175 rpm. The center distance should not exceed 44 in. Space limits the diameter of the driven sheave to 11.5 in. Determine the sheave diameters, the belt size, and the number of belts.

Table 15-3   HORSEPOWER RATINGS OF STANDARD V BELTS*

Belt section	Sheave pitch diameter, in	Belt speed, fpm				
		1000	2000	3000	4000	5000
A	2.6	0.47	0.62	0.53	0.15	
	3.0	0.66	1.01	1.12	0.93	0.38
	3.4	0.81	1.31	1.57	1.53	1.12
	3.8	0.93	1.55	1.92	2.00	1.71
	4.2	1.03	1.74	2.20	2.38	2.19
	4.6	1.11	1.89	2.44	2.69	2.58
	5.0 and up	1.17	2.03	2.64	2.96	2.89
B	4.2	1.07	1.58	1.68	1.26	0.22
	4.6	1.27	1.99	2.29	2.08	1.24
	5.0	1.44	2.33	2.80	2.76	2.10
	5.4	1.59	2.62	3.24	3.34	2.82
	5.8	1.72	2.87	3.61	3.85	3.45
	6.2	1.82	3.09	3.94	4.28	4.00
	6.6	1.92	3.29	4.23	4.67	4.48
	7.0 and up	2.01	3.46	4.49	5.01	4.90
C	6.0	1.84	2.66	2.72	1.87	
	7.0	2.48	3.94	4.64	4.44	3.12
	8.0	2.96	4.90	6.09	6.36	5.52
	9.0	3.34	5.65	7.21	7.86	7.39
	10.0	3.64	6.25	8.11	9.06	8.89
	11.0	3.88	6.74	8.84	10.0	10.1
	12.0 and up	4.09	7.15	9.46	10.9	11.1
D	10.0	4.14	6.13	6.55	5.09	1.35
	11.0	5.00	7.83	9.11	8.50	5.62
	12.0	5.71	9.26	11.2	11.4	9.18
	13.0	6.31	10.5	13.0	13.8	12.2
	14.0	6.82	11.5	14.6	15.8	14.8
	15.0	7.27	12.4	15.9	17.6	17.0
	16.0	7.66	13.2	17.1	19.2	19.0
	17.0 and up	8.01	13.9	18.1	20.6	20.7
E	16.0	8.68	14.2	17.5	18.1	15.3
	18.0	9.92	16.7	21.2	23.0	21.5
	20.0	10.9	18.7	24.2	26.9	26.4
	22.0	11.7	20.3	26.6	30.2	30.5
	24.0	12.4	21.6	28.6	32.9	33.8
	26.0	13.0	22.8	30.3	35.1	36.7
	28.0 and up	13.4	23.7	31.8	37.1	39.1

* Courtesy of Browning Manufacturing Company. These ratings are for single belts having an angle of contact of 180°.

Table 15-4   BELT-LENGTH CORRECTION FACTOR K_2*

	Nominal belt length, in				
Length factor	A belts	B belts	C belts	D belts	E belts
0.85	Up to 35	Up to 46	Up to 75	Up to 128	
0.90	38–46	48–60	81–96	144–162	Up to 195
0.95	48–55	62–75	105–120	173–210	210–240
1.00	60–75	78–97	128–158	240	270–300
1.05	78–90	105–120	162–195	270–330	330–390
1.10	96–112	128–144	210–240	360–420	420–480
1.15	120 and up	158–180	270–300	480	540–600
1.20		195 and up	330 and up	540 and up	660

* Multiply the rated horsepower per belt by this factor to obtain the corrected horsepower.

DECISIONS

1   An overload service factor of 1.2 corresponding to a 50 percent overload is selected from Table 15-5.
2   From Table 15-1, a B-section belt is selected.
3   Since the driven sheave should not exceed 11.5 in, the next smaller standard size of 11 in will be tentatively selected.
4   A center distance of 42 in is tentatively selected.

SOLUTION   The pump is to operate 24 h per day, and so 0.1 is to be added to the service factor, making it 1.3. The design horsepower of the belt is therefore

$$\text{Design hp} = (10)(1.3) = 13$$

The diameter of the small sheave is

$$d = D\frac{n_1}{n_2} = 11\frac{1175}{1750} = 7.40 \text{ in} \qquad Ans.$$

This is a standard pitch diameter for B-section belts and is also over the minimum diameter listed in Table 15-1. It will therefore be used.

From Fig. 15-3a, the angles of contact for a center distance of 42 in are

$$\theta_s = \pi - 2\sin^{-1}\frac{D-d}{2C} = \pi - 2\sin^{-1}\frac{11-7.4}{(2)(42)} = 3.056 \text{ radians}$$

$$\theta_L = \pi + 2\sin^{-1}\frac{D-d}{2C} = \pi + 2\sin^{-1}\frac{11-7.4}{(2)(42)} = 3.227 \text{ radians}$$

Table 15-5   OVERLOAD SERVICE FACTORS*

Percent overload	0	25	50	75	100	150
Service factor	1.0	1.1	1.2	1.3	1.4	1.5

* Multiply the given horsepower by these factors to obtain the design horsepower. For 16- to 24-h operation add 0.1 to these values.

The belt length is

$$L = \sqrt{4C^2 - (D - d)^2} + (\tfrac{1}{2})(D\theta_L + d\theta_s)$$
$$= \sqrt{(4)(42)^2 - (11 - 7.4)^2} + (\tfrac{1}{2})[(11)(3.227) + (7.4)(3.056)]$$
$$= 112.97 \text{ in}$$

The nearest standard size, a B112, is selected. From Table 15-2, a B112 belt has a pitch length of 113.8 in.

The belt speed is

$$V = \frac{\pi d n}{12} = \frac{(\pi)(7.4)(1750)}{12} = 3390 \text{ fpm}$$

Using Table 15-3 and interpolating, the rated horsepower per belt is 4.66. This must be corrected for the contact angle and the belt length. The contact angle for the small sheave is 3.056 radians, or 175°. From Fig. 15-4, the correction factor is 0.99. The belt-length correction factor is 1.05, from Table 15.4. Therefore the corrected horsepower per belt is

$$\text{hp per belt} = (4.66)(0.99)(1.05) = 4.85$$

and so the number of belts required is

$$N = \frac{13}{4.85} = 2.68$$

Three B-section belts must therefore be used.   ////

15-4   ROLLER CHAIN

Basic features of chain drives include a constant ratio, since no slippage or creep is involved; long life; and the ability to drive a number of shafts from a single source of power.

Roller chains have been standardized as to sizes by the ANSI.* Figure 15-5 shows the nomenclature. The pitch is the linear distance between the centers of the rollers. The width is the space between the inner link plates. These chains are manufactured in single, double, triple, and quadruple strands. The dimensions of standard sizes are listed in Table 15-6.

In Fig. 15-6 is shown a sprocket driving a chain in a counterclockwise direction. Denoting the chain pitch by p, the pitch angle by γ, and the pitch diameter of the sprocket by D, from the trigonometry of the figure we see

$$\sin \frac{\gamma}{2} = \frac{p/2}{D/2} \quad \text{or} \quad D = \frac{p}{\sin (\gamma/2)} \tag{a}$$

---

* American National Standards Institute.

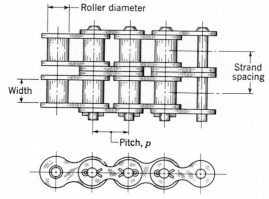

FIGURE 15-5
Portion of a double-strand roller chain.

Since $\gamma = 360/N$, where N is the number of sprocket teeth, Eq. (a) can be written

$$D = \frac{p}{\sin (180/N)} \qquad (15\text{-}7)$$

The angle $\gamma/2$, through which the link swings as it enters contact, is called the *angle of articulation*. It can be seen that the magnitude of this angle is a function of the number of teeth. Rotation of the link through this angle causes impact between the rollers and the sprocket teeth and also wear in the chain joint. Since the

Table 15-6 DIMENSIONS OF AMERICAN STANDARD ROLLER CHAIN (SINGLE STRAND)

ANSI chain number	Pitch in	Width in	Average tensile strength lb	Average weight lb per ft	Roller diameter in	Multiple strand spacing in
25	$\frac{1}{4}$	$\frac{1}{8}$	875	0.09	0.130	0.252
35	$\frac{3}{8}$	$\frac{3}{16}$	2 100	0.21	0.200	0.399
41	$\frac{1}{2}$	$\frac{1}{4}$	2 000	0.25	0.306	
40	$\frac{1}{2}$	$\frac{5}{16}$	3 700	0.42	$\frac{5}{16}$	0.566
50	$\frac{5}{8}$	$\frac{3}{8}$	6 100	0.69	0.400	0.713
60	$\frac{3}{4}$	$\frac{1}{2}$	8 500	1.00	$\frac{15}{32}$	0.897
80	1	$\frac{5}{8}$	14 500	1.71	$\frac{5}{8}$	1.153
100	$1\frac{1}{4}$	$\frac{3}{4}$	24 000	2.58	$\frac{3}{4}$	1.408
120	$1\frac{1}{2}$	1	34 000	3.87	$\frac{7}{8}$	1.789
140	$1\frac{3}{4}$	1	46 000	4.95	1	1.924
160	2	$1\frac{1}{4}$	58 000	6.61	$1\frac{1}{8}$	2.305
200	$2\frac{1}{2}$	$1\frac{1}{2}$	95 000	10.96	$1\frac{9}{16}$	2.817

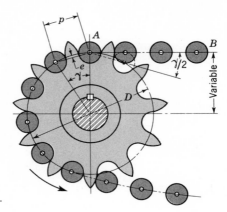

FIGURE 15-6
Engagement of a chain and sprocket.

life of a properly selected drive is a function of the wear and the surface fatigue strength of the rollers, it is important to reduce the angle of articulation as much as possible. The values of this angle have been plotted as a function of the number of teeth in Fig. 15-7a.

The number of sprocket teeth also affects the velocity ratio during the period in which the sprocket is rotating through a pitch angle. At the position shown (Fig. 15-6) the portion of the chain AB is being pulled onto the sprocket, and the pitch line of the straight portion of the chain is tangent to the pitch circle of the sprocket. However, when the sprocket has turned an angle $\gamma/2$, the pitch line AB has moved closer to the sprocket centerline by an amount e. This means not only that the straight portion AB is moving up and down as the sprocket turns, but also that the lever arm varies, and hence the velocity ratio is not constant for rotation of the sprocket through the pitch angle. This is called the *chordal speed variation*, and it is also a function of the number of teeth on the sprocket. In Fig. 15-7b this variation has been plotted as a function of the number of teeth.

The chain velocity is usually defined as the number of feet coming off the sprocket in unit time. Thus the velocity V in feet per minute is

$$V = \frac{\pi Dn}{12} = \frac{Npn}{12} \qquad (15\text{-}8)$$

where N = number of teeth on sprocket
p = chain pitch
n = speed of sprocket, rpm

Although a large number of teeth is considered desirable for the driving sprocket, in the usual case it is advantageous to obtain as small a sprocket as possible, and this requires one with a small number of teeth. For smooth operation at moderate and high speeds it is considered good practice to use a driving sprocket with at least 17 teeth; 19 or 21 will, of course, give a better life expectancy with less chain noise. Where space limitations are severe or for very slow speeds, smaller tooth numbers may be used by sacrificing the life expectancy of the chain.

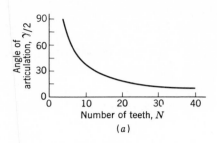

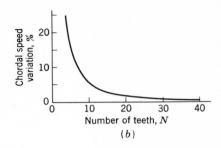

FIGURE 15-7
(*a*) Relation between the angle of articulation and the number of sprocket teeth;
(*b*) relation between the number of sprocket teeth and the chordal speed variation.

Driven sprockets are not made in standard sizes over 120 teeth because the pitch elongation will eventually cause the chain to "ride" high long before the chain is worn out. The most successful drives have velocity ratios up to 6 : 1, but higher ratios may be used at the sacrifice of chain life.

Roller chains seldom fail because they lack tensile strength but rather because they have been subjected to a great many hours of service. Actual failure may be due either to wear of the rollers on the pins or to fatigue of the surfaces of the rollers. Roller-chain manufacturers have compiled tables which give the horsepower capacity corresponding to a life expectancy of 15 kh for various sprocket speeds. These capacities are tabulated in Table 15-7 for 17-tooth sprockets. To

Table 15-7 RATED HORSEPOWER CAPACITY OF SINGLE-STRAND ROLLER CHAINS*

Sprocket speed rpm	ANSI chain number											
	25	35	41	40	50	60	80	100	120	140	160	200
50	0.08	0.139	0.193	0.322	0.620	1.05	2.44	4.67	7.91	12.3	18.0	34.2
100	0.10	0.264	0.367	0.611	1.16	1.97	4.52	8.56	14.4	22.2	32.4	60.2
150	0.12	0.379	0.523	0.870	1.65	2.82	6.39	12.0	19.9	30.6	44.2	81.3
200	0.14	0.494	0.678	1.13	2.14	3.59	8.09	15.1	24.9	38.0	54.5	98.8
300	0.21	0.705	0.954	1.59	2.99	4.98	11.1	20.3	33.2	49.9	70.5	
400	0.28	0.90	1.21	2.02	3.77	6.22	13.6	25.0	39.7	58.8		
500	0.34	1.08	1.44	2.41	4.46	7.32	15.8	28.2	44.6			
600	0.40	1.25	1.66	2.77	5.09	8.29	17.6	30.9				
800	0.51	1.56	2.04	3.41	6.17	9.91	20.5					
1000	0.61	1.83	2.37	3.95	7.05	11.1						
1200	0.70	2.08	2.65	4.41	7.75	12.1						
1400	0.79	2.29	2.88	4.80	8.31	12.7						
1600	0.87	2.48	3.06	5.10	8.70							
1800	0.94	2.65	...	5.38	8.98							
2000	1.00	2.79	...	5.57	9.13							

* The capacities shown are for a 17-tooth sprocket. For other sprocket sizes use the correction factors in Table 15-8.

obtain the capacity for a sprocket with a different number of teeth, the capacity listed in the table should be multiplied by a tooth correction factor as listed in Table 15-8.

The characteristics of the load are important considerations in the selection of roller chain. In general, extra chain capacity is required for any of the following conditions:

1 The small sprocket has less than 9 teeth for low-speed drives or less than 16 teeth for high-speed drives
2 The sprockets are unusually large
3 Shock loading occurs, or there are frequent load reversals
4 There are three or more sprockets in the drive
5 The lubrication is poor
6 The chain must operate under dirty or dusty conditions

To account for these and other conditions of operation, the horsepower of the load must be multiplied by a service factor. Suggested values for these factors are listed in Table 15-9.

The length of a chain should be determined in pitches. It is preferable to have an even number of pitches; otherwise an offset link is required. The approximate length may be obtained from the following equation:

$$\frac{L}{p} = \frac{2C}{p} + \frac{N_1 + N_2}{2} + \frac{(N_2 - N_1)^2}{4\pi^2(C/p)} \qquad (15\text{-}9)$$

where L = chain length
p = chain pitch
C = center distance
N_1 = number of teeth on small sprocket
N_2 = number of teeth on large sprocket

Table 15-8   TOOTH CORRECTION FACTORS*

Number of teeth on driving sprocket	Tooth correction factor K_T	Number of teeth on driving sprocket	Tooth correction factor K_T
11	0.53	22	1.29
12	0.62	23	1.35
13	0.70	24	1.41
14	0.78	25	1.46
15	0.85	30	1.73
16	0.92	35	1.95
17	1.00	40	2.15
18	1.05	45	2.37
19	1.11	50	2.51
20	1.18	55	2.66
21	1.26	60	2.80

* Multiply the rated horsepower from Table 15-7 by the tooth correction factor K_T.

The length of chain for a multiple-sprocket drive is most easily obtained by making an accurate scale layout and determining the length by measurement.

Lubrication of roller chains is essential in order to obtain a long and trouble-free life. Either a drip feed or a shallow bath in the lubricant is satisfactory. A medium or light mineral oil, without additives, should be used. Except for unusual conditions, heavy oils and greases are not recommended because they are too viscous to enter the small clearances in the chain parts.

EXAMPLE 15-2   A $7\frac{1}{2}$-hp speed reducer which runs at 300 rpm is to drive a conveyor at 200 rpm. The center distance is to be approximately 28 in. Select a suitable chain drive.

SOLUTION.

1 Although an odd number of sprocket teeth is to be preferred, sprockets of 20 and 30 teeth are tentatively chosen in order to obtain the proper velocity ratio. A 20-tooth sprocket will have a longer life and generate less noise than a 16- or an 18-tooth sprocket. It is chosen because space does not seem to be at a premium.

2   From Table 15-9 a service factor of 1.4 is chosen for 24-h operation with moderate shock.

The required rating is

$$\text{Required hp} = (7.5)(1.4) = 10.5$$

Examination of Table 15-7 indicates that either a No. 50 or No. 60 chain may be satisfactory. From Table 15-8 the tooth correction factor corresponding to a 20-tooth sprocket is 1.18. Therefore, from Table 15-7 the corrected capacity of a No. 50 single-strand chain at 300 rpm is

$$\text{Rated hp per strand} = (2.99)(1.18) = 3.53 \qquad \text{No. 50 chain}$$

Similarly, the capacity of a No. 60 chain is

$$\text{Rated hp per strand} = (4.98)(1.18) = 5.88 \qquad \text{No. 60 chain}$$

Table 15-9   LOAD SERVICE FACTORS K_s FOR ROLLER CHAIN*

Type of load	Service conditions	10-h day	24-h day
Uniform load	Average	1.0	1.2
Moderate shock	Abnormal	1.2	1.4
Heavy shock	Abnormal	1.4	1.7
Load reversals	Abnormal	1.5	1.9

* Courtesy of Morse Chain Company. Multiply the horsepower of the load by K_S to find the required capacity of the chain.

The number of strands required for No. 50 is

$$\text{Number of strands} = \frac{10.5}{3.53} = 2.97 \qquad \text{or 3 strands of No. 50}$$

The number required for No. 60 chain is

$$\text{Number of strands} = \frac{10.5}{5.88} = 1.79 \qquad \text{or 2 strands of No. 60}$$

The No. 60 chain would require larger sprockets, and hence it would run at a higher velocity, generate more noise, and have a shorter life. The No. 50 seems to be the better choice and is selected for this example. A comparison of the prices of the sprockets and chain for both cases might, however, make the No. 60 a better solution.

From Table 15-6, the pitch of No. 50 chain is $\frac{5}{8}$ in. Using a center distance of 28 in in Eq. (15-9), the required length of triple-strand chain in pitches is

$$\frac{L}{p} = \frac{2C}{p} + \frac{N_1 + N_2}{2} + \frac{(N_2 - N_1)^2}{4\pi^2(C/p)}$$

$$= \frac{(2)(28)}{0.625} + \frac{20 + 30}{2} + \frac{(30 - 20)^2}{4\pi^2(28/0.625)}$$

$$= 114.7 \text{ pitches}$$

The nearest even number of pitches is 114, and this will be used. A slight adjustment in the center distance is required. Substituting $L/p = 114$ into Eq. (15-9) and solving for C gives approximately $27\frac{3}{4}$ in as the new center distance.

In general, the center distance should not exceed 80 pitches; 30 to 50 pitches is a better value. In this problem the center distance is $27.75/0.625 = 44.4$ pitches, which is satisfactory. ////

15-5 ROPE DRIVES

Rope drives consisting of manila, cotton, or Dacron rope on multiple-grooved pulleys may often be a most economical form of drive over long distances and for large amounts of power. Accurate alignment of the pulleys is not necessary because they are grooved. The velocity of the rope should be quite high; 5000 fpm is a good speed for greatest economy.

15-6 WIRE ROPE

Wire rope is made with two types of winding, as shown in Fig. 15-8. The *regular lay*, which is the accepted standard, has the wire twisted in one direction to form the strands, and the strands twisted in the opposite direction to form the rope. In

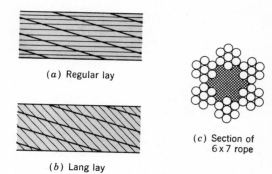

(a) Regular lay

(b) Lang lay

(c) Section of 6 x 7 rope

FIGURE 15-8
Types of wire rope.

the completed rope the visible wires are approximately parallel to the axis of the rope. Regular-lay ropes do not kink or untwist and are easy to handle.

Lang-lay ropes have the wires in the strand and the strands in the rope twisted in the same direction, and hence the outer wires run diagonally across the axis of the rope. Lang-lay ropes are more resistant to abrasive wear and failure due to fatigue than are regular-lay ropes, but they are more likely to kink and untwist.

Standard ropes are made with a hemp core which supports and lubricates the strands. When the rope is subjected to heat, either a steel center or a wire-strand center must be used.

Wire rope is designated as, for example, a $1\frac{1}{8}$-in 6 × 7 haulage rope. The first figure is the diameter of the rope (Fig. 15-8c). The second and third figures are the number of strands and the number of wires in each strand, respectively. Table 15-10 lists some of the various ropes which are available, together with their characteristics and properties. The area of the metal in standard hoisting and haulage ropes is $A_m = 0.38d$.

When a wire rope passes around a sheave, there is a certain amount of readjustment of the elements. Each of the wires and strands must slide on one another, and presumably some individual bending takes place. It is probable that in this complex action there exists some stress concentration. The stress in one of the wires of a rope passing around a sheave may be calculated as follows: From solid mechanics we have

$$M = \frac{EI}{r} \quad \text{and} \quad M = \frac{\sigma I}{c} \qquad (a)$$

where the quantities have their usual meaning. Eliminating M and solving for the stress gives

$$\sigma = \frac{Ec}{r} \qquad (b)$$

For the radius of curvature r we can substitute the radius of the sheave $D/2$. Also $c = d_w/2$, where d_w is the diameter of the wire. This substitution gives

$$\sigma = E\frac{d_w}{D} \qquad (15\text{-}10)$$

In Eq. (15-10) σ is the bending stress in the individual wires and E is the modulus of elasticity of the rope (not of the wires). The sheave diameter is represented by D. Equation (15-10) illustrates the importance of using a large-diameter sheave. The suggested minimum sheave diameters in Table 15-10 are based on a D/d_w ratio of 400. If possible, the sheaves should be designed for a larger ratio. For elevators and mine hoists, D/d_w is usually taken from 800 to 1000. If the ratio is less than 200, heavy loads will often cause a permanent set in the rope.

A wire rope may fail because the static load exceeds the ultimate strength of the rope. Failure of this nature is generally not the fault of the designer but rather that of the operator in permitting the rope to be subjected to loads for which it was not designed. On the other hand, ropes do fail because of abrasive wear and fatigue. A fatigue failure first appears as a few broken wires on the surface of the

Table 15-10   WIRE-ROPE DATA*

Rope	Weight per ft lb	Minimum sheave diameter in	Standard sizes d, in	Material	Size of outer wires	Modulus of elasticity† Mpsi	Strength‡ kpsi
6 × 7 haulage	$1.50d^2$	$42d$	$\frac{1}{4}$–$1\frac{1}{4}$	Monitor steel	$d/9$	14	100
				Plow steel	$d/9$	14	88
				Mild plow steel	$d/9$	14	76
6 × 19 standard hoisting	$1.60d^2$	$26d$–$34d$	$\frac{1}{4}$–$2\frac{3}{4}$	Monitor steel	$d/13$–$d/16$	12	106
				Plow steel	$d/13$–$d/16$	12	93
				Mild plow steel	$d/13$–$d/16$	12	80
6 × 37 special flexible	$1.55d^2$	$18d$	$\frac{1}{4}$–$3\frac{1}{2}$	Monitor steel	$d/22$	11	100
				Plow steel	$d/22$	11	88
8 × 19 extra flexible	$1.45d^2$	$21d$–$26d$	$\frac{1}{4}$–$1\frac{1}{2}$	Monitor steel	$d/15$–$d/19$	10	92
				Plow steel	$d/15$–$d/19$	10	80
7 × 7 aircraft	$1.70d^2$	——	$\frac{1}{16}$–$\frac{3}{8}$	Corrosion-resistant steel	——	——	124
				Carbon steel	——	——	124
7 × 19 aircraft	$1.75d^2$	——	$\frac{1}{8}$–$\frac{3}{8}$	Corrosion-resistant steel	——	——	135
				Carbon steel	——	——	143
19-wire aircraft	$2.15d^2$	——	$\frac{1}{32}$–$\frac{5}{16}$	Corrosion-resistant steel	——	——	165
				Carbon steel	——	——	165

* Compiled from "American Steel and Wire Company Handbook."
† The modulus of elasticity is only approximate; it is affected by the loads on the rope and, in general, increases with the life of the rope.
‡ The strength is based on the nominal area of the rope. The figures given are only approximate and are based on 1-in rope
• sizes and $\frac{1}{4}$-in aircraft-cable sizes.

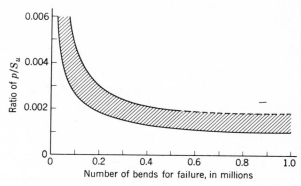

FIGURE 15-9
Experimentally determined relation between the fatigue life of wire rope and the sheave pressure; graph includes all rope sizes.

rope. Examination of the wires indicates no apparent contraction of the cross section. The failure is thus of a brittle nature and traceable to fatigue.

Drucker and Tachau* show that this failure is a function of the pressure of the rope on the sheave. This pressure is given by the equation

$$p = \frac{2F}{dD} \qquad (15\text{-}11)$$

where F = tensile force on rope
  d = rope diameter
  D = sheave diameter

Figure 15-9 is a graph showing the relation between the ratio of the pressure to the ultimate strength of the wire and the life of the rope. Examination of the graph shows that a rope is not likely to fail by fatigue if the ratio p/S_u is less than 0.001. Substitution of this ratio in Eq. (15-11) gives

$$S_u = \frac{2F}{dD} \qquad (15\text{-}12)$$

where S_u is the ultimate strength of the wire in kpsi. This strength varies considerably, because it depends upon the wire diameter as well as the material. The following values are often used:

Improved plow steel	200 kpsi
Plow steel	175 kpsi
Extra-strong cast steel	160 kpsi
Cast steel	140 kpsi
Iron	65 kpsi

* D. C. Drucker and H. Tachau, A New Design Criterion for Wire Rope, *Trans. ASME*, vol. 67, p. A-33, 1945.

Table 15-11 MINIMUM FACTORS OF SAFETY FOR WIRE ROPE

Track cables	3.2
Guys	3.5
Mine shafts, ft:	
Up to 500	8
1000–2000	7
2000–3000	6
Over 3000	5
Passenger elevators, fpm:	
50	7.50
300	9.17
800	11.25
Hoisting	5
Haulage	6
Cranes and derricks	6
Electric hoists	7

Equation (15-12) is very useful for the design of ropes and sheaves and is unique in that it contains all four variables, strength, load, rope diameter, and sheave diameter. When it is used in design, the engineer should ensure that an ample static factor of safety exists. Factors of safety listed in Table 15-11 should be applied to the tensile strength of the rope as a check on Eq. (15-12) and to ensure safety. For abnormal load conditions these values should be increased, since they are set up for the most favorable conditions of load, environment, and lubrication.

15-7 FLEXIBLE SHAFTS

One of the greatest limitations of the solid shaft is that it cannot transmit motion or power around corners. It is therefore necessary to resort to belts, chains, or gears, together with bearings and the supporting framework associated with them. The flexible shaft may often be an economical solution to the problem of transmitting motion around corners. In addition to the elimination of costly parts, its use may reduce noise considerably.

There are two main types of flexible shafts: the power-drive shaft for the transmission of power in a single direction, and the remote-control or manual-control shaft for the transmission of motion in either direction.

The construction of a flexible shaft is shown in Fig. 15-10. The cable is made by winding several layers of wire around a central core. For the power-drive shaft, rotation should be in a direction such that the outer layer is wound up. Remote-control cables have a different lay of the wires forming the cable, with more wires in each layer, so that the torsional deflection is approximately the same for either direction of rotation.

Flexible shafts are rated by specifying the torque corresponding to various radii of curvature of the casing. A 15-in radius of curvature, for example, will give

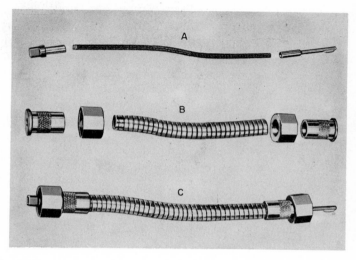

FIGURE 15-10.
Construction of a flexible shaft. The wirewound cable and end fittings are shown
at *A*. These must be enclosed by the flexible casing and end fittings, shown at *B*.
At *C* is shown the assembled flexible shaft. (*Courtesy of F. W. Stewart
Corporation.*)

from 2 to 5 times more torque capacity than a 7-in radius. When flexible shafts are
used in a drive in which gears are also used, the gears should be placed so that the
flexible shaft runs at as high a speed as possible. This permits the transmission of
the maximum amount of horsepower.

PROBLEMS

Section 15-2

15-1 A double-ply $\frac{18}{64}$-in flat leather belt is 6 in wide and transmits 15 hp. The pulley
arrangement is shown in the figure.

(*a*) Determine the tension in the tight and slack sides of the belt if the coefficient of
friction is 0.30.

(*b*) What belt tensions would result if adverse conditions caused the coefficient of
friction to drop to 0.20?

(*c*) Calculate the length of the belt.

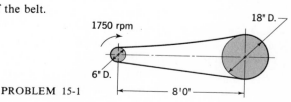

PROBLEM 15-1

15-2   A six-ply flat cotton-duck reinforced rubber belt is 200 mm wide and transmits 50 kW at a belt speed of 25 m/s. The belt has a mass of 2 kg/m of belt length. The belt is used in the crossed configuration to connect a 300-mm-diameter driving pulley to a 900-mm-diameter driven pulley at a spacing of 6 m.
(a) Calculate the belt length and the angles of wrap.
(b) Based on a coefficient of friction of 0.36, compute the belt tensions.

15-3   A belt drive is to have two 4-ft cast-iron pulleys spaced 16 ft apart. The tensile stress in the belt should not exceed 250 psi. Use $f = 0.30$ and $w = 0.035$ lb/in^3. Determine the width of $\frac{23}{64}$-in leather belt required if 75 hp is to be transmitted at 350 rpm.

15-4   A leather belt is 250 mm wide and 8 mm thick. The open belt connects a 400-mm-diameter cast-iron driving pulley to another cast-iron driven pulley 900 mm in diameter. The mass of the belt is 1.9 kg/m of belt length. The coefficient of friction of the drive is 0.28.
(a) Using a center distance of 5 m, a maximum tensile stress in the belt of 1400 kPa, and a belt velocity of 20 m/s, calculate the maximum power that can be transmitted.
(b) What are the shaft loads?

Section 15-4

15-5   A double-strand No. 60 roller chain is used to transmit power between a 13-tooth driving sprocket rotating at 300 rpm and a 52-tooth driven sprocket.
(a) What is the rated horsepower of this drive?
(b) Determine the approximate center distance if the chain length is 82 pitches.

15-6   Calculate the torque and the bending force on the shaft produced by the chain of Prob. 15-5.

15-7   A quadruple-strand No. 40 roller chain transmits power from a 21-tooth driving sprocket which turns at 1200 rpm. The velocity ratio between the two sprockets is 4 : 1.
(a) Calculate the rated horsepower of this drive.
(b) Find the tension in the chain.
(c) What is the factor of safety for the chain, based upon the average tensile strength?
(d) What should be the chain length for a center distance of approximately 20 in?

15-8   A 15-tooth No. 35 sprocket rotates at 500 rpm and transmits $\frac{3}{4}$ hp to another 15-tooth sprocket through a single-strand roller chain.
(a) Calculate the shaft torque.
(b) Find the tension in the chain.
(c) What is the pitch diameter of the sprockets?

15-9   The roller chain shown in the figure is No. 120 double-strand. The center distances shown are expressed in terms of the pitch. The 11-tooth driver rotates at 100 rpm and transmits 5 hp to the 21-tooth sprocket and 5 hp to the 17-tooth sprocket.
(a) Calculate the tension in the various portions of the chain.
(b) Find all the shaft reactions.

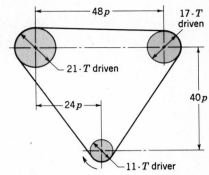

PROBLEM 15-9

15-10 The roller chain in Prob. 15-9 is single-strand No. 80. The 11-tooth sprocket is the driver and rotates at 50 rpm.

(*a*) Find the rated horsepower of the drive.

(*b*) The 17- and 21-tooth sprockets are to transfer equal torques to their shafts. For counterclockwise rotation of the 11-tooth sprocket and for transmission of the rated horsepower, what is the tension in the various portions of the chain?

(*c*) Determine the shaft reaction of each sprocket.

Section 15-6

15-11 A mine hoist uses a 2-in 6 × 19 monitor-steel wire rope. The rope is used to haul loads up to 10 tons from a shaft 480 ft deep. The drum has a diameter of 6 ft.

(*a*) Using a maximum hoisting speed of 1200 fpm and a maximum acceleration of 2 fps², determine the stress in the rope.

(*b*) What is the factor of safety?

15-12 A temporary construction elevator is to be designed to carry men and materials to a height of 90 ft. The maximum estimated load to be hoisted is 5000 lb at a velocity not exceeding 2 fps. Based on minimum sheave diameters, a minimum factor of safety, and an acceleration of 4 fps², find the number of ropes required. Use 1-in-diameter plow-steel 6 × 19 standard hoisting ropes.

15-13 Based on bending of the wires around the sheaves, find the true factor of safety for the rope of Prob. 15-12. Use $d_w = d/13$.

16

A SYSTEMS APPROACH

The systems approach is a method of engineering in which a mathematical model, representing an actual physical system, is constructed and programmed on a computer capable of man-machine interaction. The model is then "run" on the computer and the total performance observed in the computer printout or graphical display. Based on these observations, changes are made in the inputs to the physical system and the design parameters, with the objective of optimizing the design and performance of the system which is being modeled.

The engineer using the systems approach and the pilot of a modern jet aircraft have a great deal in common. The pilot observes a vast array of recording instrumentation in the cockpit and, through the aircraft control devices, makes various corrections and changes from time to time to optimize the performance of his ship. The engineer, by using a computer, monitors the performance of a simulated physical system and makes changes in the design or operation of the system to obtain the best total performance.

An important advantage of the systems approach is the reduction of lead time in the design process. A hypothetical system can be studied before prototypes are constructed. In this way, many "bugs," defects, and design changes can be foreseen and eliminated in earlier and less costly stages of design.

A second advantage of the systems approach is that it may be too expensive or impractical to instrument the real system. If the system is complex, the systems approach is the only way the performance can be observed.

A third advantage of the systems approach is that conditions can be imposed which are so strenuous that they would destroy the actual system. For example, the automobile has often been studied for safety purposes using the systems approach.

There are many kinds of systems approaches, depending upon the type of system and the kind of information desired. Since we are interested in the performance of machines, the approach we shall present is based on studies on the dynamics of mechanical systems. Obviously, you would not use this approach if you were modeling a supermarket as a system.

16-1 THE MATHEMATICAL MODEL

It is worth noting that the computer has introduced substantial changes in engineer's tasks. Where formerly an engineer expected to spend considerable time in detailed mathematical analysis of complex systems, this is now no longer necessary or even desirable. Mathematical analysis is a monotonous activity in which human beings can and usually do make errors; the analysis requires constant checking and rechecking of signs, decimal points, exponents, subscripts, and the like. Except for the simplest problems, this task should be delegated to the computer. This frees the engineer for the more important tasks of formulating problem statements and making the design decisions on the basis of the computer results.

Students, however, must perform a certain amount of analysis to understand the physics of the situation. Otherwise they are not adequately prepared to formulate the problem or to decide what to do with the solution.

The mathematical model is the set of algebraic, vector, and differential equations, and inequations which describe the behavior of the system when various inputs and design parameters are varied. Except for the simplest systems, it is not necessary actually to be able to solve these equations. Sometimes a few simplifications can be made, and the equations solved rather easily, using only a few sheets of paper. On other occasions there are so many equations and they contain so many nonlinearities that only digital or analog computation could possibly be used. The systems approach, in fact, was devised as a means of using computers for the simulation of large systems. The approach would not have been possible without such large and sophisticated computing facilities as are available today.

Experience with the systems approach in mechanical engineering shows that the simplest model gives meaningful results about 90 percent of the time. Therefore, unless something about the problem indicates otherwise, one should usually begin with a simple model. If the results obtained from the simple model are

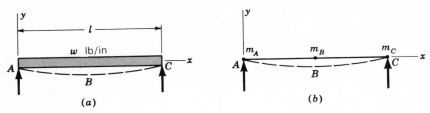

FIGURE 16-1
(a) The mass of the beam is uniformly distributed from A to C; (b) the mass of the beam is lumped at A, B, and C.

satisfactory, the task is finished, and the engineer can proceed to the next job at hand. If the results are not satisfactory, the information needed for the creation of a more sophisticated model will have been made available.

16-2   LUMPED SYSTEMS

We shall be dealing with mechanical systems containing such elements as gears, pulleys, belts, chains, rotating shafts, beams, and springs. In creating models of these and other elements it is necessary to utilize a simplification which is best described as *lumping*.

Figure 16-1 illustrates the meaning of lumping when applied to a beam. In Fig. 16-1a the real beam is seen to have a uniformly distributed weight of w lb per in of length. For this beam a partial differential equation having an infinity of solutions must be written. The resulting equation can then be solved for the dynamic response to many kinds of disturbances. By lumping the mass as shown in Fig. 16-1b, only one ordinary differential equation is written. Though the results are slightly different, there are many physical situations in which the lumped model will yield perfectly satisfactory results.

16-3   THE DYNAMIC RESPONSE OF A
DISTRIBUTED SYSTEM

It will help us to understand the effects of lumping if we compare the response of a distributed system with that of an equivalent lumped system. In the next section we shall consider the lumped system.

For the distributed system let us consider the simply supported bar in Fig. 16-1a. We wish to consider motion only in the xy plane and to eliminate gravity effects; therefore we assume the xy plane is horizontal. Every particle of the beam is to have a motion y which is a function of its location x along the beam and a function of time. Thus

$$y = f(x, t) \qquad (a)$$

Equation (3-12) for the load q on a beam will now have to be written

$$q = EI \frac{\partial^4 y}{\partial x^4} \qquad (b)$$

where q is the d'Alembert force in the y direction and is the product of the mass per unit length $\rho = w/g$ with the acceleration $\partial^2 y/\partial t^2$. Thus Eq. ($b$) becomes

$$\frac{\partial^4 y}{\partial x^4} = -\frac{\rho}{EI} \frac{\partial^2 y}{\partial t^2} \qquad (16\text{-}1)$$

Partial differential equations similar to Eq. (16-1) are sometimes solved by assuming the solution is a product-type function. For example,

$$y = xe^t \qquad \text{and} \qquad y = \sin x \log t$$

are product-type functions. If y should just happen to be this kind of a solution, we could write

$$y(x, t) = T(t)X(x) \qquad (16\text{-}2)$$

solve for X and T independently, and then put them back together to get the solution. For this example it happens that y is indeed this kind of a solution; so let us proceed.

First, take the fourth derivative of Eq. (16-2) with respect to x.

$$\frac{\partial^4 y}{\partial x^4} = T \frac{d^4 X}{dx^4} \qquad (c)$$

Next, take the second derivative of Eq. (16-2) with respect to t.

$$\frac{\partial^2 y}{\partial t^2} = X \frac{d^2 T}{dt^2} \qquad (d)$$

Now substitute Eqs. (c) and (d) into Eq. (16-1).

$$T \frac{d^4 X}{dx^4} = -\frac{\rho}{EI} X \frac{d^2 T}{dt^2} \qquad (e)$$

Separate the variables by dividing both sides by XT; the result is

$$\frac{1}{X} \frac{d^4 X}{dx^4} = -\frac{\rho}{EI} \frac{1}{T} \frac{d^2 T}{dt^2} \qquad (f)$$

But X and T were assumed to be independent functions. That being so, both sides must equal a constant. Designating that constant by the term λ^2, we have

$$\frac{1}{X} \frac{d^4 X}{dx^4} = -\frac{\rho}{EI} \frac{1}{T} \frac{d^2 T}{dt^2} = \lambda^2$$

Thus we get the two ordinary differential equations

$$\frac{d^2 T}{dt^2} + \lambda^2 \frac{EI}{\rho} T = 0 \qquad (16\text{-}3)$$

$$\frac{d^4 X}{dx^4} - \lambda^2 X = 0 \qquad (16\text{-}4)$$

From a study of ordinary differential equations we learn that the solution to (16-3) is

$$T = A_1 \cos \omega t + A_2 \sin \omega t = A \cos (\omega t - \phi) \qquad (16\text{-}5)$$

with
$$A = \sqrt{A_1^2 + A_2^2} \qquad \phi = \tan^{-1} \frac{A_2}{A_1}$$

and
$$\omega = \lambda \sqrt{\frac{EI}{\rho}} \qquad (16\text{-}6)$$

where we have not as yet assigned any meaning to the term ω. If there is doubt in your mind that Eq. (16-5) is a solution to (16-3), substitute T and its second derivative into Eq. (16-3) and prove to yourself that the two terms sum to zero.

To obtain a solution to Eq. (16-4), assume a solution of the form

$$X = Be^{ax} \qquad (g)$$

and substitute it and its fourth derivative into Eq. (16-4) to yield

$$a^4 - \lambda^2 = 0 \qquad (h)$$

When the four roots of Eq. (h) are substituted into Eq. (g) you will obtain an exponential expression containing four terms. By employing the identities

$$e^{\pm bx} = \cosh bx \pm \sinh bx$$

$$e^{\pm jbx} = \cos bx \pm j \sin bx$$

With $j = \sqrt{-1}$, you should finally get

$$X = C_1 \cosh \lambda^{1/2}x + C_2 \sinh \lambda^{1/2}x + C_3 \cos \lambda^{1/2}x + C_4 j \sin \lambda^{1/2}x \qquad (16\text{-}7)$$

Having now solved Eq. (16-3) for T and (16-4) for X, and noting that T is only a function of t and X is only a function of x, we can put these solutions together as indicated by Eq. (16-2) to get y. The result is

$$y = A \cos (\omega t - \phi)(C_1 \cosh \lambda^{1/2}x + C_2 \sinh \lambda^{1/2}x$$

$$+ C_3 \cos \lambda^{1/2}x + C_4 j \sin \lambda^{1/2}x) \qquad (16\text{-}8)$$

Our next problem is to determine the values of the constants C_1, C_2, C_3, and C_4, which we compute from the geometry of the beam. To evaluate these we use the information that the deflection and the bending moment are zero at the ends of the beam where $x = 0$ and $x = l$. Therefore

$$X(0) = X''(0) = 0 \qquad X(l) = X''(l) = 0$$

where $X'' = d^2X/dx^2$. Substituting the first pair of conditions into Eq. (16-7) gives

$$C_1 = C_3 = 0$$

From the second pair of conditions we obtain the equations

$$C_2 \sinh \lambda^{1/2}l + C_4 j \sin \lambda^{1/2}l = 0$$
$$C_2 \sinh \lambda^{1/2}l - C_4 j \sin \lambda^{1/2}l = 0 \qquad (i)$$

Of course, the solution $C_2 = C_4 = 0$ is trivial and not of interest. So set the determinant of the coefficients equal to zero.

$$\begin{vmatrix} \sinh \lambda^{1/2}l & j \sin \lambda^{1/2}l \\ \sinh \lambda^{1/2}l & -j \sin \lambda^{1/2}l \end{vmatrix} = 0$$

We obtain

$$\sinh \lambda^{1/2}l \sin \lambda^{1/2}l = 0 \qquad (j)$$

This expression can be satisfied only by making $\sin \lambda^{1/2}l = 0$. In other words,

$$\lambda^{1/2}l = n\pi \qquad n = 1, 2, 3, \ldots$$

Therefore

$$\lambda = \left(\frac{n\pi}{l}\right)^2 \qquad (16\text{-}9)$$

Substituting into Eq. (16-6) gives

$$\omega_n = \left(\frac{n\pi}{l}\right)^2 \sqrt{\frac{EI}{\rho}} \qquad n = 1, 2, 3, \ldots \qquad (16\text{-}10)$$

which is the formula for all the natural circular frequencies of the beam in radians per second. There are an infinite number of these, each corresponding to a particular mode of motion. The first three are

$$\omega_1 = \frac{\pi^2}{l^2}\sqrt{\frac{EI}{\rho}} \qquad \omega_2 = \frac{4\pi^2}{l^2}\sqrt{\frac{EI}{\rho}} \qquad \omega_3 = \frac{9\pi^2}{l^2}\sqrt{\frac{EI}{\rho}}$$

Since 2π radians constitutes one cycle of motion, the frequency in cycles per second (Hz) is

$$f = \frac{\omega}{2\pi} \qquad (16\text{-}11)$$

But we became sidetracked while attempting to find the values of C_2 and C_4. If we substitute Eq. (16-9) into either of Eqs. (i), we find, using the first, that

$$C_2 \sinh n\pi + C_4 \sin n\pi = 0$$

Therefore $C_2 = 0$ and $C_4 \neq 0$. We now substitute these constants back into Eq. (16-8), designate $B = jAC_4$, and obtain as the particular solution

$$y = B_1 \cos(\omega_1 t - \phi_1)\sin\frac{\pi x}{l} + B_2 \cos(\omega_2 t - \phi_2)\sin\frac{2\pi x}{l} + \cdots \qquad (16\text{-}12)$$

or

$$y = \sum_{n=1}^{\infty} B_n \cos(\omega_n t - \phi_n)\sin\frac{n\pi x}{l} \qquad (16\text{-}13)$$

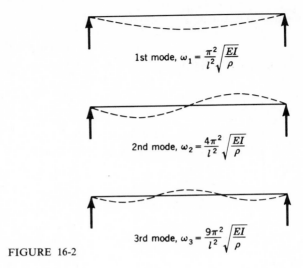

1st mode, $\omega_1 = \dfrac{\pi^2}{l^2}\sqrt{\dfrac{EI}{\rho}}$

2nd mode, $\omega_2 = \dfrac{4\pi^2}{l^2}\sqrt{\dfrac{EI}{\rho}}$

3rd mode, $\omega_3 = \dfrac{9\pi^2}{l^2}\sqrt{\dfrac{EI}{\rho}}$

FIGURE 16-2

where ω_n is not to be confused with the single natural frequency of a one-freedom system.

Suppose at some instant of time, which we shall call time zero, we visualize the beam as having a configuration

$$y = y_0 \sin \frac{\pi x}{l}$$

and that the velocity $\partial y/\partial t$ is instantaneously zero for every coordinate x along the beam. These two sets of conditions are called initial, or starting, conditions since they apply at the particular instant $t = 0$. They can be expressed as

$$y(x, 0) = y_0 \sin \frac{\pi x}{l} \qquad (k)$$

$$\frac{\partial y(x, 0)}{\partial t} = 0 \qquad (l)$$

It is information of this nature that we require in order to evaluate the two remaining sets of constants in Eq. (16-13). Equation (k) can be substituted into Eq. (16-12) directly to yield

$$y_0 \sin \frac{\pi x}{l} = B_1 \cos\left(-\phi_1\right) \sin \frac{\pi x}{l} + B_2 \cos\left(-\phi_2\right) \sin \frac{2\pi x}{l} + \cdots \qquad (m)$$

Next, taking the derivative of Eq. (16-12) with respect to time, to get the velocity, yields

$$\frac{\partial y}{\partial t} = -B_1 \omega_1 \sin\left(\omega_1 t - \phi_1\right) \sin \frac{\pi x}{l} - B_2 \omega_2 \sin\left(\omega_2 t - \phi_2\right) \sin \frac{2\pi x}{l} - \cdots \qquad (n)$$

Substituting Eq. (*l*) into Eq. (*n*) gives

$$\phi_1 = \phi_2 = \cdots = 0$$

Therefore, in Eq. (*m*), $B_1 = y_0$ and

$$B_2 = B_3 = \cdots = 0$$

and so the solution is

$$y = y_0 \cos \omega_1 t \sin \frac{\pi x}{l} \qquad (16\text{-}14)$$

We can now begin to appreciate that X is a shape function while T is a time function. The various values of λ are called the *eigenvalues*, and a particular shape corresponds to every integer n, each of which is called a *mode shape*. The shapes of the first three modes are shown in Fig. 16-2.

16-4 THE DYNAMIC RESPONSE OF A LUMPED SYSTEM

After gaining some experience with the systems approach in dealing with engineering problems, people are able to do a rather good job of predicting in advance which mode of motion will predominate. Most of the time the first mode is the important one and the effects of the higher modes are insignificant. Although this is not always the case, we shall assume it to be so in this chapter and restrict our analysis to first modes. So let us proceed to the analysis of a lumped system.

The mass m attached to the weightless spring of stiffness k is a model of a lumped system. It is called a *one-freedom system* (also called a *one-degree-of-freedom system*) because only one coordinate x is necessary to describe the motion. It is called a lumped system because the spring is weightless and the mass is assumed to be concentrated at a point.

The origin of the coordinate x in Fig. 16-3 is chosen at the equilibrium position of the mass. Thus, when the mass is given any positive displacement x, the spring force has two components. One of these is $k\delta_{st} = W$, where δ_{st} is the distance the spring was deflected when the weight was first suspended from it. The other component is kx. Summing forces acting on the mass gives

$$\sum F = -k(x + \delta_{st}) + W - m\ddot{x} = 0$$

or

$$m\ddot{x} + kx = 0 \qquad (16\text{-}15)$$

where $\ddot{x}$ is the Newtonian representation of d^2x/dt^2.

Though you may have solved equations like Eq. (16-15) many times in the past, let us solve it here to establish the approach and nomenclature. If we write Eq. (16-15) in the form

$$\ddot{x} = -\frac{k}{m} x \qquad (a)$$

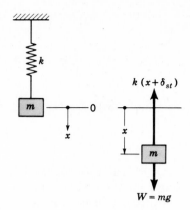

FIGURE 16-3

it can be seen that the function used for x must be the same as its second derivative. One such function is

$$x = A \sin bt \qquad (b)$$

Then

$$\dot{x} = Ab \cos bt \qquad (c)$$

and

$$\ddot{x} = -Ab^2 \sin bt \qquad (d)$$

Substituting Eqs. (b) and (d) into (a) produces

$$-Ab^2 \sin bt = -\frac{k}{m} A \sin bt \qquad (e)$$

Canceling $A \sin bt$ from both sides leaves

$$b^2 = \frac{k}{m}$$

Thus Eq. (b) is a solution provided we set $b = \sqrt{k/m}$. It can be shown in the same manner that

$$x = B \cos bt \qquad (f)$$

is also a solution to Eq. (a). A general solution is obtained by adding Eqs. (b) and (f) to get

$$x = A \sin \sqrt{\frac{k}{m}} t + B \cos \sqrt{\frac{k}{m}} t \qquad (g)$$

Physically, the constants A and B represent the manner in which the motion was started, or to put it another way, the state of the motion at $t = 0$.

We now define

$$\omega_n = \pm \sqrt{\frac{k}{m}} \qquad (16\text{-}16)$$

where the $\pm$ sign means that the solution is valid for either positive or negative time. The subscript n in Eq. (16-16) means "natural," and ω_n is called the *natural circular frequency*; it is usually measured in radians per second. Be careful not to confuse the subscript n in Eq. (16-16) with that of Eq. (16-13) where n is an integer.

Since one cycle of motion will be completed in 2π radians, the period of the motion is given by the equation

$$\tau = \frac{2\pi}{\omega_n} = 2\pi\sqrt{\frac{m}{k}} \qquad (16\text{-}17)$$

where τ is in seconds. The *frequency* is the reciprocal of the period and is

$$f = \frac{\omega_n}{2\pi} = \frac{1}{2\pi}\sqrt{\frac{k}{m}} \qquad (16\text{-}18)$$

in cycles per second.

We can now write Eq. (g) as

$$x = A \sin \omega_n t + B \cos \omega_n t \qquad (h)$$

Let us start the motion by pulling the mass downward through a positive distance x_0 and releasing it with zero velocity at the instant $t = 0$. The starting, or initial, conditions are

$$t = 0 \qquad x = x_0 \qquad \dot{x} = 0$$

Substituting the first two of these in Eq. (h) gives

$$x_0 = A(0) + B(1)$$

or $B = x_0$. Next, taking the first derivative of Eq. (h) with respect to time,

$$\dot{x} = A\omega_n \cos \omega_n t - B\omega_n \sin \omega_n t \qquad (i)$$

and substituting the first and third conditions in this equation yields

$$0 = A\omega_n(1) - B\omega_n(0)$$

or $A = 0$. Substituting A and B into Eq. (h) produces

$$x = x_0 \cos \omega_n t \qquad (16\text{-}19)$$

If we start the motion with

$$t = 0 \qquad x = 0 \qquad \dot{x} = v_0$$

we get

$$x = \frac{v_0}{\omega_n} \sin \omega_n t \qquad (16\text{-}20)$$

But if we start with

$$t = 0 \qquad x = x_0 \qquad \dot{x} = v_0$$

we get

$$x = \frac{v_0}{\omega_n} \sin \omega_n t + x_0 \cos \omega_n t \qquad (16\text{-}21)$$

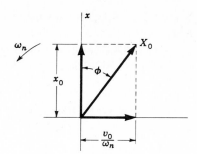

FIGURE 16-4

which is the general form of the solution. By plotting the two parts of Eq. (16-21) on the one-dimensional reference system of Fig. 16-4 it is seen that

$$x = X_0 \cos (\omega_n t - \phi) \qquad (16\text{-}22)$$

with

$$X_0 = \sqrt{x_0^2 + \left(\frac{v_0}{\omega_n}\right)^2} \qquad (16\text{-}23)$$

and

$$\phi = \tan^{-1} \frac{v_0}{x_0 \omega_n} \qquad (16\text{-}24)$$

With this background we can now return to the problem of lumping the mass of the beam of Fig. 16-1. As indicated earlier, we are interested in the simple, or first-mode, models, and so we lump the mass as shown in Fig. 16-1b. This lumping restricts our investigation to the first mode because only a single coordinate is necessary to describe the motion of the lumped mass m_B. The masses m_A and m_C are located at supports where they cannot move. The total mass of the beam is ρl, where ρ is the mass per unit length. For a beam with simple supports, a logical approach would be to place half the mass at the center and a quarter over each support. Thus $m_B = \rho l/2$. This is an assumption. Later we shall define rules for lumping the mass of other beams and structures.

Referring next to Table A-12-5, we find the deflection at the center of a simply supported beam with center load to be

$$y_{\max} = \frac{Fl^3}{48EI} \qquad (16\text{-}25)$$

Therefore the spring rate is

$$k = \frac{48EI}{l^3} \qquad (16\text{-}26)$$

Now substitute $m = m_B = \rho l/2$ and the spring rate k into Eq. (16-16) for the natural frequency ω_n. The result is

$$\omega_n = \sqrt{\frac{48EI/l^3}{\rho l/2}} = \frac{9.7980}{l^2} \sqrt{\frac{EI}{\rho}} \qquad (16\text{-}27)$$

When this frequency is compared with that of the exact solution in Sec. 16-3, we find an error of about 0.7 percent. For many engineering purposes this is completely acceptable. It also provides some measure of the kinds of errors that are introduced by the lumping procedure.

16-5 MODELING THE ELASTICITIES

Once the physical system has been represented as an abstract diagram, the equations which define the mathematical model are easy to write. The first step in creating this abstraction involves a consideration of the spring rates, or elasticities, which exist throughout the system.

In Chap. 3 linear and torsional spring rates were defined and methods of determining spring rates for simple members as well as for structures using the electric circuit analogy or Castigliano's theorem were presented. Only a few additions to the theory of Chap. 3 will be necessary here.

First, consider the system of Fig. 16-5a, where, assuming small motions, we may wish to describe the motion of the lever by the coordinate x_A, the coordinate x_B, or the coordinate θ. Thus we require formulas for transforming the spring k_A to the equivalent springs shown in Fig. 16-5c and d. Note by the coordinate θ in Fig. 16-5d that k is the stiffness of a torsional spring. Later we shall find that the equations of angular motion are easier to write when the system in abstract form is represented in the same manner as a linear-motion system, as has been done in Fig. 16-5d.

The rule for obtaining the stiffness of an equivalent spring is that the potential energy stored by the equivalent spring must be the same as the potential energy stored by the actual spring. The potential energy stored by the actual system may be obtained from Fig. 16-5a or b. It is

$$U = \tfrac{1}{2}k_A x_A^2 \qquad (a)$$

The energy stored by the system of Fig. 16-5c is

$$U = \tfrac{1}{2}kx_B^2 \qquad (b)$$

But $x_B = bx_A/a$. Using this relation and equating the energies of Eqs. (a) and (b) gives

$$k = \frac{a^2}{b^2}k_A \qquad (16\text{-}28)$$

Note, in Fig. 16-5a, that b is much larger than a; therefore k_A is a stiffer spring than its equivalent k. This knowledge makes it easier to remember Eq. (16-28). The rule is: *Multiply the actual spring rate by the square of the lever ratio to get the equivalent spring rate and remember that the stiffest spring is the one closest to the pivot.*

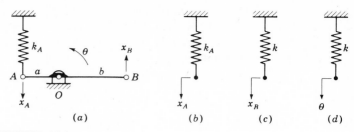

FIGURE 16-5

For Fig. 16-5d, the energy stored is

$$U = \tfrac{1}{2}k\theta^2 \qquad (c)$$

Noting that $x_A = a\theta$ and equating the energies as before gives

$$k = a^2 k_A \qquad (16\text{-}29)$$

Thus, to convert a linear-motion spring into an equivalent torsional spring, *multiply the stiffness of the linear-motion spring by the square of its radius from the lever pivot.*

A somewhat similar problem occurs in modeling rotating systems. Figure 16-6a shows two shafts of torsional stiffnesses k_1 and k_2 connected by gears having pitch diameters D_1 and D_2, respectively. Although k_2 has one end fixed, the analysis also applies if the end is free. In creating a model of Fig. 16-6a, we may wish to base the equivalent system on k_2 or on k_1. Basing the system on k_2 means that we diagram the equivalent system as in Fig. 16-6b. Both k_2 and the torque reaction at the support remain unchanged. However, k_1, T, and θ will be changed and are now denoted as k_1', T', and θ', respectively.

Designate θ_1 and θ_2, respectively, as the angle of twist of shafts 1 and 2. Then

$$\theta = \theta_1 + \frac{D_2}{D_1}\theta_2 \qquad (d)$$

and

$$\theta' = \theta_1' + \theta_2 = \frac{D_1}{D_2}\theta_1 + \theta_2 \qquad (e)$$

Thus

$$\theta_1' = \frac{D_1}{D_2}\theta_1 \qquad (16\text{-}30)$$

The reaction torque at the support will be larger than the input torque T by the amount of the gear ratio D_2/D_1. Thus

$$T' = \frac{D_2}{D_1}T \qquad (16\text{-}31)$$

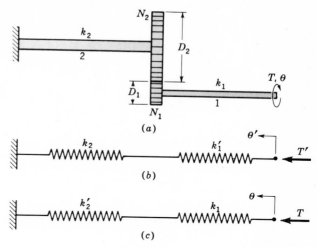

FIGURE 16-6
(a) The actual system; (b) equivalent system based on k_2; (c) equivalent system based on k_1.

Now, using Eqs. (16-30) and (16-31), we have

$$k'_1 = \frac{T'}{\theta'_1} = \frac{D_2/D_1}{D_1/D_2} \frac{T}{\theta_1} = \left(\frac{D_2}{D_1}\right)^2 k_1 \quad (16\text{-}32)$$

The ratio D_2/D_1 in these equations can always be replaced by N_2/N_1, the ratio of the tooth numbers. And for rotating systems, use the angular-velocity ratio n_1/n_2, which is the same as the gear ratio D_2/D_1.

Using Eq. (16-32), we now find, for the system of Fig. 16-6c,

$$k'_2 = \left(\frac{D_1}{D_2}\right)^2 k_2 \quad (f)$$

In our considerations of the geared shafts, the gear teeth were assumed to be rigid. Furrow and Mabie* have investigated the stiffness of gear teeth in considerable detail and provided a thorough history of previous investigations. Here, however, we shall make use of an approach due to Koževnik,† who presents a method of determining the equivalent torsional tooth stiffness such that Fig. 16-6b, say, would have three springs in series, one for shaft 1, one for the combined tooth stiffness, and the third for shaft 2.

* R. W. Furrow and H. H. Mabie, The Measurement of Static Deflection in Spur Gear Teeth, *J. Mechanisms*, vol. 5, no. 2, pp. 147–168, 1970.

† Jaroslav Koževnick, "Dynamics of Machines," p. 267, Erven P. Noordhoff, Ltd., Groningen, Netherlands, 1962. (An English translation.)

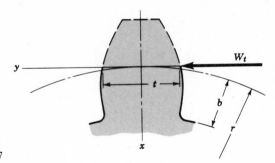

FIGURE 16-7

Kožešnik makes the following assumptions:

1 The teeth are always in contact at the pitch point
2 The entire load is carried by a single pair of teeth at all times
3 The teeth are assumed to be short rectangular cantilevers
4 No provision is made for backlash

Using as much of the gear nomenclature from Chaps. 11 and 12 as possible, denote t the tooth thickness at the pitch line, b the dedendum, r the pitch radius, W_t the tangential, or transmitted, load, and f the face width (we used F in Chaps. 11 and 12, but now we must be careful to distinguish the face width from a force). Then the moment of inertia is

$$I = \frac{ft^3}{12} \quad (16\text{-}33)$$

from Table A-30. Next, referring to Fig. 16-7, we see that the flexural deflection of the tooth is

$$y = \frac{W_t b^3}{3EI} \quad (g)$$

To this must be added the deflection due to direct shear because we are dealing with a short beam. From Eq. (2-33), the shear stress in such a beam is

$$\tau = \frac{W_t}{If} \int_y^{t/2} y\, dA = \frac{W_t}{If} \int_y^{t/2} y(f\, dy) = \frac{W_t}{2I}\left(\frac{t^2}{4} - y^2\right)$$

Now the strain energy stored in a differential volume of material $f\, dx\, dy$ will be equal to the energy in a unit volume times the differential volume. Therefore, from Eq. (3-23), we have

$$dU = \frac{\tau^2}{2G} f\, dx\, dy = \frac{W_t^2}{8GI^2}\left(\frac{t^2}{4} - y^2\right)^2 f\, dx\, dy$$

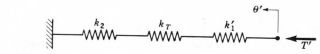

FIGURE 16-8

Integrating this expression with respect to x and y gives the total strain energy due to direct shear as

$$U = \int_0^b \int_{-t/2}^{t/2} \frac{W_t^2}{8GI^2}\left(\frac{t^2}{4} - y^2\right)^2 f\, dx\, dy = \frac{W_t^2 bt^2}{20GI}$$

Applying Castigliano's theorem, the deflection due to direct shear is

$$y = \frac{\partial U}{\partial W_t} = \frac{W_t bt^2}{10GI} = \frac{W_t b}{0.833GA} \qquad (h)$$

Adding Eqs. (g) and (h) to get the total deflection gives

$$y = W_t\left(\frac{b}{0.833GA} + \frac{b^3}{3EI}\right) \qquad (16\text{-}34)$$

The tooth stiffness is then the reciprocal of the expression in parentheses:

$$\frac{1}{k_G} = \frac{b}{0.833GA} + \frac{b^3}{3EI} \qquad (16\text{-}35)$$

Equation (16-35) is the expression developed by Koževnik for the stiffness of a single gear tooth. Note that this is a linear-motion stiffness because k is in lb/in. A pair of meshing teeth will have their stiffnesses in series, and so Eq. (3-3) applies. Also, using Eq. (16-29), we find the torsional spring rate based on shaft 2 of the gear teeth of Fig. 16-6 to be

$$k_T = \left(\frac{D_2}{2}\right)^2 \frac{k_G}{2} = \frac{D_2^2 k_G}{8} \qquad (16\text{-}36)$$

The system of Fig. 16-6b can now be diagrammed as shown in Fig. 16-8.

In many practical situations the torsional equivalent of the tooth stiffness is so large in comparison with the shaft stiffnesses that it is hardly worth considering.

For belt and rope drives Koževnik recommends Eq. (3-6) for the linear-motion stiffness, which is

$$k = \frac{AE}{l} \qquad (16\text{-}37)$$

where A is the belt cross-sectional area, E the modulus of elasticity, and l the length of one of the tangents to the pulleys plus one-third of the belt wrap. Thus, from Fig. 15-3, we have

$$l = \tfrac{1}{2}\sqrt{4C^2 - (D - d)^2} + \tfrac{1}{3}(D\theta_L + d\theta_S) \qquad (16\text{-}38)$$

where D and d are the large and small pulley diameters, C is the center distance, and θ_L and θ_S are the wrap angles for the large and small pulleys, respectively. The modulus of elasticity for leather belts ranges from 17 to 32 kpsi. For V belts E can be chosen from the range 14 to 20 kpsi. For steel rope, use Table 15-10.

If we now base the torsional stiffness of the belt on the shaft attached to the large pulley, then

$$k_B = \left(\frac{D}{2}\right)^2 \frac{AE}{l} \qquad (16\text{-}39)$$

In diagramming the equivalent stiffnesses of belts we must be careful to remember that they are one-way springs;* they can be elongated but not compressed. Also, the spring rate of a belt usually turns out to be quite soft (small) in comparison with the shaft spring rate. Note how this contrasts with gear-tooth stiffnesses. Finally, a system containing belts is usually a highly damped system due to slip and friction losses.

16-6 MODELING THE MASSES AND INERTIAS

Table A-31 contains formulas for the mass m and the mass moment of inertia I, often simply called the inertia, of simple geometric shapes.

Table 16-1 provides a few additional lumping formulas. The helical-spring formula also applies to an axially loaded bar fixed at one end and to a torsion bar fixed at one end.†

Let us first consider the weightless lever AB shown in Fig. 16-9, pivoted at O, with lumped masses m_A and m_B at each end. We are free to choose any one of the coordinates x, y, θ to describe the motion, because the system has only a single degree of freedom. The three systems shown in Fig. 16-9b are called *equivalent systems*; any one of them may be made the equivalent of Fig. 16-9a merely by choosing the proper parameters k, m, and I with respect to the corresponding describing coordinate x, y, or θ. Since the system of Fig. 16-9a has only a single natural frequency, we should obtain this same frequency no matter which of the three equivalent systems we choose for describing the motion.

If we should decide to define the motion by means of the x coordinate at A, then the equivalent system must include the mass m_A as well as an equivalent mass to replace the effect of m_B.

An equivalent system of masses or inertias is obtained by expressing in mathematical form the fact that both systems, the actual and the equivalent, have the same kinetic energies. If we choose to describe the equivalent system by the coordinate x, the kinetic energy is

$$U_{eq} = \tfrac{1}{2}m\dot{x}^2 \qquad (a)$$

---

* But see Prob. 16-7.

† For additional details concerning lumping, see Joseph E. Shigley, "Simulation of Mechanical Systems," chap. 6, McGraw-Hill Book Company, New York, 1967.

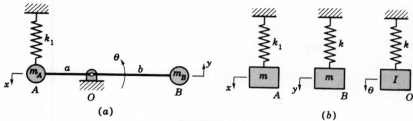

FIGURE 16-9

(a) Levered system; (b) various equivalent systems.

where m is the equivalent mass. The kinetic energy of the actual system is

$$U_{\text{act}} = \tfrac{1}{2}m_A\dot{x}^2 + \tfrac{1}{2}m_B\dot{y}^2 \qquad (b)$$

Since $\dot{y} = (b/a)\dot{x}$, Eq. (b) becomes

$$U_{\text{act}} = \frac{1}{2}\,m_A\dot{x}^2 + \frac{1}{2}\,m_B\frac{b^2}{a^2}\,\dot{x}^2 \qquad (c)$$

Upon equating Eqs. (a) and (c), we learn that

$$m = m_A + \frac{b^2}{a^2}\,m_B \qquad (d)$$

This mass m with spring k_1 then forms the equivalent system whose motion is described by the coordinate x. We see, from Eq. (d), that the same rule applies for

Table 16-1 FORMULAS FOR LUMPED SYSTEMS

m_c = concentrated mass
m_d = distributed mass
m = lumped mass of equivalent system

Name	Helical spring	Cantilever	Simply supported beam	Fixed ended beam
Actual system	m_d m_c	m_d m_c	$l/2$ m_c m_d	$l/2$ m_c m_d
Equivalent system	m	m	$l/2$ m	$l/2$ m
	$m = m_c + \tfrac{1}{3}m_d$	$m = m_c + 0.23m_d$	$m = m_c + 0.50m_d$	$m = m_c + 0.375m_d$

masses as for springs. *To obtain an equivalent mass, multiply the actual mass by the ratio of the square of the lever arms.* This rule is easy to remember if we note from Eq. (*d*) that the heaviest mass is the mass closest to the pivot point.

Now, if we wish to determine an equivalent system located at *B* and described by the coordinate *y*, the equivalent spring constant is

$$k = \frac{a^2}{b^2} k_1$$

and the equivalent mass to be located at *B* is

$$m = m_B + \frac{a^2}{b^2} m_A$$

To complete the torsional equivalent system, we compute the mass moment of inertia by means of the usual formula

$$I = m_A a^2 + m_B b^2$$

Alternatively, we can write the kinetic energies of both systems and equate them as before. Thus, for the equivalent system,

$$U_{eq} = \frac{I\dot{\theta}^2}{2} \qquad (e)$$

and for the actual system,

$$U_{act} = \frac{m_A \dot{x}^2}{2} + \frac{m_B \dot{y}^2}{2} \qquad (f)$$

If we then note that $\dot{x} = a\dot{\theta}$ and $\dot{y} = b\dot{\theta}$ and form $U_{eq} = U_{act}$, we obtain the same expression for *I*.

We must note very particularly that an equivalent system is constructed to yield a single set of results. Because of the large number of simplifying assumptions involved, we cannot expect it to yield all possible information concerning the total system behavior. In the simulation of the dynamics of the connecting rod of an internal-combustion engine, for example, MacDuff and Curreri[*] show that a two-mass equivalent of a rod with distributed mass cannot have the same mass, the same mass moment of inertia, and the same center of gravity as the actual rod. Some of these quantities can be determined satisfactorily, but not all of them at the same time. Thus, if we are interested in connecting-rod stresses, we make one set of assumptions. If we are interested in a dynamic-force analysis, we need another set. The particular system must always be formed with the actual problem and the desired results well in mind.

Next we concentrate our attention on the geared four-inertia system of Fig. 16-10a. Here an input torque drives inertia I_1, which might be the rotor of an

---

* John N. MacDuff and John R. Curreri, "Vibration Control," pp. 308–313, McGraw-Hill Book Company, New York, 1958.

electric motor, through the angle θ_1. This inertia is connected by a shaft of torsional stiffness k_{12} to a spur pinion of inertia I_2 and tooth numbers N_2. This pinion drives gear 3 of inertia I_3 and tooth numbers N_3 through the linear-motion combined tooth-mesh stiffness k_{23}. Finally, gear 3 is connected to the output inertia, which might be the impeller of a pump, through a shaft of torsional stiffness k_{34}. We now wish to create the equivalent torsional system based on the output shaft, shown in Fig. 16-10b and c. The methods to be developed would also enable us to convert to a model based on the input-shaft motion.

The work has been partially completed in the previous section. Thus, from Eq. (16-31),

$$T'_{in} = \frac{N_3}{N_2} T_{in} \qquad (g)$$

And from Eq. (16-30),

$$\theta'_1 = \frac{N_2}{N_3} \theta_1 \qquad (h)$$

Similarly,

$$\theta'_2 = \frac{N_2}{N_3} \theta_2 \qquad (i)$$

Using Eq. (16-32), we also find

$$k'_{12} = \left(\frac{N_3}{N_2}\right)^2 k_{12} \qquad (j)$$

The torsional tooth stiffness is obtained from Eq. (16-36) and is

$$k'_{23} = \frac{D_3^2 k_{23}}{8} \qquad (k)$$

where D_3 is the pitch diameter of gear 3.

Our only remaining problem is to determine the equivalent inertias I'_1 and I'_2. Since the kinetic energies must remain unchanged during the modeling, we can write

$$\tfrac{1}{2} n'^2 I' = \tfrac{1}{2} n^2 I$$

where n and I refer to the speed in rpm and inertia of the actual system and n' and I' to the speed and inertia of the equivalent system. Therefore

$$I' = \left(\frac{n}{n'}\right)^2 I \qquad (16\text{-}40)$$

which states that the equivalent inertia is equal to the actual inertia times the square of the velocity ratio. Thus, for Fig. 16-10, we have

$$I'_1 = \left(\frac{N_3}{N_2}\right)^2 I_1 \qquad (l)$$

$$I'_2 = \left(\frac{N_3}{N_2}\right)^2 I_2 \qquad (m)$$

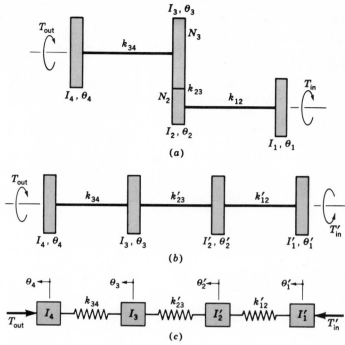

FIGURE 16-10
Modeling a rotating system. (*a*) Actual system; (*b*) equivalent torsional system;
(*c*) equivalent torsional system shown in linear form.

Equations (*l*) and (*m*) are written in terms of the ratio of the tooth numbers, since
this is the same as the velocity ratio. Note that *the heaviest inertia runs at the
slowest speed.*

16-7   MODELING FRICTION AND DAMPING

The equations describing the mathematical model can, of course, be written
without the inclusion of friction or damping effects. When this is done for actual
machines or systems, however, considerable trouble may be encountered, depending upon the model. The acceleration and the inertia forces, in particular, may
appear as wild swings in the computer output, just as if a great deal of white noise
had been introduced into the system. Introducing just a small amount of friction
will often quiet the system down so that the computer results will compare more
favorably with the instrumented output of the physical system itself.

After experiences such as these in using the systems approach, you begin to realize that Nature did indeed plan the universe with great care.

Two kinds of forces are generally introduced into the mathematical model to account for the general effect which we shall describe somewhat loosely as friction. These are *viscous damping* and *sliding*, or *Coulomb*, *friction*.

Viscous damping, also called viscous friction, is defined by the equation

$$F = -c\dot{x} \quad (16\text{-}41)$$

where $\dot{x}$ is the velocity and c is the coefficient of viscous damping. In vibration theory a particular value of the coefficient, called the *coefficient of critical damping* c_c, is defined, which delineates the transition between vibration and no vibration.* The value of this constant for a particular system may be obtained from the equation

$$c_c = 2m\omega_n \quad (16\text{-}42)$$

The ratio
$$\zeta = \frac{c}{c_c} = \frac{c}{2m\omega_n} \quad (16\text{-}43)$$

of the actual to the critical damping is very useful as an aid in the prediction of the friction in actual physical systems. There are several ways of measuring the damping in the laboratory. One widely used method is to physically disturb the system in some manner and then observe the decay of this disturbance on an oscilloscope. It is through such measurements as these that we have learned that the damping ratio ζ is usually in the range

$$0.03 \leq \zeta \leq 0.15$$

If the system contains belts, has high windage losses or very tight clearances, then choose a high damping ratio. A damping ratio of 3 percent represents structural damping, the approximate rate of decay of stress waves in a bar struck by a hammer. A free-running machine with good lubrication and no evident friction sources would probably have a damping ratio of about 5 to 7 percent. Thus, if the machine is not available for testing, making an educated guess is about the best one can do.

Since the viscous-damping force is linearly related to velocity, we often employ this kind of damping in the mathematical model even though the predominant friction forces are due to sliding, or Coulomb, friction. In many practical cases the difference in performance is barely perceptible, and so it makes little difference. It is not difficult, however, to include sliding friction in the mathematical model if the computer is going to be used to monitor the performance of the system. Assume that the coefficient of sliding friction equals the coefficient of static

* See Joseph E. Shigley, "Theory of Machines," p. 461, McGraw-Hill Book Company, New York, 1961.

friction. Though this assumption is not necessary, it makes a simpler solution. Then the required relations are

$$F = -\mu N \qquad \dot{x} > 0$$
$$-\mu N \leq F \leq +\mu N \qquad \dot{x} = 0 \qquad (16\text{-}44)$$
$$F = +\mu N \qquad \dot{x} < 0$$

where F is the friction force, μ is the coefficient of sliding friction, and N is the normal force pressing the sliding surfaces together. The friction force corresponding to $\dot{x} = 0$ will be that force between $-\mu N$ and $+\mu N$ required to maintain static equilibrium.

16-8 MATHEMATICAL MODELS FOR SHOCK ANALYSIS

Impact refers to the collision of two masses with initial relative velocity. In some cases it is desirable to achieve a known impact in design; for example, this is the case in the design of coining, stamping, and forming presses. In other cases impact occurs because of excessive deflections, or clearances between parts, and in these cases it is desirable to minimize the effects. The rattling of mating gear teeth in their tooth spaces is an impact problem caused by shaft deflection and the backlash or clearance between the teeth. This impact causes gear noise and fatigue failure of the tooth surfaces. The clearance space between a cam and follower or between a journal and its bearing may result in crossover impact and also cause excessive noise and rapid fatigue failure.

Shock is a more general term, which is used to describe any suddenly applied transient force or disturbance. Thus the study of shock includes impact as a special case. There are two general approaches to the study of shock, depending upon whether one uses only statics in the analysis or both statics and dynamics. It should be observed, however, that researchers have spent lifetimes investigating shock and impact phenomena and that entire books have been written on the subject. For this reason the material to be presented here is intended only to indicate the scope of the subject and to provide a basic understanding of what is involved.*

As a first example of the creation of a mathematical model for the study of impact, consider the collision of an automobile with a tree or a rigid obstruction. Such collisions are created deliberately on the automotive proving grounds to

---

* To avoid product liability or to achieve a more competitive and economical design may require the practicing designer to investigate further. The following books are recommended: Philip Barkan in Harold A. Rothbart (ed.), "Mechanical Design and Systems Handbook," sec. 16, McGraw-Hill Book Company, New York, 1964; Lydik S. Jacobsen and Robert S. Ayre, "Engineering Vibrations," McGraw-Hill Book Company, New York, 1958; Joseph E. Shigley, "Simulation of Mechanical Systems," McGraw-Hill Book Company, New York, 1967.

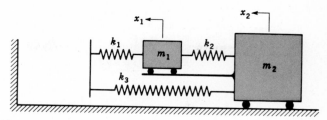

FIGURE 16-11
Two-freedom mathematical model of an automobile in collision with a rigid obstruction.

study the strength of many elements, such as door hinges and latches, seat belts and their mountings, and steering-wheel columns, for example. Various mathematical models have been, and are being, created to explain the results of these collisions. Figure 16-11 shows one of these models. Here m_1 is the lumped mass of the engine. Its displacement, velocity, and acceleration are described by the coordinate x_1 and its time derivatives. The lumped mass of the vehicle, less the engine, is described by m_2, and its motion by the coordinate x_2. Springs k_1, k_2, and k_3 represent the nonlinear stiffness of the various structural and nonstructural elements which compose the vehicle. Friction must also be included, but is not shown in this model. The determination of the variable spring rates k_1, k_2, and k_3 for such a complex structure, and where large permanent deformations are expected, can be performed only experimentally. Once these parameters, the k's, m's, and frictional coefficients, are obtained, however, a set of nonlinear differential equations can be written and solved on either the digital or the analog computer for any impact velocity.

Of course, a great many simplifying assumptions have to be made to reduce such a complex machine to the simple model of Fig. 16-11. However, the model can be made to yield very useful results, with a considerable saving in time and money.

Now consider Fig. 16-12a. Here is a mass m which drops through a distance h and collides with a beam of stiffness EI. If we can neglect the mass of the beam,

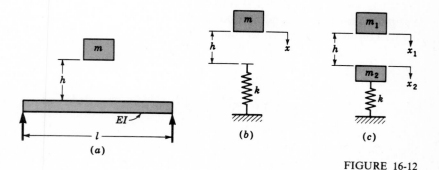

FIGURE 16-12

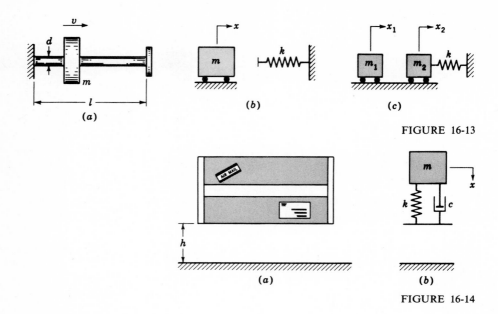

FIGURE 16-13

FIGURE 16-14

the model of this system is Fig. 16-12*b*. If the mass cannot be neglected, Fig. 16-12*c* is the model. In both cases *k* is the spring rate of the beam corresponding to the point of impact along the beam. Damping or friction can be added to either model if desired.

The abstract models of Fig. 16-12*b* and *c* can, of course, be used to represent any structure for which a spring rate can be obtained. Thus the models are quite general.

A case of impact not affected by gravity is shown in Fig. 16-13*a*. A mass *m*, propelled along a rod of length *l* and diameter *d*, strikes a stop at the end of the rod. What are the maximum stress and elongation of the rod? This problem can be solved using the models of Fig. 16-13*b* or *c*, depending upon whether the mass of the rod can be neglected.

The model of Fig. 16-14*b* is often used to simulate the dropping of a box of equipment as in *a*. Depending upon the cushioning material, the system may be highly nonlinear and have a spring force proportional to some higher power of *x*, as in Fig. 3-1*b*.

16-9 STRESS AND DEFLECTION DUE TO IMPACT

Let us now determine the stress and deflection of the beam of Fig. 16-12*a* due to the impact of the mass *m*. The analysis is quite general and can be applied to any structure for which a spring rate can be obtained. To be conservative, we shall

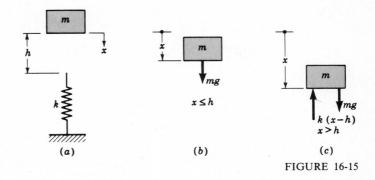

FIGURE 16-15

assume that no energy is lost. For this example we also assume that the area of the impacting mass is small, that it strikes the center of the beam, and that the mass of the beam can be neglected. Thus the model of Fig. 16-12b applies. The deflection-force relation for this beam is found in Table A-12, and from this relation we find that the spring rate is

$$k = \frac{48EI}{l^3} \quad (16\text{-}45)$$

Figure 16-15a shows the model again; we choose the origin of x corresponding to the position of the mass when time t is zero. Two free-body diagrams are necessary, one when $x \leq h$, and another when $x > h$, to account for the spring force. These are shown in Fig. 16-15b and c. For each of these free-body diagrams we can write Newton's laws by stating that the inertia force $m\ddot{x}$ is equal to the sum of the external forces acting on the mass. We then have

$$\begin{aligned} m\ddot{x} &= mg & x &\leq h \\ m\ddot{x} &= -k(x - h) + mg & x &> h \end{aligned} \quad (16\text{-}46)$$

We must also include in the mathematical statement of the problem the knowledge that the mass is released with zero initial velocity. Equation (16-46) constitutes a set of *piecewise linear differential equations*. Each equation is linear, but each applies only for a certain range of x. Thus the first equation is valid only as long as x is less than or equal to h. Similarly, the second is valid only for x's greater than h. The solution to this set of equations will yield the motion x for all values of time beginning with $t = 0$. However, we are interested in the function x only up until the time that the spring reaches its maximum deflection.

The solution to the first equation in the set is

$$x = \frac{gt^2}{2} \quad x \leq h \quad (16\text{-}47)$$

and you can verify this by direct substitution. Equation (16-47) is no longer valid after $x = h$; call this instant of time t_1. Then, from Eq. (16-47), we find

$$t_1 = \sqrt{\frac{2h}{g}} \qquad (a)$$

Differentiating to get the velocity gives

$$\dot{x} = gt \qquad x \leq h \qquad (16\text{-}48)$$

Then the velocity of the mass at $t = t_1$ is

$$\dot{x}_1 = gt_1 = \sqrt{2gh} \qquad (b)$$

Our next problem is to solve the second equation of the set (16-46). It is convenient to define a new time $t' = t - t_1$. Thus $t' = 0$ at the instant the mass strikes the spring. In solving this equation it is necessary to employ the knowledge that $x = h$ and $\dot{x} = \sqrt{2gh}$ at the instant $t' = 0$.

The second equation of the set (16-46) is an ordinary one-freedom linear differential equation which we have already studied. Applying your knowledge of differential equations, you should find that the solution is

$$x = A \cos \sqrt{\frac{k}{m}} t' + B \sin \sqrt{\frac{k}{m}} t' + h + \frac{mg}{k} \qquad x > h \qquad (c)$$

If you cannot derive this solution, you should at least prove to yourself that it is a solution by substituting it and its second time derivative into Eq. (16-46) to show that the equality is satisfied.

The constants A and B of Eq. (c) are evaluated using the known values of x and $\dot{x}$ at $t' = 0$. Substituting $x = h$ and $t' = 0$ into Eq. (c) gives

$$h = A + h + \frac{mg}{k}$$

and so

$$A = -\frac{mg}{k} \qquad (d)$$

Differentiating Eq. (c) gives the velocity as

$$\dot{x} = -A\sqrt{\frac{k}{m}} \sin \sqrt{\frac{k}{m}} t' + B\sqrt{\frac{k}{m}} \cos \sqrt{\frac{k}{m}} t' \qquad x > h \qquad (e)$$

Then substituting $\dot{x} = \sqrt{2gh}$ and $t' = 0$ gives

$$\sqrt{2gh} = B\sqrt{\frac{k}{m}}$$

or

$$B = \sqrt{\frac{2gh}{k/m}} \qquad (f)$$

Substituting the values of A and B into Eq. (c) gives the solution as

$$x = -\frac{mg}{k} \cos \sqrt{\frac{k}{m}} t' + \sqrt{\frac{2gh}{k/m}} \sin \sqrt{\frac{k}{m}} t' + h + \frac{mg}{k} \qquad x > h \qquad (16\text{-}49)$$

This solution is not really as complicated as it appears. The mass has already dropped through the distance h, which is the third term. The fourth term, mg/k, is simply the static deflection of the spring due to the weight mg. The other two terms depend upon t', and our task is to manipulate these so as to get x_{max}.

The first two terms of Eq. (16-49) are of the form

$$u = P \cos a + Q \sin a \qquad (g)$$

which is simply the sum of a sine and cosine term. Such sums can be transformed into a single sine or cosine term by the addition or subtraction of another angle from the argument. Thus we can also write Eq. (g) in the form

$$u = R \cos (a - b) \qquad (h)$$

To show that this is a valid transformation, let

$$R = \sqrt{P^2 + Q^2} \qquad \cos b = \frac{P}{R} \qquad \sin b = \frac{Q}{R} \qquad (i)$$

Then Eq. (g) can be written

$$u = R\left(\frac{P}{R} \cos a + \frac{Q}{R} \sin a\right) = R(\cos b \cos a + \sin b \sin a)$$

$$= R \cos (a - b)$$

Employing this transformation for Eq. (16-49) gives

$$x = \sqrt{\left(\frac{mg}{k}\right)^2 + \frac{2mgh}{k}} \cos \left(\sqrt{\frac{k}{m}} t' - \phi\right) + h + \frac{mg}{k} \qquad x > h \qquad (16\text{-}50)$$

where ϕ, though of no interest here, is given by

$$\phi = \frac{\pi}{2} + \tan^{-1} \sqrt{\frac{mg}{2kh}} \qquad (j)$$

The maximum deflection occurs when the cosine term is unity. We designate this spring deflection as δ and, after rearranging, find it to be

$$\delta = x - h = \frac{mg}{k} + \frac{mg}{k} \sqrt{1 + \frac{2hk}{mg}} \qquad (k)$$

The equation may be more useful if we substitute the weight W of the falling mass for mg. We then have

$$\delta = \frac{W}{k} + \frac{W}{k} \sqrt{1 + \frac{2hk}{W}} \qquad (16\text{-}51)$$

This is an exceedingly useful and general equation since it gives the deflection of any elastic structure subjected to impact. Remember that we have assumed no energy loss as well as a massless structure.

Equation (16-51) can be particularized by substituting known values of the spring rate. For the beam of Fig. 16-12a,

$$k = \frac{48EI}{l^3}$$

and the maximum deflection is found, from Eq. (16-51), to be

$$\delta = \frac{Wl^3}{48EI} + \frac{Wl^3}{48EI}\sqrt{1 + \frac{96EIh}{Wl^3}} \qquad (16\text{-}52)$$

If we turn the bar of Fig. 16-13a into a vertical direction and allow a weight W to fall through a distance h to the stop at the end of the bar, then

$$k = \frac{AE}{l}$$

from Eq. (3-6), and Eq. (16-51) becomes

$$\delta = \frac{Wl}{AE} + \frac{Wl}{AE}\sqrt{1 + \frac{2AEh}{Wl}} \qquad (16\text{-}53)$$

The maximum force in a spring occurs when the spring is deflected the greatest amount, and hence, from Eq. (16-51), we have

$$F = k\delta = W + W\sqrt{1 + \frac{2hk}{W}} \qquad (16\text{-}54)$$

This force can now be used with any of the stress equations to compute the stress caused by impact. For the bar of Fig. 16-13 in a vertical position, we have

$$\sigma = \frac{F}{A} = \frac{W}{A} + \frac{W}{A}\sqrt{1 + \frac{2AEh}{Wl}} \qquad (16\text{-}55)$$

It is interesting to note from this equation that, if $h = 0$, that is, if the weight is released while in contact with the bar but not exerting any force on it, the stress is

$$\sigma = \frac{2W}{A} \qquad (16\text{-}56)$$

This may help to explain why designers in the past have arbitrarily doubled all factors of safety when impact is expected. It is much better, of course, to compute the actual value of the impact force.

The stress and deflection equations derived above for massless elastic structures probably do not actually give conservative values. That is, the actual stresses may be higher than the predicted values. One reason for this is that the dynamic spring rate actually differs slightly from the static rate which has been used in the development. Another reason is that the systems are not really massless, but rather have a continuous or distributed mass. The solution to such a problem would require either that we solve a partial differential equation or that we divide the member into a finite number of masses and springs and then solve a finite number of ordinary simultaneous differential equations.

16-10 CAM SYSTEMS

The use of cams to convert rotary motion into reciprocating or oscillating motion is well known. In fact, one or more cam-and-follower mechanisms can be found on almost any machine we might choose to inspect. These cams perform such important functions in many cases that the behavior of the entire machine is dependent upon the design of the cam motion. For this reason it is often desirable to make use of the systems approach in designing cam-and-follower mechanisms and in analyzing their performance.

The follower system of a cam mechanism may consist of one or more rods, levers, gears, and springs, all of which are elastic. These members all have mass which is distributed in accordance with their geometry. Various kinds of friction and damping will be present because of relative sliding, air and fluid-film resistance, and internal effects. The schematic drawing of the overhead valve train in Fig. 16-16a for an automotive engine is a typical example of such a follower system.

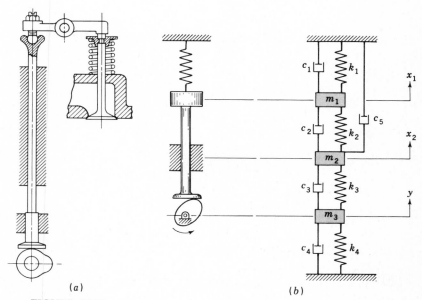

(a)   (b)

FIGURE 16-16
(a) Cross section showing the overhead valves in an automotive engine; (b) mathematical model of cam system.

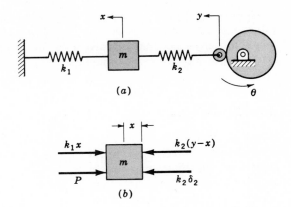

(a)

(b)

FIGURE 16-17

A mathematical model of a cam system is shown in Fig. 16-16b. This is too complicated to analyze on paper but is perfectly capable of being solved by computer. The figure suggests that m_3 is the lumped mass of the cam and part of the camshaft, and k_4 represents the camshaft stiffness. The coordinate y is the instantaneous resultant of the motion machined into the cam and the vertical motion or deflection of the camshaft.

The mass m_1 represents the load or part which is to be reciprocated by the cam motion. Spring k_1 is a compression spring whose purpose is to retain contact between the cam and follower at all times. Sometimes it is called the *retaining spring*. The coordinate x_1 describes the motion of m_1; of course, the cam profile is designed with the hope that x_1 will be a certain desired motion. The systems approach for this problem can be used to discover whether x_1 turned out to be the desired motion or not.

The follower mass is m_2, and the springs k_2 and k_3 represent the elasticity of the follower. Dashpots representing viscous damping are inserted between the masses to account for frictional effects.

The equations which define the mathematical model of a cam system are not as simple to write as one might expect. It is desirable to include in these relations the possibility that the follower need not always be in physical contact with the cam. As we shall see, it is easy enough to make this prediction but there is a pitfall in getting there.

Let us consider Fig. 16-17a, where k_1 is the retaining spring, k_2 the linearized Hertzian deflection characteristic of the cam-and-roller contact, and m is the lumped mass of the follower system. The motion y machined into the cam will be a function of the angular position of the cam θ, its derivatives, and of time t. The coordinate x describes the motion of the follower.

To assemble spring k_1, rotate the cam to its lowest position, that is, $y = 0$; squeeze k_1 through the distance δ_{st}; and insert the spring into position and release it. Upon releasing k_1, a preload P comes into existence in the system, this being the force with which the roller pushes on the cam when $y = 0$. Since springs k_1 and k_2 are in series during this preloading operation, the preload is

$$P = \frac{k_1 k_2 \delta_{st}}{k_1 + k_2} \qquad (a)$$

from Eq. (3-3). Therefore the initial compression of k_1 is

$$\delta_1 = \frac{P}{k_1} = \frac{k_2 \delta_{st}}{k_1 + k_2} \qquad (b)$$

and the initial compression of k_2 is

$$\delta_2 = \frac{P}{k_2} = \frac{k_1 \delta_{st}}{k_1 + k_2} \qquad (c)$$

   To keep track of the forces acting upon mass m, refer to Fig. 16-17b. Here we visualize mass m as being displaced through some positive distance x and the cam as having turned through some angle θ so as to produce a value of y slightly greater than x. Thus spring k_1 exerts a force made up of two components, the preload P and the force $k_1 x$, both acting in the negative x direction. Since $y > x$, spring k_2 is in compression and exerts a force $k_2(y - x)$ in the positive x direction and another force $k_2 \delta_2$ also in the positive x direction. In the usual course of solving this problem you would probably cancel the force P in Fig. 16-17b with the force $k_2 \delta_2$ because they are equal and opposite. In this example this is exactly the wrong thing to do, because you would be discarding the information you require to determine the roller-contact force. This force is

$$F = \begin{cases} k_2(y - x + \delta_2) & x - y < \delta_2 \\ 0 & x - y \geq \delta_2 \end{cases} \qquad (d)$$

Equation (d) is a mathematical statement of the fact that k_2 can only be in compression.

   Adding the d'Alembert force to those of Fig. 16-17b gives

$$\sum F = -k_1 x - P + F - m\ddot{x} = 0$$

or

$$m\ddot{x} + k_1 x = F - P \qquad (e)$$

as the equation of motion of the system. Of course, the motion $y = y(\theta, t)$ machined into the cam must be defined before the equation can be solved. And because of the piecewise nature of Eq. (d), it would be hopeless to attempt a solution of Eq. (e) by hand methods.

   In using the systems approach, students are sometimes surprised when, upon obtaining Eqs. (d) and (e), say, we stop. But after numbers are obtained, this is as far as we need to go, because these are the equations which are programmed into the computer. In view of this, it will be informative to investigate a more complex cam system. The system we shall examine is especially interesting because both rotational and linear-motion coordinates exist in the mathematical model.

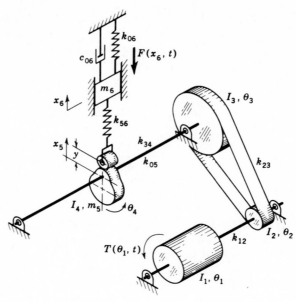

FIGURE 16-18

Figure 16-18 is a schematic drawing of a motor-driven cam system, with $T = T(\theta_1, t)$ being the torque input from the motor and $F = F(x_6, t)$ representing the force output of the cam follower. The system is further complicated by the belt-and-pulley connection between the motorshaft and camshaft, and by the fact that the camshaft has a torsional stiffness k_{34} and a linear stiffness k_{05}. Except for friction, the primary load driven by the motor is the force output F of the cam follower. If the roller should happen to leave contact with the cam for a short period of time, the possibility arises that the motor could run away with I_1, I_2, I_3, and I_4. Let us see how these problems are handled.

We first create the two subsystems shown in Fig. 16-19, one, the rotating subsystem, and the other, the linear-motion subsystem. The rotating subsystem of Fig. 16-19a is obtained using the methods of Secs. 16-5 and 16-6. Two dashpots having damping coefficients c_{01} and c_{03} have been added to account for bearing friction and windage losses. Similar dashpots could have been added to account for belt friction. The output torque $T(\theta_4, t)$ is due to the force exerted by the roller on the cam. We shall obtain an expression for this force shortly. The equations of motion of this subsystem may be written by assuming that $\theta_1' > \theta_2' > \theta_3 > \theta_4$ and that $\dot{\theta}_1'$ and $\dot{\theta}_2'$ are both positive. Summing forces on each of the inertias gives

$$T'(\theta_1', t) - c_{01}\dot{\theta}_1' - k_{12}'(\theta_1' - \theta_2') - I_1'\ddot{\theta}_1' = 0$$

$$k_{12}'(\theta_1' - \theta_2') - k_{23}'(\theta_2' - \theta_3) - I_2'\ddot{\theta}_2' = 0 \qquad (16\text{-}57)$$

$$k_{23}'(\theta_2' - \theta_3) - c_{03}\dot{\theta}_3 - k_{34}(\theta_3 - \theta_4) - I_3\ddot{\theta}_3 = 0$$

$$k_{34}(\theta_3 - \theta_4) - T(\theta_4, t) - I_4\ddot{\theta}_4 = 0$$

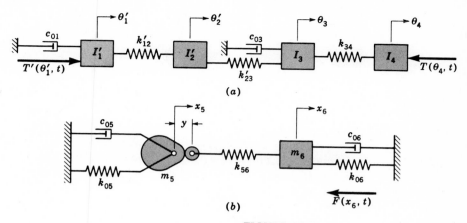

FIGURE 16-19
(a) Torsional subsystem; (b) linear subsystem.

Note that the difference in the angular coordinates appears, but not the individual coordinates themselves. If the motor runs at a high speed or for long periods of time, the individual coordinates will approach infinity. Therefore these coordinate differences should be preserved if at all possible.

Before writing the equations of motion of the linear subsystem of Fig. 16-19b, it is necessary to investigate the preloading of the retaining spring. This spring is k_{06}; it is to be assembled by compressing it through the distance δ_{st} when the cam is turned to its lowest position. The equivalent stiffness of the three series springs is

$$k = \frac{k_{05} k_{56} k_{06}}{k_{56} k_{06} + k_{05} k_{06} + k_{05} k_{56}} \qquad (f)$$

Therefore the preload is

$$P = k\delta_{st} = \frac{k_{05} k_{56} k_{06} \delta_{st}}{k_{56} k_{06} + k_{05} k_{06} + k_{05} k_{56}} \qquad (g)$$

and the compression of each spring is

$$\delta_{05} = \frac{P}{k_{05}} \qquad \delta_{56} = \frac{P}{k_{56}} \qquad \delta_{06} = \frac{P}{k_{06}} \qquad (h)$$

To write the equations of the linear-motion subsystem assume that all coordinates and their derivatives are positive and that $y > x_5 > x_6$. Then the external forces acting on m_5 and m_6 will be as shown in Fig. 16-20. The roller contact force is the same as the force exerted by spring k_{56}. This force is

$$F = \begin{cases} k_{56}(y + x_5 + \delta_{56} - x_6) & x_6 < y + x_5 + \delta_{56} \\ 0 & x_6 \geq y + x_5 + \delta_{56} \end{cases} \qquad (16\text{-}58)$$

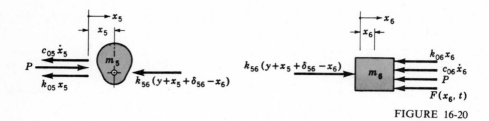

FIGURE 16-20

because this spring cannot exert a tensile force. We can now write the equations of motion for each mass by summing the external forces and adding the d'Alembert force to these. Thus, for m_5 and m_6, we have

$$P - c_{05}\dot{x}_5 - k_{05}x_5 - F - m_5\ddot{x}_5 = 0$$
$$F - k_{06}x_6 - c_{06}\dot{x}_6 - P - F(x_6, t) - m_6\ddot{x}_6 = 0 \qquad (16\text{-}59)$$

Equations (16-57) and (16-59) now constitute the general mathematical model. We have neglected rotary shear in developing these equations. If included, this would couple the linear-motion coordinate x_5 to the angular coordinate θ_4. The effect, however, is small.

As one example of the use of these equations, we might specify that the motor torque is constant and independent of speed. In fact, many motors do have a nearly constant torque-speed characteristic during starting. This would give a constant value for the term $T'(\theta'_1, t)$ in Fig. 16-19a.

For the motion y machined into the cam we might choose to use

$$y = y_0 - y_0 \cos \theta_4 t \qquad (16\text{-}60)$$

which is a harmonic motion without dwells. This equation can easily be obtained in the program during the integration of the fourth equation of the set (16-57). In the study of cam dynamics it is shown that the torque reaction to the roller-contact force is*

$$T = \frac{\dot{y}}{\dot{\theta}} F \qquad (i)$$

where $\dot{\theta}$ is the angular cam velocity. Differentiating Eq. (16-60) gives

$$\dot{y} = y_0 \dot{\theta}_4 \sin \theta_4 t$$

or

$$\frac{\dot{y}}{\dot{\theta}_4} = y_0 \sin \theta_4 t = y_0 \sin \theta_4$$

Thus the term needed for the fourth equation in the set (16-57) may be computed during the process of integrating this equation and is

$$T(\theta_4, t) = F y_0 \sin \theta_4 \qquad (16\text{-}61)$$

* See Joseph E. Shigley, "Theory of Machines," p. 570, McGraw-Hill Book Company, New York, 1961.

The output force $F(x_6, t)$ will of course depend upon the physical system and the particular task accomplished by the cam follower. In a theoretical problem one might use the function

$$F(x_6, t) = C\dot{x}_6^2 \quad (16\text{-}62)$$

which is equivalent to velocity-squared damping. With such a force the system would come up to speed and reach a steady-state condition quickly. This is especially necessary if the problem is to be run on the digital computer. Otherwise an excessive amount of computer time might be required to get through the transient part of the run.

We have accomplished our original task. Equations (16-57) to (16-62) constitute the mathematical model of the entire cam mechanism of Fig. 16-18. Once this is programmed on the computer, you can cause the computer to issue whatever information is of interest. You can "turn the motor on" and read the coordinates $\theta_1, \theta_2, \theta_3, \theta_4, x_5$, and x_6 and their derivatives as the system builds up to speed. The shaft torques $k_{12}(\theta_1 - \theta_2)$ and $k_{34}(\theta_3 - \theta_4)$ can be read out or used in the computer to obtain the torsional stresses in the shaft. The bending deflection of the camshaft is given by the coordinate x_5, and the bending force by the product $k_{05} x_5$. Thus the bending stress is easily computed.

Maximum benefits are derived from the systems approach when the engineer can monitor the performance of his mechanical system, just as the pilot monitors the performance of his aircraft. Corrections and changes are made in accordance with the observed performance. Thus the design engineer might want to try various cam motions y to learn of their effect on the performance. Shafts can be stiffened up by moving the bearings closer together and increasing the shaft diameters. A motor having a different torque-speed characteristic might be of interest. It is very desirable, therefore, that the engineer be able to interact directly with the computer.

16-11 DESIGNING WITH THE PROGRAMMABLE CALCULATOR

In addition to the speed of computation, the programmable calculator can contribute to design effectiveness as follows:

1 When iteration is required in design, an entire family of solutions can be obtained and the important performance parameters plotted to reveal the optimum. The compression spring-design program of Fig. 16-24 is a good example of such an approach.

2 Design problems frequently require decisions during the course of the analysis. In designing a column, for example, the slenderness ratio dictates whether the Euler equation or the parabolic equation should be used. The column design program of Fig. 16-25 shows how the calculator can be used to make this decision.

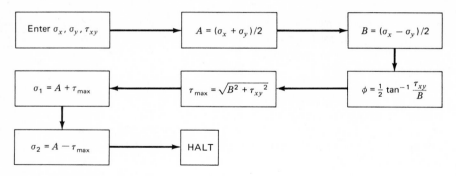

FIGURE 16-21

Mohr's circle flow chart (87 instructions). Computes σ_1, σ_2, and the angle ϕ from x to σ_1 (positive in the ccw direction). τ_{xy} is positive if cw, negative if ccw. τ_{max} is the radius of Mohr's circle, but may not be the maximum shear stress. See Secs. 2-2 and 2-3.

3 The calculator is an economical and convenient means for devising and debugging subroutines to be assembled later for use in larger digital-computer design programs.

4 Even lengthy programs can be keyed into the calculator in just a few minutes and so permanent program storage is not a vital consideration, though it is useful. The careful designer can easily build a library of design routines which will greatly improve his effectiveness.

As illustrations of the usefulness of the programmable calculator, this section contains flow charts for five useful design techniques including two detailed coded programs for the SR-52 calculator. All of these have been checked, but no effort has been made to economize on the number of instructions.

Figures 16-21 to 16-23 for the Mohr's circle diagram, the modified Goodman fatigue diagram, and the S-N diagram, respectively, are self-explanatory. In each case these programs substitute for graphical computations. Similar substitutes may be obtained by curve-fitting techniques. See Sec. 13-8, for example, where a stress-concentration chart has been replaced by an empirical equation.

The flow chart of Fig. 16-24 and the accompanying SR-52 code in Table 16-2 constitute solutions of all the compression-spring design equations of Chap. 8. The program is initiated by entering the wire diameter d, the outside coil

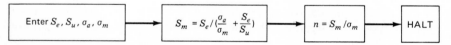

FIGURE 16-22

Flow chart for modified Goodman diagram (54 instructions). See Sec. 5-21.

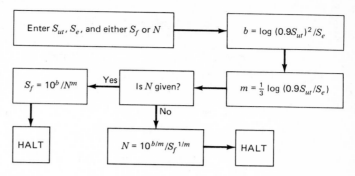

FIGURE 16-23
Flow chart for S–N diagram (105 instructions). Computes either S_f or N with S_e and S_{ut} given. See Sec. 5-11.

diameter OD, the number of active coils N, the material exponent m (see Table 8-2) and factor A, and the number of dead coils. The program computes and displays the yield strength in shear and halts. The operator then enters the desired value of the shear stress τ_{max} corresponding to the solid height. This action causes the calculator to resume the computation. The results, consisting of the mean diameter D, the spring index C, the spring rate k, and the shear-stress multiplication factor K_s, the force F and length l corresponding to the solid

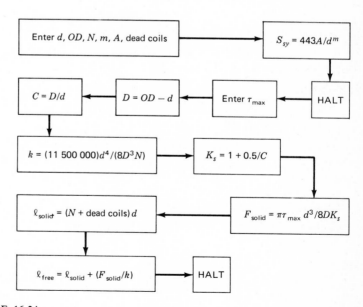

FIGURE 16-24
Flow chart for compression-spring design (184 instructions). For nomenclature, see Chap. 8.

Table 16-2 SR-52 CODE FOR COMPRESSION-SPRING DESIGN PROGRAM

Loc	Code	Key	Item		Loc	Code	Key	Item
000	46	LBL			050	04	4	
	11	A	d			54	)	
	42	STO				95	=	S_{sy}
	00	0				81	HLT	
	01	1				46	LBL	
005	81	HLT			055	17	B'	τ_{max}
	46	LBL				42	STO	
	12	B	OD			00	0	
	42	STO				07	7	
	00	0				43	RCL	
010	02	2			060	00	0	
	81	HLT				02	2	
	46	LBL				75	−	
	13	C	N			43	RCL	
	42	STO				00	0	
015	00	0			065	01	1	
	03	3				95	=	D
	81	HLT				42	STO	
	46	LBL				00	0	
	14	D	m			08	8	
020	42	STO			070	43	RCL	
	00	0				00	0	
	04	4				08	8	
	81	HLT				55	÷	
	46	LBL				43	RCL	
025	15	E	A		075	00	0	
	42	STO				01	1	
	00	0				95	=	C
	05	5				42	STO	
	81	HLT				00	0	
030	46	LBL			080	09	9	
	16	A'	ends			43	RCL	
	42	STO				00	0	
	00	0				01	1	
	06	6				45	y^x	
035	04	4			085	04	4	
	03	3				65	×	
	03	3				01	1	
	65	×				01	1	
	43	RCL				05	5	
040	00	0			090	00	0	
	05	5				00	0	
	55	÷				00	0	
	53	(				00	0	
	43	RCL				00	0	
045	00	0			095	55	÷	
	01	1				53	(	
	45	y^x				08	8	
	43	RCL				65	×	
	00	0				43	RCL	

Table 16-2 SR-52 CODE FOR COMPRESSION-SPRING DESIGN PROGRAM

Loc	Code	Key	Item	Loc	Code	Key	Item
100	00	0		145	01	1	
	08	8			01	1	
	45	y^x			54	)	
	03	3			95	=	F_{solid}
	65	×			42	STO	
105	43	RCL		150	01	1	
	00	0			02	2	
	03	3			53	(	
	54	)			43	RCL	
	95	=	k		00	0	
110	42	STO		155	03	3	
	01	1			85	+	
	00	0			43	RCL	
	01	1			00	0	
	85	+			06	6	
115	93	.		160	54	)	
	05	5			65	×	
	55	÷			43	RCL	
	43	RCL			00	0	
	00	0			01	1	
120	09	9		165	95	=	l_{solid}
	95	=	K_s		42	STO	
	42	STO			01	1	
	01	1			03	3	
	01	1			43	RCL	
125	59	π		170	01	1	
	65	×			03	3	
	43	RCL			85	+	
	00	0			43	RCL	
	07	7			01	1	
130	65	×		175	02	2	
	43	RCL			55	÷	
	00	0			43	RCL	
	01	1			01	1	
	45	y^x			00	0	
135	03	3		180	95	=	l_{free}
	55	÷			42	STO	
	54	)			01	1	
	08	8			04	4	
	65	×			81	HLT	
140	43	RCL					
	00	0					
	08	8					
	65	×					
	43	RCL					

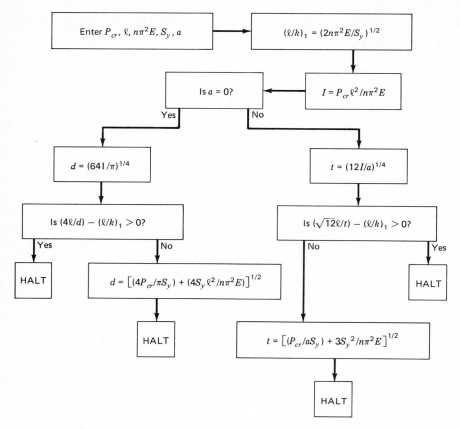

FIGURE 16-25
Column design flow chart (211 instructions). Used to design rectangular- or round-section columns of any length. See Secs. 3-10 and 3-11.

height, and the free length of the spring are stored in memory and may be displayed as desired. Other calculations concerned with the spring design can be made on the keyboard without affecting the values stored in memory until a satisfactory design is obtained. In addition, one or more items of the given data

Table 16-3   UNITS AND NOMENCLATURE FOR COLUMN DESIGN

Symbol	Name	IPS units	IPS units	SI units
P_{cr}	Critical load	lb	kip	N
l	Column length	in	in	mm
E	Elastic modulus	psi	kpsi	MPa
S_y	Yield strength	psi	kpsi	MPa
d	Diameter	in	in	mm
t	Thickness	in	in	mm
at	Width	in	in	mm

can be changed, without affecting the others, and the program rerun with the new data. While this program is in IPS units, it would be simple to construct a similar program in SI.

A program for the design of a round- or rectangular-section column is displayed in Fig. 16-25 and Table 16-4. Table 16-3 shows the symbols used together with the sets of units that the program will accept. The factor a is entered as zero if a round-section column is desired. If a rectangular-section column is desired, the program assumes that bending would occur about the axis parallel to the width of the bar. Accordingly, the factor a is entered as the ratio of the width to the thickness, and hence is a number equal to or greater than unity.

The program first determines the location of the tangency point $(l/k)_1$ joining the Euler and parabolic curves. The most of inertia I is calculated next, using the assumption that the Euler equation applies. The program then determines whether a round or a rectangular column is to be designed and calculates d or t, accordingly. This result is then entered into an equation to calculate the actual slenderness ratio l/k. This value is compared with $(l/k)_1$ to determine whether to use the Euler solution or to recalculate the dimension using the parabolic formula. If the Euler equation is found to be valid, the program halts and displays the dimension d or t. If not, the flow is to the corresponding parabolic equation. The new result then replaces the previous result in storage and the program halts. Table 16.4 shows that $(l/k)_1$ and I can also be displayed upon completion of a run.

Table 16-4 SR-52 CODE FOR COLUMN DESIGN PROGRAM

Loc	Code	Key	Item	Loc	Code	Key	Item
000	46	LBL		020	42	STO	
	11	A	P_{cr}		00	0	
	42	STO			04	4	
	00	0			81	HLT	
	01	1			46	LBL	
005	81	HLT		025	16	A'	a
	46	LBL			42	STO	
	12	B	l		00	0	
	42	STO			05	5	
	00	0			81	HLT	
010	02	2		030	46	LBL	
	81	HLT			15	E	start
	46	LBL			53	(	
	13	C	$n\pi^2 E$		02	2	
	42	STO			65	×	
015	00	0		035	43	RCL	
	03	3			00	0	
	81	HLT			03	3	
	46	LBL			55	÷	
	14	D	S_y		43	RCL	

(*continued*)

Table 16-4 SR-52 CODE FOR COLUMN DESIGN PROGRAM

Loc	Code	Key	Item	Loc	Code	Key	Item
040	00	0		090	65	×	
	04	4			43	RCL	
	54	)			00	0	
	30	$\sqrt{x}$			02	2	
	42	STO			55	÷	
045	00	0		095	43	RCL	
	06	6	$(l/k)_1$		00	0	
	43	RCL			08	8	
	00	0			75	−	
	01	1			43	RCL	
050	65	×		100	00	0	
	43	RCL			06	6	
	00	0			95	=	
	02	2			22	INV	
	40	x^2			80	if pos	
055	55	÷		105	01	1	
	43	RCL			01	1	
	00	0			02	2	
	03	3			43	RCL	
	95	=	I		00	0	
060	42	STO		110	08	8	
	00	0			81	HLT	
	07	7			53	(	
	43	RCL			43	RCL	
	00	0			00	0	
065	05	5		115	01	1	
	90	if zro			55	÷	
	01	1			43	RCL	
	04	4			00	0	
	01	1			05	5	
070	53	(		120	55	÷	
	01	1			43	RCL	
	02	2			00	0	
	65	×			04	4	
	43	RCL			85	+	
075	00	0		125	03	3	
	07	7			65	×	
	55	÷			43	RCL	
	43	RCL			00	0	
	00	0			04	4	
080	05	5		130	40	x^2	
	54	)			55	÷	
	30	$\sqrt{x}$			43	RCL	
	30	$\sqrt{x}$			00	0	
	42	STO			03	3	
085	00	0		135	54	)	
	08	8			30	$\sqrt{x}$	
	01	1			42	STO	
	02	2			00	0	
	30	$\sqrt{x}$			08	8	

Table 16-4 SR-52 CODE FOR COLUMN DESIGN PROGRAM

Loc	Code	Key	Item	Loc	Code	Key	Item
140	81	HLT		175	43	RCL	
	53	(			00	0	
	06	6			08	6	
	04	4			81	HLT	
	65	×			53	(	
145	43	RCL		180	04	4	
	00	0			65	×	
	07	7			43	RCL	
	55	÷			00	0	
	59	π			01	1	
150	54	)		185	55	÷	
	30	$\sqrt{x}$			59	π	
	30	$\sqrt{x}$			55	÷	
	42	STO			43	RCL	
	00	0			00	0	
155	08	8		190	04	4	
	04	4			85	+	
	65	×			04	4	
	43	RCL			65	×	
	00	0			43	RCL	
160	02	2		195	00	0	
	55	÷			02	2	
	43	RCL			40	x^2	
	00	0			65	×	
	08	8			43	RCL	
165	75	−		200	00	0	
	43	RCL			04	4	
	00	0			55	÷	
	06	6			43	RCL	
	95	=			00	0	
170	22	INV		205	03	3	
	80	if pos			54	)	
	01	1			30	$\sqrt{x}$	
	07	7			42	STO	
	09	09			00	0	
				210	08	8	
					81	HLT	

PROBLEMS

Section 16-4

16-1 The gears and shafts shown in the figure are made of steel. What is the spring rate of the equivalent to the geared system based on the input shaft if the gear teeth are assumed to be rigid? Compute the spring rate assuming the gear teeth are elastic.

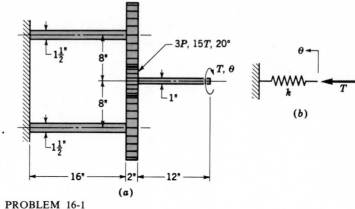

(a)

PROBLEM 16-1
(a) Geared system; (b) equivalent system.

16-2 Find the spring rate k of the equivalent of the geared system shown in the figure based on the input shaft and on rigid gear teeth. The shafts are of steel.

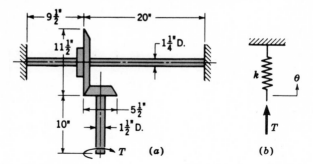

PROBLEM 16-2
(a) Geared system; (b) equivalent system.

Section 16-6

16-3 The figure illustrates a hypothetical motor-driven pump. We wish to model this system based on the pump rotor, that is, $I_4 = I_D$, $\theta_4 = \theta_D$, and $T_4 = T_D$. The mass moments of inertia in lbf·s²·in are $I_A = 0.110$, $I_B = 0.0011$, $I_C = 0.469$, and $I_D = 7.50$. The shafts and gears are of steel, and the gears have a diametral pitch of 10 teeth/in and are cut using the 20° full-depth system. When the motor is first turned on, the starting torque is $T_A = 95$ lb·in. We are interested in the response of the system to this input. The load torque is given by the relation

$$T_D = 0.246\dot{\theta}_D^2$$

Compute the spring rates of the shafts and gear teeth and convert to k_{12}, k_{23}, and

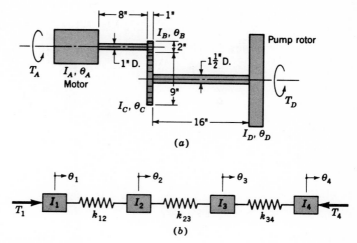

(a)

(b)

PROBLEM 16-3
(a) Schematic drawing of a geared motor-driven pump; (b) equivalent system.

k_{34}. Compute the equivalent inertias, and specify the formulas for the coordinates θ_1 to θ_4 in terms of θ_A to θ_D.

16-4 During a computer run of Prob. 16-3, the coordinates θ_i and their derivatives become available for use in calculating any information that is desired. Show how these coordinates or their derivatives can be used to compute the torsional stress in each shaft and the tangential force on the gear teeth.

16-5 The figure illustrates a motor-driven fan system in which the speed reduction is obtained using belts and sheaves. In this design study, the motor torque is to be approximated by the relation

$$T_A = 150 - 0.416\dot{\theta}_A$$

and the load torque by the relation

$$T_D = 0.108\dot{\theta}_D^2$$

The mass moments of inertia are $I_A = 0.175$, $I_B = 0$, $I_C = 1.25$, and $I_D = 3.50$, all in lbf·s²·in. The inertia of the small sheave is so small that it has been neglected, making I_B zero. Coordinate θ_2, however, is used to represent the equivalent of the pulley rotation θ_B in order that the belt pull and the input-shaft torque can be made to appear as separate terms in a computer run. Both the motor shaft and the fan shaft are made of cold-drawn steel. The equivalent system is to be based on the fan speed, that is, $\dot{\theta}_4 = \dot{\theta}_D$.

(a) Assuming a rigid system with no friction losses, determine the steady-state fan speed in rpm. What is the corresponding motor speed and motor torque?

(b) In a computer run, the input and output torques T_1 and T_4 and the coordinates θ_1 to θ_4 with their derivatives would become available for use in calculating any

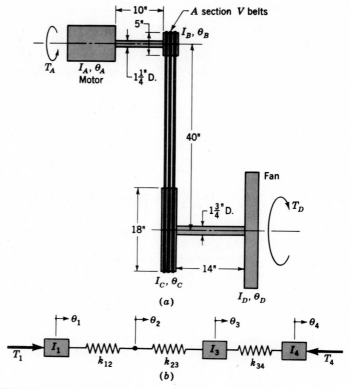

(a)

PROBLEM 16-5

(a) The physical system, a fan driven by a motor through a V belt and pulley system; (b) a mathematical model of the system; friction not shown.

information of interest. Develop formulas for computing the motor-shaft torsional stress, the belt pull, and the fan-shaft torsional stress in terms of these coordinates. The modulus of elasticity of the belts is 20 kpsi.

16-6 Compute the torsional spring rates of both shafts of the fan system of Prob. 16-5 and the combined linear-motion spring rate of the belts. The equivalent system is based on the fan speed. Find T_1, k_{12}, k_{23}, k_{34}, I_1, and I_3. Also find formulas for $\theta_1, \theta_2, \theta_3$, and θ_4 in terms of θ_A, θ_B, θ_C, and θ_D.

Section 16-8

16-7 A belt may be treated as a two-way spring by including the belt slack in the mathematical model. The figure shows how this can be done for the motor-driven fan system of Probs. 16-5 and 16-6. Note, in the figure, that the belt slack in angular

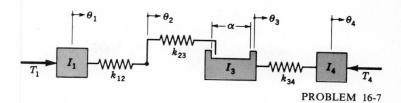

PROBLEM 16-7

units based on the fan speed is $\alpha = b/r = b/9$ radians. Write the set of equations and inequations which describe the motion of the system shown in the figure. Add viscous damping to the system by including a dashpot to ground at each origin. Designate the respective viscous-damping coefficients as c_{01}, c_{02}, c_{03}, and c_{04}.

Section 16-9

16-8 The coefficient of sliding friction for the motion of the weight W in the figure over the flat surface is μ. With the weight motionless, the end of the spring is suddenly pushed through the distance $y = a$ and held there. Determine the response x of the weight for one cycle of motion assuming that the spring force at time zero is greater than the friction force.

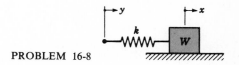

PROBLEM 16-8

16-9 Mass m_1 in the figure has an initial displacement and velocity $x(0) = -a$, $\dot{x}(0) = v_0$. Mass m_2 has zero initial displacement and velocity. The origin of each coordinate system is shown in the figure. Write the equations of motion for both masses.

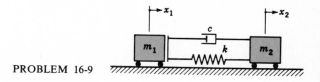

PROBLEM 16-9

16-10 In flywheel design a coefficient of speed fluctuation C_s is defined as

$$C_s = \frac{\theta_{max} - \theta_{min}}{\theta_{av}}$$

in terms of the maximum, average, and minimum angular velocities of the flywheel. In this problem we wish to define a mathematical model for the purpose of designing a flywheel to give a coefficient-of-speed fluctuation approximately equal to some desired value. The flywheel shown in the figure is intended to smooth out the speed fluctuations which would occur if it were not present or if it did not have a sufficiently large mass moment of inertia. The shafts connecting the flywheel to the driving and driven machinery are of steel. The driver exerts a torque input to the system

$$T_1 = 52\,500 + 500 \sin \dot{\theta}_1 t \qquad \text{lb·in}$$

The opposing torque T_3 exerted by the driven machinery on the shaft is exerted on every other turn of the shaft and is proportional to the square of the shaft angular velocity. Mathematically, this torque can be expressed in the form

$$T_3 = \begin{cases} 0 & \theta_3 = 0 \text{ to } 2\pi,\ 4\pi \text{ to } 6\pi,\ 8\pi \text{ to } 10\pi,\ \dots \\ -26.6\dot{\theta}_3^2 & \theta_3 = 2\pi \text{ to } 4\pi,\ 6\pi \text{ to } 8\pi,\ 10\pi \text{ to } 12\pi,\ \dots \end{cases}$$

where θ_3 is in radians and T_3 is in lb·in. Write the equations of the abstract model in terms of the given information, and state what you would do to obtain an optimum flywheel design if the desired coefficient of speed fluctuation is $C_s = 0.15$.

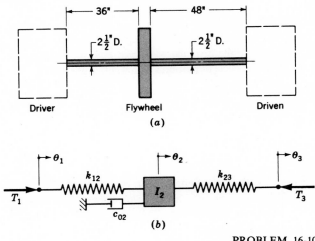

PROBLEM 16-10

ANSWERS TO SELECTED PROBLEMS

1-1 $S_y = 656$ MPa, $S_u = 1000$ MPa, 7.12 percent

2-1   (a) $\tau_{max} = 7$ kpsi;   (b) $\sigma_1 = 11.4$ kpsi, $\sigma_2 = -1.4$ kpsi, $\phi(x, \sigma_1) = 19.3°$ cw;
     (c) $\sigma_1 = 0$ kpsi, $\sigma_2 = -10$ kpsi, $\phi(x, \sigma_1) = 26.6°$ ccw; (d) $\sigma_1 = 10.2$ kpsi,
     $\sigma_2 = 4.8$ kpsi, $\phi(x, \sigma_1) = 10.9°$ cw

2-3   (a) $\sigma_1 = -11.5$ MPa, $\sigma_2 = -78.5$ MPa, $\phi(x, \sigma_1) = 31.7°$ ccw; (b) $\sigma_1 = 24$ MPa,
     $\sigma_2 = -104$ MPa, $\phi(x, \sigma_1) = 70.7°$ cw;   (c) $\sigma_1 = 88.9$ MPa, $\sigma_2 = 1.1$ MPa,
     $\phi(x, \sigma_1) = 23.4°$ ccw; (d) $\sigma_1 = 144.9$ MPa, $\sigma_2 = -24.9$ MPa, $\phi(x, \sigma_1) = 67.5°$ ccw

2-5 -0.989 GPa, $-4.78(10)^{-3}$ m/m, -0.382 mm, 0.112 mm

2-7   (a) 15.3 kpsi, 0.0204 in; (b) 0.853 in; (c) 5.24 kpsi

2-9   (a) Center bar, $\sigma_C = 1875$ psi; outer bars, $\sigma_O = -938$ psi; (b) $\sigma_C = 7180$ psi,
     $\sigma_O = 4400$ psi

2-11 $\sigma(\text{bolt}) = 0.241$ GPa, $\sigma(\text{tube}) = -0.863$ GPa

2-13 $\sigma_1 = 0.425$ GPa, $\sigma_2 = -0.0332$ GPa, $\epsilon_3 = -5.53(10)^{-4}$ m/m

2-15   (a) $R_1 = 0.3$ kN; (b) $M_1 = 55$ N·m; (c) $R_1 = 1500$ N; (d) $M_{max} = -0.6$ kN·m

2-17   (a) $M = R_1\langle x \rangle^1 - F\langle x - a \rangle^1 + R_2\langle x - l \rangle^1$;
     (b) $M = -M_1\langle x \rangle^0 + R_1\langle x \rangle^1 - (w/2)\langle x \rangle^2 + (w/2)\langle x - l \rangle^2$

2-19 (a) $M = -R_1\langle x\rangle^1 + R_2\langle x - (l/2)\rangle^1 - F\langle x - l\rangle^1$;

(b) $M = R_1\langle x\rangle^1 - (w/2)\langle x - a\rangle^2 + (w/2)\langle x - (l - a)\rangle^2 + R_2\langle x - l\rangle^1$

2-21 $1\frac{1}{2}$ in OD by $1\frac{1}{8}$ in ID

2-23 (a) $M_{max} = 180$ N·m at A; (b) 201 MPa; (c) 21 mm

2-25 At B, $\sigma_x = -16Fx/\pi d^3$; at C, $\tau_{xy} = 8F/3\pi d^2$ cw; at D, $\sigma_x = 16Fx/\pi d^3$

2-27 0.346 m

2-29 $d = 1\frac{1}{8}$ in

2-31 (a) $M_A = 420$ N·m, $M_B = 270$ N·m; (b) $\sigma_x = 131$ MPa, $\tau_{xz} = 93.3$ MPa; (c) $\phi =$ 27.4° from x to σ_1

2-33 (a) $\tau_{xz} = 71.3$ MPa, $\sigma_x = -47.5$ MPa; (b) $\sigma_1 = 51.4$ MPa, $\sigma_2 = -98.9$ MPa, $\phi = 54.2$° from x to σ_1

2-35 At A: $\sigma_x = 9.016$ kpsi, $\tau_{xz} = 12.7$ kpsi; at B: $\sigma_x = 60.9$ kpsi, $\tau_{xy} = 12.7$ kpsi; at C: $\sigma_x = -9.344$ kpsi, $\tau_{xz} = 12.7$ kpsi

2-37 (b) $\sigma_z = 18.5$ kpsi at B, $\tau_{max} = 12.9$ kpsi at the middle of the $1\frac{1}{4}$-in side; (c) $\sigma_x = 43.4$ kpsi, $\tau_{xz} = 14.5$ kpsi

2-39 At $r = a$, $\sigma_r = -4$ kpsi, $\sigma_t = 18.2$ kpsi, $\epsilon_t = 30.1(10)^{-4}$, $\epsilon_r = -15.9(10)^{-4}$, $\epsilon_l = -7.65(10)^{-4}$; at $r = b$, $\sigma_r = 0$ kpsi, $\sigma_t = 14.2$ kpsi, $\epsilon_t = 21.8(10)^{-4}$, $\epsilon_l = \epsilon_r = -7.65(10)^{-4}$

2-41 $\sigma_x = \sigma_y = p_i D/4t$, $\sigma_r = 0$

2-43 At $r = a$, $\sigma_t = 250$ MPa; at $r = b$, $\sigma_t = 400$ MPa

2-45 D(tire) = 15.980 in

2-47 $\sigma_i = 19.7$ kpsi, $\sigma_0 = -13.6$ kpsi

2-49 Maximum axial pull 161 lb; maximum bending force 98 lb

2-51 $\sigma_x = 0.42p_{max} = -31.4$ kpsi, $\sigma_y = 0.12p_{max} = -8.98$ kpsi, $\sigma_z = 0.70p_{max} = -52.4$ kpsi; the maximum shear stresses are 21.7 kpsi, 11.2 kpsi, and 10.5 kpsi all at $b = 0.0136$ in

3-1 $k = 16.6$ kip·in/rad

3-3 $k = 15.7$ kN/m

3-5 $k_A = (N_1/N_2)^2 k_1 k_2/[k_1 + (N_1/N_2)^2 k_2]$
$k_B = (N_2/N_1)^2 k_1 k_2/[(N_2/N_1)^2 k_1 + k_2]$

3-7 Use $d = 21$ mm, $\tau = 21.9$ MPa; the stress is rather small

3-9 0.215 in

3-11 Use a $2 \times 2 \times \frac{1}{8}$-in angle

3-13 $b = 21.7$ mm, $\sigma = 411$ MPa

3-15 $W = 84$ kip, $y_{max} = 0.0333$ in

3-17 0.0148 in

3-19 2.12 in

3-21 $F = 10.3$ kN

3-23 $\theta = (Tb/GJ_1) + (Ta/3EI_2)$

3-25 5.72 mm

3-27 $b = 3.79$, $(l/k)_1 = 94$

3-29 45.6 kN, 42.6 kN, 26.2 kN, 11.8 kN

3-31 53 kip

3-33 $\frac{5}{8}$ in, $\frac{3}{4}$ in, $1\frac{3}{8}$ in

3-35 $2\frac{5}{8}$ in, $1\frac{7}{8}$ in, $1\frac{3}{8}$ in

4-1 20

4-3 120, 24

4-5 6!

4-7 0.46

4-9 78.2 percent

4-11 8 ways; $\frac{1}{8}$; $\frac{15}{56}$

4-13 $\bar{S} = 49.5$ kpsi, $s_S = 3.82$ kpsi

4-15 380 psi

4-17 $c_{min} = 0.001$ in; 0.10 percent

5-1 (a) 4, 2.86, 3.2; (b) 3.51, 3.12, 3.28; (c) 4, 4, 4; (d) 3.93, 3.93, 4.52

5-3 (a) 5, 5, 5; (b) 5, 5, 5.78; (c) 5, 2.5, 2.89; (d) 5, 5, 5

5-5 410 lb

5-7 (a) $\sigma_x = 28.6$ kpsi, $\tau_{xy} = 18.33$ kpsi; (b) $n = 2.19$; (c) $n = 2.03$

5-9 8.38 MPa

5-11 $\sigma' = 81.8$ kpsi

5-13 Inner member, $\sigma' = 23.6$ kpsi; outer member, $\sigma' = 37.6$ kpsi

5-15 The factors of safety by the maximum normal stress theory, the Coulomb-Mohr theory, and the modified Mohr theory are respectively: (a) 2.34, 2.22, 2.34; (b) 6.80, 6.80, 6.80; (c) 3.25, 3.25, 3.25; (d) 3.84, 2.68, 3.27 using graphics

5-17 32.7 kpsi, $22.9(10)^3$ cycles

5-19 89 kpsi

5-21 58.2 kpsi

5-23 30.8 kN

5-25 $n(\text{fatigue}) = 1.24$; $n(\text{static}) = 1.43$

5-27 $n = 2.95$

5-29 $n = 1.63$

5-31 $d = 2\frac{1}{2}$ in

5-33 (a) 3.74; (b) 1.5; (c) 2

5-35 (a) 8.62 lb, 25.9 lb; (b) no, because $n = 1.27$

5-37 $n(\text{fatigue}) = 1.47$, $n(\text{static}) = 4.22$

5-39 $n(\text{fatigue}) = 2.84$, $n(\text{static}) = 2.66$

5-41 (a) $n(\text{fatigue}) = 1.06$, $n(\text{static}) = 1.66$; (b) $n(\text{fatigue}) = 2.31$, $n(\text{static}) = 1.15$; (c) $n(\text{fatigue}) = 1.18$, $n(\text{static}) = 1.34$; (d) $N = 46.4(10)^3$ cycles, $n(\text{static}) = 1.15$; (e) $n(\text{fatigue}) = 1.20$, $n(\text{static}) = 2$

5-43 $n = 1.47$

5-45 At a shaft angle of $60°$ $\sigma_1 = 3.9$ kpsi, $\sigma_2 = -9$ kpsi, $\phi = 56.3°$

6-1 $e_{max} \cong 0.93$ at $\psi = 35°$

6-3   (a) $T = 12.83$ lb·in; (b) $l = 4.46$ in knob to knob; $d = 0.144$ in

6-5   (a) 3.35 kW; (b) 0.125

6-7   Use 7 16-mm screws or 5 20-mm screws

6-9   Use $\frac{5}{16}''$-18UNC bolts

6-11   (a) 625 lb/bolt; (b) 1000 lb/bolt; (c) 5.93; (d) 8.66

6-13   (a) $S_e = 10.8$ kpsi; (b) $k_b = 8.83(10)^6$ lb/in, $k_m = 37.7(10)^6$ lb/in; (c) $n = 1.40$

6-15   (a) 6.75 kip based on bearing on member; (b) 13.5 kip based on bearing on member

6-17 $F = 5.85$ kN based on shear of bolt

6-19   Shear of bolt 5.35, bearing on bolt 8.85, bearing on member 5.63, strength of member 2.95

7-1   (a) 17.7 kip; (b) 35.4 kip; (c) 2.17 kip; (d) 4.33 kip

7-3   (a) 49.8 kN; (b) the weld has zero shear stress

7-5   (a) 19.5 kN; (b) 12.3 kN; (c) 90.4 kN

7-7   (a) $\tau_{max} = 10.8$ kpsi; (b) $\tau_{max} = 11$ kpsi at bottom

7-9   (a) $\tau_{max} = 2910$ psi; (b) $\tau_{max} = 17.6$ MPa

8-1   (a) 9.25 lb; (b) 1.73 in; (c) 5.35 lb/in; (d) 0.752 in; (e) 2.48 in

8-3   (a) 4.46 N; (b) 44.0 N; (c) 453 N/m; (d) 157 mm

8-5   (a) $k_o = 7.01$ lb/in, $k_i = 4.87$ lb/in; (b) 11.9 lb; (c) $\tau_o = 17.4$ kpsi, $\tau_i = 19.8$ kpsi

8-7   1.74 in, 8.82 lb

8-9 $d = 1.6$ mm, $l_F = 263$ mm, $l_S = 52.8$ mm

8-11   (a) 8.03 lb/in, 0.647 in, 75.5 kpsi; (b) 2.04

8-13 n(static) = 2.80, n(fatigue) = 1.81

8-15   (a) No, τ(solid) $< S_{sy}$; (b) 41.6 lb/in; (c) no; (d) yes

9-1   Use 25-mm bore 02-series bearings

9-3   Use 10-mm bore 02-series bearings

9-5   Use 35-mm bore 02-series bearings

9-7   4340 daN for O and 4130 daN for C

9-9 $C_R = 21.1$ kN at O; $C_R = 17.4$ kN at B

10-1   0.627 hp

10-3   1.37, 1.80, 2.31, and 2.79 hp

10-5   0.0238 Btu/s, 0.0798 in^3/s, 0.167 in^3/s, 0.000 86 in, 14.4°F

10-7 $H = 24.1$ J/s, $Q_s = 1350$ mm^3/s, $Q = 2740$ mm^3/s, $h_o = 21$ μm, $T_C = 7$°C

10-9   0.0109, 0.0351, 0.0723, 0.118 Btu/s

10-11 $T_F = 36$°F, $p_{max} = 181$ psi, $h_o = 157$ μin

10-13   54°F, 543 μin

10-15 59°C, 13.7 μm, 3360 kPa

10-17 (*a*) 0.0011 in at 71°; (*b*) 0.01; (*c*) 0.167 in³/s, 0.498 in³/s; (*d*) 321 psi at 11.7°; (*e*) about 104°; (*f*) 128°F, 156°F

10-19 78°F, $1.8(10)^{-4}$ in, 2820 psi

10-21 56°C, 3000 mm³/s, 5.63 μm

11-1 40T, 7.75 in

11-3 $p = 2\pi$ mm, $N_2 = 96$T, $C = 120$ mm

11-5 $q_a = 1.07$ in, $q_r = 0.99$ in, $p = 1.257$ in, $m_c = 1.64$

11-7 $p_b = 0.95$ in, $m_c = 1.34$

11-9 244.4 rpm ccw

11-11 114.3 rpm ccw

11-13 2530 lb

11-15 (*a*) 84T, 40T; (*b*) 491 rpm, 123 rpm; (*c*) 6.94 N·m, 23.3 N·m, 89.4 N·m

11-17 (*a*) $C_{ab} = 3\frac{1}{3}$ in, $C_{bc} = 6\frac{2}{3}$ in; (*b*) $F_{2a} = 1390$ lb, $F_{3b} = 1390$ lb, $F_{4b} = 1680$ lb; (*c*) $F_b = 2835$ lb

11-19 (*a*) 810 rpm, 778 lb·in; (*b*) 389 lb, 78 lb; (*c*) 467 lb·in; (*d*) 467 lb; (*e*) $\mathbf{R}_{Ab} = -292\mathbf{j} + 27.2\mathbf{k}$ lb, $\mathbf{R}_{Db} = 59\mathbf{j} + 81.8\mathbf{k}$ lb; (*f*) 1170 lb·in

11-21 14.1 kpsi

11-23 Use an 8-pitch 18-tooth pinion having a $2\frac{1}{8}$-in face or a 6-pitch 18-tooth pinion having a $1\frac{1}{2}$-in face

11-25 Use 8-pitch gears with a $2\frac{1}{8}$-in face width

11-27 1.27

11-29 53.0 kpsi

12-1 (*a*) $p_n = 0.393$ in, $p_t = 0.433$ in, $p_x = 0.929$ in, $p_N = 0.369$ in; (*b*) $P_t = 7.25$, $\phi_t = 21.88°$; (*c*) $a = 0.125$ in, $b = 0.156$ in, $d_P = 2.48$ in, $d_G = 4.41$ in

12-3 (*a*) $P_n = 4.14$, $p_t = 0.785$ in, $p_n = 0.758$ in, $p_x = 2.93$ in, $\phi_n = 19.4°$; (*b*) $a = 0.242$ in, $b = 0.302$ in, $d_P = 5$ in, $d_G = 16$ in

12-5 (*a*) $\mathbf{F}_{2a} = -168\mathbf{i} - 231\mathbf{j} + 400\mathbf{k}$, $\mathbf{F}_{3b} = +168\mathbf{i} + 231\mathbf{j} - 400\mathbf{k}$; (*b*) $\mathbf{F}_{2a} = -168\mathbf{i} + 231\mathbf{j} - 400\mathbf{k}$, $\mathbf{F}_{3b} = -800\mathbf{k}$, $\mathbf{F}_{4c} = +168\mathbf{i} - 231\mathbf{j} - 400\mathbf{k}$, all answers in lb

12-7 $\mathbf{F}_{2a} = -266\mathbf{i} - 96.8\mathbf{j} - 71.3\mathbf{k}$ lb, $\mathbf{F}_{3b} = 169\mathbf{i} - 169\mathbf{j}$ lb, $\mathbf{F}_{4c} = 96.8\mathbf{i} + 266\mathbf{j} + 71.3\mathbf{k}$ lb

12-9 $\mathbf{F}_C = 1230\mathbf{i} + 684\mathbf{j} - 1080\mathbf{k}$ lb, $\mathbf{F}_D = 654\mathbf{i} + 618\mathbf{j}$ lb

12-11 570 lb

12-13 140 lb

12-15 47.1 rpm cw

12-17 $\mu = 0.042$, $\mathbf{T} = 284\mathbf{k}$ lb·in, $\mathbf{F}_A = 71\mathbf{i} - 422\mathbf{j} - 1110\mathbf{k}$ lb, $\mathbf{F}_B = 71\mathbf{i} + 132\mathbf{j}$ lb

12-19 1.04 hp, 83 percent

12-21 0.343 hp, 84 percent

12-23 756 rpm

12-25 $F_E = 163i - 192j + 355k$ lb, $F_F = 110j + 145k$ lb, $T = -1030i$ lb·in

12-27 $F_C = 239i + 87k$ lb, $F_D = 563i - 239j + 1250k$ lb

12-29 Assuming gear is weakest member and using $K_v = 0.675$, the horsepower is 1.60 assuming two-way bending

12-31 32 hp using $K_v = 0.67$

13-1 Either 1 in or $1\frac{1}{16}$ in is satisfactory

13-3 $2\frac{5}{8}$ in

13-5 21 mm

13-7 $n = 1.31$

13-9 The combined moments are $M_C = 58.9$ kip·in and $M_F = 51.1$ kip·in; $n = 3.16$ at the keyway; $n = 3.21$ at the shoulder; $K_t = 1.38$ used at the keyway because the torsion is steady

14-1 (a) 538 lb; (b) 4130 lb·in; (c) 1130 lb on right shoe, 230 lb on left shoe

14-3 (a) 0.947 kN; (b) 183 N·m; (c) 2.16 kN on right shoe, 0.366 kN on left shoe

14-5 (a) 1.63 kN; (b) 151 N·m; (c) 3960 J

14-7 1.86 in, 24.9 kip·in

14-9 0.622 kN, 532 N·m

14-11 1920 lb, 627 lb, 24.6 lb

14-13 718 N·m

14-15 955 kPa

14-17 1510 lb, 1230 lb·in

14-19 (a) 140 lb, 10.7 psi; (b) 139 lb, 11 psi

14-21 $d = 3\frac{1}{2}$ in, $b = 1.11$ in, $F = 137$ lb

14-23 $\Delta T = 227°C$

14-25 $E_k = 1650$ J, $T = 310$ N·m

15-1 (a) 350 lb, 169 lb; (b) 445 lb, 264 lb; (c) 230 in

15-3 14.2 in

15-5 6.98 hp, 18 in

15-7 22.2 hp, 698 lb, 11.5, 135p

15-9 Chain tensions on 17T sprocket are 1200 and 2400 lb; shaft reaction is 3210 lb

15-11 41.4 kpsi, 2.56, $d_w = d/13$

15-13 2.15

16-1 $5.75(10)^4$ lb·in/rad

16-3 $k_{AB} = 14.1(10)^4$ lb·in/rad, $k_{12} = 286(10)^4$ lb·in/rad, $k_G = 13.25(10)^6$ lb/in, $k_{23} = 134(10)^6$ lb·in/rad, $k_{CD} = k_{34} = 35.8$ lb·in/rad, $I_1 = 2.22$ lb·s²·in, $I_2 = 0.0222$ lb·s²·in, $I_3 = I_C$, $I_4 = I_D$, $\theta_1 = 0.222\theta_A$, $\theta_2 = 0.222\theta_B$, $\theta_3 = \theta_C$, $\theta_4 = \theta_D$, $T_1 = 428$ lb·in

16-5 $n_D = 479$ rpm, $n_A = 1720$ rpm, $T_A = 75$ lb·in; motor-shaft torsional stress is
      $\tau = 2.61(10)^6(\theta_1 - \theta_2)$ psi; belt pull is $P = 1080(\theta_2 - \theta_3)$ lb; fan-shaft torsional
      stress is $\tau = 71.5(10)^4(\theta_3 - \theta_4)$ psi

16-7

$$\sum T_1 = T_1 - k_{12}(\theta_1 - \theta_2) - c_{01}\dot{\theta}_1 - I_1\ddot{\theta}_1 = 0$$

$$\sum T_2 = \begin{cases} k_{12}(\theta_1 - \theta_2) - c_{02}\dot{\theta}_2 - k_{23}(\theta_2 - \theta_3 - \alpha) = 0 & \theta_2 - \theta_3 > \alpha \\ k_{12}(\theta_1 - \theta_2) - c_{02}\dot{\theta}_2 = 0 & 0 \le \theta_2 - \theta_3 \le \alpha \\ k_{12}(\theta_1 - \theta_2) - c_{02}\dot{\theta}_2 - k_{23}(\theta_2 - \theta_3) = 0 & \theta_2 - \theta_3 < 0 \end{cases}$$

$$\sum T_3 = \begin{cases} k_{23}(\theta_2 - \theta_3 - \alpha) - c_{03}\dot{\theta}_3 - k_{34}(\theta_3 - \theta_4) - I_3\ddot{\theta}_3 = 0 & \theta_2 - \theta_3 > \alpha \\ -c_{03}\dot{\theta}_3 - k_{34}(\theta_3 - \theta_4) - I_3\ddot{\theta}_3 = 0 & 0 \le \theta_2 - \theta_3 \le \alpha \\ k_{23}(\theta_2 - \theta_3) - c_{03}\dot{\theta}_3 - k_{34}(\theta_3 - \theta_4) - I_3\ddot{\theta}_3 = 0 & \theta_2 - \theta_3 < 0 \end{cases}$$

$$\sum T_4 = k_{34}(\theta_3 - \theta_4) - c_{04}\dot{\theta}_4 - T_4 - I_4\ddot{\theta}_4 = 0$$

16-9

$$\sum F_1 = -m_1\ddot{x}_1 = 0$$
$$\sum F_2 = -m_2\ddot{x}_2 = 0 \qquad x_1 < x_2$$

$$\sum F_1 = -k(x_1 - x_2) - c(\dot{x}_1 - \dot{x}_2) - m_1\ddot{x}_1 = 0$$
$$\sum F_2 = k(x_1 - x_2) + c(\dot{x}_1 - \dot{x}_2) - m_2\ddot{x}_2 = 0 \qquad x_1 \ge x_2$$

APPENDIX

TABLES

Table A-1 STANDARD SI PREFIXES*,†

Name	Symbol	Factor
tera	T	$1\,000\,000\,000\,000 = 10^{12}$
giga	G	$1\,000\,000\,000 = 10^9$
mega	M	$1\,000\,000 = 10^6$
kilo	k	$1\,000 = 10^3$
hecto‡	h	$100 = 10^2$
deka‡	da	$10 = 10^1$
deci‡	d	$0.1 = 10^{-1}$
centi‡	c	$0.01 = 10^{-2}$
milli	m	$0.001 = 10^{-3}$
micro	μ	$0.000\,001 = 10^{-6}$
nano	n	$0.000\,000\,001 = 10^{-9}$
pico	p	$0.000\,000\,000\,001 = 10^{-12}$
femto	f	$0.000\,000\,000\,000\,001 = 10^{-15}$
atto	a	$0.000\,000\,000\,000\,000\,001 = 10^{-18}$

* If possible use multiple and submultiple prefixes in steps of 1000.
† Spaces are used in SI instead of commas to group numbers to avoid confusion with the practice in some European countries of using commas for decimal points.
‡ Not recommended but sometimes encountered.

Table A-2   CONVERSION OF ENGLISH UNITS TO SI UNITS

To convert from	To	Multiply by	
		Accurate	Common
foot (ft)	metre (m)	3.048 000*E − 01	0.305
foot-pound-force (ft·lbf)	joule (J)	1.355 818 E + 00	1.35
foot-pound-force/second (ft·lbf/s)	watt (W)	1.355 818 E + 00	1.35
inch (in)	metre (m)	2.540 000*E − 02	0.0254
gallon (gal U.S.)	metre³ (m³)	3.785 412 E − 03	0.003 78
horsepower (hp)	kilowatt (kW)	7.456 999 E − 01	0.746
mile (mi U.S. Statute)	kilometre (km)	1.609 344*E + 00	1.610
pascal (Pa)	newton/metre² (N/m²)	1.000 000*E + 00	1
pound-force (lbf avoirdupois)	newton (N)	4.448 222 E + 00	4.45
pound-mass (lbm avoirdupois)	kilogram (kg)	4.535 924 E − 01	0.454
pound-force/foot² (lbf/ft²)	pascal (Pa)	4.788 026 E + 01	47.9
pound-force/inch² (psi)	pascal (Pa)	6.894 757 E + 03	6890
slug	kilogram (kg)	1.459 390 E + 01	14.6
ton (short 2000 lbm)	kilogram (kg)	9.071 847 E + 02	907

* Exact

Table A-3   CONVERSION OF SI UNITS TO ENGLISH UNITS

To convert from	To	Multiply by	
		Accurate	Common
joule (J)	foot-pound-force (ft·lb)	7.375 620 E − 01	0.737
kilogram (kg)	slug	6.852 178 E − 02	0.0685
kilogram (kg)	pound-mass (lbm avoirdupois)	2.204 622 E + 00	2.20
kilogram (kg)	ton (short 2000 lbm)	1.102 311 E − 03	0.001 10
kilometre (km)	mile (mi U.S. Statute)	6.213 712 E − 01	0.621
kilowatt (kW)	horsepower (hp)	1.341 022 E + 00	1.34
metre (m)	foot (ft)	3.280 840 E + 00	3.28
metre (m)	inch (in)	3.937 008 E + 02	39.4
metre³ (m³)	gallon (gal U.S.)	2.641 720 E + 02	264
newton (N)	pound-force (lb avoirdupois)	2.248 089 E − 01	0.225
pascal (Pa)	pound-force/foot² (lb/ft²)	2.088 543 E − 02	0.0209
pascal (Pa)	pound-force/inch² (psi)	1.450 370 E − 04	0.000 145
watt (W)	foot-pound-force/second (ft·lb/s)	7.375 620 E − 01	0.737

Table A-4 PREFERRED SI UNITS FOR BENDING STRESS $\sigma = Mc/I$ AND TORSION STRESS $\tau = Tr/J$

M, T	I, J	c, r	σ, τ
N·m	m^4	m	Pa
N·m	cm^4	cm	MPa
N·m	mm^4	mm	GPa
kN·m	m^4	m	kPa
kN·m	cm^4	cm	GPa
kN·m	mm^4	mm	TPa
N·mm	mm^4	mm	MPa
kN·mm	mm^4	mm	GPa

Table A-5 PREFERRED SI UNITS FOR AXIAL STRESS $\sigma = F/A$ AND DIRECT SHEAR STRESS $\tau = F/A$

F	A	σ, τ
N	m^2	Pa
N	mm^2	MPa
kN	m^2	kPa
kN	mm^2	GPa

Table A-6 PREFERRED SI UNITS FOR THE DEFLECTION OF BEAMS $y = f(Fl^3/EI)$ OR $y = f(wl^4/EI)$

F, wl	l	I	E	y
N	m	m^4	MPa	μm
kN	m	m^4	GPa	μm
N	cm	mm^4	GPa	mm
N	mm	mm^4	GPa	μm

Table A-7 PHYSICAL CONSTANTS OF MATERIALS

Material	Modulus of elasticity, E		Modulus of rigidity, G		Poisson's ratio	Unit weight, w		
	Mpsi	GPa	Mpsi	GPa		lb/in^3	lb/ft^3	kN/m^3
Aluminum (all alloys)	10.3	71.0	3.80	26.2	0.334	0.098	169	26.6
Beryllium copper	18.0	124.0	7.0	48.3	0.285	0.297	513	80.6
Brass	15.4	106.0	5.82	40.1	0.324	0.309	534	83.8
Carbon steel	30.0	207.0	11.5	79.3	0.292	0.282	487	76.5
Cast iron, gray	14.5	100.0	6.0	41.4	0.211	0.260	450	70.6
Copper	17.2	119.0	6.49	44.7	0.326	0.322	556	87.3
Douglas fir	1.6	11.0	0.6	4.1	0.33	0.016	28	4.3
Glass	6.7	46.2	2.7	18.6	0.245	0.094	162	25.4
Inconel	31.0	214.0	11.0	75.8	0.290	0.307	530	83.3
Lead	5.3	36.5	1.9	13.1	0.425	0.411	710	111.5
Magnesium	6.5	44.8	2.4	16.5	0.350	0.065	112	17.6
Molybdenum	48.0	331.0	17.0	117.0	0.307	0.368	636	100.0
Monel metal	26.0	179.0	9.5	65.5	0.320	0.319	551	86.6
Nickel silver	18.5	127.0	7.0	48.3	0.322	0.316	546	85.8
Nickel steel	30.0	207.0	11.5	79.3	0.291	0.280	484	76.0
Phosphor bronze	16.1	111.0	6.0	41.4	0.349	0.295	510	80.1
Stainless steel (18-8)	27.6	190.0	10.6	73.1	0.305	0.280	484	76.0

Table A-8 PROPERTIES OF STRUCTURAL SHAPES—EQUAL ANGLES

w_a = weight per foot of aluminum sections, lb
w_s = weight per foot of steel sections, lb
A = area, in^2
I = moment of inertia, in^4
k = radius of gyration, in
y = centroidal distance, in
Z = section modulus, in^3

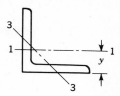

Size	w_a	w_s	A	I_{1-1}	k_{1-1}	Z_{1-1}	y	I_{3-3}	k_{3-3}
$1 \times 1 \times \frac{1}{8}$	0.28	0.80	0.23	0.02	0.30	0.03	0.30	0.008	0.19
$1 \times 1 \times \frac{1}{4}$	0.53	1.49	0.44	0.04	0.29	0.05	0.34	0.016	0.19
$1\frac{1}{2} \times 1\frac{1}{2} \times \frac{1}{8}$	0.44	1.23	0.36	0.07	0.45	0.07	0.41	0.031	0.29
$1\frac{1}{2} \times 1\frac{1}{2} \times \frac{1}{4}$	0.83	2.34	0.69	0.14	0.44	0.13	0.46	0.057	0.29
$2 \times 2 \times \frac{1}{8}$	0.59	1.65	0.49	0.18	0.61	0.13	0.53	0.08	0.40
$2 \times 2 \times \frac{1}{4}$	1.14	3.19	0.94	0.34	0.60	0.24	0.58	0.14	0.39
$2 \times 2 \times \frac{3}{8}$	1.65	4.70	1.37	0.47	0.59	0.35	0.63	0.20	0.39
$2\frac{1}{2} \times 2\frac{1}{2} \times \frac{1}{4}$	1.45	4.1	1.19	0.69	0.76	0.39	0.71	0.29	0.49
$2\frac{1}{2} \times 2\frac{1}{2} \times \frac{3}{8}$	2.11	5.9	1.74	0.98	0.75	0.56	0.76	0.41	0.48
$3 \times 3 \times \frac{1}{4}$	1.73	4.9	1.43	1.18	0.91	0.54	0.82	0.49	0.58
$3 \times 3 \times \frac{3}{8}$	2.55	7.2	2.10	1.70	0.90	0.80	0.87	0.70	0.58
$3 \times 3 \times \frac{1}{2}$	3.32	9.4	2.74	2.16	0.89	1.04	0.92	0.91	0.58
$3\frac{1}{2} \times 3\frac{1}{2} \times \frac{1}{4}$	2.05	4.9	1.69	1.93	1.07	0.76	0.94	0.80	0.69
$3\frac{1}{2} \times 3\frac{1}{2} \times \frac{3}{8}$	3.01	7.2	2.49	2.79	1.06	1.11	1.00	1.15	0.68
$3\frac{1}{2} \times 3\frac{1}{2} \times \frac{1}{2}$	3.94	11.1	3.25	3.56	1.05	1.45	1.05	1.49	0.68
$4 \times 4 \times \frac{1}{4}$	2.35	6.6	1.94	2.94	1.23	1.00	1.07	1.21	0.79
$4 \times 4 \times \frac{3}{8}$	3.46	9.8	2.86	4.26	1.22	1.48	1.12	1.75	0.78
$4 \times 4 \times \frac{1}{2}$	4.54	12.8	3.75	5.46	1.21	1.93	1.17	2.26	0.78
$4 \times 4 \times \frac{5}{8}$	5.58	15.7	4.61	6.56	1.19	2.36	1.22	2.76	0.77
$6 \times 6 \times \frac{3}{8}$	5.27	14.9	4.35	14.85	1.85	3.38	1.60	6.07	1.18
$6 \times 6 \times \frac{1}{2}$	6.95	19.6	5.74	19.38	1.84	4.46	1.66	7.92	1.17
$6 \times 6 \times \frac{5}{8}$	8.59	24.2	7.10	23.64	1.82	5.51	1.71	9.70	1.17
$6 \times 6 \times \frac{3}{4}$	10.20	28.7	8.43	27.64	1.81	6.52	1.76	11.43	1.16

Table A-9 PROPERTIES OF STRUCTURAL SHAPES—UNEQUAL ANGLES

w_a = weight per foot of aluminum sections, lb
w_s = weight per foot of steel sections, lb
A = area, in^2
I = moment of inertia, in^4
k = radius of gyration, in
x and y = respective centroidal distances, in
Z = section modulus, in^3

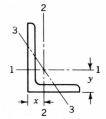

Size	w_a	w_s	A	I_{1-1}	k_{1-1}	Z_{1-1}	y	I_{2-2}	k_{2-2}	Z_{2-2}	x	I_{3-3}	k_{3-3}
$2 \times 1\frac{1}{2} \times \frac{1}{8}$	0.51	1.44	0.42	0.17	0.63	0.12	0.60	0.08	0.44	0.07	0.36	0.04	0.32
$2 \times 1\frac{1}{2} \times \frac{1}{4}$	0.98	2.77	0.81	0.31	0.62	0.23	0.66	0.15	0.43	0.14	0.41	0.08	0.32
$3 \times 2 \times \frac{3}{16}$	1.10	3.07	0.91	0.82	0.95	0.40	0.94	0.29	0.56	0.19	0.46	0.17	0.43
$3 \times 2\frac{1}{2} \times \frac{1}{4}$	1.58	4.5	1.31	1.12	0.92	0.53	0.89	0.70	0.73	0.38	0.64	0.35	0.52
$3 \times 2\frac{1}{2} \times \frac{3}{8}$	2.32	6.6	1.92	1.60	0.91	0.78	0.94	1.00	0.72	0.55	0.69	0.51	0.51
$3 \times 2\frac{1}{2} \times \frac{1}{2}$	3.02	9.4	2.49	2.03	0.90	1.01	0.99	1.26	0.71	0.72	0.74	0.65	0.51
$4 \times 3 \times \frac{1}{4}$	2.05	5.8	1.69	2.68	1.26	0.96	1.21	1.29	0.87	0.56	0.72	0.70	0.64
$4 \times 3 \times \frac{1}{2}$	3.95	11.1	3.25	4.96	1.24	1.85	1.31	2.36	0.85	1.08	0.82	1.30	0.63
$6 \times 4 \times \frac{3}{8}$	4.36	12.3	3.60	13.02	1.90	3.17	1.90	4.63	1.13	1.50	0.91	2.67	0.86
$6 \times 4 \times \frac{1}{2}$	5.74	16.2	4.74	16.95	1.89	4.19	1.96	6.01	1.13	1.98	0.97	3.47	0.86

Table A-10 PROPERTIES OF ROUND TUBING

w_a = weight per foot of aluminum tubing, lb/ft
w_s = weight per foot of steel tubing, lb/ft
A = area, in^2
I = moment of inertia, in^4
k = radius of gyration, in
Z = section modulus, in^3

Size	w_a	w_s	A	I	k	Z
$1 \times \frac{1}{8}$	0.416	1.128	0.344	0.034	0.313	0.067
$1 \times \frac{1}{4}$	0.713	2.003	0.589	0.046	0.280	0.092
$1\frac{1}{2} \times \frac{1}{8}$	0.653	1.769	0.540	0.129	0.488	0.172
$1\frac{1}{2} \times \frac{1}{4}$	1.188	3.338	0.982	0.199	0.451	0.266
$2 \times \frac{1}{8}$	0.891	2.670	0.736	0.325	0.664	0.325
$2 \times \frac{1}{4}$	1.663	4.673	1.374	0.537	0.625	0.537
$2\frac{1}{2} \times \frac{1}{8}$	1.129	3.050	0.933	0.660	0.841	0.528
$2\frac{1}{2} \times \frac{1}{4}$	2.138	6.008	1.767	1.132	0.800	0.906
$3 \times \frac{1}{4}$	2.614	7.343	2.160	2.059	0.976	1.373
$3 \times \frac{3}{8}$	3.742	10.51	3.093	2.718	0.938	1.812
$4 \times \frac{3}{16}$	2.717	7.654	2.246	4.090	1.350	2.045
$4 \times \frac{3}{8}$	5.167	14.52	4.271	7.090	1.289	3.544

Table A-11 PROPERTIES OF STRUCTURAL SHAPES—CHANNELS

w_a = weight per foot of aluminum sections, lb
w_s = weight per foot of steel sections, lb
A = area, in^2
I = moment of inertia, in^4
k = radius of gyration, in
x = centroidal distance, in
Z = section modulus, in^3

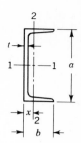

a	b	t	A	w_a	w_s	I_{1-1}	k_{1-1}	Z_{1-1}	I_{2-2}	k_{2-2}	Z_{2-2}	x
3	1.410	0.170	1.21	1.46	4.1	1.66	1.17	1.10	0.20	0.40	0.20	0.44
3	1.498	0.258	1.47	1.78	5.0	1.85	1.12	1.24	0.25	0.41	0.23	0.44
3	1.596	0.356	1.76	2.13	6.0	2.07	1.08	1.38	0.31	0.42	0.27	0.46
4	1.580	0.180	1.57	1.90	5.4	3.83	1.56	1.92	0.32	0.45	0.28	0.46
4	1.720	0.320	2.13	2.58	7.25	4.58	1.47	2.29	0.43	0.45	0.34	0.46
5	1.750	0.190	1.97	2.38	6.7	7.49	1.95	3.00	0.48	0.49	0.38	0.48
5	1.885	0.325	2.64	3.20	9.0	8.90	1.83	3.56	0.63	0.49	0.45	0.48
6	1.920	0.200	2.40	2.91	8.2	13.12	2.34	4.37	0.69	0.54	0.49	0.51
6	2.034	0.314	3.09	3.73	10.5	15.18	2.22	5.06	0.87	0.53	0.56	0.50
6	2.157	0.437	3.82	4.63	13.0	17.39	2.13	5.80	1.05	0.52	0.64	0.51
7	2.090	0.210	2.87	3.47	9.8	21.27	2.72	6.08	0.97	0.58	0.63	0.54
7	2.194	0.314	3.60	4.36	12.25	24.24	2.60	6.93	1.17	0.57	0.70	0.52
7	2.299	0.419	4.33	5.24	14.75	27.24	2.51	7.78	1.38	0.56	0.78	0.53
8	2.260	0.220	3.36	4.10	11.5	32.30	3.10	8.10	1.30	0.63	0.79	0.58
8	2.343	0.303	4.04	4.89	13.75	36.11	2.99	9.03	1.53	0.61	0.85	0.55
8	2.527	0.487	5.51	6.67	18.75	43.96	2.82	10.99	1.98	0.60	1.01	0.57
9	2.430	0.230	3.91	4.74	13.4	47.68	3.49	10.60	1.75	0.67	0.96	0.60
9	2.485	0.285	4.41	5.34	15.0	51.02	3.40	11.34	1.93	0.66	1.01	0.59
9	2.648	0.448	5.88	7.11	20.0	60.92	3.22	13.54	2.42	0.64	1.17	0.58
10	2.600	0.240	4.49	5.43	15.3	67.37	3.87	13.47	2.28	0.71	1.16	0.63
10	2.739	0.379	5.88	7.11	20.0	78.95	3.66	15.79	2.81	0.69	1.32	0.61
10	2.886	0.526	7.35	8.89	25.0	91.20	3.52	18.24	3.36	0.68	1.48	0.62
10	3.033	0.673	8.82	10.67	30.0	103.45	3.43	20.69	3.95	0.67	1.66	0.65
12	3.047	0.387	7.35	8.89	25.0	144.37	4.43	24.06	4.47	0.78	1.89	0.67
12	3.170	0.510	8.82	10.67	30.0	162.08	4.29	27.01	5.14	0.76	2.06	0.67

Table A-12 SHEAR, MOMENT, AND DEFLECTION OF BEAMS

1. Cantilever—end load

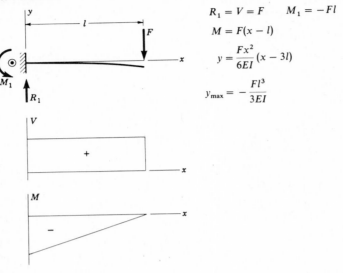

$$R_1 = V = F \qquad M_1 = -Fl$$

$$M = F(x - l)$$

$$y = \frac{Fx^2}{6EI}(x - 3l)$$

$$y_{max} = -\frac{Fl^3}{3EI}$$

2. Cantilever—intermediate load

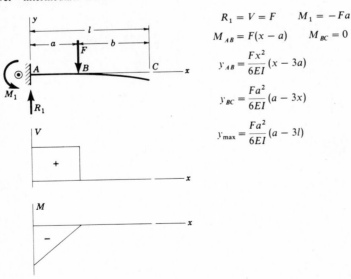

$$R_1 = V = F \qquad M_1 = -Fa$$

$$M_{AB} = F(x - a) \qquad M_{BC} = 0$$

$$y_{AB} = \frac{Fx^2}{6EI}(x - 3a)$$

$$y_{BC} = \frac{Fa^2}{6EI}(a - 3x)$$

$$y_{max} = \frac{Fa^2}{6EI}(a - 3l)$$

Table A-12 SHEAR, MOMENT, AND DEFLECTION OF BEAMS (*continued*)

3. Cantilever—uniform load

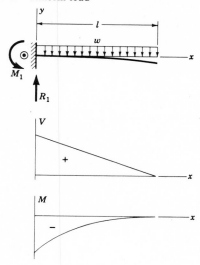

$$R_1 = wl \qquad M_1 = -\frac{wl^2}{2}$$

$$V = w(l - x) \qquad M = -\frac{w}{2}(l - x)^2$$

$$y = \frac{wx^2}{24EI}(4lx - x^2 - 6l^2)$$

$$y_{max} = -\frac{wl^4}{8EI}$$

4. Cantilever—moment load

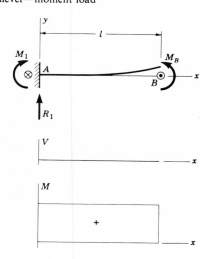

$$R_1 = 0 \qquad M_1 = M_B \qquad M = M_B$$

$$y = \frac{M_B x^2}{2EI} \qquad y_{max} = \frac{M_B l^2}{2EI}$$

Table A-12 SHEAR, MOMENT, AND DEFLECTION OF BEAMS (*continued*)

5. Simple supports—center load

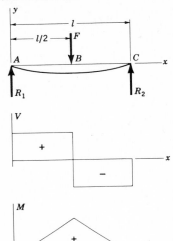

$$R_1 = R_2 = \frac{F}{2} \qquad V_{AB} = R_1$$

$$V_{BC} = -R_2$$

$$M_{AB} = \frac{Fx}{2} \qquad M_{BC} = \frac{F}{2}(l - x)$$

$$y_{AB} = \frac{Fx}{48EI}(4x^2 - 3l^2)$$

$$y_{max} = -\frac{Fl^3}{48EI}$$

6. Simple supports—intermediate load

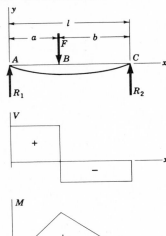

$$R_1 = \frac{Fb}{l} \qquad R_2 = \frac{Fa}{l} \qquad V_{AB} = R_1$$

$$V_{BC} = -R_2$$

$$M_{AB} = \frac{Fbx}{l} \qquad M_{BC} = \frac{Fa}{l}(l - x)$$

$$y_{AB} = \frac{Fbx}{6EIl}(x^2 + b^2 - l^2)$$

$$y_{BC} = \frac{Fa(l - x)}{6EIl}(x^2 + a^2 - 2lx)$$

Table A-12 SHEAR, MOMENT, AND DEFLECTION OF BEAMS (*continued*)

7. Simple supports—uniform load

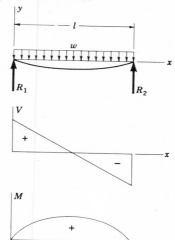

$$R_1 = R_2 = \frac{wl}{2} \qquad V = \frac{wl}{2} - wx$$

$$M = \frac{wx}{2}(l - x)$$

$$y = \frac{wx}{24EI}(2lx^2 - x^3 - l^3)$$

$$y_{max} = -\frac{5wl^4}{384EI}$$

8. Simple supports—moment load

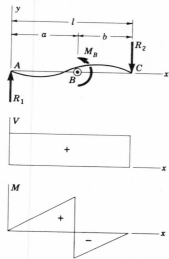

$$R_1 = -R_2 = \frac{M_B}{l} \qquad V = \frac{M_B}{l}$$

$$M_{AB} = \frac{M_B x}{l} \qquad M_{BC} = \frac{M_B}{l}(x - l)$$

$$y_{AB} = \frac{M_B x}{6EIl}(x^2 + 3a^2 - 6al + 2l^2)$$

$$y_{BC} = \frac{M_B}{6EIl}[x^3 - 3lx^2 + x(2l^2 + 3a^2) - 3a^2l]$$

Table A-12   SHEAR, MOMENT, AND DEFLECTION OF BEAMS (*continued*)

9. Simple supports—twin loads

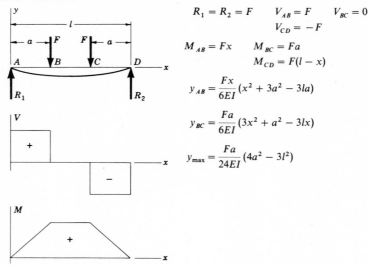

$$R_1 = R_2 = F \qquad V_{AB} = F \qquad V_{BC} = 0$$
$$V_{CD} = -F$$

$$M_{AB} = Fx \qquad M_{BC} = Fa$$
$$M_{CD} = F(l - x)$$

$$y_{AB} = \frac{Fx}{6EI}(x^2 + 3a^2 - 3la)$$

$$y_{BC} = \frac{Fa}{6EI}(3x^2 + a^2 - 3lx)$$

$$y_{max} = \frac{Fa}{24EI}(4a^2 - 3l^2)$$

10. Simple supports—overhanging load

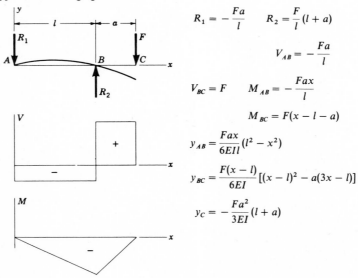

$$R_1 = -\frac{Fa}{l} \qquad R_2 = \frac{F}{l}(l + a)$$

$$V_{AB} = -\frac{Fa}{l}$$

$$V_{BC} = F \qquad M_{AB} = -\frac{Fax}{l}$$

$$M_{BC} = F(x - l - a)$$

$$y_{AB} = \frac{Fax}{6EIl}(l^2 - x^2)$$

$$y_{BC} = \frac{F(x - l)}{6EI}[(x - l)^2 - a(3x - l)]$$

$$y_C = -\frac{Fa^2}{3EI}(l + a)$$

Table A-12   SHEAR, MOMENT, AND DEFLECTION OF BEAMS (*continued*)

11. One fixed and one simple support
    —center load

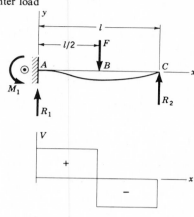

$$R_1 = \frac{11F}{16} \qquad R_2 = \frac{5F}{16} \qquad M_1 = -\frac{3Fl}{16}$$

$$V_{AB} = R_1 \qquad V_{BC} = -R_2$$

$$M_{AB} = \frac{F}{16}(11x - 3l) \qquad M_{BC} = \frac{5F}{16}(l - x)$$

$$y_{AB} = \frac{Fx^2}{96EI}(11x - 9l)$$

$$y_{BC} = \frac{F(l - x)}{96EI}(5x^2 + 2l^2 - 10lx)$$

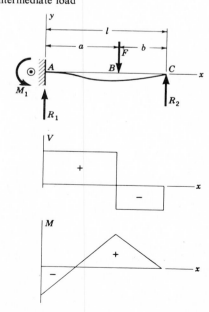

12. One fixed and one simple support
    —intermediate load

$$R_1 = \frac{Fb}{2l^3}(3l^2 - b^2) \qquad R_2 = \frac{Fa^2}{2l^3}(3l - a)$$

$$M_1 = \frac{Fb}{2l^2}(b^2 - l^2) \qquad V_{AB} = R_1$$

$$V_{BC} = -R_2$$

$$M_{AB} = \frac{Fb}{2l^3}[b^2l - l^3 + x(3l^2 - b^2)]$$

$$M_{BC} = \frac{Fa^2}{2l^3}(3l^2 - 3lx - al + ax)$$

$$y_{AB} = \frac{Fbx^2}{12EIl^3}[3l(b^2 - l^2) + x(3l^2 - b^2)]$$

$$y_{BC} = y_{AB} - \frac{F(x - a)^3}{6EI}$$

Table A-12 SHEAR, MOMENT, AND DEFLECTION OF BEAMS (*continued*)

13. One fixed and one simple support
    —uniform load

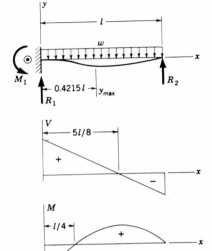

$$R_1 = \frac{5wl}{8} \qquad R_2 = \frac{3wl}{8} \qquad M_1 = -\frac{wl^2}{8}$$

$$V = \frac{5wl}{8} - wx$$

$$M = \frac{w}{8}(4x^2 + 5lx - l^2)$$

$$y = \frac{wx^2}{48EI}(l - x)(2x - 3l)$$

$$y_{max} = -\frac{wl^4}{185EI}$$

14. Fixed supports—center load

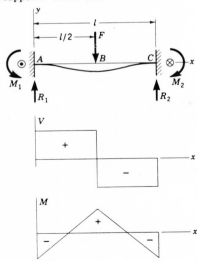

$$R_1 = R_2 = \frac{F}{2} \qquad M_1 = M_2 = -\frac{Fl}{8}$$

$$V_{AB} = -V_{BC} = \frac{F}{2}$$

$$M_{AB} = \frac{F}{8}(4x - l) \qquad M_{BC} = \frac{F}{8}(3l - 4x)$$

$$y_{AB} = \frac{Fx^2}{48EI}(4x - 3l)$$

$$y_{max} = -\frac{Fl^3}{192EI}$$

Table A-12 SHEAR, MOMENT, AND DEFLECTION OF BEAMS (*continued*)

15. Fixed supports—intermediate load

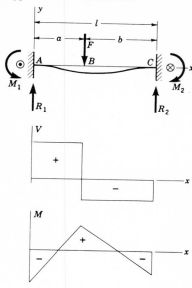

$$R_1 = \frac{Fb^2}{l^3}(3a + b) \qquad R_2 = \frac{Fa^2}{l^3}(3b + a)$$

$$M_1 = -\frac{Fab^2}{l^2} \qquad M_2 = -\frac{Fa^2b}{l^2}$$

$$V_{AB} = R_1$$

$$V_{BC} = -R_2$$

$$M_{AB} = \frac{Fb^2}{l^3}[x(3a + b) - la]$$

$$M_{BC} = M_{AB} - F(x - a)$$

$$y_{AB} = \frac{Fb^2x^2}{6EIl^3}[x(3a + b) - 3al]$$

$$y_{BC} = \frac{Fa^2(l - x)^2}{6EIl^3}[(l - x)(3b + a) - 3bl]$$

16. Fixed supports—uniform load

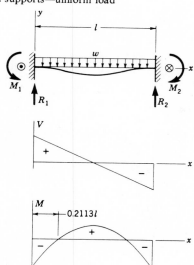

$$R_1 = R_2 = \frac{wl}{2} \qquad M_1 = M_2 = -\frac{wl^2}{12}$$

$$V = \frac{w}{2}(l - 2x)$$

$$M = \frac{w}{12}(6lx - 6x^2 - l^2)$$

$$y = -\frac{wx^2}{24EI}(l - x)^2$$

$$y_{max} = -\frac{wl^4}{384EI}$$

Table A-13 ORDINATES OF THE STANDARD NORMAL DISTRIBUTION

$$f(z) = \frac{1}{\sqrt{2\pi}} e^{-z^2/2}$$

z	0	1	2	3	4	5	6	7	8	9
0.0	0.3989	0.3989	0.3989	0.3988	0.3986	0.3984	0.3982	0.3980	0.3977	0.3973
0.1	0.3970	0.3965	0.3961	0.3956	0.3951	0.3945	0.3939	0.3932	0.3925	0.3918
0.2	0.3910	0.3902	0.3894	0.3885	0.3876	0.3867	0.3857	0.3847	0.3836	0.3825
0.3	0.3814	0.3802	0.3790	0.3778	0.3765	0.3752	0.3739	0.3725	0.3712	0.3697
0.4	0.3683	0.3668	0.3653	0.3637	0.3621	0.3605	0.3589	0.3572	0.3555	0.3538
0.5	0.3521	0.3503	0.3485	0.3467	0.3448	0.3429	0.3410	0.3391	0.3372	0.3352
0.6	0.3332	0.3312	0.3292	0.3271	0.3251	0.3230	0.3209	0.3187	0.3166	0.3144
0.7	0.3123	0.3101	0.3079	0.3056	0.3034	0.3011	0.2989	0.2966	0.2943	0.2920
0.8	0.2897	0.2874	0.2850	0.2827	0.2803	0.2780	0.2756	0.2732	0.2709	0.2685
0.9	0.2661	0.2637	0.2613	0.2589	0.2565	0.2541	0.2516	0.2492	0.2468	0.2444
1.0	0.2420	0.2396	0.2371	0.2347	0.2323	0.2299	0.2275	0.2251	0.2227	0.2203
1.1	0.2179	0.2155	0.2131	0.2107	0.2083	0.2059	0.2036	0.2012	0.1989	0.1965
1.2	0.1942	0.1919	0.1895	0.1872	0.1849	0.1826	0.1804	0.1781	0.1758	0.1736
1.3	0.1714	0.1691	0.1669	0.1647	0.1626	0.1604	0.1582	0.1561	0.1539	0.1518
1.4	0.1497	0.1476	0.1456	0.1435	0.1415	0.1394	0.1374	0.1354	0.1334	0.1315
1.5	0.1295	0.1276	0.1257	0.1238	0.1219	0.1200	0.1182	0.1163	0.1145	0.1127
1.6	0.1109	0.1092	0.1074	0.1057	0.1040	0.1023	0.1006	0.0989	0.0973	0.0957
1.7	0.0940	0.0925	0.0909	0.0893	0.0878	0.0863	0.0848	0.0833	0.0818	0.0804
1.8	0.0790	0.0775	0.0761	0.0748	0.0734	0.0721	0.0707	0.0694	0.0681	0.0669
1.9	0.0656	0.0644	0.0632	0.0620	0.0608	0.0596	0.0584	0.0573	0.0562	0.0551
2.0	0.0540	0.0529	0.0519	0.0508	0.0498	0.0488	0.0478	0.0468	0.0459	0.0449
2.1	0.0440	0.0431	0.0422	0.0413	0.0404	0.0396	0.0387	0.0379	0.0371	0.0363
2.2	0.0355	0.0347	0.0339	0.0332	0.0325	0.0317	0.0310	0.0303	0.0297	0.0290
2.3	0.0283	0.0277	0.0270	0.0264	0.0258	0.0252	0.0246	0.0241	0.0235	0.0229
2.4	0.0224	0.0219	0.0213	0.0208	0.0203	0.0198	0.0194	0.0189	0.0184	0.0180
2.5	0.0175	0.0171	0.0167	0.0163	0.0158	0.0154	0.0151	0.0147	0.0143	0.0139
2.6	0.0136	0.0132	0.0129	0.0126	0.0122	0.0119	0.0116	0.0113	0.0110	0.0107
2.7	0.0104	0.0101	0.0099	0.0096	0.0093	0.0091	0.0088	0.0086	0.0084	0.0081
2.8	0.0079	0.0077	0.0075	0.0073	0.0071	0.0069	0.0067	0.0065	0.0063	0.0061
2.9	0.0060	0.0058	0.0056	0.0055	0.0053	0.0051	0.0050	0.0048	0.0047	0.0046
3.0	0.0044	0.0043	0.0042	0.0040	0.0039	0.0038	0.0037	0.0036	0.0035	0.0034
3.1	0.0033	0.0032	0.0031	0.0030	0.0029	0.0028	0.0027	0.0026	0.0025	0.0025
3.2	0.0024	0.0023	0.0022	0.0022	0.0021	0.0020	0.0020	0.0019	0.0018	0.0018
3.3	0.0017	0.0017	0.0016	0.0016	0.0015	0.0015	0.0014	0.0014	0.0013	0.0013
3.4	0.0012	0.0012	0.0012	0.0011	0.0011	0.0010	0.0010	0.0010	0.0009	0.0009
3.5	0.0009	0.0008	0.0008	0.0008	0.0008	0.0007	0.0007	0.0007	0.0007	0.0006
3.6	0.0006	0.0006	0.0006	0.0005	0.0005	0.0005	0.0005	0.0005	0.0005	0.0004
3.7	0.0004	0.0004	0.0004	0.0004	0.0004	0.0004	0.0003	0.0003	0.0003	0.0003
3.8	0.0003	0.0003	0.0003	0.0003	0.0003	0.0002	0.0002	0.0002	0.0002	0.0002
3.9	0.0002	0.0002	0.0002	0.0002	0.0002	0.0002	0.0002	0.0002	0.0001	0.0001

Table A-14   AREAS UNDER THE STANDARD NORMAL DISTRIBUTION CURVE

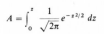

$$A = \int_0^z \frac{1}{\sqrt{2\pi}} e^{-z^2/2} \, dz$$

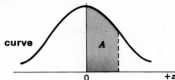

z	0	1	2	3	4	5	6	7	8	9
0.0	0.0000	0.0040	0.0080	0.0120	0.0160	0.0199	0.0239	0.0279	0.0319	0.0359
0.1	0.0398	0.0438	0.0478	0.0517	0.0557	0.0596	0.0636	0.0675	0.0714	0.0754
0.2	0.0793	0.0832	0.0871	0.0910	0.0948	0.0987	0.1026	0.1064	0.1103	0.1141
0.3	0.1179	0.1217	0.1255	0.1293	0.1331	0.1368	0.1406	0.1443	0.1480	0.1517
0.4	0.1554	0.1591	0.1628	0.1664	0.1700	0.1736	0.1772	0.1808	0.1844	0.1879
0.5	0.1915	0.1950	0.1985	0.2019	0.2054	0.2088	0.2123	0.2157	0.2190	0.2224
0.6	0.2258	0.2291	0.2324	0.2357	0.2389	0.2422	0.2454	0.2486	0.2518	0.2549
0.7	0.2580	0.2612	0.2642	0.2673	0.2704	0.2734	0.2764	0.2794	0.2823	0.2852
0.8	0.2881	0.2910	0.2939	0.2967	0.2996	0.3023	0.3051	0.3078	0.3106	0.3133
0.9	0.3159	0.3186	0.3212	0.3238	0.3264	0.3289	0.3315	0.3340	0.3365	0.3389
1.0	0.3413	0.3438	0.3461	0.3485	0.3508	0.3531	0.3554	0.3577	0.3599	0.3621
1.1	0.3643	0.3665	0.3686	0.3708	0.3729	0.3749	0.3770	0.3790	0.3810	0.3830
1.2	0.3849	0.3869	0.3888	0.3907	0.3925	0.3944	0.3962	0.3980	0.3997	0.4015
1.3	0.4032	0.4049	0.4066	0.4082	0.4099	0.4115	0.4131	0.4147	0.4162	0.4177
1.4	0.4192	0.4207	0.4222	0.4236	0.4251	0.4265	0.4279	0.4292	0.4306	0.4319
1.5	0.4332	0.4345	0.4357	0.4370	0.4382	0.4394	0.4406	0.4418	0.4429	0.4441
1.6	0.4452	0.4463	0.4474	0.4484	0.4495	0.4506	0.4515	0.4525	0.4535	0.4545
1.7	0.4554	0.4564	0.4573	0.4582	0.4591	0.4599	0.4608	0.4616	0.4625	0.4633
1.8	0.4641	0.4649	0.4656	0.4664	0.4671	0.4678	0.4686	0.4693	0.4699	0.4706
1.9	0.4713	0.4719	0.4726	0.4732	0.4738	0.4744	0.4750	0.4756	0.4761	0.4767
2.0	0.4772	0.4778	0.4783	0.4788	0.4793	0.4798	0.4803	0.4808	0.4812	0.4817
2.1	0.4821	0.4826	0.4830	0.4834	0.4838	0.4842	0.4846	0.4850	0.4854	0.4857
2.2	0.4861	0.4864	0.4868	0.4871	0.4875	0.4878	0.4881	0.4884	0.4887	0.4890
2.3	0.4893	0.4896	0.4898	0.4901	0.4904	0.4906	0.4909	0.4911	0.4913	0.4916
2.4	0.4918	0.4920	0.4922	0.4925	0.4927	0.4929	0.4931	0.4932	0.4934	0.4936
2.5	0.4938	0.4940	0.4941	0.4943	0.4945	0.4946	0.4948	0.4949	0.4951	0.4952
2.6	0.4953	0.4955	0.4956	0.4957	0.4959	0.4960	0.4961	0.4962	0.4963	0.4964
2.7	0.4965	0.4966	0.4967	0.4968	0.4969	0.4970	0.4971	0.4972	0.4973	0.4974
2.8	0.4974	0.4975	0.4976	0.4977	0.4977	0.4978	0.4979	0.4979	0.4980	0.4981
2.9	0.4981	0.4982	0.4982	0.4983	0.4984	0.4984	0.4985	0.4985	0.4986	0.4986
3.0	0.4987	0.4987	0.4987	0.4988	0.4988	0.4989	0.4989	0.4989	0.4990	0.4990
3.1	0.4990	0.4991	0.4991	0.4991	0.4992	0.4992	0.4992	0.4992	0.4993	0.4993
3.2	0.4993	0.4993	0.4994	0.4994	0.4994	0.4994	0.4994	0.4995	0.4995	0.4995
3.3	0.4995	0.4995	0.4995	0.4996	0.4996	0.4996	0.4996	0.4996	0.4996	0.4997
3.4	0.4997	0.4997	0.4997	0.4997	0.4997	0.4997	0.4997	0.4997	0.4997	0.4998
3.5	0.4998	0.4998	0.4998	0.4998	0.4998	0.4998	0.4998	0.4998	0.4998	0.4998
3.6	0.4998	0.4998	0.4999	0.4999	0.4999	0.4999	0.4999	0.4999	0.4999	0.4999
3.7	0.4999	0.4999	0.4999	0.4999	0.4999	0.4999	0.4999	0.4999	0.4999	0.4999
3.8	0.4999	0.4999	0.4999	0.4999	0.4999	0.4999	0.4999	0.4999	0.4999	0.4999
3.9	0.5000	0.5000	0.5000	0.5000	0.5000	0.5000	0.5000	0.5000	0.5000	0.5000

Table A-15 LIMITS AND FITS FOR CYLINDRICAL PARTS*

The limits shown in the accompanying tabulations are in thousandths of an inch. The size ranges include all sizes *over* the smallest size in the range, up to and *including* the largest size in the range. The letter symbols are defined as follows:

RC   *Sliding and running fits.* Running and sliding fits are intended to provide a similar running performance, with suitable lubrication allowance, throughout the range of sizes. The clearance for the first two classes, used chiefly as slide fits, increase more slowly with diameter than the other classes, so that accurate location is maintained even at the expense of free relative motion.

RC1 *Close sliding fits* are intended for the accurate location of parts which must assemble without perceptible play.

RC2 *Sliding fits* are intended for accurate location but with greater maximum clearance than class RC1. Parts made to this fit move and turn easily, but are not intended to run freely, and in the larger sizes may seize with small temperature changes.

RC3 *Precision running fits* are about the closest fits which can be expected to run freely, and are intended for precision work at slow speeds and light journal pressures, but are not suitable where appreciable temperature differences are likely to be encountered.

RC4 *Close running fits* are intended chiefly for running fits on accurate machinery with moderate surface speeds and journal pressures, where accurate location and minimum play is desired.

RC5—RC6 *Medium running fits* are intended for higher running speeds or heavy journal pressures, or both.

RC7 *Free running fits* are intended for use where accuracy is not essential or where large temperature variations are likely to be encountered, or under both of these conditions.

RC8—RC9 *Loose running fits* are intended for use where wide commercial tolerances may be necessary, together with an allowance, on the external member.

L   *Locational fits.* Locational fits are fits intended to determine only the location of the mating parts; they may provide rigid or accurate location, as with interference fits, or provide some freedom of location, as with clearance fits. Accordingly, they are divided into three groups: clearance fits, transition fits, and interference fits.

LC *Locational clearance fits* are intended for parts which are normally stationary but which can be freely assembled or disassembled. They run from snug fits for parts requiring accuracy of location, through the medium clearance fits for parts such as ball, race, and housing, to the looser fastener fits where freedom of assembly is of prime importance.

LT *Locational transition fits* are a compromise between clearance and interference fits, for application where accuracy of location is important but either a small amount of clearance or interference is permissible.

LN *Locational interference fits* are used where accuracy of location is of prime importance and for parts requiring rigidity and alignment with no special requirements for bore pressure. Such fits are not intended for parts designed to transmit frictional loads from one part to another by virtue of the tightness of fit, since these conditions are covered by force fits.

FN *Force fits.* Force and shrink fits constitute a special type of interference fit, normally characterized by maintenance of constant bore pressures throughout the range of sizes. The interference therefore varies almost directly with diameter, and the difference between its minimum and maximum value is small so as to maintain the resulting pressures within reasonable limits.

FN1 *Light drive fits* are those requiring light assembly pressures and produce more or less permanent assemblies. They are suitable for thin sections or long fits or in cast-iron external members.

FN2 *Medium drive fits* are suitable for ordinary steel parts or for shrink fits on light sections. They are about the tightest fits that can be used with high-grade cast-iron external members.

FN3 *Heavy drive fits* are suitable for heavier steel parts or for shrink fits in medium sections.

FN4—FN5 *Force fits* are suitable for parts which can be highly stressed or for shrink fits where the heavy pressing forces required are impractical.

* Extracted from American Standard Limits and Fits for Cylindrical Parts, USAS B4.1-1967, with the permission of the publishers, The American Society of Mechanical Engineers, United Engineering Center, 345 East 47th Street, New York 10017. Limit dimensions are tabulated in this standard for nominal sizes up to and including 200 in.

Table A-15-1   RUNNING AND SLIDING FITS

Class		0–0.12	0.12–0.24	0.24–0.40	0.40–0.71	0.71–1.19	1.19–1.97	1.97–3.15	3.15–4.73
RC1	Hole	+0.2 / −0	+0.2 / −0	+0.25 / −0	+0.3 / −0	+0.4 / −0	+0.4 / −0	+0.5 / −0	+0.6 / −0
	Shaft	+0.1 / −0.25	−0.15 / −0.3	−0.2 / −0.35	−0.25 / −0.45	−0.3 / −0.55	−0.4 / −0.7	−0.4 / −0.7	−0.5 / −0.9
RC2	Hole	+0.25 / −0	+0.3 / −0	+0.4 / −0	+0.4 / −0	+0.5 / −0	+0.6 / −0	+0.7 / −0	+0.9 / −0
	Shaft	−0.1 / −0.3	−0.15 / −0.35	−0.2 / −0.45	−0.25 / −0.55	−0.3 / −0.7	−0.4 / −0.8	−0.4 / −0.9	−0.5 / −1.1
RC3	Hole	+0.4 / −0	+0.5 / −0	+0.6 / −0	+0.7 / −0	+0.8 / −0	+1.0 / −0	+1.2 / −0	+1.4 / −0
	Shaft	−0.3 / −0.55	−0.4 / −0.7	−0.5 / −0.9	−0.6 / −1.0	−0.8 / −1.3	−1.0 / −1.6	−1.2 / −1.9	−1.4 / −2.3
RC4	Hole	+0.6 / −0	+0.7 / −0	+0.9 / −0	+1.0 / −0	+1.2 / −0	+1.6 / −0	+1.8 / −0	+2.2 / −0
	Shaft	−0.3 / −0.7	−0.4 / −0.9	−0.5 / −1.1	−0.6 / −1.3	−0.8 / −1.6	−1.0 / −2.0	−1.2 / −2.4	−1.4 / −2.8
RC5	Hole	+0.6 / −0	+0.7 / −0	+0.9 / −0	+1.0 / −0	+1.2 / −0	+1.6 / −0	+1.8 / −0	+2.2 / −0
	Shaft	−0.6 / −1.0	−0.8 / −1.3	−1.0 / −1.6	−1.2 / −1.9	−1.6 / −2.4	−2.0 / −3.0	−2.5 / −3.7	−3.0 / −4.4
RC6	Hole	+1.0 / −0	+1.2 / −0	+1.4 / −0	+1.6 / −0	+2.0 / −0	+2.5 / −0	+3.0 / −0	+3.5 / −0
	Shaft	−0.6 / −1.2	−0.8 / −1.5	−1.0 / −1.9	−1.2 / −2.2	−1.6 / −2.8	−2.0 / −3.6	−2.5 / −4.3	−3.0 / −5.2
RC7	Hole	+1.0 / −0	+1.2 / −0	+1.4 / −0	+1.6 / −0	+2.0 / −0	+2.5 / −0	+3.0 / −0	+3.5 / −0
	Shaft	−1.0 / −1.6	−1.2 / −1.9	−1.6 / −2.5	−2.0 / −3.0	−2.5 / −3.7	−3.0 / −4.6	−4.0 / −5.8	−5.0 / −7.2
RC8	Hole	+1.6 / −0	+1.8 / −0	+2.2 / −0	+2.8 / −0	+3.5 / −0	+4.0 / −0	+4.5 / −0	+5.0 / −0
	Shaft	−2.5 / −3.5	−2.8 / −4.0	−3.0 / −4.4	−3.5 / −5.1	−4.5 / −6.5	−5.0 / −7.5	−6.0 / −9.0	−7.0 / −10.5
RC9	Hole	+2.5 / −0	+3.0 / −0	+3.5 / −0	+4.0 / −0	+5.0 / −0	+6.0 / −0	+7.0 / −0	+9.0 / −0
	Shaft	−4.0 / −5.6	−4.5 / −6.0	−5.0 / −7.2	−6.0 / −8.8	−7.0 / −10.5	−8.0 / −12.0	−9.0 / −13.5	−10.0 / −15.0

Table A-15-2   LOCATIONAL CLEARANCE FITS

Size range

Class		0–0.12	0.12–0.24	0.24–0.40	0.40–0.71	0.71–1.19	1.19–1.97	1.97–3.15	3.15–4.73
LC1	Hole	+0.25 −0	+0.3 −0	+0.4 −0	+0.4 −0	+0.5 −0	+0.6 −0	+0.7 −0	+0.9 −0
	Shaft	+0 −0.2	+0 −0.2	+0 −0.25	+0 −0.3	+0 −0.4	+0 −0.4	+0 −0.5	+0 −0.6
LC2	Hole	+0.4 −0	+0.5 −0	+0.6 −0	+0.7 −0	+0.8 −0	+1.0 −0	+1.2 −0	+1.4 −0
	Shaft	+0 −0.25	+0 −0.3	+0 −0.4	+0 −0.5	+0 −0.6	+0 −0.7	+0 −0.9	+0 −1.0
LC3	Hole	+0.6 −0	+0.7 −0	+0.9 −0	+1.0 −0	+1.2 −0	+1.6 −0	+1.8 −0	+2.2 −0
	Shaft	+0 −0.4	+0 −0.5	+0 −0.6	+0 −0.7	+0 −0.8	+0 −1.0	+0 −1.2	+0 −1.4
LC4	Hole	+1.6 −0	+1.8 −0	+2.2 −0	+2.8 −0	+3.5 −0	+4.0 −0	+4.5 −0	+5.0 −0
	Shaft	+0 −1.0	+0 −1.2	+0 −1.4	+0 −1.6	+0 −2.0	+0 −2.5	+0 −3.0	+0 −3.5
LC5	Hole	+0.4 −0	+0.5 −0	+0.6 −0	+0.7 −0	+0.8 −0	+1.0 −0	+1.2 −0	+1.4 −0
	Shaft	−0.1 −0.35	−0.15 −0.45	−0.2 −0.6	−0.25 −0.65	−0.3 −0.8	−0.4 −1.0	−0.4 −1.1	−0.5 −1.4
LC6	Hole	+1.0 −0	+1.2 −0	+1.4 −0	+1.6 −0	+2.0 −0	+2.5 −0	+3.0 −0	+3.5 −0
	Shaft	−0.3 −0.9	−0.4 −1.1	−0.5 −1.4	−0.6 −1.6	−0.8 −2.0	−1.0 −2.6	−1.2 −3.0	−1.4 −3.6
LC7	Hole	+1.6 −0	+1.8 −0	+2.2 −0	+2.8 −0	+3.5 −0	+4.0 −0	+4.5 −0	+5.0 −0
	Shaft	−0.6 −1.6	−0.8 −2.0	−1.0 −2.4	−1.2 −2.8	−1.6 −3.6	−2.0 −4.5	−2.5 −5.5	−3.0 −6.5
LC8	Hole	+1.6 −0	+1.8 −0	+2.2 −0	+2.8 −0	+3.5 −0	+4.0 −0	+4.5 −0	+5.0 −0
	Shaft	−1.0 −2.0	−1.2 −2.4	−1.6 −3.0	−2.0 −3.6	−2.5 −4.5	−3.0 −5.5	−4.0 −7.0	−5.0 −8.5
LC9	Hole	+2.5 −0	+3.0 −0	+3.5 −0	+4.0 −0	+5.0 −0	+6.0 −0	+7.0 −0	+9.0 −0
	Shaft	−2.5 −4.1	−2.8 −4.6	−3.0 −5.2	−3.5 −6.3	−4.5 −8.0	−5.0 −9.0	−6.0 −10.5	−7.0 −12.0
LC10	Hole	+4.0 −0	+5.0 −0	+6.0 −0	+7.0 −0	+8.0 −0	+10.0 −0	+12.0 −0	+14.0 −0
	Shaft	−4.0 −8.0	−4.5 −9.5	−5.0 −11.0	−6.0 −13.0	−7.0 −15.0	−8.0 −18.0	−10.0 −22.0	−11.0 −25.0
LC11	Hole	+6.0 −0	+7.0 −0	+9.0 −0	+10.0 −0	+12.0 −0	+16.0 −0	+18.0 −0	+22.0 −0
	Shaft	−5.0 −11.0	−6.0 −13.0	−7.0 −16.0	−8.0 −18.0	−10.0 −22.0	−12.0 −28.0	−14.0 −32.0	−16.0 −38.0

Table A-15-3 LOCATIONAL TRANSITION FITS

Class		Size range							
		0–0.12	0.12–0.24	0.24–0.40	0.40–0.71	0.71–1.19	1.19–1.97	1.97–3.15	3.15–4.73
LT1	Hole	+0.4 / −0	+0.5 / −0	+0.6 / −0	+0.7 / −0	+0.8 / −0	+1.0 / −0	+1.2 / −0	+1.4 / −0
	Shaft	+0.10 / −0.10	+0.15 / −0.15	+0.2 / −0.2	+0.2 / −0.2	+0.25 / −0.25	+0.3 / −0.3	+0.3 / −0.3	+0.4 / −0.4
LT2	Hole	+0.6 / −0	+0.7 / −0	+0.9 / −0	+1.0 / −0	+1.2 / −0	+1.6 / −0	+1.8 / −0	+2.2 / −0
	Shaft	+0.2 / −0.2	+0.25 / −0.25	+0.3 / −0.3	+0.35 / −0.35	+0.4 / −0.4	+0.5 / −0.5	+0.6 / −0.6	+0.7 / −0.7
LT3	Hole			+0.6 / −0	+0.7 / −0	+0.8 / −0	+1.0 / −0	+1.2 / −0	+1.4 / −0
	Shaft			+0.5 / +0.1	+0.5 / +0.1	+0.6 / +0.1	+0.7 / +0.1	+0.8 / +0.1	+1.0 / +0.1
LT4	Hole			+0.9 / −0	+1.0 / −0	+1.2 / −0	+1.6 / −0	+1.8 / −0	+2.2 / −0
	Shaft			+0.7 / +0.1	+0.8 / +0.1	+0.9 / +0.1	+1.1 / +0.1	+1.3 / +0.1	+1.5 / +0.1
LT5	Hole	+0.4 / −0	+0.5 / −0	+0.6 / −0	+0.7 / −0	+0.8 / −0	+1.0 / −0	+1.2 / −0	+1.4 / −0
	Shaft	+0.5 / +0.25	+0.6 / +0.3	+0.8 / +0.4	+0.9 / +0.5	+1.1 / +0.6	+1.3 / +0.7	+1.5 / +0.8	+1.9 / +1.0
LT6	Hole	+0.4 / −0	+0.5 / −0	+0.6 / −0	+0.7 / −0	+0.8 / −0	+1.0 / −0	+1.2 / −0	+1.4 / −0
	Shaft	+0.65 / +0.25	+0.8 / +0.3	+1.0 / +0.4	+1.2 / +0.5	+1.4 / +0.6	+1.7 / +0.7	+2.0 / +0.8	+2.4 / +1.0

Table A-15-4 LOCATIONAL INTERFERENCE FITS

Class		Size range							
		0–0.12	0.12–0.24	0.24–0.40	0.40–0.71	0.71–1.19	1.19–1.97	1.97–3.15	3.15–4.73
LN1	Hole	+0.25 / −0	+0.3 / −0	+0.4 / −0	+0.4 / −0	+0.5 / −0	+0.6 / −0	+0.7 / −0	+0.9 / −0
	Shaft	+0.45 / +0.25	+0.5 / +0.3	+0.65 / +0.4	+0.8 / +0.4	+1.0 / +0.5	+1.1 / +0.6	+1.3 / +0.7	+1.6 / +1.0
LN2	Hole	+0.4 / −0	+0.5 / −0	+0.6 / −0	+0.7 / −0	+0.8 / −0	+1.0 / −0	+1.2 / −0	+1.4 / −0
	Shaft	+0.65 / +0.4	+0.8 / +0.5	+1.0 / +0.6	+1.1 / +0.7	+1.3 / +0.8	+1.6 / +1.0	+2.1 / +1.4	+2.5 / +1.6
LN3	Hole	+0.4 / −0	+0.5 / −0	+0.6 / −0	+0.7 / −0	+0.8 / −0	+1.0 / −0	+1.2 / −0	+1.4 / −0
	Shaft	+0.75 / +0.5	+0.9 / +0.6	+1.2 / +0.8	+1.4 / +1.0	+1.7 / +1.2	+2.0 / +1.4	+2.3 / +1.6	+2.9 / +2.0

Table A-15-5 FORCE AND SHRINK FITS

Size range

Class		0–0.12	0.12–0.24	0.24–0.40	0.40–0.56	0.56–0.71	0.71–0.95	0.95–1.19	1.19–1.58
FN1	Hole	+0.25, −0	+0.3, −0	+0.4, −0	+0.4, −0	+0.4, −0	+0.5, −0	+0.5, −0	+0.6, −0
	Shaft	+0.5, +0.3	+0.6, +0.4	+0.75, +0.5	+0.8, +0.5	+0.9, +0.6	+1.1, +0.7	+1.2, +0.8	+1.3, +0.9
FN2	Hole	+0.4, −0	+0.5, −0	+0.6, −0	+0.7, −0	+0.7, −0	+0.8, −0	+0.8, −0	+1.0, −0
	Shaft	+0.85, +0.6	+1.0, +0.7	+1.4, +1.0	+1.6, +1.2	+1.6, +1.2	+1.9, +1.4	+1.9, +1.4	+2.4, +1.8
FN3	Hole							+0.8, −0	+1.0, −0
	Shaft							+2.1, +1.6	+2.6, +2.0
FN4	Hole	+0.4, −0	+0.5, −0	+0.6, −0	+0.7, −0	+0.7, −0	+0.8, −0	+0.8, −0	+1.0, −0
	Shaft	+0.95, +0.7	+1.2, +0.9	+1.6, +1.2	+1.8, +1.4	+1.8, +1.4	+2.1, +1.6	+2.3, +1.8	+3.1, +2.5
FN5	Hole	+0.6, −0	+0.7, −0	+0.9, −0	+1.0, −0	+1.0, −0	+1.2, −0	+1.2, −0	+1.6, −0
	Shaft	+1.3, +0.9	+1.7, +1.2	+2.0, +1.4	+2.3, +1.6	+2.5, +1.8	+3.0, +2.2	+3.3, +2.5	+4.0, +3.0

Size range

Class		1.58–1.97	1.97–2.56	2.56–3.15	3.15–3.94	3.94–4.73	4.73–5.52	5.52–6.30	6.30–7.09
FN1	Hole	+0.6, −0	+0.7, −0	+0.7, −0	+0.9, −0	+0.9, −0	+1.0, −0	+1.0, −0	+1.0, −0
	Shaft	+1.4, +1.0	+1.8, +1.3	+1.9, +1.4	+2.4, +1.8	+2.6, +2.0	+2.9, +2.2	+3.2, +2.5	+3.5, +2.8
FN2	Hole	+1.0, −0	+1.2, −0	+1.2, −0	+1.4, −0	+1.4, −0	+1.6, −0	+1.6, −0	+1.6, −0
	Shaft	+2.4, +1.8	+2.7, +2.0	+2.9, +2.2	+3.7, +2.8	+3.9, +3.0	+4.5, +3.5	+5.0, +4.0	+5.5, +4.5
FN3	Hole	+1.0, −0	+1.2, −0	+1.2, −0	+1.4, −0	+1.4, −0	+1.6, −0	+1.6, −0	+1.6, −0
	Shaft	+2.8, +2.2	+3.2, +2.5	+3.7, +3.0	+4.4, +3.5	+4.9, +4.0	+6.0, +5.0	+6.0, +5.0	+7.0, +6.0
FN4	Hole	+1.0, −0	+1.2, −0	+1.2, −0	+1.4, −0	+1.4, −0	+1.6, −0	+1.6, −0	+1.6, −0
	Shaft	+3.4, +2.8	+4.2, +3.5	+4.7, +4.0	+5.9, +5.0	+6.9, +6.0	+8.0, +7.0	+8.0, +7.0	+9.0, +8.0
FN5	Hole	+1.6, −0	+1.8, −0	+1.8, −0	+2.2, −0	+2.2, −0	+2.5, −0	+2.5, −0	+2.5, −0
	Shaft	+5.0, +4.0	+6.2, +5.0	+7.2, +6.0	+8.4, +7.0	+9.4, +8.0	+11.6, +10.0	+13.6, +12.0	+13.6, +12.0

Table A-16 AMERICAN STANDARD PIPE

Nominal size, in	Outside diameter, in	Threads per inch	Standard No. 40	Wall thickness, in	
				Extra strong No. 80	Double extra strong
$\frac{1}{8}$	0.405	27	0.070	0.098	
$\frac{1}{4}$	0.540	18	0.090	0.122	
$\frac{3}{8}$	0.675	18	0.093	0.129	
$\frac{1}{2}$	0.840	14	0.111	0.151	0.307
$\frac{3}{4}$	1.050	14	0.115	0.157	0.318
1	1.315	$11\frac{1}{2}$	0.136	0.183	0.369
$1\frac{1}{4}$	1.660	$11\frac{1}{2}$	0.143	0.195	0.393
$1\frac{1}{2}$	1.900	$11\frac{1}{2}$	0.148	0.204	0.411
2	2.375	$11\frac{1}{2}$	0.158	0.223	0.447
$2\frac{1}{2}$	2.875	8	0.208	0.282	0.565
3	3.500	8	0.221	0.306	0.615
$3\frac{1}{2}$	4.000	8	0.231	0.325	
4	4.500	8	0.242	0.344	0.690
5	5.563	8	0.263	0.383	0.768
6	6.625	8	0.286	0.441	0.884
8	8.625	8	0.329	0.510	0.895

Table A-17 MECHANICAL PROPERTIES OF STEELS*

UNS number	Processing	Yield strength, kpsi	Tensile strength, kpsi	Elongation in 2 in, %	Reduction in area, %	Brinell hardness H_B
G10100	HR	26	47	28	50	95
	CD	44	53	20	40	105
G10150	HR	27	50	28	50	101
	CD	47	56	18	40	111
G10180	HR	32	58	25	50	116
	CD	54	64	15	40	126
G10350	HR	39	72	18	40	143
	CD	67	80	12	35	163
	Drawn 800°F	81	110	18	51	220
	Drawn 1000°F	72	103	23	59	201
	Drawn 1200°F	62	91	27	66	180
G10400	HR	42	76	18	40	149
	CD	71	85	12	35	170
	Drawn 1000°F	86	113	23	62	235
G10500	HR	49	90	15	35	179
	CD	84	100	10	30	197
	Drawn 600°F	180	220	10	30	450
	Drawn 900°F	130	155	18	55	310
	Drawn 1200°F	80	105	28	65	210
G15216§	HR‡	81	100	25	57	192
G41300	HR‡	60	90	30	45	183
	CD‡	87	98	21	52	201
	Drawn 1000°F	133	146	17	60	293
G41400	HR‡	63	90	27	58	187
	CD‡	90	102	18	50	223
	Drawn 1000°F	131	153	16	45	302
G43400	HR‡	69	101	21	45	207
	CD‡	99	111	16	42	223
	Drawn 600°F	234	260	12	43	498
	Drawn 1000°F	162	182	15	40	363
G46200	Core†	89	120	22	55	248
	Drawn 800°F	94	130	23	66	256
G61500	HR‡	58	91	22	53	183
	Drawn 1000°F	132	155	15	44	302
G87400	HR‡	64	95	25	55	190
	CD‡	96	107	17	48	223
	Drawn 1000°F	129	152	15	44	302
G92550	HR‡	78	115	22	45	223
	Drawn 1000°F	160	180	15	32	352

* Tabulated in accordance with the Unified Numbering System for Metals and Alloys (UNS), Society of Automotive Engineers, Warrendale, Pa., 1975. This reference contains the cross reference numbers for AISI, ASTM, FED, MIL SPEC, and SAE specifications.

The values shown for hot-rolled (HR) and cold-drawn (CD) steels are *estimated minimum values* which can usually be expected in the size range of $\frac{3}{4}$ to $1\frac{1}{4}$ in. A minimum value is roughly several standard deviations below the arithmetic mean. The values shown for heat-treated steels are so-called *typical* values. A typical value is neither the mean nor the minimum. It can be obtained by careful control of the purchase specifications and the heat-treatment, together with continuous inspection and testing. The properties shown in this table are from a variety of sources and are believed to be representative. There are so many variables which affect these properties, however, that their approximate nature must be clearly recognized.

Multiply strength in kpsi by 6.89 to get the strength in MPa.

† Case hardened, core properties. ‡ Annealed. § Same as AISI 52100.

Table A-18 MECHANICAL PROPERTIES OF WROUGHT ALUMINUM ALLOYS

These are *typical* properties for sizes of about $\frac{1}{2}$ in. A typical value may be neither the mean nor the minimum. It is a value which can be obtained when the purchase specifications are carefully written and with continuous inspection and testing. The values given for fatigue strength S_f correspond to $50(10)^7$ cycles of completely reversed stress. Aluminum alloys do not have an endurance limit. The yield strength is the 0.2 percent offset value.

UNS alloy number	Temper	Yield strength S_y, kpsi	Tensile strength S_u, kpsi	Shear modulus of rupture S_{su}, kpsi	Fatigue strength S_f, kpsi	Elongation in 2 in, %	Brinell hardness H_B
A91100	-O	5	13	9.5	5	45	23
	-H12	14	15.5	10	6	25	28
	-H14	20	22	14	9	16	40
	-H16	24	26	15	9.5	14	47
	-H18	27	29	16	10	10	55
A93003	-O	6	16	11	7	40	28
	-H12	17	19	12	8	20	35
	-H14	20	22	14	9	16	40
	-H16	24	26	15	9.5	14	47
	-H18	27	29	16	10	10	55
A93004	-O	10	26	16	14	25	45
	-H32	22	31	17	14.5	17	52
	-H34	27	34	18	15	12	63
	-H36	31	37	20	15.5	9	70
	-H38	34	40	21	16	6	77
A92011	-T3	48	55	32	18	15	95
	-T8	45	59	35	18	12	100
A92014	-O	14	27	18	13	18	45
	-T4	40	62	38	20	20	105
	-T6	60	70	42	18	13	135
A92017	-O	10	26	18	13	22	45
	-T4	40	62	38	18	22	105
A92018	-T61	46	61	39	17	12	120
A92024	-O	11	27	18	13	22	47
	-T3	50	70	41	20	16	120
	-T4	48	68	41	20	19	120
	-T36	57	73	42	18	13	130
A95052	-O	13	28	18	17	30	45
	-H32	27	34	20	17.5	18	62
	-H34	31	37	21	18	14	67
	-H36	34	39	23	18.5	10	74
	-H38	36	41	24	19	8	85
A95056	-O	22	42	26	20	35	
	-H18	59	63	34	22	10	
	-H38	50	60	32	22	15	
A96061	-O	8	18	12.5	9	30	30
	-T4	21	35	24	13.5	25	65
	-T6	40	45	30	13.5	17	95
A97075	-T6	72	82	49	24	11	150

Table A-19 MECHANICAL PROPERTIES OF ALUMINUM ALLOY CASTINGS

These are typical properties for $\frac{1}{2}$-in sizes and may be neither the mean nor the minimum, but are properties which can be attained with reasonable care. The fatigue strength is for $50(10)^7$ cycles of reversed stress. Both yield strengths are obtained by the 0.2 percent offset method. Multiply strengths in kpsi by 6.89 to get strength in MPa.

| UNS number and temper | Tension | | | Compressive yield strength, kpsi | Shear modulus of rupture, kpsi | Fatigue strength, kpsi | Brinell hardness H_B |
	Yield strength, kpsi	Ultimate strength, kpsi	Elongation in 2 in, %				
A03190*	18	27	2.0	19	22	10	70
A03190-T6*	24	36	2.0	25	29	10	80
A03330†	19	34	2.0	19	27	14.5	90
A03330-T5†	25	34	1.0	25	27	12	100
A03330-T6†	30	42	1.5	30	33	15	105
A03330-T62†	40	45	1.5	40	36	10	105
A03550-T6*	25	35	3.0	26	28	9	80
A03550-T7*	36	38	0.5	38	28	9	85
A03550-T71*	29	35	1.5	30	26	10	75
A03560-T51*	20	25	2.0	21	20	7.5	60
A03560-T6*	24	33	3.5	25	26	8.5	70
A03560-T7*	30	34	2.0	31	24	9	75

* Sand casting.
† Permanent-mold casting.

Table A-20 TYPICAL PROPERTIES OF GRAY CAST IRON

The American Society for Testing Materials (ASTM) numbering system for gray cast iron is established such that the numbers correspond to the *minimum tensile strength* in kpsi. Thus an ASTM No. 20 cast iron has a minimum tensile strength of 20 kpsi. Note particularly that the tabulations are *typical* values.

| ASTM number | Tensile strength S_{ut}, kpsi | Compressive strength S_{uc}, kpsi | Shear modulus of rupture S_{su}, kpsi | Modulus of elasticity, Mpsi | | Endurance limit S_e, kpsi | Brinell harness, H_B |
				Tension	Torsion		
20	22	83	26	9.6–14	3.9–5.6	10	156
25	26	97	32	11.5–14.8	4.6–6.0	11.5	174
30	31	109	40	13–16.4	5.2–6.6	14	201
35	36.5	124	48.5	14.5–17.2	5.8–6.9	16	212
40	42.5	140	57	16–20	6.4–7.8	18.5	235
50	52.5	164	73	18.8–22.8	7.2–8.0	21.5	262
60	62.5	187.5	88.5	20.4–23.5	7.8–8.5	24.5	302

Table A-21 TYPICAL PROPERTIES OF SOME COPPER-BASE ALLOYS

All yield strengths are by 0.5 percent offset method. Multiply strength in MPa by 0.145 to get strength in kpsi.

UNS number	Alloy name	Form	Temper	Yield strength MPa	Tensile strength MPa	Elongation in 50 mm %	Rockwell hardness
C17000	Beryllium	Rod	Hard	515	790	5	98B
		Rod	Soft	170	415	50	77B
		Sheet	Hard	1000	1240	2	—
C21000	Gilding brass	Sheet	Hard	345	385	5	64B
		Sheet	Soft	70	235	45	46F
C22000	Commercial bronze	Sheet	Hard	370	420	5	70B
		Sheet	Soft	70	255	45	53F
		Rod	Hard	380	415	20	60B
		Rod	Soft	70	275	50	55F
C23000	Red brass	Sheet	Hard	395	480	5	77B
		Sheet	Soft	85	275	47	59F
		Rod	Hard	360	395	23	75B
		Rod	Soft	70	275	55	55F
C26000	Cartridge brass	Sheet	Hard	435	525	8	82B
		Sheet	Soft	105	325	62	64F
		Rod	Hard	360	480	30	80B
		Rod	Soft	110	330	65	65F
C27000	Yellow brass	Sheet	Hard	415	510	8	80B
		Sheet	Soft	105	325	62	64F
		Rod	Hard	310	415	25	80B
		Rod	Soft	110	330	65	65F
C28000	Muntz metal	Sheet	Hard	415	550	10	85B
		Sheet	Soft	145	370	45	80F
		Rod	Hard	380	515	20	80B
		Rod	Soft	145	370	50	80F
		Tube	Hard	380	510	10	80B
		Tube	Soft	160	385	50	82F
C33000	Low-leaded brass	Tube	Hard	415	515	7	80B
		Tube	Soft	105	325	60	64F
C33200	High-leaded brass	Sheet	Hard	415	510	7	80B
		Sheet	Soft	115	340	52	68F
C46200	Naval brass	Sheet	Hard	480	620	5	90B
		Rod	Hard	365	515	20	82B
		Tube	Hard	455	605	18	95B

Table A-22 TYPICAL MECHANICAL PROPERTIES OF WROUGHT STAINLESS STEELS*

All yield strengths are obtained using the 0.2 percent offset method.

UNS number	Processing	Yield strength, kpsi	Tensile strength, kpsi	Elongation in 2 in, %	Reduction in area, %	Brinell hardness H_B
S20100	Annealed	55	155	55		
	$\frac{1}{4}$ hard	75	125	20		
	$\frac{1}{2}$ hard	110	150	10		
	$\frac{3}{4}$ hard	135	175	5		
	Full hard	140	185	4		
S20200	Annealed	55	110	55		
	$\frac{1}{4}$ hard	75	125	12		
S30100	Annealed	40	110	60		165
	$\frac{1}{4}$ hard	75	125	25		
	$\frac{1}{2}$ hard	110	150	15		
	$\frac{3}{4}$ hard	135	175	12		
	Full hard	140	185	8		
S30200	Annealed	37	90	55	65	155
	$\frac{1}{4}$ hard	75	125	12		
S30300	Annealed	35	90	50	55	160
S30400	Annealed	35	85	55	65	150
S31000	Annealed	40	95	45	65	170
S31400	Annealed	50	100	45	60	170
S41400	Annealed	95	120	17	55	235
	Drawn 400°F	150	200	15	55	415
	Drawn 600°F	145	190	15	55	400
	Drawn 800°F	150	200	16	58	415
	Drawn 1000°F	120	145	20	60	325
	Drawn 1200°F	105	120	20	65	260
S41600	Annealed	40	75	30	65	155
	Drawn 400°F	145	190	15	55	390
	Drawn 600°F	140	180	15	55	375
	Drawn 800°F	150	195	17	55	390
	Drawn 1000°F	115	145	20	65	300
	Drawn 1200°F	85	110	23	65	225
	Drawn 1400°F	60	90	30	70	180
S43100	Annealed	95	125	20	60	260
	Drawn 400°F	155	205	15	55	415
	Drawn 600°F	150	195	15	55	400
	Drawn 800°F	155	205	15	60	415
	Drawn 1200°F	95	125	20	60	260
S50100	Annealed	30	70	28	65	160
S50200	Annealed	30	70	30	75	150

* *By permission*, "Metals Handbook," 8th ed., vol. 1, p. 414, American Society for Metals, Metals Park, Ohio, 1961.

Table A-23 TYPICAL PROPERTIES OF MAGNESIUM ALLOYS*

Since magnesium does not have an endurance limit, the fatigue strength shown is for $50(10)^7$ cycles of reversed stress.

Magnesium alloy	Tensile yield strength, kpsi	Tensile strength, kpsi	Elonga-tion in 2 in, %	Compres-sive yield strength, kpsi	Brinell hardness H_B	Shear strength, kpsi	Fatigue strength, kpsi
Cast AM 265C	11	27	6	11	48	...	11
Cast AM 240-T4	12	35	9	12	52	20	11
Cast AM 260-T6	20	38	3	20	78	22	11.5
Die-cast AM 263	22	34	3	...	...	...	14
Wrought AM 3S	30	40	7	11	40–52	16.7	11
Wrought AM C52S	30	40	17	20	50–71	19	15
Wrought AM C57S	32	44	14	20	55–74	20.5	17
Wrought AM 59S	38	51	9	27	70	22	18

* Courtesy of American Magnesium Corporation.

Table A-24 DECIMAL EQUIVALENTS OF WIRE AND SHEET-METAL GAUGES*

Name of gauge:	American or Brown & Sharpe	Birmingham or Stubs iron wire	United States Standard	Manu-facturers Standard†	Steel wire or Washburn & Moen	Music wire	Stubs steel wire	Twist drill
Prin-cipal use:	Nonferrous sheet, wire, and rod	Tubing, ferrous strip, flat wire, and spring steel	Ferrous sheet and plate, 480 lb per ft³	Ferrous sheet	Ferrous wire except music wire	Music wire	Steel drill rod	Twist drills and drill steel
7/0	——	——	0.500	——	0.490 0			
6/0	0.580 0	——	0.468 75	——	0.461 5	0.004		
5/0	0.516 5	——	0.437 5	——	0.430 5	0.005		
4/0	0.460 0	0.454	0.406 25	——	0.393 8	0.006		
3/0	0.409 6	0.425	0.375	——	0.362 5	0.007		
2/0	0.364 8	0.380	0.343 75	——	0.331 0	0.008		
0	0.324 9	0.340	0.312 5	——	0.306 5	0.009		
1	0.289 3	0.300	0.281 25	——	0.283 0	0.010	0.227	0.228 0
2	0.257 6	0.284	0.265 625	——	0.262 5	0.011	0.219	0.221 0
3	0.229 4	0.259	0.25	0.239 1	0.243 7	0.012	0.212	0.213 0
4	0.204 3	0.238	0.234 375	0.224 2	0.225 3	0.013	0.207	0.209 0
5	0.181 9	0.220	0.218 75	0.209 2	0.207 0	0.014	0.204	0.205 5

* Reproduced by courtesy of the Reynolds Metal Company. Specify sheet, wire, and plate by stating the gauge number, the gauge name, and the decimal equivalent in parentheses.
† Reflects present average unit weights of sheet steel.

Table A-24 DECIMAL EQUIVALENTS OF WIRE AND SHEET-METAL GAUGES* (*concluded*)

Name of gauge:	American or Brown & Sharpe	Birmingham or Stubs iron wire	United States Standard	Manu-facturers Standard†	Steel wire or Washburn & Moen	Music wire	Stubs steel wire	Twist drill
Prin-cipal use:	Nonferrous sheet, wire and rod	Tubing, ferrous strip, flat wire, and spring steel	Ferrous sheet and plate, 480 lb per ft^3	Ferrous sheet	Ferrous wire except music wire	Music wire	Steel drill rod	Twist drills and drill steel
6	0.162 0	0.203	0.203 125	0.194 3	0.192 0	0.016	0.201	0.204 0
7	0.144 3	0.180	0.187 5	0.179 3	0.177 0	0.018	0.199	0.201 0
8	0.128 5	0.165	0.171 875	0.164 4	0.162 0	0.020	0.197	0.199 0
9	0.114 4	0.148	0.156 25	0.149 5	0.148 3	0.022	0.194	0.196 0
10	0.101 9	0.134	0.140 625	0.134 5	0.135 0	0.024	0.191	0.193 5
11	0.090 74	0.120	0.125	0.119 6	0.120 5	0.026	0.188	0.191 0
12	0.080 81	0.109	0.109 357	0.104 6	0.105 5	0.029	0.185	0.189 0
13	0.071 96	0.095	0.093 75	0.089 7	0.091 5	0.031	0.182	0.185 0
14	0.064 08	0.083	0.078 125	0.074 7	0.080 0	0.033	0.180	0.182 0
15	0.057 07	0.072	0.070 312 5	0.067 3	0.072 0	0.035	0.178	0.180 0
16	0.050 82	0.065	0.062 5	0.059 8	0.062 5	0.037	0.175	0.177 0
17	0.045 26	0.058	0.056 25	0.053 8	0.054 0	0.039	0.172	0.173 0
18	0.040 30	0.049	0.05	0.047 8	0.047 5	0.041	0.168	0.169 5
19	0.035 89	0.042	0.043 75	0.041 8	0.041 0	0.043	0.164	0.166 0
20	0.031 96	0.035	0.037 5	0.035 9	0.034 8	0.045	0.161	0.161 0
21	0.028 46	0.032	0.034 375	0.032 9	0.031 7	0.047	0.157	0.159 0
22	0.025 35	0.028	0.031 25	0.029 9	0.028 6	0.049	0.155	0.157 0
23	0.022 57	0.025	0.028 125	0.026 9	0.025 8	0.051	0.153	0.154 0
24	0.020 10	0.022	0.025	0.023 9	0.023 0	0.055	0.151	0.152 0
25	0.017 90	0.020	0.021 875	0.020 9	0.020 4	0.059	0.148	0.149 5
26	0.015 94	0.018	0.018 75	0.017 9	0.018 1	0.063	0.146	0.147 0
27	0.014 20	0.016	0.017 187 5	0.016 4	0.017 3	0.067	0.143	0.144 0
28	0.012 64	0.014	0.015 625	0.014 9	0.016 2	0.071	0.139	0.140 5
29	0.011 26	0.013	0.014 062 5	0.013 5	0.015 0	0.075	0.134	0.136 0
30	0.010 03	0.012	0.012 5	0.012 0	0.014 0	0.080	0.127	0.128 5
31	0.008 928	0.010	0.010 937 5	0.010 5	0.013 2	0.085	0.120	0.120 0
32	0.007 950	0.009	0.010 156 25	0.009 7	0.012 8	0.090	0.115	0.116 0
33	0.007 080	0.008	0.009 375	0.009 0	0.011 8	0.095	0.112	0.113 0
34	0.006 305	0.007	0.008 593 75	0.008 2	0.010 4	——	0.110	0.111 0
35	0.005 615	0.005	0.007 812 5	0.007 5	0.009 5	——	0.108	0.110 0
36	0.005 000	0.004	0.007 031 25	0.006 7	0.009 0	——	0.106	0.106 5
37	0.004 453	——	0.006 640 625	0.006 4	0.008 5	——	0.103	0.104 0
38	0.003 965	——	0.006 25	0.006 0	0.008 0	——	0.101	0.101 5
39	0.003 531	——	——	——	0.007 5	——	0.099	0.099 5
40	0.003 145	——	——	——	0.007 0	——	0.097	0.098 0

* Reproduced by courtesy of the Reynolds Metal Company. Specify sheet, wire, and plate by stating the gauge number, the gauge name, and the decimal equivalent in parentheses.
† Reflects present average unit weights of sheet steel.

Table A-25 **CHARTS OF THEORETICAL STRESS-CONCENTRATION FACTORS K_t***

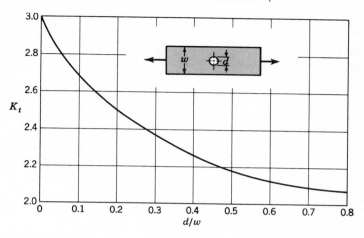

FIGURE A-25-1

Bar in tension or simple compression with a transverse hole. $\sigma_0 = F/A$, where $A = (w - d)t$, and where t is the thickness.

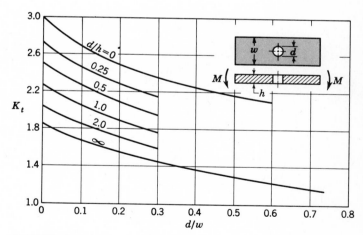

FIGURE A-25-2

Rectangular bar with a transverse hole in bending. $\sigma_0 = Mc/I$, where $I = (w - d)h^3/12$.

---

* Unless otherwise stated, these factors are from R. E. Peterson, Design Factors for Stress Concentration, *Machine Design*, vol. 23, no. 2, p. 169, February 1951; no. 3, p. 161, March 1951; no. 5, p. 159, May 1951; no. 6, p. 173, June 1951; no. 7, p. 155, July 1951; reproduced with the permission of the author and publisher.

Table A-25 **CHARTS OF THEORETICAL STRESS-CONCENTRATION FACTORS** K_t (*continued*)

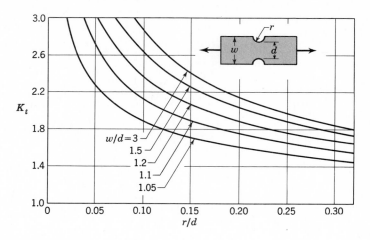

FIGURE A-25-3

Notched rectangular bar in tension or simple compression. $\sigma_0 = F/A$, where $A = dt$ and t is the thickness.

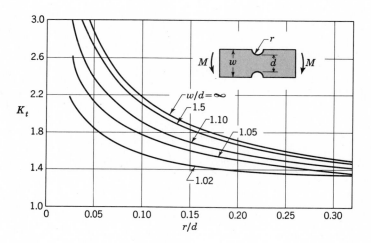

FIGURE A-25-4

Notched rectangular bar in bending. $\sigma_0 = Mc/I$, where $c = d/2$ and $I = td^3/12$. The thickness is t.

Table A-25 CHARTS OF THEORETICAL STRESS-CONCENTRATION FACTORS K_t *(continued)*

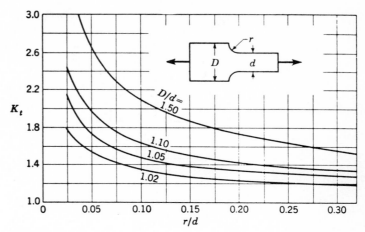

FIGURE A-25-5
Rectangular filleted bar in tension or simple compression. $\sigma_0 = F/A$, where $A = dt$ and t is the thickness.

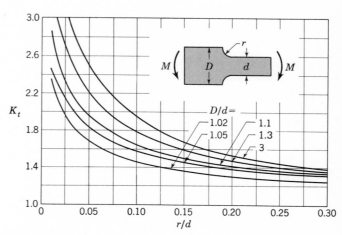

FIGURE A-25-6
Rectangular filleted bar in bending. $\sigma_0 = Mc/I$, where $c = d/2$, $I = td^3/12$, and t is the thickness.

Table A-25 **CHARTS OF THEORETICAL STRESS-CONCENTRATION FACTORS** K_t *(continued)*

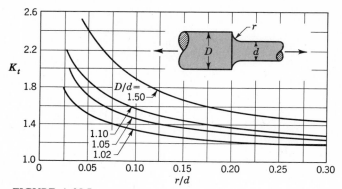

FIGURE A-25-7
Round shaft with shoulder fillet in tension. $\sigma_0 = F/A$, where $A = \pi d^2/4$.

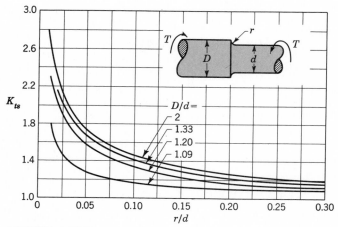

FIGURE A-25-8
Round shaft with shoulder fillet in torsion. $\tau_0 = Tc/J$, where $c = d/2$ and $J = \pi d^4/32$.

Table A-25   **CHARTS OF THEORETICAL STRESS-CONCENTRATION FACTORS** K_t (*continued*)

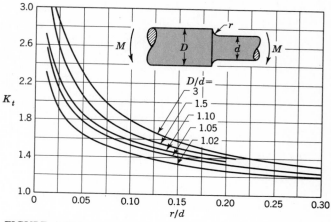

FIGURE A-25-9
Round shaft with shoulder fillet in bending. $\sigma_0 = Mc/I$, where $c = d/2$ and $I = \pi d^4/64$.

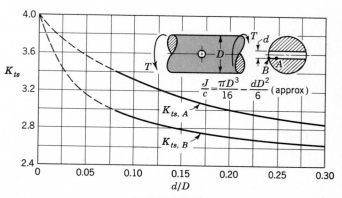

FIGURE A-25-10
Round shaft in torsion with transverse hole.

Table A-25   CHARTS OF THEORETICAL STRESS-CONCENTRATION FACTORS K_t *(continued)*

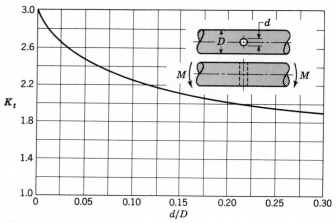

FIGURE A-25-11

Round shaft in bending with a transverse hole. $\sigma_0 = M/[(\pi D^3/32) - (dD^2/6)]$, approximately.

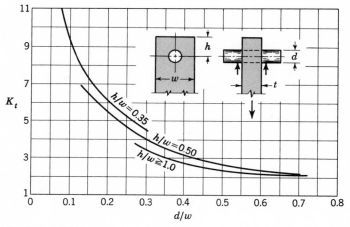

FIGURE A-25-12

Plate loaded in tension by a pin through a hole. $\sigma_0 = F/A$, where $A = (w - d)t$. When clearance exists, increase K_t 35 to 50 percent. (*M. M. Frocht and H. N. Hill, Stress Concentration Factors around a Central Circular Hole in a Plate Loaded through a Pin in Hole, J. Appl. Mechanics, vol. 7, no. 1, p. A-5, March 1940.*)

Table A-25 CHARTS OF THEORETICAL STRESS-CONCENTRATION FACTORS K_t (*continued*)

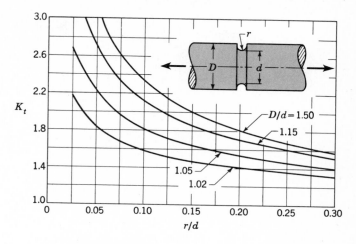

FIGURE A-25-13
Grooved round bar in tension. $\sigma_0 = F/A$, where $A = \pi d^2/4$.

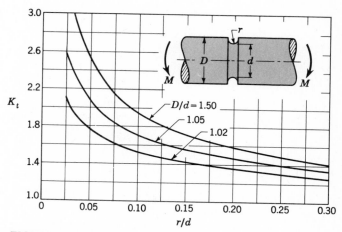

FIGURE A-25-14
Grooved round bar in bending. $\sigma_0 = Mc/I$, where $c = d/2$ and
$I = \pi d^4/64$.

Table A-25 CHARTS OF THEORETICAL STRESS-CONCENTRATION FACTORS K_t *(concluded)*

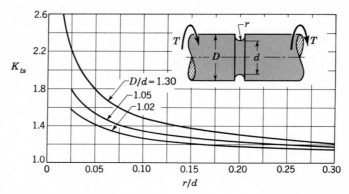

FIGURE A-25-15
Grooved round bar in torsion. $\tau_0 = Tc/J$, where $c = d/2$ and $J = \pi d^4/32$.

GREEK ALPHABET

Alpha	A	α		Nu	N	ν	
Beta	B	β		Xi	Ξ	ξ	
Gamma	Γ	γ		Omicron	O	o	
Delta	Δ	δ	∂	Pi	Π	π	
Epsilon	E	ϵ		Rho	P	ρ	
Zeta	Z	ζ		Sigma	Σ	σ	ς
Eta	H	η		Tau	T	τ	
Theta	Θ	θ	ϑ	Upsilon	Υ		
Iota	I	ι		Phi	Φ	ϕ	φ
Kappa	K	κ	$\varkappa$	Chi	X	χ	
Lambda	Λ	λ		Psi	Ψ	ψ	
Mu	M	μ		Omega	Ω	ω	

Table A-26   DIMENSIONS OF ROUND-HEAD MACHINE SCREWS (ASA B18.6-1947)*

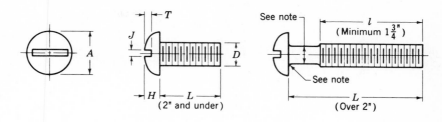

Nominal size	D Max diameter of screw	A Head diameter		H Height of head		J Width of slot		T Depth of slot	
		Max	Min	Max	Min	Max	Min	Max	Min
0	0.060	0.113	0.099	0.053	0.043	0.023	0.016	0.039	0.029
1	0.073	0.138	0.122	0.061	0.051	0.026	0.019	0.044	0.033
2	0.086	0.162	0.146	0.069	0.059	0.031	0.023	0.048	0.037
3	0.099	0.187	0.169	0.078	0.067	0.035	0.027	0.053	0.040
4	0.112	0.211	0.193	0.086	0.075	0.039	0.031	0.058	0.044
5	0.125	0.236	0.217	0.095	0.083	0.043	0.035	0.063	0.047
6	0.138	0.260	0.240	0.103	0.091	0.048	0.039	0.068	0.051
8	0.164	0.309	0.287	0.120	0.107	0.054	0.045	0.077	0.058
10	0.190	0.359	0.334	0.137	0.123	0.060	0.050	0.087	0.065
12	0.216	0.408	0.382	0.153	0.139	0.067	0.056	0.096	0.072
$\frac{1}{4}$	0.250	0.472	0.443	0.175	0.160	0.075	0.064	0.109	0.082
$\frac{5}{16}$	0.3125	0.590	0.557	0.216	0.198	0.084	0.072	0.132	0.099
$\frac{3}{8}$	0.375	0.708	0.670	0.256	0.237	0.094	0.081	0.155	0.117
$\frac{7}{16}$	0.4375	0.750	0.707	0.328	0.307	0.094	0.081	0.196	0.148
$\frac{1}{2}$	0.500	0.813	0.766	0.355	0.332	0.106	0.091	0.211	0.159
$\frac{9}{16}$	0.5625	0.938	0.887	0.410	0.385	0.118	0.102	0.242	0.183
$\frac{5}{8}$	0.625	1.000	0.944	0.438	0.411	0.133	0.116	0.258	0.195
$\frac{3}{4}$	0.750	1.250	1.18ᶜ	0.547	0.516	0.149	0.131	0.320	0.242

Notes:

All dimensions are given in inches.

Head dimensions for sizes $\frac{7}{16}$ in and larger are in agreement with round-head cap screw dimensions except the minimum values have been decreased to provide tolerances in proportion to balance of table.

The diameter of the unthreaded portion of machine screws shall not be less than the minimum pitch diameter nor more than the maximum major diameter of the thread.

The radius of the fillet at the base of the head shall not exceed one-half the pitch of the screw thread.

* By permission from Vallory H. Laughner and Augustus D. Hargan, "Handbook of Fastening and Joining of Metal Parts," McGraw-Hill Book Company, New York, 1956.

Table A-27 DIMENSIONS OF HEXAGON-HEAD CAP SCREWS (ASA B18.2-1952)*

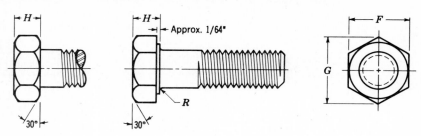

Nominal size or basic major diameter of thread	Body diameter minimum (maximum equal to nominal size)	Width across flats F			Width across corners G		Height H			Radius of fillet R	
		Max	(basic)	Min	Max	Min	Nom	Max	Min	Max	Min
$\frac{1}{4}$	0.2500	0.2450	$\frac{7}{16}$	0.4375	0.428	0.505	0.488	$\frac{5}{32}$	0.163	0.150	0.023	0.009
$\frac{5}{16}$	0.3125	0.3065	$\frac{1}{2}$	0.5000	0.489	0.577	0.557	$\frac{13}{64}$	0.211	0.195	0.023	0.009
$\frac{3}{8}$	0.3750	0.3690	$\frac{9}{16}$	0.5625	0.551	0.650	0.628	$\frac{15}{64}$	0.243	0.226	0.023	0.009
$\frac{7}{16}$	0.4375	0.4305	$\frac{5}{8}$	0.6250	0.612	0.722	0.698	$\frac{9}{32}$	0.291	0.272	0.023	0.009
$\frac{1}{2}$	0.5000	0.4930	$\frac{3}{4}$	0.7500	0.736	0.866	0.840	$\frac{5}{16}$	0.323	0.302	0.023	0.009
$\frac{9}{16}$	0.5625	0.5545	$\frac{13}{16}$	0.8125	0.798	0.938	0.910	$\frac{23}{64}$	0.371	0.348	0.041	0.021
$\frac{5}{8}$	0.6250	0.6170	$\frac{15}{16}$	0.9375	0.922	1.083	1.051	$\frac{25}{64}$	0.403	0.378	0.041	0.021
$\frac{3}{4}$	0.7500	0.7410	$1\frac{1}{8}$	1.1250	1.100	1.299	1.254	$\frac{15}{32}$	0.483	0.455	0.041	0.021
$\frac{7}{8}$	0.8750	0.8660	$1\frac{5}{16}$	1.3125	1.285	1.516	1.465	$\frac{35}{64}$	0.563	0.531	0.062	0.047
1	1.0000	0.9900	$1\frac{1}{2}$	1.5000	1.469	1.732	1.675	$\frac{39}{64}$	0.627	0.591	0.062	0.047
$1\frac{1}{8}$	1.1250	1.1140	$1\frac{11}{16}$	1.6875	1.631	1.949	1.859	$\frac{11}{16}$	0.718	0.658	0.125	0.110
$1\frac{1}{4}$	1.2500	1.2390	$1\frac{7}{8}$	1.8750	1.812	2.165	2.066	$\frac{25}{32}$	0.813	0.749	0.125	0.110
$1\frac{3}{8}$	1.3750	1.3630	$2\frac{1}{16}$	2.0625	1.994	2.382	2.273	$\frac{27}{32}$	0.878	0.810	0.125	0.110
$1\frac{1}{2}$	1.5000	1.4880	$2\frac{1}{4}$	2.2500	2.175	2.598	2.480	$\frac{15}{16}$	0.974	0.902	0.125	0.110

* By permission from Vallory H. Laughner and Augustus D. Hargan, "Handbook of Fastening and Joining of Metal Parts," McGraw-Hill Book Company, New York, 1956.

Notes:

All dimensions given in inches.

Bold type indicates products unified dimensionally with British and Canadian Standards.

Taper of head (angle between one side and axis) shall not exceed 2 deg, specified width across flats being the largest dimension.

Top of head shall be flat and chamfered. Diameter of top circle shall be maximum width across flats within a tolerance of minus 15 percent.

Bearing surface shall be flat and either washer-faced or with chamfered corners. Diameters of bearing surface shall be 95 percent of maximum width across flats within a tolerance of plus or minus 5 percent.

Bearing surface shall be at right angles to axis of body within a tolerance of 2 deg for sizes up to and including 1 in, and within a tolerance of 1 deg for sizes larger than 1 in. The bearing surface shall be concentric with axis of body within a tolerance of 3 percent of the maximum width across flats.

Minimum thread length shall be twice the diameter plus $\frac{1}{4}$ in for lengths up to and including 6 in; twice the diameter plus $\frac{1}{2}$ in for lengths over 6 in. The tolerance shall be plus $\frac{3}{16}$ in or $2\frac{1}{4}$ threads, whichever is greater. On products that are too short for minimum thread lengths the distance from the bearing surface of the head to the first complete thread shall not exceed the length of $2\frac{1}{2}$ threads, as measured with a ring thread gauge, for sizes up to and including 1 in and $3\frac{1}{2}$ threads for sizes larger than 1 in.

Threads shall be coarse, fine, or 8-thread series, class 2A for plain (unplated) cap screws. For plated cap screws, the diameters may be increased by the amount of class 2A allowance. Thickness or quality of plating shall be measured or tested on the side of the head.

Point shall be flat and chamfered, length of point to first full thread not to exceed $1\frac{1}{2}$ threads.

Tolerance on length for sizes up to and including $\frac{3}{4}$ in shall be minus $\frac{1}{32}$ in for lengths up to and including 1 in; minus $\frac{1}{16}$ in for lengths over 1 in to and including 2 in; and minus $\frac{3}{32}$ in for lengths over 2 in to 6 in, inclusive. The tolerance shall be doubled for sizes over $\frac{3}{4}$ in and lengths longer than 6 in.

Total runout (eccentricity and angularity) of thread in relation to body for sizes up to and including $\frac{3}{4}$ in shall not exceed 0.010 in for each inch of length when measured in a sleeve gauge; the deviation of shank from a surface plate on which it is rolled shall not exceed 0.0312 in. A suggested gauge is shown in the Appendix. For sizes over $\frac{3}{4}$ in total runout shall be subject to negotiation.

Unless otherwise specified, physical properties of steel cap screws shall correspond to SAE Grades 2 or 5. Cap screws may also be made from alloy steel, brass, bronze, corrosion-resisting steel, aluminum alloy, or such other material as specified.

Table A-28 DIMENSIONS OF FINISHED HEXAGON BOLTS (ASA B18.2-1952)*

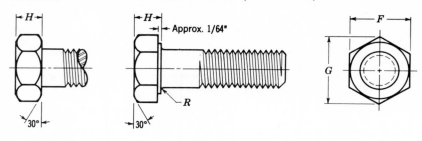

Nominal size or basic major diameter of thread	Body diameter minimum (maximum equal to nominal size)	Width across flats F			Width across corners G		Height H			Radius of fillet R	
		Max	(basic)	Min†	Max	Min	Nom	Max	Min	Max	Min
$\frac{1}{4}$	0.2500	0.2450	$\frac{7}{16}$	0.4375	0.428	0.505	0.488	$\frac{5}{32}$	0.163	0.150	0.023	0.009
$\frac{5}{16}$	0.3125	0.3065	$\frac{1}{2}$	0.5000	0.489	0.5777	0.557	$\frac{13}{64}$	0.211	0.195	0.023	0.009
$\frac{3}{8}$	0.3750	0.3690	$\frac{9}{16}$	0.5625	0.551	0.650	0.628	$\frac{15}{64}$	0.243	0.226	0.023	0.009
$\frac{7}{16}$	0.4375	0.4305	$\frac{5}{8}$	0.6250	0.612	0.722	0.698	$\frac{9}{32}$	0.291	0.272	0.023	0.009
$\frac{1}{2}$	0.5000	0.4930	$\frac{3}{4}$	0.7500	0.736	0.866	0.840	$\frac{5}{16}$	0.323	0.302	0.023	0.009
$\frac{9}{16}$	0.5625	0.5545	$\frac{13}{16}$	0.8125	0.798	0.938	0.910	$\frac{23}{64}$	0.371	0.348	0.041	0.021
$\frac{5}{8}$	0.6250	0.6170	$\frac{15}{16}$	0.9375	0.922	1.083	1.051	$\frac{25}{64}$	0.403	0.378	0.041	0.021
$\frac{3}{4}$	0.7500	0.7410	$1\frac{1}{8}$	1.1250	1.100	1.299	1.254	$\frac{15}{32}$	0.483	0.455	0.041	0.021
$\frac{7}{8}$	0.8750	0.8660	$1\frac{5}{16}$	1.3125	1.285	1.516	1.465	$\frac{35}{64}$	0.563	0.531	0.062	0.047
1	1.0000	0.9900	$1\frac{1}{2}$	1.5000	1.469	1.732	1.675	$\frac{39}{64}$	0.627	0.591	0.062	0.047
$1\frac{1}{8}$	1.1250	1.1140	$1\frac{11}{16}$	1.6875	1.631	1.949	1.859	$\frac{11}{16}$	0.718	0.658	0.125	0.110
$1\frac{1}{4}$	1.2500	1.2390	$1\frac{7}{8}$	1.8750	1.812	2.165	2.066	$\frac{25}{32}$	0.813	0.749	0.125	0.110
$1\frac{3}{8}$	1.3750	1.3630	$2\frac{1}{16}$	2.0625	1.994	2.382	2.273	$\frac{27}{32}$	0.878	0.810	0.125	0.110
$1\frac{1}{2}$	1.5000	1.4880	$2\frac{1}{4}$	2.2500	2.175	2.598	2.480	$\frac{15}{16}$	0.974	0.902	0.125	0.110
$1\frac{5}{8}$	1.6250	1.6130	$2\frac{7}{16}$	2.4375	2.356	2.815	2.686	1	1.038	0.962	0.125	0.110
$1\frac{3}{4}$	1.7500	1.7380	$2\frac{5}{8}$	2.6250	2.538	3.031	2.893	$1\frac{3}{32}$	1.134	1.054	0.125	0.110
$1\frac{7}{8}$	1.8750	1.8630	$2\frac{13}{16}$	2.8125	2.719	3.248	3.100	$1\frac{5}{32}$	1.198	1.114	0.125	0.110
2	2.0000	1.9880	3	3.0000	2.900	3.464	3.306	$1\frac{7}{32}$	1.263	1.175	0.125	0.110
$2\frac{1}{4}$	2.2500	2.2380	$3\frac{3}{8}$	3.3750	3.262	3.897	3.719	$1\frac{3}{8}$	1.423	1.327	0.188	0.173
$2\frac{1}{2}$	2.5000	2.4880	$3\frac{3}{4}$	3.7500	3.625	4.330	4.133	$1\frac{17}{32}$	1.583	1.479	0.188	0.173
$2\frac{3}{4}$	2.7500	2.7380	$4\frac{1}{8}$	4.1250	3.988	4.763	4.546	$1\frac{11}{16}$	1.744	1.632	0.188	0.173
3	3.0000	2.9880	$4\frac{1}{2}$	4.5000	4.350	5.196	4.959	$1\frac{7}{8}$	1.935	1.815	0.188	0.173

* By permission from Vallory H. Laughner and Augustus D. Hargan, "Handbook of Fastening and Joining of Metal Parts," McGraw-Hill Book Company, New York, 1956.
† In sizes $\frac{1}{4}$ to 1 in a tolerance of minus 0.050 D may be used when the product is hot-made.

Notes:

All dimensions given in inches. *Bold type indicates products unified dimensionally with British and Canadian Standards.* "Finished" in the title refers to the quality of manufacture and the closeness of tolerance and does not indicate that surfaces are completely machined. Taper of head (angle between one side and axis) shall not exceed 2 deg, specified width across flats being the largest dimension. Top of head shall be flat and chamfered. Diameter of top circle shall be maximum width across flats within a tolerance of minus 15 percent. Bearing surface shall be flat and either washer-faced or with chamfered corners. Diameter of bearing surface shall be 95 percent of maximum width across flats within a tolerance of plus or minus 5 percent. Bearing surface shall be at right angles to axis of body within a tolerance of 2 deg for sizes up to and including 1 in, and within a tolerance of 1 deg for sizes larger than 1 in. The bearing surface shall be concentric with axis of body within a tolerance of 3 percent of the maximum width across flats. Minimum thread length shall be twice the diameter plus $\frac{1}{4}$ in for lengths up to and including 6 in; twice the diameter plus $\frac{1}{2}$ in for lengths over 6 in. The tolerance shall be plus $\frac{3}{16}$ in or $2\frac{1}{2}$ threads, whichever is greater. On products that are too short for minimum thread lengths, the distance from the bearing surface of the head to the first complete thread shall not exceed the length of $2\frac{1}{2}$ threads, as measured with a ring thread gauge, for sizes up to and including 1 in and $3\frac{1}{2}$ threads for sizes larger than 1 in. Threads shall be coarse-, fine-, or 8-thread series, class 2A for plain (unplated) bolts. For plated bolts, the diameters may be increased by the amount of class 2A allowance. Thickness or quality of plating shall be measured or tested on the side of the bolt head. Point shall be flat and chamfered, length of point to first full thread not to exceed $1\frac{1}{2}$ threads. Tolerance on bolt length for sizes up to and including $\frac{3}{4}$ in shall be minus $\frac{1}{32}$ in for lengths up to and including 1 in; minus $\frac{1}{16}$ in for lengths over 1 in to and including 2 in; and minus $\frac{3}{32}$ in for lengths over 2 in to 6 in, inclusive. The tolerance shall be doubled for sizes over $\frac{3}{4}$ in and lengths longer than 6 in. Total runout (eccentricity and angularity) of thread in relation to body for sizes up to and including $\frac{3}{4}$ in shall not exceed 0.010 in for each inch of length when measured in a sleeve gauge; the deviation of shank from a surface plate on which it is rolled shall not exceed 0.0312 in. A suggested gauge is shown in the Appendix. For sizes over $\frac{3}{4}$ in total runout shall be subject to negotiation. Unless otherwise specified, physical properties of steel bolts shall correspond to SAE Grades 2 or 5. Bolts may also be made from alloy steel, brass, bronze, corrosion-resisting steel, aluminum alloy, or such other material as specified.

Table A-29 DIMENSIONS OF FINISHED HEXAGON AND HEXAGON JAM NUTS (ASA B18.2-1952)*

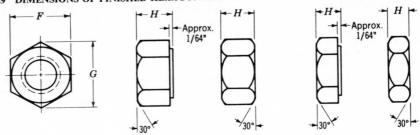

Nominal size or basic major diameter of thread		Width across flats F			Width across corners G		Thickness, nuts H			Thickness, jam nuts H		
		Max	(basic)	Min	Max	Min	Nom	Max	Min	Nom	Max	Min
$\frac{1}{4}$	0.2500	$\frac{7}{16}$	0.4375	0.428	0.505	0.488	$\frac{7}{32}$	0.226	0.212	$\frac{5}{32}$	0.163	0.150
$\frac{5}{16}$	0.3125	$\frac{1}{2}$	0.5000	0.489	0.577	0.557	$\frac{17}{64}$	0.273	0.258	$\frac{3}{16}$	0.195	0.180
$\frac{3}{8}$	0.3750	$\frac{9}{16}$	0.5625	0.551	0.650	0.628	$\frac{21}{64}$	0.337	0.320	$\frac{7}{32}$	0.227	0.210
$\frac{7}{16}$	0.4375	$\frac{11}{16}$	0.6875	0.675	0.794	0.768	$\frac{3}{8}$	0.385	0.365	$\frac{1}{4}$	0.260	0.240
$\frac{1}{2}$	0.5000	$\frac{3}{4}$	0.7500	0.736	0.866	0.840	$\frac{7}{16}$	0.448	0.427	$\frac{5}{16}$	0.323	0.302
$\frac{9}{16}$	0.5625	$\frac{7}{8}$	0.8750	0.861	1.010	0.982	$\frac{31}{64}$	0.496	0.473	$\frac{11}{32}$	0.365	0.323
$\frac{5}{8}$	0.6250	$\frac{15}{16}$	0.9375	0.922	1.083	1.051	$\frac{35}{64}$	0.559	0.535	$\frac{3}{8}$	0.387	0.363
$\frac{3}{4}$	0.7500	$1\frac{1}{8}$	1.1250	1.088	1.299	1.240	$\frac{41}{64}$	0.665	0.617	$\frac{27}{64}$	0.446	0.398
$\frac{7}{8}$	0.8750	$1\frac{5}{16}$	1.3125	1.269	1.516	1.447	$\frac{3}{4}$	0.776	0.724	$\frac{31}{64}$	0.510	0.458
1	1.0000	$1\frac{1}{2}$	1.5000	1.450	1.732	1.653	$\frac{55}{64}$	0.887	0.831	$\frac{35}{64}$	0.575	0.519
$1\frac{1}{8}$	1.1250	$1\frac{11}{16}$	1.6875	1.631	1.949	1.859	$\frac{31}{32}$	0.999	0.939	$\frac{39}{64}$	0.639	0.579
$1\frac{1}{4}$	1.2500	$1\frac{7}{8}$	1.8750	1.812	2.165	2.066	$1\frac{1}{16}$	1.094	1.030	$\frac{23}{32}$	0.751	0.687
$1\frac{3}{8}$	1.3750	$2\frac{1}{16}$	2.0625	1.994	2.382	2.273	$1\frac{11}{64}$	1.206	1.138	$\frac{25}{32}$	0.815	0.747
$1\frac{1}{2}$	1.5000	$2\frac{1}{4}$	2.2500	2.175	2.598	2.480	$1\frac{9}{32}$	1.317	1.245	$\frac{27}{32}$	0.880	0.808
$1\frac{5}{8}$	1.6250	$2\frac{7}{16}$	2.4375	2.356	2.815	2.686	$1\frac{25}{64}$	1.429	1.353	$\frac{29}{32}$	0.944	0.868
$1\frac{3}{4}$	1.7500	$2\frac{5}{8}$	2.6250	2.538	3.031	2.893	$1\frac{1}{2}$	1.540	1.460	$\frac{31}{32}$	1.009	0.929
$1\frac{7}{8}$	1.8750	$2\frac{13}{16}$	2.8125	2.719	3.248	3.100	$1\frac{39}{64}$	1.651	1.567	$1\frac{1}{32}$	1.073	0.989
2	2.0000	3	3.0000	2.900	3.464	3.306	$1\frac{23}{32}$	1.763	1.675	$1\frac{3}{32}$	1.138	1.050
$2\frac{1}{4}$	2.2500	$3\frac{3}{8}$	3.3750	3.262	3.897	3.719	$1\frac{59}{64}$	1.970	1.874	$1\frac{13}{64}$	1.251	1.155
$2\frac{1}{2}$	2.5000	$3\frac{3}{4}$	3.7500	3.625	5.330	4.133	$2\frac{9}{64}$	2.193	2.089	$1\frac{22}{64}$	1.505	1.401
$2\frac{3}{4}$	2.7500	$4\frac{1}{8}$	4.1250	3.988	4.763	4.546	$2\frac{23}{64}$	2.415	2.303	$1\frac{37}{64}$	1.634	1.522
3	3.0000	$4\frac{1}{2}$	4.5000	4.350	5.196	4.959	$2\frac{37}{64}$	2.638	2.518	$1\frac{45}{64}$	1.643	1.643

* By permission from Vallory H. Laughner and Augustus D. Hargan, "Handbook of Fastening and Joining of Metal Parts," McGraw-Hill Book Company, New York, 1956.

Notes:

All dimensions given in inches.

Bold type indicates products unified dimensionally with British and Canadian Standards.

"Finished" in the title refers to the quality of manufacture and the closeness of tolerance and does not indicate that surfaces are completely machined.

Taper of the sides of nuts (angle between one side and the axis) shall not exceed 2 deg, the specified width across flats being the largest dimension.

Tops of nuts shall be flat and chamfered. Diameter of top circle shall be the maximum width across flats within a tolerance of minus 15 percent for washer-faced nuts and within a tolerance of minus 5 percent for double-chamfered nuts.

Bearing surface shall be washer-faced or with chamfered corners. Diameter of circle of bearing surface shall be the maximum width across flats within a tolerance of minus 5 percent. Tapped hole shall be counter-sunk $\frac{1}{64}$ in over the major diameter of thread for nuts up to and including $\frac{1}{2}$ in and $\frac{1}{32}$ in over the major diameter of thread for nuts over $\frac{1}{2}$ in size.

Bearing surface shall be at right angles to the axis of the threaded hole within a tolerance of 2 deg for $\frac{5}{8}$-in nuts or smaller and 1 deg for nuts larger than $\frac{5}{8}$ in; therefore the maximum total runout of bearing face would equal the tangent of specified angle times the distance across flats.

Thread shall be coarse-, fine-, or 8-thread series; class 2B.

Suitable material for steel nuts is covered by ASTM A-307; other materials will be as agreed upon by manufacturer and user.

Tolerance on width across flats may be increased 0.015 in for hot-formed nuts $\frac{5}{8}$ in and smaller.

Table A-30 PROPERTIES OF SECTIONS

A = area, in^2
I = moment of inertia, in^4
J = polar moment of inertia, in^4
Z = section modulus, in^3
k = radius of gyration, in
$\bar{y}$ = centroidal distance, in

Rectangle

$$A = bh$$

$$I = \frac{bh^3}{12}$$

$$Z = \frac{bh^2}{6}$$

$$k = 0.289h$$

$$\bar{y} = \frac{h}{2}$$

Triangle

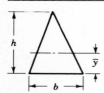

$$A = \frac{bh}{2}$$

$$I = \frac{bh^3}{36}$$

$$Z = \frac{bh^2}{24}$$

$$k = 0.236h$$

$$\bar{y} = \frac{h}{3}$$

Circle

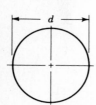

$$A = \frac{\pi d^2}{4}$$

$$I = \frac{\pi d^4}{64}$$

$$Z = \frac{\pi d^3}{32}$$

$$J = \frac{\pi d^4}{32}$$

$$k = \frac{d}{4}$$

$$\bar{y} = \frac{d}{2}$$

Hollow circle

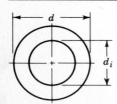

$$A = \frac{\pi}{4}(d^2 - d_i^2)$$

$$I = \frac{\pi}{64}(d^4 - d_i^4)$$

$$Z = \frac{\pi}{32d}(d^4 - d_i^4)$$

$$J = \frac{\pi}{32}(d^4 - d_i^4)$$

$$k = \sqrt{\frac{d^2 + d_i^2}{16}}$$

$$\bar{y} = \frac{d}{2}$$

Table A-31 MASS AND MASS MOMENTS OF INERTIA OF GEOMETRIC SHAPES

ρ = density, weight/unit volume

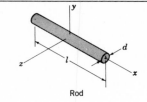

Rod

$$m = \frac{\pi d^2 l \rho}{4g}$$

$$I_y = I_z = \frac{ml^2}{12}$$

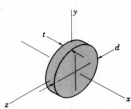

Round disk

$$m = \frac{\pi d^2 t \rho}{4g}$$

$$I_x = \frac{md^2}{8}$$

$$I_y = I_z = \frac{md^2}{16}$$

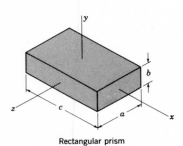

Rectangular prism

$$m = \frac{abc\rho}{g}$$

$$I_x = \frac{m}{12}(a^2 + b^2)$$

$$I_y = \frac{m}{12}(a^2 + c^2)$$

$$I_z = \frac{m}{12}(b^2 + c^2)$$

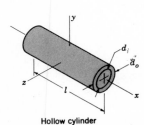

Cylinder

$$m = \frac{\pi d^2 l \rho}{4g}$$

$$I_x = \frac{md^2}{8}$$

$$I_y = I_z = \frac{m}{48}(3d^2 + 4l^2)$$

Hollow cylinder

$$m = \frac{\pi l \rho}{4g}(d_o^2 - d_i^2)$$

$$I_x = \frac{m}{8}(d_o^2 + d_i^2)$$

$$I_y = I_z = \frac{m}{48}(3d_o^2 + 3d_i^2 + 4l^2)$$

Index

Index

A frame, 269
Absolute systems, 17
Acceleration, 19
Acme thread, 230
Actuating force for brakes and clutches, 526
Actuator, linear, 271
Addendum, 399
AFBMA (Anti-Friction Bearing Manufacturers Association), 323
AGMA (American Gear Manufacturers Association), 416
AISC (American Institute of Steel Construction), 12, 124n.
AISC formula, 119
AISI (American Iron and Steel Institute), 165
Allowable stress, 12
Allowance, 154
Alloys, 165
Alternating stresses, 177
Aluminum alloys, numbering system, 166
Aluminum Association, 166
AND operation, 136
Angle of action, 406
Angle of recess, 406
Angle of twist, 58
Angular acceleration, 19
Angular velocity, 19
Annular gear, 408
ANSI (American National Standards Institute), 416, 564
Area, 19
ASM (American Society for Metals), 160n., 166, 204n.
ASME (American Society of Mechanical Engineers), 20, 350
ASTM (American Society for Testing Materials), 248
Average life, 323
Average value (see Mean value)
AWS (American Welding Society), 287
Axial pitch of worm gearsets, 469

Axle, meaning of, 505
Ayre, Robert S., 600

Backlash in gear teeth, 400
Ball bushing, 323
Barkan, Phillip, 600
Barth equation, 428
Base circle in gearing, 402
Base circles, construction of, 405
Base pitch:
  in gearing, 407
  normal, 465
Base units, 19
Basic dynamic capacity of bearings, 327
Basic size, 154
Basic units, 17
Beam:
  slope of, 98
  spot-welded, 85
Beam sections, 49
Beams:
  curvature of, 98
  curved, 68
  sign conventions, 98
  solid round, 49
  with unsymmetrical sections, 47
Bearing, journal: nomenclature, 355
  sleeve, 347
Bearing characteristic, 353
Bearing characteristic number, 362
Bearing groove, 379
Bearing materials, 393
  operating limits, 394
Bearing performance, optimization of, 378
Bearing stress, 258
Bearings:
  alloys used, 393
  application factors, 332
  ball: angular-contact, 320
    load ratings, 330, 331
    nomenclature, 320
    types, 321

Bearings:
  collar, 235
  dimensions, 329
  equivalent loads, 329
  film thickness, 354
  flanged, 391
  grooves in, 390
  heat-dissipation, 385
  hydrodynamic theory, 359
  life, 323
  load rating, 327
  loading, 387
  lubricant temperature, 384
  mounting, 339
  needle, 322
  partial, 355
  pedestal, 384
  permissible loads, 388
  pillow-block, 384
  power loss in, 384
  pressure distribution, 357, 373, 381
  pressure-fed, 379
  recommended life, 332
  reliability, 325
  roller, 321
    load ratings, 332
    types of, 322
  sealing of, 338
  sleeve, unit loads, 388
  slider, 357
  tapered roller, nomenclature, 333
  thrust, 390
  two-piece, 389
  useful life, definition of, 323
  velocity distribution in, 382
Belleville spring, 310
Belt lengths, 559
Belt tension, 558
Belts:
  flat, 556
  materials for, 558
  stiffness of, 593
  toothed, 557
Bending, sign conventions, 38
Bending moment, 37
  sign of, 40
Bending stresses, 43, 45
Benjamin, Jack R., 190n.
Biaxial stress, 27

BIPM (International Bureau of Weights
  and Measures), 20
Bit brace, 90
Blake, J. C., 247n.
Boas, Mary L., 132
Bolt, 237
Bolt grades, head markings, 249
Bolt preload, distribution of, 246
Bolted joints:
  factor of safety, 250
  gasketed, 253
  shear loaded, 256
  stiffness of, 242
  stiffness ratio, 253
  tightening of, 245
Bolts:
  materials for, 247, 248
  preloading of, 244
  stresses in, 253
Bonding, 289
Booser, E. R., 394
Boundary lubrication, 348
Boundary values, 99
Box beam, 85
Boyd, John, 362
Brackets, 270
Brakes:
  actuating forces, 529
  band-type, 539
  disk, 541
  force analysis: external shoe, 535
    internal shoe, 529
  pivoted-shoe, 537
  self-energizing, 527
  (*See also* Clutches)
Brazing, 289
Breakeven point, 13
Brinell hardness, 164
British system, 17
Brittle materials:
  characteristics of, 173
  in fatigue, 196
Brittleness, 164
Broghamer, E. I., 431
Bryan, Joseph G., 146
Buckingham, Earle, 479
Buckling, 97, 114
Bushing, ball, 323
Butt weld, 276

C clamp, 267
Cam system, 607
Cameron, A., 349*n*.
Cams, torque requirements, 612
Cap screws, 239
Carlson, Harold C. R., 303*n*.
Cast iron, 10
Castigliano's theorem, 109
CBBI (Cast Bronze Bearing Institute),
  388
Centrifugal force in gear wheels, 446
Centroid of bolt groups, 258
Centroidal axis, 46
Chain (*see* Roller chain)
Channel, 50
Chester, Louis B., 190*n*., 512
CIPM (International Committee of
    Weights and Measures), 350
Circular frequency, 19, 587
Circular pitch, 399
Clamp frames, 71
Class interval, 143
Class limits, 143
Claussen, G. E., 278*n*.
Clearance, 154
  in bearings, 355, 387
  in gear teeth, 400
  radial, 352
Clips, 112
Clockwise-counterclockwise rule, 40
Clutches:
  axial, 540
  disk, 540
  overload release, 545
  overrunning, 545
  rim-type, 527
  square-jaw, 545
  (*See also* Brakes)
Coefficients:
  of friction: in belting, 559
    for worm gearing, 474
  of thermal expansion, 68
Coffin, L. F., 175*n*.
Coleman, J. J., 204*n*.
Collar bearing, 235
Column design, 618
Columns:
  computer analysis, 618
  eccentric, 119

Colwell, L. V., 212*n*.
Combination, 133
Compound event, 136
Compression:
  pure, 34
  simple, 114
Compression tests, 10
Cone clutch, 543
Cone distance in bevel gearing, 483
Conical spring, 311
Connecting rod, 219
  built-up, 80
Construction code, 12
Contact pressure, 63
Contact ratio in gearing, 410
Contact stress, 74
Continuity, 142
Conversion, 24
Coping saw, 93
Cornell, C. Allin, 190*n*.
Cornet, I., 175*n*.
Corrosion, 197
Cost, 12
Coulomb friction, 599
Coulomb-Mohr theory, 174
Countershafts, 86
Cram, W. D., 212*n*.
Crandall, S. H., 40
Crane hook, 91
Crankshaft, 88
Critical damping, 599
Critical load, 115
Crown gear, 491
Curreri, John R., 596
Curvature:
  of beams, 45
  center of, 70
  of springs, 297
Curvature formula, 98
Curved beams, 68
  formulas for, 73

Dahl, Norman C., 40
d'Alembert force, 581, 609
Damping:
  kinds, 599
  velocity-squared, 613
Dao-Thien, My, 190*n*.
Datsko, Joseph, 165

Dead coils in springs, 302
Decimal point, 20
Dedendum, 399
Deflection:
  of curved parts, 112
  due to impact, 606
  due to shear, 592
Deformation, 34
  angular, 58
Degrees of freedom, 585
Dennison, E. S., 383
Density, 19
Derived units, 17, 19
Design, 4
  definition, 3
Design chart for columns, 119
Design factor, 8, 11
Diametral pitch, 399
  values of, 418
Differential, automotive, 501
Differential equations, piecewise, 603
Directions, principal, 28
Distortion energy, 171
Distortion-energy theory, 170, 512
Distribution:
  normal, 146
  of strengths, 166
Distribution function, 141
Dixon, Wilfred J., 145, 146
Dodge, Thomas M., 190n., 512
Dolan, Thomas J., 207, 431
Drucker, D. C., 573
Ductility, 10, 164
Dudley, Darle W., 438, 441, 447, 466, 480, 481
Dummy force, 109
Duplex bearings, 341
Dynamic factor in gearing, 430
Dynamic load rating, 327

Eccentric loading of fasteners, 259
Eccentricity, 119, 355
Eccentricity ratio, 355, 365
  for columns, 120
Edel, Henry D., Jr., 15
Efficiency:
  of screws, 267
  of worm gearing, 472

Eigenvalue, 585
Elastic coefficient in gearing, 440
Elastic constants, 35
Elastic limit, 9
Elasticity, 35, 95
Elastohydrodynamic lubrication, 348
Elongation, 34, 97
End conditions, 115, 116
Endurance limit:
  definition, 180
  mean values, 181
  surface, 211
  torsional, 208
Endurance limits:
  of cast iron, 181
  of steel, 181
Energy, 19, 106
  potential, 589
Ensign, C. R., 187
Epicyclic gear trains, 418
Equality, 21
Equivalency, 21
Euler formula, 115
Event, 135
Exerciser, finger, 318
Eye bolt, 92

Face of gears, 400
Face width of worm gearing, 471
Factor of safety, 11
  of bolted joints, 250
  of columns, 114
  in gearing, 437
  of wire rope, 574
Failure of riveted joints, 257
Failure theories, comparison, 173
Fatigue of springs, 305
Fatigue damage, 185
Fatigue failure, 177
Fatigue strength, 179
  definition, 180
  formulas, 184
  modifying factors, 188
  of nonferrous alloys, 182
  torsional, 206
Fatigue-strength reduction factor (*see* Stress concentration factor)
Fatigue stress, 200

Fatigue testers, 179
Fillet weld, 276
Findley, W. N., 204*n*.
Finite life, 183
Fish-scale analogy, 244
Fishhook, 92
Flexibility, 94
Flow variable, 368
Force, 19, 21
  concentrated, 6
Force analysis of worm gearing, 472
Force couple, 19
Force fit, stresses in, 216
Four-bar mechanism, 159
FPS system, 17
Frames, 112
Freche, J. C., 187
Frequency, 19
  natural, 587
Frequency function, 141
Frequency polygon, 143
Friction:
  coefficient of, 352, 392, 394
  in screw threads, 232
  sliding, 599
  starting, 319
Friction coefficients for clutch materials,
  547
Friction materials for brakes and
    clutches, 547
Friction variable, 367
Frictional forces in worm gearing, 472
Frictional radius, 541
Fuller, Dudley D., 348*n*.
Furrow, R. W., 591

Gagne, A. F., Jr., 547
Gaskets:
  pressure, 253
  stiffness of, 241
Gaussian distribution, 146
Gear materials, 443
Gear teeth:
  bending deflection, 593
  bevel: stresses in, 487
    tooth proportions for, 482
  conjugate, 401
  contact of, 407, 433

Gear teeth:
  cutting methods, 415
  endurance limit of, 435
  form factors, 426
  generation of, 413
  helical: contact of, 457
    proportions, 460
  long-and-short addendum system, 416
  standards for, 416
  stiffness of, 593
  strength of, 424
  stress distribution in, 427
  template for, 405
  terminology, 399
  thickness, 400
Gear trains, notation for, 421
Gears:
  bevel, 482
    contact stresses, 488
    force analysis, 484
    mounting methods, 487
    pitch angles, 482
    terminology, 482
    tooth forces, 484
  crossed helical, 467
  fabrication, 444
  helical, 457
    face width, 460
    force analysis, 461
    hand of, 468
    nomenclature, 458
    stresses in, 463
  herringbone, 458
  internal, 408
  nomenclature, 399
  power losses in, 443
  spiral, 467
  spiral bevel, 489
  spiroid, 495
  worm, 469
    correction factors, 481
    materials, 480
    nomenclature, 469
    power losses in, 480
    power rating, 479
    stresses in, 479
  Zerol bevel, 493
Generating line in gearing, 405
Geometry factor, gears, 432, 441

Geometry factors:
  bevel gearing, 488
  helical gearing, 464
  spiral bevel gearing, 492
  spur gears, 434
Gib-head key, 264
Gitchel, K. R., 433
Gluing, 289
Goodman diagram, 201
  computer solution, 614
Grassi, R. C., 175n.
Gravitational systems, 17
Gravity, standard, 18
Grip, 240
  of riveted joints, 256
Grooves in bearings, 379
Gun barrels, 60
  stresses in, 216

Hanley, B. C., 204n.
Hardness, 164
Haugen, Edward B., 190n.
Heat, 19
Heat dissipation in bearings, 385
Heat loss, 384
Heating of clutches and brakes, 548
Helix angle, 231, 456
Hertz, 74
Hertz stresses, 212
Histogram, 143
Hoersch, V. A., 75
Hooke's law, 9, 35
Hoops, 60
Horger, Oscar J., 207
Horsepower formulas, 59
Hub, 65
Hydraulic cylinders, 60, 271
Hydrodynamic lubrication, 348
Hydrostatic lubrication, 348
Hydrostatic stress, 170
Hypoid gearing, 493

Impact, definition, 600
Integration:
  boundaries of, 99
  graphical, 103

Interference, 154
  in gearing, 411
Interference fit, 64
Internal friction theory, 174
International System of Units, 18
Involute, 402
Involute curve, generation of, 403
IPS system, 17
ISO (International Standardization
    Organisation), 229
Iteration, 6

Jacobsen, Lydik S., 600
Jennings, C. H., 288
Johnson formula, 118
Joints, bolted, 240
Journal, 347
Juvinall, Robert C., 108n., 181, 183, 204n.

Kececioglu, Dimitri, 190n., 209, 512
Kelvin, 19
Keys:
  square, 265
  strength of, 264
  types, 263
Keyways, types, 265
Kilogram mass, 21
Kinematic chain, 158
Kinetic energy, 548, 594
Knoop hardness, 164
Kožešnick, Jaroslav, 591
Kurtz, H. J., 247n.

Lardner, Thomas J., 40
Law of gearing, 402
Lead, 228, 470
Lead angle of worms, 471
Leibensperger, R. L., 337
Le Système International d'Unitès (SI),
    18
Lewis, Wilfred, 424
Lewis form factor for gearing, 425
Libra, 17
Life, 183
  of bearing, 323
Life expectancy of bearing, 324
Life factor for gears, 442

Limits, 154
Line of action, 401
  definition, 405
Lipson, Charles, 181, 183, 190n., 192, 212n., 324
Little, Robert E., 204n., 209
Load distribution factors:
  for bevel gearing, 487
  for helical gears, 466
  for spur gears, 439
Load intensity, 38
Load-stress factor, 212
Loading diagram, 38
Lubricant, specific heat of, 374
Lubricant film, 354
Lubrication:
  of antifriction bearings, 337
  boundary, 392
  elastohydrodynamic, 337
  film thickness, 354
  reason for, 347
  stability of, 353
  thin-film, 353
  types of, 348
Lubrication systems, 376, 385
Lumped systems, formulas for, 595
Lumping, definition, 580

Mabie, H. H., 591
MacCullough, Gleason H., 59
MacDuff, John N., 596
Machine screws, 238
McKee, S. A., 353
McKee, T. R., 353
Magnaflux inspection, 177
Maier, Karl W., 314
Major diameter, 228
Manson, S. S., 187
Manson's method, 187
Margin of safety, 11, 176
Marin, J., 209
Mass, 21
Massey, Frank J., Jr., 145, 146
Massoud, M., 190n.
Materials:
  for bolts, 247
  for springs, 303
  statistical properties, 166
Maximum film pressure variable, 369

Maximum normal stress theory, 167
Maximum shear stress theory, 169, 505
Maximum strain-energy theory, 170
Maximum stresses, 193
Mean strength, 206
Mean value, 142
Mechanical equivalent of heat, 374
Median, 143
Median life, 323
Merritt, H. E., 416
Metric threads, specification of, 231
Miller, W. R., 209
Miner, M. A., 185
Miner's rule, 185
Minimum film-thickness variable, 365
Minimum life of bearings, 323
Minor diameter, 228
Mischke, C., 190n., 192, 325
Mitchell, L. D., 204n., 512
Mnemonic notation, 177
Mode, 143
Mode shape, 585
Modified Goodman diagram, 202
Modified Mohr theory, 175
Module of gear teeth, 399
Modulus:
  of elasticity, 35
    of belts, 594
  of rigidity, 35
  of rupture, 11, 174, 207, 306
Mohr's circle, 27
  computer solution, 614
Mohr's circle diagram, 29
Moment:
  external, 45
  internal, 45
  sign of, 43
  torsional, 58
Moment of inertia, 57, 263
  area, 45
  computation of, 57
  polar, 58, 59
  unit, 285
  unit polar, 280
Moment connection, 280
Moment load, 259
Moments, resultants of, 54
Morton, Hudson T., 323n.
Multiplication, theorem of, 137

Nachtigall, A. J., 187
Neutral axis, 43
  location of, 45
Neutral plane, 43
Newton, units of, 18
Newton's laws, 16
  of viscous flow, 349
Nominal size, 154
Nominal stress, 193
Nonlinearity, 95
Normal distribution, 146
  unit, 147
Normal pitch, 458
Normal stress, 27
Norris, C. H., 278
Notch sensitivity, 194
  of cast iron, 196
  charts, 195
Numbers, groups of, 20
Nuts:
  selection of, 251
  strength of, 251

Octahedral stresses, 513
Offset method, 9
Ohji, K., 209
One-way bending of gear teeth, 437
Optimum section, 71
OR operation, 136
Outboard mounting, 487
Outcome, 135
Overload factor for gears, 437

Palmgren, A., 185
Palmgren-Miner theory, 185
Parabolic formula, 118
Parallel-axis theorem, 46
Partial differential equations, 581
Paul, Burton, 175
Percentage elongation, 164
Period, 587
Permutation, 131
Peterson, R. E., 193, 252, 264
Petroff's law, 352
Photoelasticity, 427
Pinion, definition, 399
Pinkus, Oscar, 348n.
Pins, 263
Pipe, 60

Pitch, 228
  axial, 458
  diametral, 399
  kinds, 399
  normal, 458
  standards for, 418
  transverse, 458
Pitch angle in bevel gearing, 483
Pitch circle, 399
Pitch circles, definition of, 401
Pitch point of gears, 401
Pitting, 212
Plane stress, 27
Planet gears, 419
Planetary gear trains, 418
Plating, 197
Poisson, 35
Poisson's ratio, 35, 37
Pole, 104
Pole distance, 104
Popov, Egor P., 108
Population, 142
Position vector, 54
Power, 19
Power formula in SI, 59
Power loss in bearings, 371
Precision, 21
Prefixes, 18
  in denominators, 20
  use of, 20
Preload, minimum, 248
Presentation, 8
Press, 79
  power-driven, 231
Press fit, 63
Pressure, 19, 60
Pressure angle:
  in gearing, 405
  normal, 459
  for worm gearing, 470
Pressure line in gearing, 405
Pressure ratio, 373
Pressure vessels, 60, 268
Primary shear, 259
  in welds, 284
Principal directions, 28
Principal stresses, 28
Probability:
  conditional, 137
  cumulative, 141
  definition, 133

Probability density, 141
Probability distribution, 141
Product of inertia, 48
Programmable calculator, 613
Proof load, 247
Proof strength, 247
Proportional limit, 9
Pure, meaning for stress, 34
pV values for brakes and clutches, 549
Pythogorean theorem, 153

R. R. Moore machine, 179
Rack, 407, 491
Radius:
  of curvature, 98
    of gear teeth, 440
  of gyration, 115
Raimondi, A. A., 362
Random experiment, 140
Rating life, 323
Reduction in area, 164
Reliability:
  of bearings, 325
  meaning, 191
Reliability factor, 192
Repeated stress, 177
Retaining rings, 266
Revolute, 158
Reynolds, Osborne, 356
Rigid body, 6
Rigidity, 35, 94
Ring gear, 419
Rippel, Harry C., 391n.
Riveted joints, 256
Rockwell hardness, 164
Roller chain, 564
  correction factors, 568, 569
  dimensions, 565
  ratings, 567
Rope, 570
Rotating-beam machine, 179
Rothbart, Harold A., 75, 600
Rounding, rules for, 23
Rounding off, 21

SAE (Society of Automotive Engineers),
    165, 248
Safety, 11
  of columns, 114
  factor of (see Factor of safety)

Salakian, A. G., 278
Sample, 142
Sample point, 135
Sample space, 135
Sample variance, 144
Saybolt viscosity, 351
Scale of integral, 104
Screw, 237
Screw threads:
  diameter of, 228
  tensile-stress areas, 229
  (See also Threads)
Screws:
  efficiency of, 234
  head types, 238
  machine, 238
  power, 231
  self-locking, 234
  stress concentration factors, 252
  torque relations, 233
Sear, Arthur W., 380
Secant formula, 119
Secondary shear, 259
  in welded joint, 280
Section modulus, 45
Self-locking effect in brakes and clutches,
    526
Series springs, 96
Shaft:
  definition, 504
  overhanging, 89
Shaft angle, 467
Shafting:
  flexible, 575
  stresses in, 505
Shapes, hot-rolled, 49
Shear:
  in bending, 48
  pure, 34
Shear center, 50
Shear deflection, 592
Shear-energy theory, 170
Shear flow, 56
Shear force, 37, 56
  in bolted joints, 260
  sign of, 39
Shear-force diagram, 38
Shear modulus of elasticity, 35
Shear strain, 35
Shear stress, 27
  in bending, 48

Shear stress:
  in hollow beams, 49
  maximum, 33, 76, 505
  octahedral, 513
  sense of, 27
Sheth, Narendra J., 190*n*., 192, 324*n*.
Shock, definition, 600
Shrink fit, 63
Shube, Eugene, 326*n*.
SI units, 18
  rules, 20
Side-flow variable, 369
Sidebottom, O. M., 513
Simple, meaning for stress, 34
Sines, George, 195, 204*n*., 506
Singularity functions, 40, 99
  integrals of, 41
Size, 154
Size factors, 190
Skills, 7
Sleeve bearings (*see* Bearings)
Slenderness ratio, 115
Slope of beams, 98
Slug, 17
Smith, J. O., 513
S-N diagram, 179
  for cast iron, 196
  computer solution, 614
Soap-film analogy, 59
Soderberg, C. R., 507
Soderberg diagram, 202
Soldering, 289
Solid-film lubrication, 348
Sommerfeld number, 362
Spangenberg, Craig, 130
Specific dynamic capacity of bearings, 327
Specific heat, 374
Specifications, 5
Speed, 19
Speed ratio in gearing, 440
Spindle, definition, 505
Spiral angle, 491, 493
Splines, 447
Spoke, design of, 446
Spot welding, 288
Spotts, M. F., 75, 146
Spring:
  definition, 95
  equivalent, 96
  latching, 221

Spring constant, 96
  of bolts, 240
Spring index, 296
Spring rate, 96
  of a bar, 97
  torsional, 97
Spring steels, strength of, 304
Springs:
  active coils, 302
  cantilever, 220
  capacity of, 314
  combinations of, 96
  compression: buckling of, 300
    types of ends, 302
  computer analysis, 615
  equivalent, 589
  extension: preload, 301
    types of ends, 300
  fatigue strengths, 306
  fatigue stresses, 305
  form coefficient, 314
  frequency of, 313
  helical, 296
    deflection of, 298
  on levers, 590
  materials for, 303
  nested, 298
  one-way, 594
  in parallel, 96
  resonance of, 313
  retaining, 608
  in series, 96
  stiffness of, 95
  stress-correction factors, 297
  stress distribution, 298
  torsion, 308
    deflection of, 309
    rate of, 309
Sprockets, 566
Spur gears, definition, 399
Square threads, 230
Standard deviation, 145
Standard gravity, 21
Standard sizes, 13
Standard units, 147
Standardized variable, 191
Starting conditions, 586
Static moment, 56
Steel, alloy, 165
Sternlicht, Beno, 348*n*.

Stiffness, 35
  of bolts, 240
Stirling's approximation, 132
Stochastic variable, 140
Strain, 34
  measurement, 35
  principal, 36
  shear, 35
  triaxial, 36
Strain energy, 106, 170
  due to shear, 593
Strength, 9
  of gear teeth, 424
  of spring steels, 304
  surface, 442
  tensile, 10
  torsional, 10
  ultimate, 10
  ultimate torsional, 11
Strength amplitude, 206
Strength distribution, 191
Stress, 9, 19, 27
  allowable, 12
  bending, 45
  biaxial, 27, 36
  distribution in curved beams, 71
  Hertzian, in gears, 440
  hoop, 60
  kinds, 200
  longitudinal, 60
  maximum shear, 76
  minimum, 200
  nominal, 193
  plane, 27
  due to press fit, 63
  pure, 34
  radial, 60
  in screw threads, 237
  shear, 48
    maximum, 33
  tangential, 60
  thermal, 67
  torsion: distribution of, 58
    in various sections, 59
  triaxial, 37
  uniaxial, 36
  von Mises, 512
Stress amplitude, 200
Stress components, 27
  signs of, 27

Stress concentration:
  in gear teeth, 431
  in riveted joints, 257
Stress-concentration factors, 193
  for extension springs, 301
  formula for, 517
  in gears, 437
  in hubs, 65
  for keyways, 264
  for screws and bolts, 252
  for springs, 297
  for welds, 288
  for wire rope, 571
Stress distribution, 34, 191
  in gear teeth, 427
  in welds, 278
Stress raisers, 193
Stress range, 200
Stress-strain diagram, 9
Stresses:
  bending, 43
  components of, 27
  conventions, 27
  in gear teeth, 75
  Hertzian, 75
  permissible, for welds, 287
  in pressure vessels, 60
  principal, 28
  repeated, 177
  residual, 275
  in rolling bearings, 75
  senses of, 27
  in shafts, 505
  in springs, 296
  triaxial, 33
Stud, 237
Sun gear, 419
Superposition, 36, 103
Surface endurance limit, 213
Surface factor, 189
Surface-finish factors for gear teeth, 436
Surface strength, 212
  of gear teeth, 442
SUV (Saybolt Universal Viscosity), 351
Swanson, W. M., 349n.
Symmetry, axis of, 43
Synthesis, 6

T section, 73
Tachau, H., 573
Tanks, 60
Temperature factor, 193
  for gearing, 442
Temperature gradient, 68
Tensile strength, 10
  distribution of, 166
Tension, pure, 34
Tension test, 10, 164, 167
Thermal stress, 67
Thread standards, 228
Threads:
  American National, 229
  areas of, 229
  gauge sizes, 229
  helix angle, 231
  kinds, 230
  lead of, 228
  metric sizes, 229
  multiple, 229
  specifying, 231
  stresses in, 237
  terminology, 228
  Unified (American National), 229
Throat of weld, 277
Thrust bearings, 390
  (*See also* Bearings)
Thrust collar, 235
Timing belt, 557
Timken bearings, catalogue sheet, 336
Timoshenko, S., 59
Tolerances, 13, 154
  bilateral, 154
  of springs, 305
Tooth contact, lowest point of, 433
Torque, 19
  pure, 58
Torque coefficient, 246
Torque-twist diagram, 10
Torsion, 58
  of sections, 59
  of welded joint, 280
Torsion test, 10, 167
Torsional strength, 10, 174
Tower, Beauchamp, 356
Train value, definition of, 418
Transfer formula, 263
Transmitted load in gearing, 422
Transverse pitch, 458

Trapezoidal section, 73
Tredgold's approximation, 483
Triaxial stress, 32
Twisting moment, 58

U frames, 71
Ultimate strength, 10
Undercutting of gear teeth, 411
Uniaxial stress, 36
Uniform stress, 34
Unit strain, 34
Units, 17
  prefixes of, 18
  spelling, 18
  systems of, 17
UNS (Unified Numbering System), 165
USASI (United States of America
    Standards Institute), 154

V belts, 556
  correction factors, 563
  ratings, 562
  standards, 560
Value, 15
Value analysis, 14
Valves, overhead, 607
Variables, kinds, 103
Variance, 144
  population, 146
Vaughn, D. T., 204*n*., 512
Vector method, 52
Vectors, 51
  cross products, 51
Velocity, 19
Velocity factor:
  in gearing, 428
  for helical gears, 465
Velocity ratio, 418
Vickers hardness, 164
Viscosimeter, Saybolt Universal, 350
Viscosity:
  absolute, definition, 349
  dynamic, 350
  kinematic, 351
  units, 350
  of various fluids, 351
Viscosity-temperature chart, 363, 364
Viscous damping, 599

Volume, 19, 21
Volute spring, 311
von Mises-Hencky theory, 170
von Mises stress, 171, 209
    maximum, 210

Wadsworth, George P., 146
Wahl, A. M., 297
Waisman, J. L., 195, 204n.
Wear:
    of clutches and brakes, 540
    of gear teeth, 439
Wear factor, 212
Weibull distribution, 192, 324
Weight, 21
Weld groups:
    bending properties, 285
    centroids of, 282
Weld symbols, 276
Welded joint, shear of, 280
Welds:
    failure theories, 286
    stress distributions, 278
    torsional properties, 282
    types, 276

Westinghouse formula, 510
Wire rope, 570
    lay of, 571
    life, 573
    standards, 572
    strengths, 573
    stresses, 572
Wirsching, Paul H., 190n.
Woodruff key, 264
Work, 19
Worm, 469
Wrench torque, 245

Yield point, 9
Yield strength, 9
    distribution of, 166
    in shear, 169, 171
    torsional, 11
Yih, Chia-Shun, 351n.

Zimmerli, F. P., 306
Zodiac, sign of, 17